For
MOTHER AND SUSAN,
who loved;

PAUL AND JIM,
who nurtured;

AND ERIC,
who first encouraged me
to come to terms with Darwin

CONTENTS

Where does Circumstance end, and Providence, where begins it?
What are we to resist, and what are we to be friends with?
If there is battle, 'tis battle by night: I stand in the darkness,
Here in the mêlée of men, Ionian and Dorian on both sides,
Signal and password known; which is friend and which is foeman?
Is it a friend? I doubt, though he speak with the voice of a brother.
Still you are right, I suppose; you always are, and will be;
Though I mistrust the Field-Marshal, I bow to the duty of order.
Yet is my feeling rather to ask, where *is* the battle?
Yes, I could find in my heart to cry, notwithstanding my Elspie,
O that the armies indeed were arrayed! O joy of the onset!
Sound thou Trumpet of God, come forth, Great Cause, to array us,
King and leader appear, thy soldiers sorrowing seek thee.
Would that the armies indeed were arrayed, O where is the battle!
Neither battle I see, nor arraying, nor King in Israel...

Arthur Hugh Clough, *The Bothie
of Tober-na-Vuolich*

PREFACE

For one hundred years it has been fashionable to employ military metaphors to characterise the religious debates over evolution in the later nineteenth century. Implicit in this historiography of 'conflict' and 'warfare' is the positivistic assumption that science and metaphysics, evolutionary theory and Christian theology, can or should be divorced. This book undertakes a revision of the received historiography by describing its polemical origins and baneful effects and by offering an interpretation of Protestant responses to Darwin that shows their affinities with the metaphysical and theological traditions from which Darwinism and post-Darwinian evolutionary thought derived.

In offering a non-violent interpretation of the post-Darwinian controversies this study supports and enlarges the standard revisionist thesis that Christian theology has been congenial to the development of modern science. What M. B. Foster, R. K. Merton, and R. Hooykaas *inter alia* have argued concerning the rise of physical science and technology in the sixteenth and seventeenth centuries – namely, that the Christian (and especially the Reformed) doctrine of a contingent creation, ordered and superintended by a perpetual Providence, has led to the adoption of empirical methods in science and the extension of causo-mechanical explanations of nature – is here applied to the rise and spread of theories of biological evolution in the later nineteenth century. By considering the views of twenty-eight Christian controversialists in Great Britain and America, it is argued that Darwin's theory of evolution by natural selection could be accepted in substance only by those whose theology was distinctly orthodox; that this was so because the theory itself presupposed a cosmology and a causality which, owing much to orthodox doctrines of creation and providence, could be made consonant *a priori* with orthodox theistic beliefs; and that, conversely, other theories of evolution, rationalist and immechanical alike, were embraced by those whose theology was notably liberal because such theories, themselves the product of heterodox theologies of nature, promised to secure theistic beliefs which Darwinism seemed bound to offend.

These conclusions will certainly be regarded by some critics as dictated by 'Darwinian' prejudice and by others as dictated by prejudice in favour of orthodoxy. The only critics from whom *prima facie* sympathy can be expected are those, on the one hand, who would be theologically orthodox Darwinians, and those on the other who radically question the metaphysical foundations of both Protestant orthodoxy and modern evolutionary science. If the sympathisers predominate and the conclusions summarised above are substantially accepted, then let it be said that the way forward will likely lie with those who perceive the orthodoxy of radicalism. They alone have begun to show why the Kingdom of God has yet to be realised in the science of any society.

Prejudice, however, is inescapable, for all history begins in biography. The historian's *Sitz im Leben* conditions the subject of his work no less than its treatment and conclusions. I am grateful to Craig Massey, who first impressed me fifteen years ago with the challenge of evolution to Christian faith, and to Kenneth Arneson, who awakened my interest in history at the same time. In subsequent years Wilbur Applebaum introduced me to the history of science, John Montgomery exemplified lofty standards of historical research, David Wells increased my knowledge of Victorian ecclesiastical history, and Basil Hall encouraged me to write the book which the thesis he supervised has now become. Others who have read and criticised drafts of this study in part or in whole are Richard Aulie, Michael Bartholomew, Muriel Blaisdell, A. J. Cain, Owen Chadwick, John Durant, John Greene, Udo Krolzik, Bernard Norton, Henry Rack, Gerrylynn Roberts, and Richard Ziemacki. To these esteemed teachers, colleagues, and friends I impute none of my prejudices or mistakes. Beyond every material and cognitive debt, I owe them the inspiration which has enabled me to persevere in writing, despite the problems and perils of the interpretive task at hand.

My indebtedness to other friends and colleagues is deeper and more various than I can hope to acknowledge. Edward Dowey, Jr and Glenn Wittig made possible a summer's research in Princeton, New Jersey. Michael McGiffert arranged for his excellent dissertation to be put at my disposal. Jimmy McGill and Eric Korn shamelessly abetted my bibliomania, furnishing me with many of the books that form the basis of this study. The librarians of the University of Manchester, the Open University, and Princeton Theological Seminary borrowed other books and literature on my behalf. Chris Clark, Susan Darling, Gaynor Hollington, Patricia Hutchinson, and Teresa Philp kindly set aside more challenging responsibilities in order to type and retype the manuscript. A host of friends on both sides of the Atlantic gave of their time, their advice, and their hospitality in so many ways and on so many occasions during my years of research.

In the last result this book owes its existence to the generosity of the British people, who established and sustain the Marshall Aid Commemoration Commission, which in 1972 elected me to a scholarship of ample means to support my work. To the chairman and officers of the Commission, to Geraldine Cully, and to all whom they represent, I am, and shall always be, profoundly grateful.

Lidlington, Bedfordshire   J.R.M.
13 March 1978

PREFACE TO
THE PAPERBACK IMPRESSION

The call for a new and more accessible impression has afforded the opportunity to correct several errors and minor misprints and to make a very few stylistic changes in the text. Purchasers of the first edition may be interested to know that the present volume, apart from the enlarged bibliographic addendum, is substantially identical with the one they possess.

This is not of course to imply that I could now write the same book, or that I am fully satisfied with it as it stands. My remarks about 'prejudice' in the original Preface have only to be reaffirmed. Both in its treatment of the post-Darwinian controversies, by reference to individual psychology, as a 'non-violent' episode of intellectual history, and in its conclusions, which have been taken too readily as apologetic, the book is an artifact of my concerns and preconceptions in the early 1970s. More recently I have been struggling to anchor the conclusions in their historic social contexts for reasons crucially (though not completely) dissimilar to those which had actuated me at the start. Were the book being re-written, I would acknowledge a greater debt to Frank Turner and Bob Young (see pp. 13, 366 below) through a revised estimate of social Darwinism and a new emphasis on the military metaphor as mediating a genuine conflict between the theodicies of competing social élites.

Still I offer the new impression without hesitation, not least because of encouragement from generous and incisive reviewers such as Dick Aulie, David Hollinger, Ron Numbers, Roger Smith, and Frank Turner. The book holds numerous leads for further research and an updated bibliography through which to pursue them. If my interpretations should also attract attention, now from a wider audience, I trust this will be in part because readers have learnt to substitute 'Christian Evolutionism' for 'Christian Darwinisticism', a term more problematic than I had anticipated (p. 15), and because they have discerned my intentions in the epigraphs fore and aft, and in the final paragraphs of the Introduction and the Conclusion.

London   J.R.M.

3 April 1981

INTRODUCTION:
THE TERRAIN OF REVISION

[Darwin's 'Origin of Species'] was badly received by the generation to which it was addressed. . . . But the present generation will probably behave just as badly if another Darwin should arise, and inflict upon them that which the generality of mankind most hate – the necessity of revising their convictions. Let them, then, be charitable to us ancients. . . . Let them as speedily perform a strategic right-about-face, and follow the truth wherever it leads. . . . It may be, that, as history repeats itself, their happy ingenuity will. . .discover that the new wine is exactly of the same vintage as the old, and that (rightly viewed) the old bottles prove to have been expressly made for holding it.

T. H. Huxley[1]

It is an agreeable irony that T. H. Huxley, the incubus of late-Victorian theology, concluded his pioneering essay, 'On the reception of the "Origin of Species"', with an allusion to the biblical metaphor of wine and wineskins. It is at once more ironic and less agreeable that in this context he obliquely credited the 'happy ingenuity' of his generation for discovering that the new Darwinian wine was of the same vintage as older causo-mechanical explanations of natural phenomena, and thus that older theological bottles, 'rightly viewed', had been made expressly to contain it. Less agreeable, this latter irony, because it violates most of the preconceptions which, until quite recently, have fostered by accounts of Christian responses to Darwin in the later nineteenth century. Was there not a massive 'conflict between religion and science', a full-scale 'warfare of science with theology'? Did not Huxley himself serve valiantly as the 'gladiator-general' of the evolutionary troops? Were not the 'fundamentalists' compelled 'to retire from the lists, bleeding and crushed, if not annihilated; scotched, if not slain'? Did not the 'Darwinian revolution' overthrow the orthodox *ancien régime*, with its hoary dogmas of creation, providence, and design?

Following Darwin's dictum that 'all observation must be for or against some view if it is to be of any service',[2] the present study undertakes to revise the traditional historiography in the light of Huxley's seminal suggestion by examining Protestant responses to Darwin in Great Britain and America as a constitutive part of the history of post-Darwinian evolutionary thought.[3] The post-Darwinian controversies, as we shall refer to these responses so conceived, comprise one of the most frequently discussed and least adequately understood aspects of nineteenth-century intellectual and social history. More than forty years have elapsed since the subject was first critically studied; twenty years and a Darwin centenary have passed since historians first began to call for an in-depth treatment of the religious debates over evolution.[4] The time would therefore seem opportune to cut through the teeming overgrowth of secondary and tertiary literature on the post-Darwinian controversies in order to give interpretation fresh access to the luxuriant undergrowth of primary sources.

The scope of the literature

Although commentaries on the relations of evolutionary thought and Christian faith have appeared with some regularity since the turn of the century,[5] the historical study of the post-Darwinian controversies can be said to date from 1932, when Arthur M. Schlesinger, Sr, professor of history in Harvard University, published his well-known essay, 'A critical period in American religion, 1875–1900'. Schlesinger pointed out that 'Darwinism' was one of three 'threats to orthodoxy' in the last quarter of the nineteenth century and left it to his student, Bert James Loewenberg, to describe more fully how this was so.[6] In 1934, seventy-five years after the *Origin of Species* appeared, Loewenberg submitted a dissertation which contained the first critical account of the post-Darwinian controversies. Within seven years he had published three articles based on his dissertation, a trilogy which has remained standard ever since. But by this time the subject had been pursued in greater depth by others.

In the decade between the submission of Loewenberg's dissertation and the end of the Second World War, six more studies of the post-Darwinian controversies were completed: dissertations by Sister Mary Frederick Eggleston (1934), Windsor Hall Roberts (1936), and Samuel Regester Neel, Jr (1942); and articles by Sidney Ratner (1936), William Ebenstein (1939), and Herbert W. Schneider (1945).[7] Of these studies, Eggleston's, Ratner's, and Ebenstein's retain the least value (being, respectively, diffuse, intemperate, and prosaic); Schneider's alone may be regarded as seminal; and Roberts' and Neel's are without doubt the most substantial of all, the former for its handling of the American periodical literature, the latter for its systematic comparison of the views of religious writers on evolution with Darwin's theory itself. If Loewenberg's four contributions are included, this outpouring – ten works in a little over a decade – was not to be exceeded until historians had passed through the Darwin centenary of 1959.

Meanwhile, studies of the post-Darwinian controversies were enriched by a growing general literature on the inter-relations of science, philosophy, religion, and society in the later nineteenth century. A useful dissertation by Herbert J. Kramer (1948), an important essay by Stow Persons (1950), and a well-researched chapter by Walter P. Metzger (1955) contained the only notable discussions of religious responses to evolution in the period. Contributions to the general literature, on the other hand, came from Jacques Barzun (1941), Avery Milton Church (1943), David F. Bowers (1944), Richard Hofstadter (1944), Robert E. D. Clark (1948), Philip P. Wiener (1949), Paul B. Sears (1950), Edward A. White (1952), and William Irvine (1955).[8] Church's meagre dissertation is merely suggestive of a comprehensive, critical and statistical study which might be made of the American pulpit's response to Victorian scientific naturalism; Clark's pioneering work is frankly apologetic. Barzun and Sears wrote tracts for their times – the Second World War and the Cold War – which, with all their limitations, are still readable as examples of cultural critique. White's study of naturalism in American thought and Irvine's joint biography of Darwin and Huxley are more historical in approach, the latter remaining perhaps the most generally accessible *entrée* to the Darwinian debates. The essay by Bowers, a parallel account of Hegel and Darwin in America, is less distinguished as an example of historical writing, but Hofstadter's interpretation of social Darwinism and Wiener's study of evolution in American philosophy deserve to rank as classics.

The year 1956 was the calm before the storm. No significant discussion of the post-Darwinian controversies appeared, not even in a work of general scope. Then in 1957 the centennial deluge of Darwiniana began with Edward J. Pfeifer's dissertation on the reception of Darwinism in the United States. In 1958 there was Alvar Ellegård's magisterial

study of the British periodical press and a dissertation by Michael
McGiffert on the Christian Darwinism of Asa Gray and George
Frederick Wright. The centennial year saw publication of Gertrude
Himmelfarb's comprehensive, though controversial, interpretation of
the 'Darwinian revolution' and an essay by Basil Willey on the
response of 'clerical orthodoxy' in Britain. Among general and related
works that appeared in the same period, the more important were
Loren Eiseley's elegant study of 'evolution and the men who discovered
it' (1958), an essay by Noel Annan on science and religion in Britain
(1959), and John C. Greene's wide-ranging investigation of 'evolution
and its impact on western thought' (1959).[9]

During the early 1960s historians of the post-Darwinian controversies paused to take their bearings, guided by Loewenberg, who
alone could speak of Darwin and Darwin studies with thirty years'
experience.[10] Then in 1967, with the publication of Richard H. Overman's monograph on evolution and the Christian doctrine of creation,
the most recent decade of writing on the post-Darwinian controversies
got under way. As in the period before the Darwin centenary, there
also appeared a useful corpus of general and background literature
related to the subject: works on the history of sociology and anthropology by John Burrow (1966), John S. Haller, Jr (1971), and Anthony
Leeds (1974); surveys of evolutionary thought in American culture by
Paul F. Boller, Jr (1969) and Cynthia Eagle Russett (1976); studies of
the impact of scientific naturalism in Victorian Britain by Frank
Miller Turner (1974); and outstanding, if conflicting, interpretations
of the intellectual and ideological context of nineteenth-century
scientific thought by Maurice Mandelbaum (1971), Robert M. Young
(1973), and Owen Chadwick (1975).[11]

In 1968 the post-Darwinian controversies in the United States were
interpreted by George H. Daniels in a short anthology of primary
sources and by John Angus Campbell in his dissertation on the response
of 'American Christianity' to Darwin's rhetorical strategy in the
Origin of Species. Thereafter, beginning in 1970, the British controversies received almost exclusive attention. In that year the subject
figured prominently in the second volume of Chadwick's history of the
Victorian Church and in an original essay by Young. Again, in 1971,
Young added importantly to the literature, examining responses to
Darwin's metaphor of natural selection; Simonsson meanwhile sought
to improve on Ellegård's analysis of the periodical press. The British
controversies were discussed in 1974 by John Hedley Brooke, Alan

Richardson, and Colin A. Russell in correspondence texts published by the Open University. In the same year M. J. S. Hodge and Frederick Burkhardt touched on the subject in their contributions to a volume on the comparative reception of Darwinism, a volume which also contained a relevant essay on the United States by Edward J. Pfeifer. In 1975 Michael Ruse interpreted early British responses to Darwin in light of the controversies before 1859. Peter J. Bowler broke new ground in 1977 by suggesting a re-evaluation of Darwin's impact on the argument from design.[12]

Other bodies of literature have enlarged the study of the post-Darwinian controversies, though individually their subjects are narrower or their aims more limited than those of the works mentioned above. Apart from the numerous discussions of science and religion in the pre-Darwinian nineteenth century,[13] there are biographical studies of the post-Darwinian controversialists,[14] descriptions of denominational responses to evolutionary thought,[15] interpretations of the impact of evolution on Christian doctrines,[16] and analyses of the mutual bearings of evolution and philosophy.[17] For the present, however, most of these works must be disregarded as we examine the contours of the principal literature of the subject.

The shape of the literature

Like historical writing in any field, the literature of the post-Darwinian controversies is best represented, not as lying on an elevated qualitative plateau, but as occupying a mountainous region with its hills and valleys, its lofty peaks and yawning chasms. Four eminent features of this region greet the historiographer's eye.

First, although the number of works discussing the controversies in Great Britain nearly equals the number dealing with the United States, only the studies by Eggleston and Overman begin to offer a transatlantic perspective.

Second, interpretations of the post-Darwinian controversies have in most cases been made largely on the basis of secondary sources or from very restricted inductions. The only studies which rise appreciably above these limitations are those of Loewenberg, Roberts, Metzger, Pfeifer, Ellegård, and Campbell.

Third, historians who have examined Protestant responses to Darwin in the greatest detail have concentrated on the period before 1880, neglecting the last decades of the century. The only such historians who take a significantly larger view are Roberts, Neel, and Campbell.

Fourth and finally, studies of the post-Darwinian controversies have been conducted with little regard for the history of evolutionary thought after Darwin. The only works which begin to interpret the religious controversies as constitutive of the development of scientific and philosophical doctrines of evolution are those of Neel, Schneider, Persons, Pfeifer, Ellegård, Daniels, and Young.

Now it would not only be presumptuous but plainly self-defeating to disparage the existing literature of the post-Darwinian controversies because much of it lies beneath these four qualitative eminences or because no single work is represented at them all. Limited aims are legitimate aims, and are the kind most likely to be fulfilled. The literature includes studies of great interest and profound scholarship, works full of curious facts, bibliographical leads, and seminal interpretations, without which a basic reassessment of the post-Darwinian controversies could hardly begin. Yet, granting all this, there are substantial reasons for thinking that a study which undertakes this reassessment will have to measure up at the four points established above – to emulate the merits of the most eminent works and minimise their deficiencies.

There is first of all the matter of international coverage. The post-Darwinian controversies in Great Britain and in America, considered as segments of the history of ideas, cannot be understood in isolation from each other. A strong case might be made for separate treatment of the controversies elsewhere,[18] but to neglect one half of the trans-atlantic English-speaking community would be to overlook the fact that in matters scientific and religious, or more particularly, in matters related to the debate over evolution, Britain and America did practically constitute a single community of thought during the later nineteenth century.[19] Darwin's works were published simultaneously on both sides of the Atlantic; so were those of Huxley, Herbert Spencer, John Tyndall, Charles Lyell, and numerous other scientists. What was not printed with authorisation was summarily pirated by enterprising Americans, at least until the introduction of international copyright legislation in the 1890s. British publishers were perhaps more circumspect in this regard but they managed nevertheless to treat their customers to a wide selection of American works on science and religion.[20] Meanwhile the principal controversialists travelled back and forth across the Atlantic, finding a place in the public eye and securing a wider readership for their books. By 1882, the year of Darwin's death,

Huxley, Spencer, Tyndall, Lyell, and the Duke of Argyll had been welcomed in the United States; Henry Drummond, whose *Natural Law in the Spiritual World* (1883) was pirated no less than fifteen times, had visited twice by 1893. By this date in Britain, on the other hand, Asa Gray, John William Dawson, George Frederick Wright, James McCosh, John Fiske, Henry Ward Beecher, and Minot Judson Savage were personally known. Others, such as Thomas Rawson Birks in England and Charles Hodge in America, gained recognition through association with international bodies like the Evangelical Alliance. Therefore the historical instincts of Eggleston and Overman, who alone offer a transatlantic perspective on the post-Darwinian controversies, are quite inerrant, though the research which underlies their studies leaves much to be desired.

This deficiency brings to mind the second of the points made above: namely, that comparatively few historians of the post-Darwinian controversies have set forth their interpretations with a very great concern for thoroughness. Typically, several major writers are taken as 'representative', a handful of periodicals are quoted as barometers of scientific and religious opinion, or secondary sources which supply the needed interpretations are simply pressed into service. Little if any notice is taken of the thousands of books and articles on evolution and religion that were published in the wake of the *Origin of Species*, and no effort, therefore, is made to ensure that the literature which informs the interpretations is in some sense outstanding or representative within the whole. Yet this is precisely what a full and discerning account of the post-Darwinian controversies requires. Loewenberg acknowledged this implicitly in his handling of the American monographic literature, Roberts, Pfeifer, and Campbell in their coverage of the American periodical literature, Metzger in his use of a wide range of published and unpublished sources, and Ellegård in his peerless analysis of the periodical press in Britain. None of these writers, however, offers a transatlantic perspective, nor do any but Roberts and Campbell give much attention to developments in the last two decades of the century.

And this, it will be recalled, is a third distinguishing feature of the literature of the post-Darwinian controversies. The most detailed studies of Protestant responses to Darwin concentrate on the period up to about 1880. Few interpretations venture far beyond. There are of course some fairly good reasons for this limitation, besides the obvious one of practicability. If the years before 1872 are considered alone, as Ellegård has done, then there is the dual advantage of studying British reactions to both the *Origin of Species* and the *Descent of Man* (1871) without having to take much account of the controversies in the United States, where the Civil War and Reconstruction were major pre-occupations.[21] Or if 1880 is taken as the *terminus ad quem*, then one may be sure, as Huxley pointed out, that the *Origin of Species* had 'come of age': transmutation had won wide acceptance, bolstered by some telling fossil discoveries, and a generation of evolutionists, who themselves had recently 'come of age', were advancing the biological sciences at an unprecedented rate. At the same time, as has often been observed, a large number of religious leaders in Britain and America had begun to make unambiguous overtures to evolution.[22]

The problem with these demarcations, logical and convenient though they may be, is that they exclude a vast body of literature, apart from which Protestant responses to Darwin are but partially understood. Research into the bibliography of monographs on science and religion published between 1860 and 1900 has shown that more than half of these works – most dealt with evolution in one form or another – appeared in the latter two decades of the period: in Britain about fifty-five per cent and in America somewhat more than sixty per cent. The controversies did not dwindle away by the mid-eighties but, on the contrary, judging from the frequency with which monographs were being published, Protestant efforts to accommodate evolution were then at a peak.[23] This may be explained by the fact that the intellectual climate which fostered religious accommodations, a climate pervaded by Neo-Hegelian, Spencerian, and pragmatic philosophies, Lamarckian and Neo-Darwinian biologies, and liberal theologies deriving in part from the critical study of the Bible, did not become widespread until the last two decades of the century. Among those who have discussed the post-Darwinian controversies in the greatest detail only Roberts, Neel, and Campbell have undertaken significant discussions of Protestant responses in this period. Each, however, writes only of the United States, Roberts and Campbell alone ground their work in wide-ranging research, and Neel alone concerns himself seriously with the context of post-Darwinian evolutionary thought.

Here, finally, is the fourth and perhaps the most important of the points established above. For if historians had been careful to view Protestant responses to Darwin in relation to contemporary scientific and philosophical doctrines of evolution, then there would be no need to insist either on a transatlantic perspective or on a discussion of responses after 1880. In other words, the main currents of evo-

lutionary thought, which both informed and transported the controversialists, were continuous through space and time. Theories of
evolution developed local varieties, to be sure, but from the sixties
through the early decades of the twentieth century neither Britain nor
America was without prominent advocates of what we shall refer to
generally as 'Lamarckian evolution'. Spencer, whose *System of
Synthetic Philosophy* appeared in instalments between 1860 and 1896,
was Britain's leading Lamarckian and America's most popular one.
The representatives of American 'Neo-Lamarckianism' – a term first
used in 1884 (later shortened to 'Neo-Lamarckism') to designate a
movement which had been forming since 1866 – were less successful
in exporting their speculations but in Britain their spirit prevailed after
1870 in naturalists such as Richard Owen, George Henslow, and
St George Mivart. Meanwhile 'Darwinism', as we shall employ the
word, was defended by Darwin until his death in 1882 and thereafter
by Huxley, Francis Galton, and George John Romanes in Britain and
by Gray and Wright in the United States. 'Neo-Darwinism' likewise
had its roots in Darwin's lifetime, most notably in the writings of Alfred
Russel Wallace, and after 1883 it gained a wider currency through the
publications of the German zoologist August Weismann. No study of
the post-Darwinian controversies concerns itself with all these schools
of evolutionary thought, much less distinguishes carefully among them.
However, Neel, Schneider, Persons, Pfeifer, Ellegård, Daniels, and
especially Young do variously testify that the controversies are best
understood when seen as constitutive of the development of scientific
and philosophical doctrines of evolution in the later nineteenth century.

The sense of the literature

If the bibliographic scope of the literature of the post-Darwinian
controversies may be likened to a dense and sprawling jungle, and if its
qualitative shape can be compared to a range of mountains, then its
historiographic sense is best represented by a flat and arid desert. In
this region there are attractive features as well, but the interpretations
in question do not generally venture beyond the first level of abstraction. The meaning of Protestant responses to Darwin is sought chiefly
through narration, description, and classification.

The first and most obvious concern of historians is the dating and
ordering of past events. Historians of the post-Darwinian controversies
are certainly no exception. Starting with some basic chronological

questions – How long was evolution resisted? When was evolution finally accepted? What events made acceptance possible for scientists and for theologians? – they have divided the controversies into stages. For example, Schlesinger writes that in America 'the religious controversy over biological evolution reached its most critical stage in the late 1870s. . . . An analysis of the turbid flood of argument which poured forth from the press during the eighties reveals a steady advance – or retreat – from a position of pure emotional obscurantism to one of concession and accommodation.' Similarly, Loewenberg identifies the period from 1860 to 1880 as one of 'acrid polemics' and the decades from 1880 to 1900 as one of evolution's infiltration through thought and culture. The turning-point for acceptance of evolution, he says, was the death of Louis Agassiz in 1873.[24] Roberts, while concurring for the most part with the Schlesinger–Loewenberg periodisation, locates the turning-point in 1874 with the publication of Hodge's *What Is Darwinism?* Pfeifer, however, finds that 'the turning of the tide' coincided with the emergence of an American school of Neo-Lamarckism in 1866, that the 'softening' of anti-Darwinian arguments continued after 1868, that a 'generally calmer mood' prevailed through 1871, notwithstanding the publication of the *Descent of Man*, and, finally, that some form of 'Christian evolution' had gained wide acceptance by 1880.[25]

The British controversies, on the other hand, have been carefully partitioned by Ellegård and Chadwick. Ellegård discerns three periods during which supernaturalistic accounts of creation were gradually abandoned: from 1859 to 1863, when the debate over evolution peaked, declined, and flared up again; from 1864 to 1869, a quieter period on the whole; and from 1870 to 1872, when Wallace, Mivart, and the *Descent of Man* renewed the debate and brought it to focus on natural selection and human evolution. Chadwick, taking a broader view, asserts that the compatibility of evolution and Christian doctrine was gradually acknowledged 'among more educated Christians' between 1860 and 1885, though after 1876 acceptance of evolution was 'both permissible and respectable'. Darwin's burial in Westminster Abbey in 1882 and Frederick Temple's 1884 Bampton Lectures, *The Relations between Religion and Science*, were among the events which highlighted the progressive accommodation, and Temple's consecration in 1896 as archbishop of Canterbury 'may be taken to mark the final acceptance of the doctrine of evolution among the divines, clergy and leading laity of the established church'.[26]

A second approach to interpreting the post-Darwinian controversies is largely a refinement of chronological narration, for it involves the detailed description of individual responses to evolution or the responses of individual denominations. The former is the concern not only of biographers but of many historians who take a larger interest in the controversies. Thus Neel writes of Fiske, McCosh, Joseph Le Conte, Lyman Abbott, and Francis Howe Johnson. Persons describes the views of McCosh, Le Conte, Savage, Paul Ansel Chadbourne, and Francis Ellingwood Abbot. Benz (1967) discusses Fiske, McCosh, Abbott, Savage, Drummond, and Wright. Campbell gives special attention to McCosh, Le Conte, Mivart, and Joseph Cook. The responses of such individuals are usually considered for the purpose of determining when and how each one came to accept or reject evolution, but in a few cases there is a greater design: Neel and Campbell compare the responses with Darwin's views on evolution; Benz undertakes by his expositions to qualify the work of Pierre Teilhard de Chardin.

Likewise, the object in examining the denominational responses to evolution is only occasionally more profound. Apart from studies of the Roman Catholic Church, a literature of wide and often acute scholarship, which lies for the most part beyond the purview of the present study,[27] most works that deal with the subject describe when and how individual denominations accepted or rejected evolution, or offer general comparisons of denominational responses. In the first instance Dietz (1958) and Street (1959) furnish good examples and in the latter, Eggleston, Roberts, and Pfeifer. The consensus seems to be that Unitarians in America were the most receptive to evolution, Congregationalists the most influential in interpreting and propagating it, Presbyterians alternately very hostile or quite accommodating, and Methodists, Baptists, and Lutherans reluctant but generally uninvolved. Ellegård finds a similar pattern in the British controversies, grouping Methodism and the Low Church as 'strongly anti-Darwinian', Baptists, Presbyterians, Quakers, and the High Church as taking a 'less adverse stand', especially the High Church party, and Unitarians and the Broad Church as 'much less anti-Darwinian than the other religious groups'. At the same time, however, he points out that 'the attitudes towards Darwinism in the various sects reflect with remarkable accuracy their ideological position' and thus joins with Schneider and Persons in suggesting a theological interpretation of denominational responses which rises above mere description.[28]

The third major historiographic ploy that appears in the literature of the post-Darwinian controversies follows on from the second, for it involves the classification of responses which are described. Historians, like most human beings, prefer order to confusion and efficient expressions to tedious enumerations. Instinctively, therefore, in discussing the welter of Christian responses to Darwin, they establish various conventions for speaking of the responses – conventions which may in turn become categories or canons for understanding the post-Darwinian controversies. For example, Overman describes the responses of 'Rational Supernaturalistic Orthodoxy', 'Romantic Liberalism', and 'Scientific Modernism'. McGiffert identifies a 'scientific left' composed of Huxley, Spencer, Tyndall, and the 'German Darwinians', a 'theological left' composed of Congregationalists and Unitarians such as Beecher, Johnson, Savage, John White Chadwick, and Octavius B. Frothingham, a 'theological right' represented by Hodge and Enoch Fitch Burr among others, and a community of thought committed to 'reconciliationism', which counted McCosh, Le Conte, Cook, Gray, Wright, and Alexander Winchell among its members. Such terminology is certainly useful, and, like the narration and description with which it is invariably combined, it may help to interpret the 'when' and the 'how' of Protestant responses to Darwin. But the more interesting and difficult question of 'why', the problem of constructing a 'phylogeny' of the post-Darwinian controversies rather than a mere descriptive 'taxonomy', is left almost untouched. This question, this problem, cannot be overlooked if the limitations of the extant literature are to be transcended.

Charting a revision

In the present study we endeavour to penetrate the bibliographic jungle surrounding the post-Darwinian controversies, to surmount those qualitative features which distinguish the most eminent literature of the subject, and to cultivate in the historiographic desert an interpretation of Protestant responses to Darwin that shows their affinities with the metaphysical and theological traditions from which Darwinism and post-Darwinian evolutionary thought derived.

Our point of departure is the traditional historiography of 'conflict' or 'warfare' between a variously hypostatised 'religion' and 'science'. This 'military metaphor' first flourished in the last decades of the nineteenth century chiefly through the influence of the American historians John William Draper and Andrew Dickson White. Enriched and enlarged by the vocabulary of Victorian politics, the polemics of T. H. Huxley, and the tactics of American Fundamentalism in the twentieth century, the military metaphor has remained the standard source of imagery for speaking of the post-Darwinian controversies. Its acceptance in conjunction with narrative, descriptive, and taxonomic treatments of the subject has thus been largely tacit. Its effect, however, has been to predetermine historical understanding and thereby, we argue, to perpetuate false conceptualisation – much as William Whewell's vision of 'two antagonist doctrines of geology', catastrophes and uniformity, has affected interpretations of the pre-Darwinian debates over the theories of Charles Lyell. 'To arrive at a just interpretation of the controversy over Darwinism', as McGiffert rightly points out, '. . .the inquiring historian must cut through the cake of metaphor which encrusts the subject. The analogies of warfare are peculiarly seductive, especially as they were so commonly used by those who were engaged in the contest.'[29]

Yet the very fact that the military metaphor was 'commonly used' within the post-Darwinian controversies testifies to its symbolic importance, whatever the failings of 'warfare' as an historical analogy. The sense of 'conflict' symbolised in the military metaphor can be explicated at various levels and the sociological approaches of Turner and Young would appear to hold great promise in this regard.[30] In the present study we concentrate on the personal level, examining the 'crisis of faith' which Darwin precipitated in numerous Victorian minds. Our aim is not to furnish psychological evidence of a 'Darwinian revolution' but, on the contrary, to qualify this interpretation by showing how largely the crisis arose and was resolved within the framework of established religious beliefs. Far from asserting that Darwinism implied 'the replacement of one entire *Weltanschauung* by a different one' and required 'a new concept of God and a new basis for religion', we maintain, with Young, that such 'orthodox accounts which stress the growth of scientific naturalism as a development away from traditional theological and social doctrines, must be fundamentally reconsidered. In their place we require an interpretation which shows the deeper continuities.'[31]

In a scientific 'revolution', according to Thomas Kuhn, the 'paradigm', or conceptual and methodological framework, of 'normal science' encounters certain anomalous facts, enters a period of 'crisis', and ends up being discarded in favour of an incommensurable new paradigm which becomes the framework for the resumption of normal science. Although Kuhn intends this as an analysis of the internal development of science and its sociology, he does weakly acknowledge that 'external social, economic, and intellectual...conditions outside the sciences may influence the range of alternatives' available to one who undertakes to terminate a crisis by proposing some 'revolutionary reform'. He makes no attempt to analyse the crisis-state of the individual scientist or the role of 'external...conditions' in the creative process by which the crisis is resolved – omissions which have not gone unnoticed by admirers and critics. But he does point out that these questions 'need far more investigation' and that they 'demand the competence of the psychologist even more than that of the historian'.[32]

Now the post-Darwinian controversies did not take place over communally held paradigms nor did they result in any revolutionary reforms. Nothing in fact is so notable about the science of the period as the absence of a new evolutionary paradigm and the failure of a normal science of natural history to re-emerge.[33] However, while the impact of Darwinism was not 'revolutionary' in the Kuhnian sense, it nevertheless involved a 'crisis': a crisis of belief in creation, providence, and design, of belief in the reality of the divine purposes in nature and the omnipotence and beneficence of the divine character which they reveal. For many individuals this crisis found its resolution in a quite mundane way, through the writing of a book. To understand the crisis is, therefore, in some measure to understand each book; and to understand each book, or some representative portion of them, is to arrive at an interpretation of the most tangible historical evidence of the post-Darwinian controversies. Clearly, then, 'the competence of the psychologist' in explaining the crisis and its resolution must affect the interpretation at least as much as the historian's competence in handling the published evidence. This is why we ground our historical revision in Leon Festinger's 'theory of cognitive dissonance', perhaps the most influential general theory of attitude change, and one which has had a large and beneficial influence on the study of social psychology.[34]

Festinger's exposition of 'conflict', 'dissonance', and 'dissonance reduction' equips us to investigate the 'deeper continuities' between pre-Darwinian and post-Darwinian theologies of nature. Taking as our touchstone the response to Darwinism itself – a theory which is distinguished carefully from Lamarckian, Spencerian, and Neo-Darwinian versions of evolution – we show that 'Christian Anti-Darwinism' emerged in the last result from a conflict between Darwinian doctrines and certain fundamental philosophical, rather than specifically Christian, beliefs: namely, the perennial belief that full and final certainty can be obtained through inductive inference and must be obtained for a scientific theory to be thoroughly credible; and the belief, lately indebted to the Neo-Platonism of German romantic philosophy, that every form of life is essentially fixed by the divine will. If a 'Darwinian revolution' occurred at all it was these beliefs about certainty and fixity which were primarily overthrown.

For Christians whose faith was not allied to a philosophy of certainty and fixity, and who therefore could countenance some theory of evolution, Darwinism created distinctly theological conflicts: conflicts regarding the purposes and character of God as manifested in the creation of the world and mankind. The dissonance involved in arbitrating between Darwinian and theological truths could be variously reduced, and 'Christian Darwinisticism' emerged – as Morse Peckham's awkward but appropriate fusion of 'Darwinism' and 'romanticism' suggests[35] – in efforts to transform the Darwinian theory with metaphysics of providence and progress which, by supplanting causomechanical explanations, could secure a teleology and a theodicy on an evolutionary basis. The metaphysics in question were those of Lamarckian evolution on the one hand, of theological liberalism on the other. Their lineage, as we point out, extended far into the pre-Darwinian past.

It was possible, however, to achieve dissonance reduction without substantially modifying the Darwinian theory. 'Christian Darwinism' was also metaphysically sophisticated but not in a manner inimical to a causo-mechanical account of biological evolution. Its metaphysics were those of a theological orthodoxy which extended back through the Protestant Reformation to the Fathers of the Church. The paradox of orthodox Darwinians and liberal Darwinists (save for two whose views on evolution most nearly resembled Darwin's) demands an explanation and in conclusion we endeavour to provide it through an application of Festinger's theory. First, we argue that Darwinism was steeped in the orthodoxy of the Anglican clergymen William Paley and Thomas Robert Malthus; second, that Darwin, though deeply influenced by their respective teleology and theodicy, could not reconcile their theology proper with the phenomena his theory presupposed; third, that the orthodox theology of the Christian Darwinians was relevant to an acceptance of Darwinism not only by virtue of its consonance with the orthodoxy of Paley and Malthus, but because of

the ability of its Calvinistic and Trinitarian doctrines to reduce the
dissonance which led Darwin to abandon the Christian faith; and
finally, that the liberal theology of Christian Darwinists could not co-
exist with Darwinism because it lacked both the consonance with the
orthodoxy of Paley and Malthus, and the means of dissonance reduction inherent in Calvinistic and Trinitarian doctrines.

To argue that Darwinism was the legitimate offspring of an orthodox theology of nature; to assert with Huxley that Darwin raised no
problem for the 'philosophical Theist' which had not existed 'from
the time that philosophers began to think out the logical grounds and
logical consequences of Theism';[36] and therefore to agree with the
'happy ingenuity' of Huxley's generation that, 'rightly viewed', orthodox theological bottles proved to have been made expressly for holding
the new Darwinian wine – to reason thus may well invite the accusation that one is doing scarcely veiled apologetics. Whether it be possible
to write history without defending some 'ultimate concern' is of course
highly debatable, but an endeavour to vindicate the particular orthodoxy of the Christian Darwinians would doubtless be as unacceptably
whiggish as the more familiar attempt to co-opt the Darwinism of
Darwin in behalf of the prevailing orthodoxy in evolutionary biology.
In general, moreover, it does not at present seem obvious that Christian
theology is best defended by historical accounts of its formative influence on modern science. At the end of this study, as at the beginning, it
will thus remain 'a matter of controversy...whether evolutionary
theory demonstrates the need for a new religion to include the new
idea of an evolving Universe or whether nothing more is needed than a
transformed – or for the first time clearly understood – Christianity'.[37]

PART I

HISTORIANS AND HISTORIOGRAPHY

I

DRAPER, WHITE, AND THE
MILITARY METAPHOR

When historians accept the statement of a partisan as a truth of history they often put themselves at the mercy of a bias which is not their own, the word of a controversialist is received without criticism because it is convenient to do so and not because the historian shares the controversialist's passions; indeed he may never have given much thought about the direction in which those passions might have pulled the statements he accepts.

G. Kitson Clark[1]

Clever metaphors die hard. Their tenacity of life approaches that of the hardiest micro-organisms. Living relics litter our language, their *raisons d'être* forever past, ignored if not forgotten, and their present fascination seldom impaired by the confusions they may create. In politics and religion, where name-calling is at a peak, each generation labels its mugwumps, levellers, quislings, whigs, and tories, its anabaptists, Calvinists, fundamentalists, papists, and puritans. Even a phrase common to Roman history, 'barbarian assault', has been widely employed as a slapdash argument for dealing firmly with the lower classes.[2] When once a catchy phrase, a memorable name, or a colourful concept enters the common language, it never fails to make history. In so doing it often takes on a history of its own.

Such is the case of the military metaphor. Through constant repetition in historical and philosophical exposition of every kind, from pulpit, platform, and printed page, the idea of science and religion at 'war' has become an integral part of Western intellectual culture. Like other clever metaphors, this one shows few signs of dying out. Historians, no less than the antagonists and apologists of Christian faith, continue to find it irresistible.[3] Had they but found it so engaging as an object of historical research a different situation would doubtless obtain. For a study of the origins of the military metaphor and of its influence over the past one hundred years inspires little confidence in its utility as an historiographic device. Rather, this captivating idea

remains the one best guide to the sentiments of its embattled founders,
John William Draper and Andrew Dickson White.

John William Draper and his 'conflict'

In 1871 Edward Livingston Youmans entered his second decade as
America's chief purveyor of popular science. Already, like a voice
crying in the wilderness, he had prepared the way for evolution and
made straight the paths of publication for Charles Darwin, T. H.
Huxley, and the one whose latchet he felt unworthy to loose, Herbert
Spencer. Now, with a host of converts to his credit and evolution an
established truth, he approached his favourite publisher, Daniel
Appleton of New York City, with a new prophetic proposal – the idea
of issuing a library of scientific tracts written by the foremost authori-
ties of the day and prepared for simultaneous release in the United
States, Great Britain, France, and Germany. It was an unwieldy
scheme, to be sure, but Youmans' evangelical fervour triumphed over
every objection and within a year volume one, *Forms of Water* by
John Tyndall, had appeared. One after another they came – Bagehot's
Physics and Politics, Bain's *Mind and Body*, Spencer's *Study of
Sociology* among them – little red octavos stamped 'International
Scientific Series' on their spines.[4]

If Youmans' enterprise was to succeed he had to solicit manuscripts
from accomplished authors. Only known names would sell and only
those who had already contributed to scientific literature would be in a
position to compose their books on such short notice as he was granting. Thus Tyndall had just delivered at the Royal Institution the
annual Christmas lectures on which his volume was based; Bagehot
could draw on his articles which had been appearing since 1867 in the
Fortnightly Review; Bain was the author of three major works on
mental philosophy; and Spencer (thanks to Youmans' promotional
work) had completed several volumes of his *Synthetic Philosophy* and
was already engaged in research for the *Principles of Sociology*. Who
then would undertake a contribution – the thirteenth of the Series – on
the historical relations of science and religion? An historian he should
be and a scientist as well. Neither W. E. H. Lecky nor Leslie Stephen
were scientists; the former, in any event, was well occupied with his
history of eighteenth-century England and the latter had not yet
divested himself of holy orders. If only the scientific historian Henry
Thomas Buckle were still alive, Youmans may have thought, this

volume might help bring his great *History of Civilization* a step closer
to completion.

Youmans did not have to look far to find his author. The New York
of Daniel Appleton boasted a university on whose faculty sat John
William Draper (1811–1882), professor of chemistry and president of
the medical school, a scientist who claimed to have made the first
photographic portrait by Daguerre's process in 1838, who took the
first known photograph of the moon two years later, and who had
achieved no small recognition for his researches in chemistry, spectro-
graphy, and photographic techniques. Draper, moreover, was a well-
known historian. The mantle of Buckle, next to Auguste Comte the
man whose temper was nearest his own, had fallen at his feet, and
within a year of Buckle's death in 1862 Draper had published *A History
of the Intellectual Development of Europe.*[5]

The book was not good history by any stretch of the imagination. Its
plot gathers linear momentum from Comte's assurance that history
moves from theology to metaphysics to positive thought and angular
momentum from the rise and fall of civilisations according to Draper's
own notion that societies recapitulate the stages of human life. Lubri-
cating this lurching interpretation is a pre-Darwinian environmentalism
that makes national history a record of adaptation to nature, while
over all presides the great rationalist draughtsman Gibbon, warning
against 'the triumph of barbarism and religion'. 'The emotional
climax of the book', writes Draper's biographer, Donald Fleming, 'is
the vain effort of Roman Catholicism to hold back the universal
dominion of the scientific spirit. The popes appear as the heads of an
enormous bureaucracy tyrannizing over the minds of men, and
sacrificing the advance of reason and science to the cause of continued
faith in the supernatural. The chapters dealing with earlier history are
best approached as a kind of backward extension of this conflict.'[6]

But if Draper's *History* was inadequate those who read the book
were for the most part inadequately equipped to judge it. Youmans,
who was a chemist turned publicist, not an historian, must certainly be
numbered among them. Draper's work appealed to him because it
flattered the growing faith in the omnicompetence of science, in the
power and prerogative of scientists to embrace all of life within the
framework of 'law'. When Youmans told Spencer, 'What we want are
ideas – large, organizing ideas', he had exactly this in mind. And when
in 1873 he asked Draper to contribute the volume on science and
religion to his thriving library of popular science, one of those 'large,

organizing ideas' was conceived.[7] The book was not altogether a new one. Under the editorial circumstances its period of gestation was rather limited. Draper culled large portions from his *History*, condensed them, and spliced on a preface and three concluding chapters. What new material the book contained, however, was most significant, and can be taken into account only after a digression into Draper's somewhat clouded religious and educational past.

John William Draper was born at St Helens near Liverpool, the son of a Methodist minister. In 1822, at the age of eleven, he entered a Methodist grammar school where he continued to manifest an early penchant for books and science, undeterred by a school-master who 'thought it to be his business to drive the nail of knowledge through his pupils [*sic*] head and clinch it with repeated blows of his rattan at the other end'.[8] On graduation he entered the newly opened University of London, immersed himself in Benthamism and Positivism, and began to drift from the precepts of his Methodist youth. The drift was only a slow one, however, for in 1832, with his wife and three sisters, Draper sailed for America to become a teacher of natural history in a small Methodist college in Virginia. Arriving too late to take the job, he found himself supported by an even smaller Methodist educational institution, The Misses Draper Seminary for Girls, which his sisters opened on the property adjoining his laboratory. Four years, eight published papers, and a doctor's degree later, Draper finally obtained an academic post. He was appointed professor of chemistry and natural philosophy at Hampden–Sydney College in Prince Edward County, Virginia. The school was founded in 1775 by the presbytery of Hanover County and, unlike the University of London, which was also founded in religious interests, it retained a strong ecclesiastical spirit. Each day began with chapel at sunrise and concluded with another service at four in the afternoon. Unfortunately the interim was too occupied with remedial instruction and too often interrupted by ritual discipline to offer a very congenial atmosphere for a professor who aspired to do serious scientific research.

Moving with his family to the University of the City of New York in 1839, Draper assumed the chair of chemistry which he held for the remainder of his career and began the researches on which his scientific reputation chiefly depends. In the absence of correspondence destroyed at Draper's request at the time of his death, it is difficult to reconstruct his continuing religious development in this period. One incident that took place in 1853 and survived in the memories of two of Draper's grandchildren therefore takes on considerable biographical importance. It involved 'the family rebel', Draper's sister Elizabeth.

When her eight-year-old nephew William was dying, she hid a devotional book which he cried for, and after his death laid it on Draper's breakfast plate. He met this cool challenge by ordering her out of the house. Though he never forgave her, she passed a happy, unrepentant life as a Catholic convert in Bridgeport, Connecticut.[9]

Draper's biographer suggests that religious controversy may have formed the backdrop of this sad scene. Perhaps it would not be impertinent also to intrude the hypothesis that, if we assume Elizabeth to have been a devout Roman Catholic at the time, she might have taken every measure to ensure that nothing would interfere with the reception by her beloved nephew of the last sacrament.

Apart from this incident just enough is known of Draper's mature religious outlook to say that it was about as full of anomalies as his *History of the Intellectual Development of Europe*. On the one hand Draper never entirely abandoned the values of his Methodist heritage: theism was indispensable; individualism, liberty, and scientific progress were the best basis for the future happiness of mankind; private philanthropy and the judicious use of scientific innovations for the public good were undeniable obligations. But on the other hand Draper was an incurable rationalist: his God was the aloof cosmic architect of Leibniz; the human body, according to his *Human Physiology* (1856), is a perpendicular machine; and evolution long before 1859 described for him the orderly manner in which matter had progressed ever since the Deity had set it perfectly into perpetual motion. Why the decrees of science never seemed to affect the dictates of faith remains part of the enigma of Draper's religion. 'Time and again he ignored their implications', says Fleming. 'It was his way of crossing the tight rope from Wesleyanism to Darwinism.'[10]

Darwin's *Origin of Species* appeared in 1859 and the next year marked a turning-point in Draper's life. It was then, midway between the completion of his *History* in 1858 and its publication four years later, that the chemist in him began to be rivalled by the would-be scientific historian. Nothing signifies the change more distinctly and dramatically than the one incident for which Draper has most often been remembered by historians of the nineteenth century. On 30 June, addressing the convocation of the British Association at Oxford, Draper droned on for more than an hour in a crowded stuffy room and thereby postponed the main event of the day, that inevitable encounter between Bishop Samuel Wilberforce and T. H. Huxley. Doubtless Draper would rather have made his mark with the address, 'On the intellectual development of Europe considered with reference to the views of Mr Darwin and others that the progression of organisms is determined by law', and with the *History* on which it was based. But greater men and larger issues were in attendance and with these, historians, like Draper's audience, have been mainly concerned. Indeed, since Draper himself had become something of an historian, the confrontation that he witnessed must have remained vivid in his own memory as well.

'Somewhere he...picked up a fear of arbitrary power', observes Fleming.[11] Did it come from the discipline of Methodism in a preacher's family? from authoritarian educational experiences in the religious hinterlands of academia? from witnessing the relentless grip of Roman Catholicism on his own sister? from a deistic religious philosophy? from the spectacle of a slick-spoken bishop compensating scientific ignorance with ecclesiastical prestige?[12] Doubtless each of these factors made some contribution. Above all, however, it was the pretensions of the Roman Catholic Church and its pontiff in the decade after 1860 which at once filled Draper with trepidation and stirred his passion for intellectual liberty and the progress of science to new heights of intensity. The arbitrary aggression against science of the world's most powerful religious and political body had to be repulsed. The force of argument marshalled in an old book would meet the challenge, augmented by a brigade of fresh chapters newly informed by the enemy's latest intrigues. Thus in 1874 Draper fortified Youmans' International Scientific Series by deploying the *History of the Conflict between Religion and Science*.

'The history of Science', wrote Draper in his preface, 'is not a mere record of isolated discoveries; it is a narrative of the conflict of two contending powers, the expansive force of the human intellect on one side, and the compression arising from traditionary faith and human interests on the other'.[13] Lest there be some question, the reader is not left long to guess the enemy's identity. Insisting on his complete impartiality, Draper passes over the Protestant and Eastern churches and fixes his sights on the Mother of all.

It has not been necessary to pay much regard to more moderate or intermediate opinions for, though they may be intrinsically of great value, in conflicts of this kind it is not with the moderates but with the extremists that the impartial reader is mainly concerned. Their movements determine the issue. . . . In speaking of Christianity, [therefore,] reference is generally made to the Roman Church, partly because its adherents compose the majority of Christendom, partly because its demands are the most pretentious, and partly because it has commonly sought to enforce those demands by the civil power.[14]

Now in 1864 the intellectual world had been shocked and confused at the promulgation of the encyclical *Quanta cura*, with its appended 'Syllabus of errors' condemning in eighty propositions what seemed to be every contemporary challenge to Roman supremacy. Not the least among them were the 'evil opinions' that 'the Decrees of the Apostolic See and of Roman Congregations interfere with the free progress of science' (Prop. 13), that 'public institutions generally which are devoted to teaching literature and science and providing for the education of youth, be exempted from all authority of the Church' (Prop. 47), and that 'the Roman Pontiff can and ought to reconcile and harmonize himself with progress, with liberalism, and with modern civilization' (Prop. 80). Some liberal Catholics, who were as disturbed by the Syllabus as their brethren in other communions, took consolation in the Pope's subsequent summoning of a General Council. There, it was thought, explanations might be given and moderating statements issued. But it was not to be. Before the Council opened in December 1869 the Church's old guard let it be known that the Syllabus would acquire teeth in the dogmatic definition of papal infallibility. Liberal protests were to no avail and in July 1870 the Council's First Dogmatic Constitution on the Church of Christ, *Pastor Aeternus*, declared that 'the Roman Pontiff, when he speaks *ex cathedra*, . . .is possessed of that infallibility with which the divine Redeemer willed that His Church should be endowed in defining doctrine regarding faith or morals'. As for the errors specified in the Syllabus, whereas there they had only been condemned, in the Dogmatic Constitution on the Catholic Faith, *Dei Filius*, those who commit them receive the anathema: those who 'shall say that human sciences are to be so freely treated, that their assertions, although opposed to revealed doctrine, can be held as true, and cannot be condemned by the Church', or those who 'shall assert it to be possible that sometimes, according to the progress of science, a sense is to be given to doctrines propounded by the Church different from that which the Church has understood and understands'.[15]

So here was the 'extremism' with which, Draper believed, 'the impartial reader is mainly concerned'. 'Roman Christianity and Science', he asserted, 'are recognized by their respective adherents as being absolutely incompatible; they cannot exist together; one must yield to the other; mankind must make its choice – it cannot have both'.[16] Unfortunately, Draper himself was far from impartial. His statements gained plausibility by emphasising the Vatican's magisterial pronouncements and ignoring the historical circumstances which attended the development of ultramontanism after 1858. These circumstances vitiated much of what he had to say.

That the papal régime was thoroughly obsolete, that the Papal States were hardly significant in an era of secular power politics, and that Pio Nono was a relatively unintelligent, if beloved, pope, inept at diplomacy and largely unexposed to the changing world about his small dominion, are historical judgements whose omission can perhaps be excused. They have been left for later historians to make.[17] What cannot be excused is Draper's serious overreaction to the minor threat which the Vatican in fact posed to the modern world. This much could have been known in 1873. Temporal power had retreated before the army of Victor Emmanuel II, King of Piedmont and Sardinia, and in 1860 clung tenaciously to that tiny strip of land along Italy's western shore, the Patrimony of St Peter. No sooner had infallibility been promulgated, twelve years later, than Napoleon III declared war on Prussia and withdrew his protectionary forces from Civita Vecchia, thereby enabling Victor Emmanuel to complete his sweep of the papal lands. The pope remained at the mercy of the Italian parliament, practically imprisoned in his tiny sacred province. His standing in certain quarters of the Church may have been very high but it had reached its nadir among the ruling bodies of the world. His spiritual power over all those loyal to Rome and his infallibility when speaking *ex cathedra* may have been declared in stern uncompromising terms but his political power had gone from him.

One would never have thought so from reading Draper's little red book. The pope is only 'feigning that he is a prisoner'. In reality

the Roman Catholic Church is the most widely diffused and the most powerfully organized of all modern societies. It is far more a political than a religious combination.... Its movements are guided by the highest intelligence and skill. Catholicism obeys the orders of one man, and has therefore a unity, a compactness, a power, which Protestant denominations do not possess. Moreover, it derives inestimable strength from the souvenirs of the great name of Rome.[18]

In his concluding chapter Draper revealed that he was not unaware of the pope's loss of temporal power and the widespread discontent among the Catholic clergy resulting from the Syllabus and the dogma of infallibility. Indeed, while in Munich on Christmas day 1870 he had a 'long and interesting conversation' with the Catholic historian Ignaz von Döllinger, whose vociferous denial of papal infallibility led to his excommunication three months later.[19] But such knowledge did little to mitigate Draper's polarising rhetoric. The spectre of 'arbitrary power' was too fearful for him.

Draper not only failed to estimate accurately the extent of the pope's political power; he also failed to achieve an historian's understanding of the pronouncements which he took as evidence of papal aggression against modern science. 'Since the endangering of her position had been mainly brought about by the progress of science', Draper declared, the Roman Church 'presumed to define its boundaries, and prescribe limits to its authority. Still more, she undertook to denounce modern civilization.'[20] This was a grand oversimplification. It said far too much for science and nothing at all for the affairs of the Italian peninsula, which offered the immediate occasion for the harsh language of the Syllabus.[21] Furthermore, Draper nowhere indicated that the Syllabus was primarily a non-infallible classified index to pronouncements made in encyclicals or briefs issued from Rome since the time of Pius VI. 'Since these, for the most part, were concerned with particular people, books, and occasions, it was not to be supposed that they... fulfilled the conditions of an infallible pronouncement.'[22] As for the dogma of infallibility itself, Draper was equally imperceptive. The pope 'cannot claim infallibility in religious affairs, and decline it in scientific', he said. 'Infallibility embraces all things. It implies omniscience. If it holds good for theology, it necessarily holds good for science.'[23] Needless to say, a careful reading of *Pastor Aeternus* would have precluded this kind of criticism.

Finally, if neither the temporal power nor the dogmatic pronouncements of the Vatican were, as Draper conceived them, 'absolutely incompatible' with science, then one may well enquire whether there was reason to believe that the Roman Church had in some measure come to terms with 'modern civilization' or had perhaps even fostered science in the century of its greatest practical advance. For surely if such evidence existed Draper must be held accountable for neglecting to temper his bellicose spirit with it. It would not have been too difficult to discover that Pius IX, for all his failure to read the signs of the times, was not averse to employing the latest technological innovations. His government was the first in Italy to use adhesive postage stamps. In 1853 gas lamps made their appearance on the streets of Rome and the next year brought the telegraph to the Papal States. Pio Nono was almost anxious to have railways. In 1856 he granted a concession for the construction of a railway of fifty miles' length, from Rome to Civita Vecchia, and in 1858 another concession was given for fifteen miles of track between Rome and Frascati.[24] Among those who provided the theoretical basis for these technological advances were not a few sons of the Church: Dumas in chemistry, Galvani in electricity, Fresnel and Fraunhofer in optics, Leverrier in astronomy. A consistent unbeliever, 'so passionately hostile to Christianity as to reject in science and in practical life all aid or help' that comes from the hand of the Roman Church, wrote the Jesuit historian Kneller in 1910,

will have to light his house with tallow candles, for stearine comes to him from the Catholic hands of Chevreul; and he cannot use electricity without tribute, in the very quantitive terminology in which his bill is calculated, to the Catholic names Ampère and Volta. Aluminium he must refuse and abandon, for he owes it to the Catholic Sainte-Claire Deville. He cannot continue to pasteurize his wine; he cannot use Schönbein's collodium in photography, nor can he use water-glass or cement. His medicine will have to manage without Pelletier's quinine, Laënnec's ausculation, and Pasteur's whole fabric of bacteriology.[25]

But, after all, the *History of the Conflict between Religion and Science* was a tract for its times, not a history of them. It comes as no surprise that Draper had nothing constructive to say about the Roman Church and science. For this he is not culpable, nor for writing a tract, but for calling it a 'history' and masquerading his deep-seated fears under a solemn declaration of impartiality.

'As a contribution to the intellectual atmosphere of that age', Owen Chadwick observes, 'the book was powerful'.[26] Its author spoke, not as the Scribes and the Pharisees, but as one having authority. No other title in the hundred-odd volume International Scientific Series sold as well. In the United States the *Conflict* passed through fifty printings in about as many years.[27] In Great Britain there were twenty-one editions in fifteen years. Elsewhere it achieved comparable notoriety in French, German, Italian, Dutch, Spanish, Polish, Japanese, Russian, Portuguese, and Servian translations. The Spanish edition of 1876 was entered in the *Index librorum prohibitorum*, a distinction Draper shared with Copernicus, Galileo, Kepler, Locke, and John Stuart

Mill.[28] Of Draper's critics Fleming summarises, 'Most...went angrily about in circles from the premise that there could be no conflict to the conclusion that there was none. But it was of no use, they were beating him over the head with his own problem. They allowed him to set the terms of the debate.'[29] How ironic that religion was being made to fight a new war at the very time its citadel had been captured. And how much more ironic was the truth of Draper's *ex cathedra* pronouncement on behalf of the church-militant scientific, possessed of no less infallibility than that which he denied the pope: 'This is only as it were the preface, or forerunner, of a body of literature, which the events and wants of our times will call forth.'[30]

The 'warfare' of Andrew Dickson White

A body of polemical literature was indeed called forth by the 'events and wants' of American life in the years after 1865. In the wake of the Civil War another war began – the 'warfare of science with theology' – and this due to causes as complex and divisive as those of the earlier conflict. While politicians and sundry reconstructionists endeavoured to weld together their country's fragile loyalties, industrialists plundered its human and natural resources, compounding the social fractures that already existed. While Protestant clergymen surveyed their divided communions, new religious currents eroded their churches' faith, adding the problem of heresy to that of unity. And all the while, in the midst of disturbing social and religious uncertainties, echoes of evolution resounded from the college halls of America, stirring suspicion and resentment on every hand. Among laymen, at any rate, the war had postponed all but hearsay knowledge of the *Origin of Species*. But by 1870 there was no lack of evidence that a new and dangerous doctrine was taking the scientific world by storm.[31]

To religious folk it was not always clear what should be opposed and with what priority. The proliferation of heterodox opinions of every shade and hue is difficult enough for the historian to comprehend, let alone the average contemporary. There was Spencer's *First Principles* in 1864 and Darwin's *Descent of Man* in 1871, which of course had to be condemned out of hand. But in 1866 Mary Baker Patterson (later Eddy) began promulgating 'Christian Science' and in the following year Francis Ellingwood Abbot organised the Free Religious Association. In 1870 the Positivist David Goodman Croly began to issue *The Modern Thinker*, Lester Ward's National Liberal

Reform League first published *The Iconoclast*, and the Free Religious Association started its influential paper *The Index*. The *Popular Science Monthly*, edited by E. L. Youmans, appeared in 1872 and within two years fulfilled all that its title promised by reaching a circulation of 12 000. In 1873 the free-thinker De Robigne Mortimer Bennett launched *The Truth-Seeker* from Paris, Illinois. By 1874 a dozen volumes of the International Scientific Series had been published, spilling newfangled ideas onto the bookshelves of the average man. Meanwhile Robert Ingersoll, the leading infidel of his generation, was ranting about the country proclaiming the benefits of agnosticism and British unbelievers were making their more refined opinions heard. John Tyndall, the physicist critic of the physical efficacy of prayer, visited in 1872. Charles Bradlaugh, the notorious Secularist, conducted lecture tours in 1873, 1874, and 1875. And T. H. Huxley, Darwin's 'bulldog', made his *début* in 1876, lecturing on evolution at New York, at Nashville, and at the christening of Johns Hopkins University (the last *sans* invocation, an ominous omission). To make matters worse, in 1873 the cause of a waning scientific orthodoxy suffered an irreparable loss in the death of Louis Agassiz, the famous Harvard zoologist and the foremost opponent of evolution on the continent. Soon, in 1878, the foremost theological anti-evolutionist, Charles Hodge of Princeton Theological Seminary, would join Agassiz in a new tenure.[32]

Quite understandably, the effect of these developments on the Christian public was to produce widespread confusion and strife. Condemnations tended to be categorical, novelty was frequently mistaken for infidelity, and opinion turned to prejudice on every hand. In the post-war years the attitudes of religious leaders and laymen were, on the whole, hardly conducive to change. Yet in this illiberal period it befell a small fraternity of intellectual leaders to inaugurate a fundamental and far-reaching change in American life: the reform of higher education. They were all university presidents: Daniel Coit Gilman of Johns Hopkins, who began twenty-five years' leadership in 1876, on the crest of Huxley's visit; James Burrill Angell of the University of Michigan, who served for thirty-eight years beginning in 1871; Charles William Eliot of Harvard, who took the helm in 1869 and steered masterfully through four decades; and Andrew Dickson White (1832–1918), the pioneer president of the new breed, who began his seventeen-year experiment at Cornell University in 1868.[33]

Before the Civil War college presidents were fatherly figures whose task was to preserve the educational past in an unruffled status quo. Typically they were clergymen, charged by their denominations and clerical trustees with the teaching of controversial subjects – ethics, metaphysics, and natural philosophy – in which, typically, they distinguished themselves as defenders of the faith.[34] Gilman, Angell, Eliot, and White were neither clergymen nor apologists and they certainly held no brief for the classical status quo. They were prototypes of the university president of today. With swelling enrolments, the influx of irrepressible new knowledge, and the expansion of institutional premises, the old time college president had become obsolete. Student affairs, curricular decisions, and fiscal matters were delegated to specialists and committees, while the task of the new breed of university presidents became increasingly one of administrative coordination. The patriarch and teacher gave way to the publicist and business executive.[35]

However, in the expectation of hidebound constituencies and especially in the eyes of a religious public who were scandalised by the bewildering variety of post-bellum developments in religion and science, the president's role changed hardly at all. Indeed, if anything it was confirmed. He was to remain the chief *conservateur* of a young man's mind and morals. Even at liberal Unitarian Harvard, for example, Eliot encountered considerable resistance to his proposals to expand the system of elective courses; and before he could secure a temporary appointment in 1870 for the young philosophical evolutionist John Fiske, he had to overcome the deep misgivings of the Overseers.[36] Meanwhile in central New York state, far from enlightened New England, Andrew Dickson White was waging 'war' on behalf of Cornell University, an institution he hoped would be full of course offerings and devoid of sectarian trammels. To appreciate fully White's reaction to the obstacles he faced we must look briefly into his religious and educational background.

Horace and Clara Dickson White were High Church Episcopalians and they set before their eldest son an example of devout and pious churchmanship. Andrew's earliest ideal was the surpliced parish priest; later he was delighted to play the organ for services at St Paul's Church in Syracuse, New York, where the family attended.[37] But there came a pinch in what seemed otherwise to be a tolerably religious upbringing. Horace White, zealous churchman that he was, took a leading part in the establishment of a parish school at St Paul's. The good offices of the bishop and the rector being thereby secured, it was determined that Andrew should complete his final preparation for college under the instruction of the school's first master and that he should subsequently attend the college favoured by the bishop, the rector, the master (whose Alma Mater it was), and his own father. It mattered not that he aspired to attend a New England university. To little Episcopally controlled Hobart College in Geneva, New York he must go, and so he did – reluctantly.[38]

If his experience at Hobart had been favourable Andrew might easily have forgiven those who 'knew better'. As it happened he was nearly victimised by their well-meaning but visionless plan. The seventeen-year-old youth was a serious student, greatly impressed by a large collection of books and not in the least hesitant to read them every one. But how could he do this in the midst of the tumult that was Hobart? Later in life White stated that in all the colleges and universities with which he had been associated *together* he had not seen 'so much carousing and wild dissipation' as he saw in 'this little "Church College" of which the especial boast was that, owing to the small number of its students it was "able to exercise a direct Christian influence upon every young man committed to its care"'. One year of the uproar was all that young Andrew could stand and he appealed to his father to send him elsewhere. But Horace White said No, in deference to his ecclesiastical friends. Whereupon Andrew fled home and school to live with his former Syracuse tutor until, three months later, his father's convictions had changed. The boy returned home and shortly after Christmas of the same year, 1850, he departed with his father for Yale.[39]

Andrew was not yet out of danger. His academic future was again nearly sabotaged when, aboard the train, his father met a student returning to Trinity College in Hartford, Connecticut. The student spoke glowingly of his Alma Mater, a leading institution of the Protestant Episcopal Church, and in response the elder White produced letters of recommendation from the rector of St Paul's and others, addressed not to the officials of Yale University, but to the officials of Trinity College. Horace White was determined to go to Hartford, not New Haven. Andrew would not hear of it. He argued that Yale had much the better library. His father in turn tempted him with a gift of the best private library in the United States. But the young scholar was adamant. They went on to Yale where in 1854 Andrew completed the B.A. and in 1856, after studies at the Sorbonne and the University of Berlin, he was awarded the master's degree.[40]

In 1857, at the age of twenty-five, Andrew Dickson White accepted the professorship of history in the University of Michigan. The president of the University, Henry Philip Tappan, was a forerunner of the new breed of presidents of the sixties and seventies, a former Presbyterian clergyman on the one hand and an admirer of the German research university on the other. As his efforts in urging the German system on the eastern colleges had been to no avail, Tappan looked west and undertook to build at Ann Arbor a non-sectarian institution in which students would be permitted to choose freely among the various courses of instruction. His enterprise not only stood 'practically at the beginning of the transition from the old sectarian college to the modern university';[41] it also earned him the undying opposition of religious and educational traditionalists, every feature of which made an indelible impression on the new professor of history. White shared Tappan's enthusiasm for the German system of education, having so recently studied in it himself. Thus he shared Tappan's disgust at the chauvinists in the Michigan legislature who demanded, 'We want an American, not a Prussian, system.' Moreover, White knew intimately – perhaps better than Tappan – the ungodly condition into which denominational education in the United States had too often fallen and so felt a good deal of righteous indignation at the opposition of the state's small religious colleges to the 'Godless' university in their midst. Little did the young professor realise at the time how soon he himself would be subjected to the same kind of harassment in his own efforts to establish a modern university in the state of New York.[42]

The dream of a university 'worthy of the commonwealth and of the nation', full of course offerings and devoid of religious or sectarian trammels, first came to White during his nightmare experience at Hobart College. Travel, reading, and, above all, witnessing liberal education under fire at Ann Arbor, transformed the idea into an agenda; and after his election to the state senate of New York in 1863 the agenda began to issue in concrete action. Under the provisions of the federal Land Grant Act of 1862 New York had received the title to public lands, the income from which was to be employed for the foundation and maintenance of colleges where the industrial classes might obtain both a liberal and a practical education. Since the entire land fund had been placed at the disposal of a single second-rate institution which could not, in the time allotted, meet the conditions for retaining it, it was suggested that the fund be divided this way or that among the state's colleges. Not surprisingly, a land-rush ensued as representatives from nearly all of the smaller denominational schools clamoured for pieces of property. But White, deploring the idea of a state subsidy for sectarian education, resisted every attempt to divide the fund. By keeping the land intact, he argued, its income could be employed to subsidise a single first-class university along the lines of his old dream. In time White persuaded one of his senate colleagues, a wealthy philanthropist named Ezra Cornell, to support his far-sighted plan. Cornell in turn offered to endow the university with one-half million dollars and assisted White in drafting its legislation. Their bill provided that the school would combine the vocational education stipulated by the Land Grant Act with the best features of a German research university; that it would be an institution, free from partisan dominance, where 'any person can find instruction in any study'.[43]

'The introduction of this new bill into the legislature', said White, 'was a signal for war. Nearly all the denominational colleges girded themselves for the fray, and sent their agents to fight us at Albany; they also stirred up the secular press, without distinction of party, in the regions where they were situated, and the religious organs of their various sects in the great cities.'[44] After no mean struggle the bill passed both the assembly and the senate, and at length, in the autumn of 1868, Cornell University opened its doors. However, the slighted colleges and their constituencies were thoroughly exasperated; a scandal-hungry press remained only too eager to exploit whatever rumour it could obtain about the controversial new institution; and religious people everywhere took offence that a university should be founded which did not have as its aim the protection and propagation of the Christian faith. Thus from the very outset of his administration White was besieged by self-righteous and vindictive attacks.

It stood to reason that the same rhetoric which taunted evolution and evolutionists was now flung at the new university. Critics were not far mistaken in thinking that White had established at Cornell 'a school which in its main outlines conformed to Spencer's ideas'; nor was their recollection of his frequent verbal references to the English philosopher entirely amiss.[45] Had not White in fact early advocated that 'truth shall be sought for truth's sake' in the university, that 'it shall not be the purpose of the Faculty to stretch or cut Science exactly to fit

"Revealed Religion" '?[46] But no matter how the evidence is stacked, the punishment cannot be said to have fitted the crime. Cornell was a 'Godless institution', established for the propagation of 'atheism' and 'infidelity'; its students were 'raw recruits of Satan'; its faculty stood accused of everything 'from atheism to pocket-picking'. Agassiz, the foremost anti-evolutionist in America, who served as a non-resident professor, was said to be promulgating 'atheism and Darwinism'; Goldwin Smith, professor of history, who had only replied to an article in the periodical, was labelled a 'Westminster Reviewer'; and President White, who had the audacity to recommend that his students read Buckle, Lecky, and Draper, was nothing but an 'atheist'.[47] And all this notwithstanding the prominence of the school's interdenominational chapel, the presence of a vigorous Young Men's Christian Association on the campus, and the sweet reasonableness of the Cornell administration.[48]

White was irritated and dismayed. Was he not himself a sincerely religious person? Had he not done his utmost to deal impartially with all the religious interests represented in his institution? Had he not even denied his own religious preference by refusing to permit the endowment of a chaplaincy occupied by a clergyman of the Protestant Episcopal Church?[49] 'For a long time I stood on the defensive', White later reflected, 'but as. . .this seemed only to embitter our adversaries, I finally determined to take the offensive, and having been invited to deliver a lecture in the great hall of the Cooper Institute at New York, took as my subject "The Battle-fields of Science" '.[50] The immediate object was to show on the basis of a few carefully chosen historical examples that every interference with the freedom of scientific enquiry in the supposed interests of religious dogma has brought ill on both science and religion, and that in every battle which has resulted from such interference science has emerged victorious. The larger objective, as White told Cornell, was to give their religious opponents 'a lesson which they will remember'.[51]

The lecture caught on instantly, much to White's delight. It was printed in full the next day, 19 December 1869, in the *New York Daily Tribune*; extracts from it appeared everywhere in the press; and White repeated it often – at Boston, New Haven, Ann Arbor and elsewhere. Subsequently the lecture grew into two articles which Youmans was only too happy to publish in the *Popular Science Monthly*, and in 1876, after revision and enlargement, these articles appeared on both sides of the Atlantic as *The Warfare of Science*.[52] Although the book was at first rather poorly received, it soon became apparent that continuing attacks served to elicit more and more favourable responses, many of them from eminent Christian men. Thus White could consider his offensive on behalf of academic freedom to be an outstanding success.[53]

But it was not simply a strategic blow on behalf of academic freedom. The *Warfare of Science* was the characteristic reaction of a liberal scholar whose liberality had too long been abused. White had endured a year at Hobart College, the attempts of religious men to dissuade him from the training he knew he required in favour of a church-sponsored education, the sight of a university president frustrated by the vested interests of denominational Christianity, and repeated denunciations of his own educational dream-come-true. An explosion was inevitable and it is a tribute to White that it took the temperate form of historical argument. There is cause to regret only that the circumstances which occasioned his indignation are set forth, not at the outset, but at the end of the book, and there in a condensed form.[54] For, believing themselves to be under the impact of simple historical fact, readers have too often succumbed to the force of the book's opening paragraphs, little realising their origin on President White's personal battlefield at Cornell.

I purpose to present an outline of the great, sacred struggle for the liberty of science – a struggle which has lasted for so many centuries, and which yet continues. A hard contest it has been; a war waged longer, with battles fiercer, with sieges more persistent, with strategy more shrewd than in any of the comparatively transient warfare of Caesar or Napoleon or Moltke.

I shall ask you to go with me through some of the most protracted sieges, and over some of the hardest-fought battle-fields of this war. We will look well at the combatants; we will listen to the battle-cries; we will note the strategy of leaders, the cut and thrust of champions, the weight of missiles, the temper of weapons; we will look also at the truces and treaties, and note the delusive impotency of all compromises in which the warriors for scientific truth have consented to receive direction or bias from the best of men uninspired by the scientific spirit, or unfamiliar with scientific methods.

My thesis, which, by an historical study of this warfare, I expect to develop, is the following: *In all modern history, interference with science in the supposed interest of religion, no matter how conscientious such interference may have been, has resulted in the direst evils both to religion and to science – and invariably. And, on the other hand, all untrammeled scientific investigation, no matter how dangerous to religion some of its stages may have seemed, for the time, to be, has invariably resulted in the highest good of religion and of science.* I say 'invariably.' I mean exactly that. It is a rule to which history shows not one exception.[55]

White was 'essentially a crusader', observed the Cornell historian Carl Becker. 'His crusading spirit extended to every aspect of life on campus.'[56] In this spirit White brought the university into existence and fought its religious enemies; in this spirit he spent himself thereafter in the cause of liberty, reason, and toleration. For twenty years the 'warfare' of science grew obsessively in his mind, nourished, he said, by 'my main reading, even for my different courses of lectures', enriched by the researches of his close friend and former student, George Lincoln Burr, and committed to writing in the various parts of the world which called him from the presidency of Cornell in 1885.[57] The result was the appearance in 1896 of 'a landmark in general liberal historiography', *A History of the Warfare of Science with Theology in Christendom.*[58]

Like its prototype, the two-volume *magnum opus* shunned Draper's blatant anti-Catholicism. Although White had some admiration for Draper's *Conflict*, he did not himself succumb to the pressure of current events and find his 'warfare' in the encounter between modern science and the Roman Church. Doubtless he knew European history too well, and contemporary European history in particular, to make such a mistake. Rather, his warfare took place between the liberality of the scientific outlook and the constraints imposed by sectarian dogmatic theology.[59] It was less a political war than an intellectual one, and it was a war in which science has to contend, not with an incorrigible foe, but with a friend that oversteps its bounds. In making these distinctions White showed himself to be a better historian than Draper. In applying them he distinguished his work with a fairness and balance that has no place in Draper's polemical tract. Thus White freely acknowledged that the 'theological war against a scientific method in geology was waged more fiercely in Protestant countries than Catholic' because, in opposing science, 'the older Church had learned by her costly mistakes, especially in the cases of Copernicus and Galileo'. He blessed Nicholas Wiseman, later Cardinal Wiseman, whose conduct in the geological controversies, he said, 'contrasts admirably with that of timid Protestants, who were filling England with shrieks and denunciations'.[60]

The *Warfare*'s virtue was its purpose as well as its point of view. White took pains to stress in the introduction that he was 'bred a churchman', that he had been 'elected a trustee of one church college and a professor in another', and that his 'greatest sources of enjoyment were ecclesiastical architecture, religious music, and the more devout forms of poetry'. As the effects of the book in the decade after its publication must have caused many to doubt these statements, White emphasised in his *Autobiography* that many of his 'closest associations and dearest friendships' were with clergymen. 'Clergymen are generally', he said, '. . .among the best and most intelligent men that one finds'.[61] The problem was that the character of some clergymen was less desirable, and just in proportion, it seemed, to their alignment with sectarian creeds. So far Draper and White were agreed. But whereas Draper declared that 'the ecclesiastic must learn to keep himself within the domain he has chosen, and cease to tyrannize over the philosopher', while at the same time making tyrannising statements such as 'faith must render an account of herself to Reason' and 'mysteries must give place to facts', White proclaimed that 'in the field left to them – their proper field – the clergy will more and more, as they cease to struggle against scientific methods and conclusions, do work even nobler and more beautiful than anything they have heretofore done'.[62] Clearly, his aim in writing had been not only 'to aid in freeing science from trammels', but to 'strengthen religious teachers' by enabling them to see some of the lessons of history.[63]

The last battle in White's lifelong warfare on behalf of intellectual liberty was the publication of his lectures on 'the evolution of humanity in criminal law'. They appeared first between 1904 and 1908 as a series of biographical essays in the *Atlantic Monthly* and then in 1910 under the title *Seven Great Statesmen in the Warfare of Humanity with Unreason*. Although the studies of Sarpi, Grotius, Turgot, Stein, Cavour, and Bismarck serve individually to illustrate White's continuing crusade, his essay on Christian Thomasius, the German jurist and publicist, contains its most striking manifestation. The 'most permanent of all' the blessings which Thomasius bestowed on mankind, in the judgement of Cornell's ex-president, was 'his general influence on higher education. . .in favour of FREEDOM FROM SECTARIAN INFLUENCE OR CONTROL'. Of Thomasius, indeed, White wrote what might well have been his own epitaph: 'From first to last he was a warrior. . . . Only a man who could fling himself, and all that he was, and all that he hoped to be, into the fight. . .could really be of use.'[64]

The warfare of Andrew Dickson White is susceptible of many criticisms and no aspect of it more so than his *magnum opus*. However, rather than belabour the faults and limitations made obvious by generations of hostile and friendly critics,[65] we may perhaps be permitted some remarks *ad hominem*, in criticism of the spirit that gave rise to the book and to a great extent vitiates it. For there is evidence to suggest that the oppositions of sectarian dogmatic theology were not all that aroused in White a certain 'irascibility of temperament' in the defence of his beloved university.[66]

In 1874 White himself secured the appointment of a young Jewish scholar, Felix Adler, to lecture at Cornell in Hebrew and oriental literature. Some thought that Adler's lectures were calculated to inspire rationalistic views, but such objections could be overlooked – overlooked, that is, until Adler ventured to suggest in a public lecture that some of the doctrines of Christianity are found in other religions. This was too much. The Cornell administration warned Adler to cease promulgating his religious opinions and at the end of his three-year appointment his contract was not renewed. William Channing Russell, vice-president of the University, contrived to make it appear that the trustees had not objected to Adler's teachings but only to the desire of the philanthropist who had endowed Adler's post to control nominations. President White had his own explanation: he implied that Adler voluntarily withdrew. What actually happened is clarified by a letter from Russell to Adler which has been uncovered by Walter P. Metzger: Russell, whose first love after his family was Cornell, refused to jeopardise the school in any way. If, as Adler wished, there were to be a fight for reappointment, 'victory would not be of greatest importance, but...defeat would be lasting injury'. Ironically, Russell himself was dismissed by the trustees four years later for reasons related to lack of religious conviction.[67] By presiding over both these episodes of academic intolerance and by omitting any reference to them in his *Autobiography*, White lays himself open to the charge that institutionalism may simply be the unacceptable face of ecclesiasticism, from which in turn it might be inferred that the 'warfare of science with theology' is a vast over-simplification of the process by which new ideas are accepted and intellectual liberty is achieved.[68]

To be sure, White's *Warfare* contains an exhaustive compilation of cases relating to the history of toleration, frequent exact citations of primary sources, and numerous references to bodies of literature otherwise explored only with difficulty – all this in addition to the relative virtues we signalised earlier, an historical viewpoint and a constructive purpose. For this we must not fail to express our gratitude to White's memory. But in the light of the confusion and misunderstanding perpetrated by the work, it would certainly be remiss not to add – in the words of John Dillenberger – that 'there is no reason to regard it as a scholarly book today, much less as an adequate interpretation'.[69]

The irony of Draper's *Conflict* was its war-cry in the face of a defeated ecclesiastical opposition. Certainly there is also a touch of irony in the fact that the co-founder and first president of a university intended by him to 'battle mercantile morality and temper military passion' should present to it 'a sort of *Festschrift* – a tribute to Cornell University as it enters the second quarter-century of its existence' – which, probably more than any other single book, has made military passions normative in the interpretation of the relations of science and Christian faith.[70] The greater irony of Draper's work was that it predicted infallibly the body of literature that would follow it. Surely it is no less ironic that White should have succeeded infallibly in that which he so deplored, 'in thrusting still deeper into the minds of thousands of men that most mistaken of all mistaken ideas: the conviction that religion and science are enemies'.[71]

Prisoners of war

The *Conflict* and the *Warfare* had an enormous appeal. Each book was the product of forces in religion and science which converged in the life of a well-known American educator; each was published when these same forces were buffeting a sensitive religious public. The one became the best-seller of the International Scientific Series. The other 'captured the anticlerical sentiment of the *fin de siècle* "nineties"'.[72] Eventually the forces dissipated, public sentiments moderated, and the books' popularity declined. But the *Conflict* and the *Warfare* had left their mark. In capturing the imaginations of their authors' contemporaries, they employed a metaphor to describe the historical relations of science and Christian faith which has captivated writers on the subject, without respect of persons, for one hundred years. There is no better evidence of this continuing appeal than the pervasive use of the military metaphor to describe the forces of religion and science which first prompted Draper and White to write their books. For although the *Conflict* says nothing about evolution and religion, and the *Warfare* devotes only twenty of nine hundred pages to 'the final effort of theology' to defeat a scientific account of origins, historians have found little but 'conflict' and 'warfare' in the post-Darwinian controversies.

'The books by Draper and White were welcomed eagerly by all who were on the scientific side of the argument', writes Sidney Warren in his history of American free-thought. 'They found in these works new intellectual weapons with which to assail the citadel of Christianity.'[73] A chief weapon, as Warren's statement suggests, was the military metaphor itself. Pamphlets such as *The Irresistible Surrender of Orthodoxy* (1895) by J. E. Roberts and *The Struggle between Religion and Science* (1923) by Marshall J. Gauvin show from their titles alone how the metaphor could be deployed. More influential than these polemics in militarising the post-Darwinian controversies, though no less enamoured of the military metaphor, have been the vintage histories of rationalism. In *The History of English Rationalism in the Nineteenth Century* (1906) A. W. Benn states that Darwin's theory of natural selection 'seemed to promise the greatest victory ever yet won by science over theology' and so was 'turned to account in the warfare between science and theology'. He describes the 'retreat of theology' from natural religion and external evidences, arguing that final causes, 'the last entrenchment of theism', could 'hardly be held against the new artillery' and that the 'attempts to disarm and capture evolutionary science in the interests of theism or of Biblical religion' proved wholly unsuccessful.[74] Similarly, in his *History of Freethought in the Nineteenth Century* (1929), J. M. Robertson states that the *Origin of Species* was an 'irresistible arsenal of arguments'. Against it 'many of the clergy kept up the warfare of ignorance; but the battle was won within twenty years'. Bertrand Russell, the eminent philosopher, admits indebtedness to White in his historical survey, *Religion and Science* (1935), first published in the Home University Library. In the war waged continuously by traditional religion on medieval grounds, says Russell, 'science has invariably proved victorious'. 'Darwinism was as severe a blow to theology' as Copernican astronomy, over which the war's 'first pitched battle' was fought. Homer W. Smith, a scientist–historian, also relies on White's *Warfare*, which he heralds as 'one of the outstanding American contributions to rationalism'. In *Man and His Gods* (1953) Smith maintains that, after Darwin, 'medieval artillery was no longer effective against the "destroyers of the church", who were. . .too heavily armed with facts and reinforced by popular sympathy to be driven under cover'.[75]

Rationalists and free-thinkers might be expected to escalate the conflict between science and theology with a hostile metaphor. But surely those without anti-religious battle-axes to grind, historians of science and historians of ideas, animated by a spirit of objectivity, have written in a more dispassionate manner. Unfortunately, this has not always been the case. The military metaphor is not merely exemplified in general historiographical statements – those, for example, of John C. Greene, who endeavours to view Darwin and his writings 'in the broad perspective of the historical conflict between science and religion', and William Coleman, who states that 'evolution. . .provided a prominent battleground for the ongoing contention between science and religion for the allegiance of the European mind'.[76] The metaphor is extended *carte blanche* to describe the post-Darwinian controversies.

It was a 'battle royal. . .over the creation story', says F. Sherwood Taylor in a section headed 'The conflict of science and religion in the nineteenth century'. It was a 'fight for evolution' in which Huxley, possessed of a 'true fighting temperament', bore the 'brunt of the attack', according to William Cecil Dampier. It was an 'arms clash on Parnassus' following the 'catastrophic war' over geology, in the view of Julian M. Drachman. As the impact of the *Origin of Species* would 'take too long for the necessities of warfare', Drachman adds, Darwin's 'forces' had to be 'mobilized more quickly'. Huxley's 'embattled sentences marched toward a climax' in his encounter with Bishop Wilberforce at the British Association in 1860 – 'one of the most stirring incidents in the early war over evolution' – while 'reports from America showed the campaign raging there too under the generalship of Asa Gray'. And in a chapter entitled 'Attacks from all sides' Garrett Hardin features the 'first and most dramatic battle Darwinism had to fight', the Huxley–Wilberforce confrontation, which 'began a war that was vigorously waged without let-up for at least ten years'.[77]

Bert James Loewenberg, the pioneer historian of evolution in nineteenth-century America, writes that 'Darwin fired a shot heard round the theological world.' 'To question the source of species', he says, 'was to train the guns of speculation upon the last citadel of creationism'. Warren, the historian of American free-thought, explains that if the Church were to 'continue to battle the Darwinian legions, it would alienate many adherents who were beginning to see in those forces the standard bearers of true rational thought'. Herbert W. Schneider, the historian of American philosophy, asserts that 'the great majority of dogmatic theologians' in the United States 'adopted a militant attitude toward evolutionary theory'. After 1871, when the implications of

Darwinism had been fully drawn out, 'a bitter struggle ensued, and the theologians gave way very slowly, point by point, and only after waging a hopeless battle against superior forces'. Jacques Barzun, the American historian and *littérateur*, seeks to refine the metaphor by arguing that the storm over the *Origin of Species* was, above all, 'a major incident...in the dispute between the believers in consciousness and the believers in mechanical action; the believers in purpose and the believers in pure chance'. Thus, he declares, 'the so-called warfare between science and religion...comes to be seen as the warfare between two philosophies and perhaps two faiths'. Carlton J. H. Hayes, a social historian and a colleague of Schneider's and Barzun's at Columbia University, dignifies White's work in the heading of a chapter. Under the rubric ' "Warfare between Science and Theology" ' he writes that 'the fight began in earnest in the decade of the '60's over evolution and biblical criticism, and from 1871 to 1900 it raged on a wide front. The offensive passed early from "theology" to "science", whose heavy artillery was manned by such embattled Darwinians as Huxley, Tyndall, and Haeckel'. Huxley was well known as 'Darwin's bulldog' but Tyndall, according to Hayes, was 'Huxley's chief lieutenant' and Haeckel 'the outstanding artilleryman' in Germany. Even George John Romanes, Darwin's distinguished disciple, 'interspersed amateurish biological studies with cannon shots at basic religious beliefs'. 'While the big guns boomed', Hayes adds, taking up new aspects of the warfare, 'line after line of infantry – "higher critics", anthropologists, sociologists, psychologists – advanced unwaveringly with brand-new weapons against the old citadels of Christianity'.[78]

Religious historians also have perpetuated the military metaphor, though it would hardly seem to have been in their interests to do so. Some of course, being apologists, have had to deal with the relations of science and faith in the established phraseology, if only to correct its erroneous implications. But the majority, by falling prey to a captivating metaphor, have allowed Draper and White to 'set the terms of the debate'. The choice between adopting and questioning the military metaphor has existed from the outset, and is best illustrated by the two most substantial contemporary responses to Draper and White, both of which appeared in 1877: *The Final Philosophy* by Charles Woodruff Shields (1825–1904) and *Geschichte der Beziehungen zwischen Theologie und Naturwissenschaft* by Otto Zöckler (1833–1906).

If ambition be the criterion Shields was the nineteenth century's outstanding reconciler of science and Christianity. In a short treatise published in 1861 under the title *Philosophia Ultima* he advocated as a desirable and attainable object the production of a book, in conjunction with a scheme of academic studies, that would survey all the sciences and restate Christian theology in such a way as to eliminate their apparent conflicts. The idea impressed some wealthy friends and through their offices a fund was established which provided for Shields' appointment in 1865 as professor of the harmony of science and revealed religion in the College of New Jersey (later Princeton University). Within the classroom and without, for forty years Shields endeavoured to fulfil his vision. He was 'one of the last of that venerable band of clerical professors...who regarded themselves and were regarded by others as no less defenders of Christian orthodoxy than teachers of literature, philosophy and science'. The outcome of his labours was *The Final Philosophy* and two additional volumes, which together were published as *Philosophia Ultima* (1888–1905). Like its author, the work was an anachronism, another harmony of science and revelation.[79]

But it was a most anomalous anachronism. For with plans drawn from the pre-Darwinian past, where science and revelation were to be harmonised, and with a sense of urgency conditioned by the evolutionary present, where revelation seemed about to surrender to the 'warfare of science', Shields foresaw the 'final' harmony of science and revealed religion from a perspective as embattled as that of Draper and White.

That conflict which is raging in the bosom of this age between the reason of man and the word of God...is here to be viewed by us in the calm region of abstraction, in the cool mood of philosophy, and in the clear light of prophecy. As from the loopholes of a retreat, wherein we are being drilled for the actual warfare, we look forth on a battle-field, bounded only by the horizon of thought, covered all over with the smoke of controversy, and whereon not kings and peoples alone, but great ideas and principles are struggling for the mastery, with lasting interests of humanity staked upon the issue.[80]

After attending briefly to the 'early conflicts and alliances between science and religion' Shields takes up the great contemporary conflict in which the 'lines' are drawn between the natural and the supernatural, the 'weapons' are reason and revelation, and the 'issues' are civilisation and Christianity. Chapter by chapter he surveys the opposing ranks, summarising the effects of their movements on each of the sciences, on philosophy, and on civilisation. There are the 'battles of infidels and apologists' (which include Draper's *Conflict* and White's *Warfare of Science*), the 'truces of sciolists and dogmatists', the 'exploits of religious eclectics', and the 'surrenders of religious scepticism'.

None of these manoeuvres meets with Shields' approval. The warfare, he declares, must cease. Philosophy must arbitrate the dispute, a philosophy which is neither Positive nor Absolute, but 'Final.' From all appearances this was a philosophical proposal and Shields therefore could be misunderstood as one who fancied himself a modern-day Aquinas or Bacon. That it was not a philosophical but an eschatological statement is perhaps more apparent to those who live a century later. Shields believed that the millennial economy would be the orderly and ultimate outgrowth of the present world system. In the analogy of prophecy, in the historical philosophy of progress, and in society about him he saw remarkable signs of advance towards that Divine consummation. 'The same view', he says, 'harmonizes the otherwise conflicting interests which science and religion have fostered. . . . It begins at once to practically unite the natural and the supernatural, the terrestrial and the celestial, the human and the divine.'[81]

The ultimate philosophy may rise under another name and in other ways; but whenever, wherever, and however inaugurated it is itself inevitable. . . . It is that perfect system of knowledge and of society which both logically and providentially results from the whole previous development of humanity. . . . It is the millennium projected upon rational sequence as well as divine decree.[82]

As some laboured to bring in God's kingdom through social activism so Shields strove to articulate its philosophy, latent and emerging in the human mind moved by the Holy Spirit. Science, he believed, would thereby come progressively into harmony with revelation 'until the reason of man shall stand forth coincident with the word of God'.[83] The 'warfare of science' was not destined to purge faith of its myths and superstitions but to issue in a philosophy of millennial peace.

Far different is the approach of Zöckler, whose two lavishly documented 800-page volumes retain a considerably greater historical value than the works of Draper, White, and Shields. Although the Greifswald professor of theology shared with Shields the burden of harmonising science and scripture, he knew history too well to become entangled in a captivating metaphor.[84] 'It would perhaps be opportune, and would lend our work a greater attraction in the eyes of many, if we were simply to write a history of the conflicts between theology and natural science', he says at the outset of the book. 'But popular though such a history of conflicts might be, it would not be true.' Zöckler therefore refuses 'to support such writers of the Buckle school as J. W. Draper, G. H. Lewes, A. D. White, et al.' and that school's audience, 'with their glorifications of the warfares or the irresistible triumphs of science'. However, he does linger over White's *Warfare of Science* in one of his lengthy notes, showing that 'elegantly written little book' absolutely no mercy: 'Besides numerous unfounded rhetorical exaggerations, it contains many traditional misconceptions, ...chronological discrepancies, and otherwise annoying blunders. It is difficult to comprehend how a scholar of the stature of Tyndall [who contributed an introductory note to the English edition] can recommend such a superficial and bungling piece of work.' As for Draper's work, says Zöckler, it is 'of a similar calibre'. In his *Geschichte* the emphasis would be different. 'If the relations between theology and natural science are to be presented with objective, historical fidelity and free from onesidedness', he declares, 'then it is just as important to emphasise the helpful, as the hindering, influences of the former on the latter'.[85]

But it was Shields, not Zöckler, who struck the keynote for religious historians writing on science and religion. Again the military metaphor is not confined to general historiographical statements – those, for example, of A. C. McGiffert, who writes of 'the age-long conflict between science and theology', and James Y. Simpson, who submits the chapters of his conciliatory *Landmarks in the Struggle between Science and Religion* (1925) as 'in a sense guides to...old battlefields'. The metaphor is extended willy-nilly to describe the post-Darwinian controversies.[86]

First to employ this interpretation were of course the controversialists themselves. The last quarter of the nineteenth century saw a veritable explosion of military-minded literature on the relations of science and religion – titles such as *Science, Her Martyrdom and Victory* (1877) by William Sharman, *The Present Conflict with Unbelief* (1887) by John Kelly, and *Religion and Science as Allies* (1889) by James Thompson Bixby. Typical of works of this genre were books by two English clergymen, Gavin Carlyle and Nevison Loraine. In *The Battle of Unbelief* (1878) Carlyle sees the troops of that 'great system of unbelief', which masquerades under the name of evolutionary science, 'invading every home, alluring many of the young'. 'Its great enemy, which it opposes at all hazards', he asserts, 'is the word of God'. Loraine, on the other hand, is more optimistic. In *The Battle of Belief* (1891) he reviews 'the disposition of the forces engaged, the weapons used, and the modes of attack', finding to his relief that while Christianity 'holds the field' and 'advances all along the line', the 'opposing forces', whose 'only alliance is their common hostility to the Christian faith', are beset with 'confusion and internal strife'. 'No weapons from the armoury of physical nature', he says, 'can reach, much less destroy, the citadel of our divine faith'.[87]

Twentieth-century historians of British religious thought have followed the lead of the nineteenth-century commentators. Hector Macpherson relies on White's *Warfare* in describing two 'camps', scientists and theologians, and the 'violence' of their more 'reactionary' members in the controversy over evolution. F. R. Tennant writes, with evident indebtedness to White, that Darwin's teaching 'constituted the final and irresistible onslaught of science on the old view as to the nature of Biblical authority'. Similarly, H. D. A. Major declares that 'the religious opponents of Evolution were driven from the field' and, in consequence, 'the Divine fiats of *Genesis* were. . .withdrawn into the impregnable fortress of allegory or religious myth'. After evolutionists, abetted by Bishop Colenso, had made 'warfare with the Traditional-ists', he adds, 'the ark of Jehovah was carried forth no more to battle against the Philistines – at least, not in England'.[88]

In his *Religion in the Victorian Era* (1936) L. E. Elliott-Binns dis-cusses the post-Darwinian controversies under the heading, 'The Conflict between Science and Religion'. 'The conflict waged [*sic*] most fiercely during the second half of the century', he says; 'the battle was really joined' over the *Origin of Species*; and the 'first serious clash' was the Huxley–Wilberforce confrontation. 'In making its conquests, science, like every other conqueror, had. . .to destroy much.' Elliott-Binns continues in the same vein in his history of English theo-logical thought, explaining that 'the conflict between religion and science. . .was due to a great variety of causes, and perhaps above all to the failure of both sets of combatants to understand the aims and objects of the other'. The same vision of polarisation animates a pas-sage in Bernard Reardon's excellent study, *From Coleridge to Gore* (1971). 'Both sides were in the mood for conflict', says Reardon; 'the combatants on either hand were making larger assumptions than a cool concern for truth would have warranted'; and 'a truce was possible only when it was recognized that the spheres of both were limited'.[89]

The most important study of Darwin's reception in the American religious periodical press comes fully armed with the imagery of war. On 'chevaux de bataille', writes Windsor Hall Roberts, Darwin's theological opponents rode forth in a 'furious assault' on evolution. They sought out the weakest points in the 'Darwinian armor', anticipating an 'easy victory over this unworthy foe'. 'Any weapon which would aid them in beating back the new heresy was considered legitimate.' The idea was to 'shell the Darwinians out of their trenches' without wasting 'scattered firing...on obsolete forts of skepticism'. The second decade of battle, in Roberts' opinion, 'witnessed the most intense phase of the struggle', both sides alternately 'claiming the victory'. Another American contribution to the history of the relations of science and religion, *The Christian View of Science and Scripture* (1955) by the neo-evangelical apologist Bernard Ramm, was written to call a moribund Fundamentalism back to the 'noble tradition' of the 'great and learned evangelical Christians' of the later nineteenth century. It begins with a description of 'the battle of the Bible and science', a battle in which 'the victors were on the side of modernism and unbelief'. 'The orthodox had little time to develop a strategy to combat the critic', Ramm concedes, and so 'had to pitch in and fight' the best they could.[90]

This montage of war-like quotations could be extended without the slightest difficulty. We might add the remarks of lay theologians such as C. A. Coulson, who speaks of two phases in the relationship of science and religion after Darwin, the first in which Christianity was 'losing almost every battle', the second in which there has been an 'uneasy peace'. We might also enlist the views of philosophical theologians such as Richard H. Overman, who tells of 'the great battle between orthodox Christian theology and the Darwinian science'.[91] To continue, however, would be to belabour a point which by now should be quite clear: that the captivating metaphor of Draper and White has made historians prisoners of war. One looks almost in vain for some new interpretation, some better understanding, amid the tedious terminology of battles, truces, surrenders, combatants, and armaments which is supposed to describe Christian responses to Darwin. From fear of Roman Catholic power and revulsion at sectarian control of higher education Draper and White composed their highly-coloured histories – this much is easy to comprehend. It only remains to account for the uncritical extension of their historiography by writers who stand at a more or less dispassionate distance from the offending events. Some explanations will emerge as we consider the milieu in which historical studies of the post-Darwinian controversies have been written.

2

∞∞∞∞∞∞∞∞∞∞∞∞∞∞∞∞∞∞∞∞∞∞∞∞∞∞∞∞∞∞∞∞∞∞∞

POLITICS, POLEMICS, AND THE
MILITARY MILIEU

> E'er since the world evolved to form
>   And creatures moved upon its face,
> The fittest have survived the storm
>   Of beasts and of the human race.
> But as the world grows old 'tis seen
>   The fight grows fierce – 'tis hard to live,
> War to the knife rages between
>   Mankind, and none do quarter give.
> In learning, commerce, trade, and war,
> Such battling, struggling, strife, was never seen before.
> 'Psychosis'[1]

Military metaphors are as old as war itself. From Roman militarism came the imagery of battle contained in the New Testament, which has ever been at the disposal of Christians who believe themselves to be combating evil. From Christians, in large measure, this imagery has passed into common usage, so that even the opponents of faith have learnt to wage metaphorical war against their adversary. Thus it would not be surprising to discover that Draper, White and the prisoners of their historiography have drawn heavily on the vocabulary of a milieu in which military metaphors were rife. In particular it would appear that mid-Victorian politics, the polemics of T. H. Huxley, and the militant tactics of American Fundamentalists in the twentieth century have together furnished no small amount of unforgettably vivid military imagery to writers on science and religion. As influence can only occasionally be assigned to one or another of these aspects of the military milieu, Draper and White must still be regarded as the principal *casus belli*. But the influences which bore upon them and the enrichment of an entrenched historical tradition that stems from their writings are better understood with this background in mind.

Political provenances

Wars and rumours of wars occupied the thoughts of Christians on both shores of the Atlantic no less in the middle decades of the nineteenth century than at its close. The Darwinian 'revolution' was preceded in Britain by the Crimean War and the Indian Mutiny and followed by the Boer War; in the United States it was postponed by the Civil War and completed during the war with Spain. The earlier conflicts, however, were determinative for the rise and spread of military sentiments among Christians of both nations. On either hand believers sought scriptural sanctions for their government's action. If the hostilities were not seen as a crusade, or God's war, then they could at least be considered a just war. The military imagery of the Pauline epistles – the Christian as soldier, clothed in the 'whole armour of God' – and the eschatological conflict of good and evil described in the Apocalypse were pressed into service, heedless of the fact that 'the use of military metaphors was a part of the Romanizing of the Gospel': that, in other words, such paradoxical language was at once an intelligible and a transformed use of the first-century idiom.[2]

In Great Britain many Christians idealised the Crimean conflict, one motive for which had been 'a desire for warlike adventure that seized the English people after forty years of peace'.[3] Of the hagiographical literature that appeared after the Treaty of Paris in 1856 no work enjoyed a greater vogue than Catherine Marsh's novel, *Memorials of Captain Hedley Vicars, 97th Regiment*. Its effect was to convince the public that with little difficulty one might be both a spiritual and a secular soldier. The Indian Mutiny in 1857–8 furnished proof historical, not fictitious, that soldiery and Christianity were thoroughly compatible. Amid the massacres and reprisals of that iniquitous affair, which so enflamed the spirit of revenge in the homeland, stood General Henry Havelock, a hero of the highest rank and a Baptist of the most uncompromising breed. Again and again in the summer of 1857 his troops defeated overwhelming forces, winning for him the K.C.B. and a major-generalship. Fraught with the anxiety and fatigue of his last campaign, however, he succumbed to dysentery in the autumn of the same year, before news of his last promotion could reach him. Havelock became a martyr at once, and soon, on a surfeit of Christian eulogies, a virtual saint of the dissenting churches. For many the conclusion was thenceforth inescapable that a thoroughly evangelical Christian makes the best of all possible soldiers. 'The army', observes Olive Anderson, 'had been accepted by many different sections of the religious public as part of the church militant on earth'.[4]

The political cataclysms of the fifties were only the prelude to a series of intellectual tremors which jarred large sectors of the British churches into militant defence of the faith once delivered to the saints. Beginning in 1857 each year for the greater part of a decade seemed to bring a new and deliberate assault on established religious truth: Buckle's *History of Civilization in England*, Mansel's *Limits of Religious Thought*, the *Origin of Species*, *Essays and Reviews* and its legal aftermath, Bishop Colenso's *Pentateuch and Book of Joshua Critically Examined*, Lyell's *Antiquity of Man*, Huxley's *Man's Place in Nature*, and Lecky's *History of the Rise and Influence of the Spirit of Rationalism in Europe*. Now at last, midway between the atheist G. J. Holyoake's conviction for blasphemy in 1841 and the onset of the Secularist Charles Bradlaugh's successful parliamentary campaign to 'affirm' forty years later, it was becoming (in the words of Henry Sidgwick) 'impossible. . .to conceal from anybody the extent to which rationalistic views are held, and the extent of their deviation from traditional opinion'.[5]

The intimate British politics of church and state did not always tryst behind closed doors. Rowland Williams, who contributed the article 'Bunsen's biblical researches' to *Essays and Reviews*, was on that account indicted for heresy by the Tractarian bishop of Salisbury, W. K. Hamilton, condemned by the Court of Arches, but acquitted in 1864 by the Judicial Committee of the Privy Council. Likewise his partner in theological crime, Henry Bristow Wilson, the author of the article on the National Church, though convicted for denying the inspiration of scripture and eternal punishment, was cleared in the same judgement. The foremost defeat for the politics of religious conservatism came five years later when another of the 'septem contra Christum' was nominated to the see of Exeter by a new prime minister, the Tractarian and politically liberal Gladstone. Withstanding pressure and protest from every quarter, Frederick Temple refused to clear his name from association with *Essays and Reviews*. Only after his consecration did he withdraw his article from the book, and then not to imply censure of the other writers, but because he wished to acknowledge that some expressions allowed to the headmaster of Rugby School were not therefore permissible to the bishop of Exeter.[6]

Christian defences against modern unbelief of such proportions were not always of the bare-fisted variety encountered by Bradlaugh, who was swept from his platform by a mob of believers at Burnley in 1861 and at Wigan had to endure an assault upon his full lecture hall led by a local clergyman, whose secretary forced himself through a window while other belligerents threw lime into the building and poured water through its ventilators.[7] Generally churchmen were militant more in metaphor than in reality as they endeavoured to neutralise the political challenges of infidelity by securing the election of people's souls. 'The use of military imagery to describe the individual's spiritual life and the church's work in the world is as ancient as the church itself; but the popular hymns, sermons and to some extent the devotional literature of the sixties very often leave the impression that in the decade it acquired a new lease on life.'[8] During those tumultuous years, when unorthodoxy in science and religion was assailing heart and mind, Sunday School children first sang Baring-Gould's 'Onward Christian Soldiers, marching as to war' and their parents learnt the words of 'Fight the good fight' by J. S. B. Monsell and 'For all the saints' by Bishop William Walsham How. Later came para-military organisations – the Salvation Army, the Church Army, and the Boy's Brigade – and later still the 'Student Movement', embracing the Student Volunteer Missionary Union in Britain and the Student Volunteer Movement for Foreign Missions in the United States. All during the flourishing years of these latter organisations 'military phraseology and viewpoint were prevalent' in projecting an evangelical 'conquest' of the world.[9]

Yet behind the metaphor in its every form lay a real political hope, variously expressed no doubt, but as unforgettable as it has ever been in times when the Church has believed itself to be under attack by the forces of Antichrist. Writing in 1878, the Reverend Gavin Carlyle found himself living in 'a crisis in the world's history' during which, he believed, the hieroglyphs of the much-neglected Apocalypse would soon be realised in historical events. 'Is there not a great conflict between light and darkness, increasing in intensity every day, – likely to be more fierce? On the one hand the Church of Christ, reviving and strengthening, and full of the buoyancy of life; on the other, her sceptical opponents, ready to destroy her, not with physical weapons, ...but with the weapons of universal doubt.' The forces being thus arrayed in his mind, Carlyle concluded: 'Many events indicate that we are now at the beginning at least of the great struggle which is not to end until the ushering in of the millennial period.'[10]

The 'chief encouragement to the increased use of military metaphors' in mid-Victorian Britain stemmed from 'the long-standing trans-Atlantic connections of evangelical and revivalist circles'.[11] Militant evangelical and revivalist faith, in turn, was largely the product of the American Civil War. The absence of an established church in the United States had seldom caused its citizens to question whether their country was 'Christian'. Nor did it matter to them that the convictions of the founding fathers owed rather more to the Enlightenment than to orthodox Christianity. For it was only too obvious that God – the God of the Protestant majority – had led this people from the fleshpots of Europe and had brought them into a land flowing with 'inalienable rights' under the bold and obedient leadership of General George Washington, who, with the rest of the nation's elders, had by divine guidance bestowed on them sacred writings for the instruction of national life and character, the Declaration of Independence, and the Constitution.[12] But with brother pitted against brother, with Methodists, Baptists, and Presbyterians hopelessly divided over the moral issue of slavery, 'Christian' America entered a four-year identity crisis, a slough of despond, and a spiritual quandary. Could 'that nation, or any nation so conceived, and so dedicated', to use President Lincoln's famous words, '. . .long endure' the unmitigated sin of slavery (as one side saw it) or the divisive aggressions of abolitionism (as their brethren perceived the issue) without incurring the wrath of Almighty God?

The answer became apparent as each side took upon itself the burden of divine judgement and retribution. 'The Protestant forces rallied confidently to battle for the Christianization of American life', says Robert T. Handy. In North and South during the Civil War this was the dominant spirit, the citizens of both sides believing from their leaders that they were the 'true defenders of Christian America'. In his remarkable study, *Revivalism and Social Reform*, Timothy L. Smith calls the northern campaign 'the spiritual warfare against slavery'. On the front lines of abolitionism were the revivalists and revival clergymen of America's 'second evangelical awakening' in 1857–8 – Charles Grandison Finney, Elder Jacob Knapp, George B. Cheever, Henry Ward Beecher, Edward N. Kirk, Francis Wayland, and Albert Barnes. 'No avenue of propaganda could have been devised more effectively to harden Northern antipathy toward slavery than the pulpits and pens of such men', observes Smith. 'Revival religion's war on slavery sustained the theocratic ideal that God must rule American society.' In 1861 Horace Bushnell, the leading

Congregationalist theologian, declared the Civil War to be a great 'moral regeneration' which, when complete, would be 'a kind of religious crowning of our nationality'. Lincoln's Gettysburg Address in 1863 revealed its atoning basis in the 'honored dead' who had made possible a 'new birth of freedom'. And when the Great Emancipator himself joined their ranks at the hand of an assassin two years later, the symbolic equation of him with Jesus entered the civil religion once and for all. The American *Heilsgeschichte* had begun its Christian dispensation.[13]

Julia Ward Howe's 'Battle Hymn of the Republic' is only one of the more durable examples of the militant fervour which inspired American evangelicals during the Civil War. At the time there were numerous compositions expressing similar sentiments, such as 'Stand up, stand up for Jesus! Ye soldiers of the Cross' by George Duffield, 'Hold the Fort' by P. P. Bliss, and the famous war songs of Philip Phillips, 'Battling for the Lord', 'Won't you Volunteer?' and 'Recruit for the Army above'. William E. Boardman, executive secretary of the United States Christian Commission during the war, became a full-time evangelist after his duties ended. His first book, *The Higher Christian Life*, published at the height of the 1858 revival, was a phenomenal success on both sides of the Atlantic, assuring a sizable circulation in the same regions for his post-war production, *He That Overcometh; or, A Conquering Gospel* (1869). But the 'conquering Gospel' was exported chiefly by Dwight L. Moody, a former volunteer agent of the wartime organisation which Boardman served. Evangelistic work among the northern troops had equipped Moody with a vividly militaristic vocabulary that was fresh on his tongue when he first visited the British Isles in 1867. Its phrases – a 'campaign' was a local mission and 'press the fight' meant testify boldly – were augmented by the voice and harmonium of Ira Sankey in later missions (1873–5, 1881–4, 1891–2), popularising much of the martial hymnody which had taken the American churches by storm.[14]

Their hymnody and rhetoric notwithstanding, Victorian evangelicals should not be held directly accountable for describing the relations of science and faith in a military metaphor. Many there were spurred into the militant defence of religious truth who did see a threat in the advance of physical discovery, but doubtless the great majority of these Christian thinkers also believed with the aged Dr Pusey that 'unscience, not science', is 'adverse to faith'. Even the Reverend

Carlyle, whom we have met as a prophet of an intellectual Armageddon, held that 'nothing can be more perilous to the influences of religion than the idea that there is any conflict between her and genuine science'.[15] Rather, with ample justification, evangelicals may be held accountable indirectly for the rise and spread of the military metaphor. By outspokenly admiring and imitating the military, Christians unwittingly established a strong association of themselves with those whose task it was to make war, and thereby gave rise to a certain interpretation of their opposition to various cultural and intellectual trends. Certainly, at any rate, the muted horror with which T. H. Huxley reacted to General Booth's 'Darkest England' scheme and the political ramifications of the Salvation Army was not entirely without justification.[16] Nor was President White's connexion of warfare with sectarian political attacks on Cornell University purely coincidental. Perhaps then we can to some extent pardon, or at least better understand, individuals who chose to abide by the metaphor variously expressed in religious life by those with whom they strongly disagreed.

But this is only one side of the story. The political provenances in which believers took up a military metaphor were susceptible of another interpretation which historians of the Buckle–Lecky school, and free-thinkers generally, did not hesitate to exploit to the full. The rationalist historian A. W. Benn expresses it in an unusually revealing manner:

The English revolution in speculative opinion went hand in hand with a world-wide revolution in politics, each receiving inspiration from the other. Simultaneously with the appearance of Buckle's volume came the first step toward emancipation in Russia, and the appearance of his second volume coincided with the completion of that great work. Among the memorable events of 1859 Mill's 'Liberty' and Darwin's 'Origin of Species' rank together with the expulsion of Austria from Lombardy and Central Italy. Next year, within three months after Wilson and Jowett led their little band against the fortress of conservative theology, Garibaldi set sail with his thousand for Marsala; and the last stand of the Bourbons at Gaeta played a sinister accompaniment to the protest of our Bishops against 'Essays and Reviews.' Then the conflict between freedom and slavery, civilisation and barbarism, took shape on a scale of wholly unprecedented magnitude in the American War of Secession, to be fought out simultaneously with the war of opinion in England, leading to the same triumphant issue for the cause of reason on both sides of the Atlantic, the fall of Richmond in April, 1865, being followed by what we now know to have been the defeat of English clericalism in the general election

of July in that same year. Then, twelve months later, as though anticipating the new Atlantic cable, the lightnings of deliverance crossed
from the battle-fields of Virginia to the battle-fields of Bohemia, where
the champion of Jesuitism, rising in her fall, went down at Sadowa before
the champion of North German science and culture.[17]

This 'vast tide of feeling in favour of emancipation from constituted
authority', as Benn called it, also flowed from the pens of Draper and
White. Among their liberal admirers in the United States was Minot
Judson Savage, a Unitarian clergyman and one of the first Christian
leaders in America to declare himself in favour of evolution. Taking
over the phrase employed by the politician William Henry Seward to
characterise the Civil War, Savage associated a pre-Darwinian world-
view with southern slavery in the title of one of his later books – which
we give in full – *The Irrepressible Conflict between Two World-
Theories: Five Lectures Dealing with Christianity and Evolutionary
Thought, to Which Is Added 'The Inevitable Surrender of Orthodoxy'* (1892). These examples show clearly that evangelical Christians
were not alone in adopting the prevailing political vocabulary.[18]

To place an embargo on every military metaphor would be impos-
sibly pedantic, especially when one considers the extent to which both
religion and science participate in the political process. When, for
example, Charles E. Raven, who otherwise shuns the military metaphor, describes the aftermath of the post-Darwinian controversies as
'ravages of war' he is not an uncritical follower of Draper and White.
Writing during the Second World War, he employs the inter-personal
and intellectual consequences of disagreement over evolution to pre-
figure the human trauma of the global conflict through which his
readers were living. 'Warfare', he declares, 'is always disastrous as a
method of solving problems; for both parties to it emerge with their
ideas narrowed and distorted and their characters inevitably warped'.[19]
In fact, Raven, who was himself a pacifist, intends to turn the metaphor against itself by an understanding of the political responsibilities
of science and religion. 'To talk about the triumphant march of
science', he says, 'would be as ironical as to sing "Like a mighty army,
moves the church of God". Both science and religion must take a share
of blame for the appalling catastrophes which they ought to have been
able to prevent.'[20]

Furthermore, contemporary political catchwords often lend colour
to discussions of religion and science, and in so doing create interest
and insight. At a time when the words of Winston Churchill's famous

speech at Fulton, Missouri were still ringing in people's ears, John Baillie stated, 'Nowhere is it possible to draw an iron curtain across the field of our experience and proclaim either the exclusion of science from the one side or its totalitarian right on the other.' And several years later, at the height of the Cold War, H. G. Wood wrote of the 'amicable coexistence' of science and religion.[21] Although one might fault some implications of these metaphors – that, for example, there are hostile powers to be separated by a political boundary – there is no need to do so. The statements were not historical in character. It is just when the vocabulary of war predominates in historical interpretations that the task of revision must be undertaken.

The onslaught of T. H. Huxley

In *Windyhaugh*, a novel set in the 1860s, Margaret Todd has one of her characters comment that 'the strange determined resolution to go out in pursuit of the Truth', which seemed so typical of the time, 'reminds one almost of the crusades, of the search for the Holy Grail. ...Here was a whole army with its enthusiasts, its raw recruits, its mercenaries, its troop of mere camp-followers.'[22] This vivid picture may or may not represent historical reality, but there was certainly one public figure of the sixties who would have found it congenial: the self-styled 'gladiator-general' of evolutionary science, Thomas Henry Huxley (1825–1895).[23] Truth was his obsession, together with its corollaries, intellectual liberty – the freedom to seek the truth – and personal morality – the obligation to speak the truth. Warfare was his favourite metaphor by which to describe the scientific pursuit of truth in the face of ecclesiastical constraints on truth-seeking and failures at truth-telling. With objectives clear, strategy defined, and the pen as his weapon, Huxley produced a polemical discharge which has won him decorations for the highest valour in settling the Victorian controversies in favour of Darwinism and evolution.

An omnivorous reader, Huxley could not have failed to digest Draper's *Conflict* and White's *Warfare of Science*. What little evidence that he did, however, should not be construed to mean that his polemical style was somehow an outgrowth of the metaphor developed in these books.[24] As he awaited the birth of his first son on the last evening of 1856, an occasion twice hallowed for the making of resolutions, Huxley projected in his diary fifteen to twenty 'Meisterjahre', beginning in 1860, during which he would undertake 'to smite all humbugs, however big; to give a nobler tone to science; to set an example of abstinence from petty personal controversies, and of toleration for everything but lying'.[25] How far he succeeded in giving science a nobler tone and in abstaining from petty personal controversies is largely a matter of opinion. But it is a matter of historical record that from 1860 until his death Huxley missed hardly a chance to smite humbugs of every variety and to condemn untruthfulness in every form. 'Battles, like hypotheses', he admonished Ray Lankester, 'are not to be multiplied beyond necessity'. He hastened to add, 'No use to *tu quoque* me. Under the circumstances of the time, warfare has been my business and duty.'[26]

From the outset Huxley knew these circumstances well. He may not have had much contact with the geological controversies over the age of the earth and the Deluge instigated by Charles Lyell's *Principles of Geology* (1830–3) – in later years he would learn from Lyell of the 'social ostracism' which had pursued him after its publication[27] – but the outcry against the factual and theological errors contained in the anonymous *Vestiges of the Natural History of Creation* (1844) was well within recollection.[28] Realising, therefore, the pressures that could be brought to bear on advocates of evolution, and realising that the work in preparation by his naturalist friend, Charles Darwin, was hardly less heretical than *Vestiges*, Huxley expected that its religious opponents would be no less vociferous. Indeed, there was every reason to think they would be all the more vociferous in view of the book's expert presentation of evidence and the established reputation of its author. But though humbug were manifested with all the prestige of ecclesiastical authority, Huxley would quash it. 'You must recollect', he wrote to Darwin on reading a pre-publication copy of the *Origin of Species*, 'that some of your friends. . .are endowed with an amount of combativeness which. . .may stand you in good stead. I am sharpening up my claws and beak in readiness.'[29]

In an address delivered at the Royal Institution in February 1860 Huxley anticipated those among Darwin's immediate opposition who would forbid the investigation of phenomena presumed to be directly dependent on the Divine will, and attacked them with ferocity. 'There is a wonderful tenacity of life about this sort of opposition to physical science', he said. 'Crushed and maimed in every battle, it yet seems never to be slain; and after a hundred defeats it is at this day as rampant, though happily not so mischievous, as in the time of Galileo.'[30] Then in April of the same year came one of Huxley's most memorable literary moments, two glorious, swaggering, inflammatory paragraphs, doubtless believed by many to have emanated from the depths of hell.

In this nineteenth century, as at the dawn of modern physical science, the cosmogony of the semi-barbarous Hebrew is the incubus of the philosopher and the opprobrium of the orthodox. Who shall number the patient and earnest seekers after truth, from the days of Galileo until now, whose lives have been embittered and their good name blasted by the mistaken zeal of Bibliolaters? Who shall count the host of weaker men whose sense of truth has been destroyed in the effort to harmonize impossibilities – whose life has been wasted in the attempt to force the generous new wine of Science into the old bottles of Judaism, compelled by the outcry of the same strong party?

It is true that if philosophers have suffered, their cause has been amply avenged. Extinguished theologians lie about the cradle of every science as the strangled snakes beside that of Hercules; and history records that whenever science and orthodoxy have been fairly opposed, the latter has been forced to retire from the lists, bleeding and crushed, if not annihilated; scotched, if not slain. But orthodoxy is the Bourbon of the world of thought. It learns not, neither can it forget; and though, at present, bewildered and afraid to move, it is as willing as ever to insist that the first chapter of Genesis contains the beginning and the end of sound science; and to visit, with such petty thunderbolts as its half-paralysed hands can hurl, those who refuse to degrade Nature to the level of primitive Judaism.[31]

Thus it may not have been without cause that, two months later, the bishop of Tractarian Oxford chose to launch a semi-official assault on the Darwinian party, gathered in his own diocese for the annual meeting of the British Association for the Advancement of Science.

No battle of the nineteenth century, save Waterloo, is better known. No encounter between science and religion has been more often described. And at no time in history have the iniquities of Pope Urban VIII and the curia which condemned Galileo been visited on their spiritual descendants so thoroughly as on Saturday, 30 June 1860. For when Samuel Wilberforce (1805–1873) had fully ventured forth against Darwin with ridicule and mock politeness, he 'suffered a sudden and involuntary martyrdom'. Huxley 'committed forensic murder with a wonderful artistic simplicity, grinding orthodoxy between the facts and the supreme Victorian value of truth-telling'.[32] The entire episode – from Draper's sonorous address, to the bishop's intimation of Huxley's ancestry, to the scientist's grave expression of preference for the ape over one who used gifts of eloquence to obscure the truth – has been recounted elsewhere in full, if conflicting, detail.[33] Few scenes in modern history contain the drama and fewer still the humour and human interest of the Huxley–Wilberforce confrontation. This alone might explain its recurrence in the literature of the post-Darwinian controversies were it not for the lure of the military metaphor. Draper does not mention the confrontation. White passes over it in four un-warlike sentences. 'The famous clash...was not reported by a single London daily newspaper at the time, and...of the few weekly reviews that mentioned it none brought out the force of Huxley's remark.' But later commentators seldom neglect to make it a major 'battle'.[34]

It is true that one of the earliest first-hand accounts of the scene calls it 'the battle of the "Origin"'; that Francis Darwin describes Huxley's tilt against the anatomist Richard Owen on 28 June, together with the Wilberforce confrontation, as 'two pitched battles'; and that Darwin himself refers to the latter incident as 'the battle royal at Oxford'.[35] Some historians therefore may owe their terminology to these sources.[36] But the ritual recountings of other writers – the following are only a representative sample – seem to depend more on a fertile metaphor than on historical investigation. Francis Warre Cornish, a church historian, claims that the attempt by Wilberforce 'to destroy the Darwinian theory by theological weapons damaged the current theology more than the theory', despite the well-known fact that the bishop had, for the most part, simply served up scientific arguments learnt from Owen, seasoned with his own acidulous wit. Cyril Bibby, Huxley's biographer, states that 'the crowd soon tired of the grape-shot and called out for the heavy artillery', falsely implying that Draper's address, which immediately preceded the bishop's 'heavy artillery', was hostile towards the Darwinians. Vilhelm Grønbech, the Danish historian, asserts that here, in 'one of the great battles' of a war which 'ended in an overwhelming victory for science', the 'sharp sword of debate was swung with such vigor that listeners fainted and had to be carried out', misleading one to believe that more than one poor lady expressed her sentiments in that bygone idiom. Philip G. Fothergill, a Roman Catholic botanist, disregarding Huxley's premeditated attacks prior to the June gathering of the British Association and claiming that 'a violent controversy ensued' on that occasion, says: 'As far as opposi-tion from clergymen is concerned, it would have been better if they had held their peace until the scientific critics at least had had time to load their guns for an attack.'[37]

Perhaps the most flagrant use of the Huxley–Wilberforce con-frontation to support the military historiography occurs in volume

three of the English translation of the *Histoire général des sciences*.
The French tells how 'des controverses passionées et violentes se
déchaînèrent' under the general rubric 'L'accueil fait au darwinisme'.
But in English, under the heading 'Religious Opposition to Darwinism', the 'bitter and violent controversies' become a 'war between
science and religion' which 'may be illustrated' by the Owen–Wilberforce attempt to ridicule Darwin. The single sentence given to describing for French readers the resulting scene is expanded into three
agonising paragraphs.[38] Were it not that this distortion of history
appears in a standard work of reference its importance would pale
before a statement which exceeds all other statements in its imaginative
use of the Huxley–Wilberforce confrontation to substantiate the
military metaphor. 'That exchange', says Reginald Stackhouse,

> was the beginning of what became not only the religious battle of the
> century but a war which fought some of its fiercest engagements in the
> next century and which is still going on as a kind of 'spiritual cold war.'
> From the outset the conflict has followed much the same lines as the
> debate between Wilberforce and Huxley. On one side have appeared
> champions of religion arguing against evolution not so much with
> scientific reasons as with appeals to the Bible, and deriding the advocates
> of evolution as infidels. The other side has been dominated by the
> scientists who have treated Darwin's hypothesis as an absolutely proved
> fact and have regarded all critics as narrow-minded bigots. For a century
> no Christian communion has been free from this debate, and just as both
> sides retired on that Sunday [*sic*] in Oxford in 1860 without either achiev-
> ing a clear victory, so both sides are still sending doctrinal salvos at each
> other.[39]

Nothing could be farther from the truth. But nothing so well illustrates
the deleterious effects of the military metaphor.[40]

'If men like Samuel of Oxford are to have the guidance of her
destinies', wrote Huxley to Charles Kingsley, 'that great and powerful
instrument for good or evil, the Church of England', will be 'shivered
into fragments by the advancing tide of science'. Only 'by the efforts
of men who, like yourself, see your way to the combination of the
practice of the Church with the spirit of science', he added, could that
disaster be averted.[41] Such a man, in Huxley's opinion, was not his
former student, the Roman Catholic author of *On the Genesis of
Species*, St George Mivart (1827–1900). By publishing his critique of
Darwinism in 1871 Mivart gave his own critics every opportunity to
evaluate it in the light of the 'Syllabus of errors' and the Vatican

Council. By claiming the support of Augustine, Aquinas, and Suarez in an elaboration of his own version of evolution by internal innate forces, he assured that they would do it. Huxley pounced on him with glee. The casuistical practice of Mivart's church had destroyed the spirit of science in him, and there was nothing Darwin's bulldog enjoyed more than fights with two-faced religious rationalisers of science.

Except fights with Roman Catholic rationalisers. The 'great antagonist' of science, Huxley believed, was the Church of Rome – 'the one great spiritual organization which is able to resist, and must, as a matter of life and death, resist, the progress of science and modern civilization'. Evidence for such a conviction was obtained in an encounter which Huxley relates with typically militant candour.

It was my fortune some time ago to pay a visit to one of the most important of the institutions in which the clergy of the Roman Catholic Church in these islands are trained; and it seemed to me that the difference between these men and the comfortable champions of Anglicanism and of Dissent, was comparable to the difference between our gallant Volunteers and the trained veterans of Napoleon's Old Guard.

The Catholic priest is trained to know his business, and do it effectually. The professors of the college in question, learned, zealous, and determined men, permitted me to speak frankly with them. We talked like outposts of opposed armies during a truce – as friendly enemies: and when I ventured to point out the difficulties their students would have to encounter from scientific thought, they replied: 'Our Church has lasted many ages, and has passed safely through many storms. The present is but a new gust of the old tempest, and we do not turn out our young men less fitted to weather it, than they have been, in former times, to cope with the difficulties of those times. The heresies of the day are explained to them by their professors of philosophy and science, and they are taught how these heresies are to be met.'

I heartily respect an organization which faces its enemies in this way; and I wish that all ecclesiastical organizations were in as effective a condition. I think it would be better, not only for them, but for us. The army of liberal thought is, at present, in very loose order; . . .we should be the better for a vigorous and watchful enemy to hammer us into cohesion and discipline.[42]

But General Huxley himself was ever ready to smite the 'great antagonist'. Physical science was the 'irreconcilable enemy' of the scholastic system, he declared in 1874, the year Draper's *Conflict* was published. 'The College of Cardinals has not distinguished itself in Physics or Physiology; and no Pope has, as yet, set up public laboratories in the Vatican.' As late as 1889 Huxley felt sure the Inquisition could burn again: 'The wolf would play the same havoc now, if it could only get its blood-stained jaws free from the muzzle imposed by the secular arm.' The ultimate rebuke at Huxley's hand was to be compared to the Church of Rome. Seizing on Comte's denial of liberty to conscience in matters scientific in his *Philosophie positive*, and the philosopher's proposal of a 'nouveau pouvoir spirituel' in the general appendix to his *Politique positive*, Huxley declared Positivism to be little more than 'Catholicism *minus* Christianity'. 'The logical practical result of this part of his doctrine', he said, 'would be the establishment of something corresponding with the eminently Catholic, but admittedly anti-scientific, institution – the Holy Office'.[43]

Mivart's fatal error was not his criticism and misquotation of Darwin, nor even his insistence that mankind's rational soul originated through divine intervention in the evolutionary process. For such disloyalty Huxley could merely accuse him of being a traitor to science. It was his attempt to show that the last great scholastic, Suarez, following Augustine, allowed that creation may have occurred by the development of primordial matter through the powers with which it was divinely endowed – a concept compatible with organic evolution – that destroyed him. For lesser offences men have gone to the stake. Rushing to the library at St Andrews University, Huxley plunged into a folio edition of Suarez and soon (he believed) had evidence sufficient to satisfy any ecclesiastical court that, in the 'progress of science' (to use the words of *Dei Filius*), Mivart had indeed given a doctrine of the Church 'a sense. . .different from that which the Church has understood and understands'. He prosecuted his case in the *Contemporary Review*, demonstrating considerable ability in the exegesis of Latin texts. 'If Suarez has rightly stated Catholic doctrine', Huxley summed up, 'then is evolution utter heresy. And such I believe it to be. . . . One of its greatest merits in my eyes is the fact that it occupies a position of complete and irreconcilable antagonism to that vigorous and consistent enemy of the highest intellectual, moral, and social life of mankind – the Catholic Church.' Turning to the convicted, Huxley exhorted, 'Let him not imagine he is, or can be, both a true son of the Church and a loyal soldier of science.' To Darwin, on the other hand, he exulted, 'I have come out in the new character of a defender of Catholic orthodoxy, and upset Mivart out of the mouth of his own prophet.'[44]

Only two positions, it seemed, were compatible with the supreme virtue of intellectual honesty: strict orthodoxy and agnosticism. And since a 'declaration of war to the knife against secular science' was the only position 'logically reconcilable with the axioms of orthodoxy' – Protestant or Roman Catholic – Huxley would not allow its representatives the moral luxury of occupying neutral ground. Fight or surrender; there was no other option. 'A hell of honest men', he told Mivart, 'will...be more endurable than a paradise full of angelic shams'.[45] It is not too difficult therefore to imagine the effect on Huxley of an essay which deplored the French theologian, Albert Réville's, surrender of Genesis to science, found comforting hints of the nebular hypothesis in the book's first verses, and a thoroughgoing compatibility of its account of creation with the facts of paleontology.

The patent falsity of such a harmony was not all that sent Huxley 'blaspheming about the house with the first healthy expression of wrath known for a couple of years'.[46] There amid a cloud of verbiage, imposing on its readers from the pages of the November 1885 issue of *The Nineteenth Century*, was the name of the Liberal prime minister himself, William Ewart Gladstone (1809–1898). On only two occasions in his career did Huxley later admit to taking the offensive, and this was one of them. 'It was for good reasons', he said.[47] After five years in office Gladstone had just lost a general election to the Conservatives. Far from defeated, however, he was making strong his alliance with Charles Parnell and laying plans to introduce his Home Rule Bill for the first time. But he lacked the support of late-Victorian intellectuals, most of whom regarded him as a demagogue.[48] Unionism prevailed among the leaders of literature and science, and Huxley was no exception. The Irish had once been to him 'a pack of Hibernian jobbers'; now he believed them to be 'ingrained liars'. 'That', he maintained, 'is at the bottom of all Irish trouble'.[49] Any plan to grant the Irish home rule thus ran counter not only to his political beliefs but to his most fundamental of moral convictions. In this matter, Huxley declared, 'the ignorance of the so-called educated classes in this country is stupendous;...in the hands of people like Gladstone it is a political force'. What better reason, then, to display him as 'nothing but a copious shuffler'?[50]

'The interpreters of Genesis and the interpreters of Nature' appeared in the December issue of the *Nineteenth Century*. Gladstone graciously ceded a few points in the January number while assuming new harmonistic contortions. Huxley shot back with 'Mr Gladstone and Genesis' in February, but not before 'taming my wild cat' (as he put it) at the editor's request.[51] Then the controversy lagged as each party engaged himself on other fronts. The defeat of the Home Rule Bill and his own defeat in the next election sent Gladstone back to the drawing boards. There, however, he found time to employ a familiar metaphor in a famous review of *Robert Elsmere* (1888), that sensational novel of tradition-bound religion and scientific biblical criticism in which, to use Gladstone's words, 'there is a great inequality in the distribution of the arms. Reasoning is the weapon of the new scheme; emotion the sole resource of the old.'[52] Meanwhile Huxley rebuked the philosopher W. S. Lilly, who had claimed to show that as a consequence of his materialism Huxley could not be moral, romped over Canon Liddon's repristination of mediaeval realism, had the last word with the Duke of Argyll in the ensuing discussion, tangled over agnosticism with the evangelical Henry Wace in three articles, and returned to Liddon long enough to obtain from him an excuse to liquidate the Deluge. But a swipe at Jesus for vandalism, for destroying the swine of Gadara, in the controversy with Wace brought Huxley back on to a collision course with the Grand Old Man. Gladstone upbraided him in one of the articles of his series in *Good Words*, arousing Huxley from 'dreams of peace' by 'the noise of approaching battle'.

The old warrior braced himself for the struggle.

I had fondly hoped that Mr Gladstone and I had come to an end of disputation, and that the hatchet of war was finally superseded by the calumet, which, as Mr Gladstone, I believe, objects to tobacco, I was quite willing to smoke for both. But I have had, once again, to discover that the adage that whoso seeks peace will ensue it, is a somewhat hasty generalisation. The renowned warrior with whom it is my misfortune to be opposed in most things has dug up the axe and is on the war-path once more. The weapon has been wielded with all the dexterity which long practice has conferred on a past master in craft, whether of wood or state. And I have reason to believe that the simpler sort of the great tribe which he heads, imagine that my scalp is already on its way to adorn their big chief's wigwam. I am glad therefore to be able to relieve any anxieties which my friends may entertain without delay. I assure them that my skull retains its normal covering, and that though, naturally, I may have felt alarmed, nothing serious has happened. My doughty adversary has merely performed a war dance, and his blows have for the most part cut the air. I regret to add, however, that by misadventure, and I am afraid I must say carelessness, he has inflicted one or two severe contusions on himself.[53]

Gladstone responded warmly in the conclusion of his *Impregnable Rock of Holy Scripture* (1890), which collected the articles of the

Good Words series. Stimulated afresh by his lack of moderation, Huxley struck back in the March 1891 issue of the *Nineteenth Century*, warning that 'when some chieftain, famous in political warfare, adventures into the region of letters or of science, in full confidence that the methods which have brought fame and honour in his own province will answer there, he is apt to forget that he will be judged by...people on whom rhetorical artifices have long ceased to take effect; and to whom mere dexterity in putting together cleverly ambiguous phrases, and even the great art of offensive misrepresentation, are unspeakably wearisome'.[54] The words spoke volumes. Gladstone, now over eighty, had had to put away Parnell in November after the revelation of his scandalous adultery with Kitty O'Shea. His Irish policy was at a crisis and another election was on the horizon. 'As to Gladstone and his *Impregnable Rock*, it wasn't worth attacking for themselves', Huxley told his son Leonard several years later; 'but it was most important at that time to shake him in the minds of sensible men'.[55]

And so Huxley waged war unto the very end. In 1892 he introduced the collection of a decade's polemical writing, *Essays on Some Controverted Questions*, with a fifty-three page history of conflicts over *the* 'controverted question' of Nature and Supernature. The alliance of Protestant and humanist at the time of the Reformation was 'bound to be of short duration and, sooner or later, to be replaced by internecine warfare' because intellectual liberty for Protestants did not extend to the supernaturalism of the Bible. In the eighteenth century, however, Bishop Butler 'captured the guns of the free-thinking array, and turned their batteries upon themselves'.[56] Not until the nineteenth century have supernaturalists had to 'cope with an enemy whose full strength is only just beginning to be put out, and whose forces, gathering strength year by year, are hemming them round on every side'. 'This enemy', proclaimed Huxley, 'is Science'.[57] The gladiator-general made his last assault on the Right Honourable Arthur James Balfour, who had laid himself open to aggression in his *Foundations of Belief* (1895). 'I think the cavalry charge in this month's *Nineteenth* will amuse you', Huxley wrote to a daughter. 'The heavy artillery and the bayonets will be brought into play next month.'[58] Not the next month or the next year but in 1932 the second part appeared.[59] Influenza had outflanked the old warrior and his revision could not be completed in the weakened condition which led to his first and final surrender on

29 June 1895. For once, at least, he did not fire the parting shot.

'Huxley was capable of shedding light, but he too much enjoyed engendering heat; he could construct a consistent ideology, but he preferred to demolish the shaky structures erected by his opponents.'[60] Charming and genial, able to lead and labour with, as well as to annihilate, those with whom he disagreed, Huxley nevertheless frequently fell victim to his own passions. 'It is not surprising that his language again and again suggests the stark words of the old Puritans', observes Grønbech. 'In him there is the spirit of those warriors who were outwardly resolute because they had fought courageously against the enemies within.'[61] 'Controversy is as abhorrent to me as gin to a reclaimed drunkard', Huxley confessed to John Morley; 'but oh dear! it would be so nice to squelch that pompous imposter'. When Herbert Spencer sought to induce him to reply to an article on the grounds that 'intellectual warfare' with Gladstone had been good for his health, he retorted, 'Your stimulation of my combative instincts is downright wicked. I will not look at the *Fortnightly* article lest I fall into temptation.' Though he himself might resist, if another should succumb he would allow him, 'once at any rate in his life, to perform a public war-dance against all sorts of humbug and imposture', but then admonish him, as he did Ernst Haeckel, to remove the war-paint as soon as possible. 'It has no virtue except as a sign of one's own frame of mind and determination, and when that is once known, is little better than a distraction.'[62]

In a lesser figure the polemics would indeed be a distraction. But as Huxley was the arch-antagonist of faith and certainly one of the most colourful contributors to the Victorian 'conflict' between science and religion, his irrepressible onslaught is rather a deception. Too easily historians and contemporaries alike have mistaken the prominent for the prevalent, and the outspoken for the ordinary. The essays of an episcophagous scientist have thus become a thesaurus for speaking of the post-Darwinian controversies.[63]

The anti-evolution crusade

The military metaphor was conceived *au milieu* of wars and war-like sentiments. Two decades of political turmoil and the christianisation of its terminology preceded Draper's *Conflict* and White's *Warfare of Science*. Throughout the last third of the nineteenth century Christian doctrines and their defenders were besieged by the polemics of T. H.

Huxley. And when at last the first generation of protagonists – Draper and White; Huxley, Wilberforce, Mivart, and Gladstone – had left the scene of strife, a fresh outburst of military passions revived and refurbished the metaphor for later historians. We refer not to the Great War but to the American religious phenomenon which immediately followed it: the emergence of militant Fundamentalism. Like the Huxley–Wilberforce confrontation but in reverse, militant Fundamentalism has deeply biased interpretations of the post-Darwinian controversies.

Not surprisingly, those who lived through the Fundamentalist period seem to have been the most prone to writing history in its light. The pioneer interpreter of evolution in America, Bert James Loewenberg, went to university in the heyday of militant Fundamentalism and completed his doctoral studies soon after its close. He writes in an influential article based on his Harvard dissertation that the hypothesis of evolution 'opened another battle in the perennial warfare between science and theology which was destined to rage until almost the close of the century', a battle in which 'those who owed their allegiance to that variety of orthodoxy known as fundamentalism would not long remain silent on the issues presented by science and philosophy'. Henry Steele Commager, a social historian for whom the controversies of the 1920s were well within recollection, thinks that 'the strength and persistence of fundamentalism well into the twentieth century is one of the curiosities of the history of American thought'. Completing his bilateral anachronism, which is also perhaps a curiosity of American thought, he refers to the famous 'monkey trial' of the Tennessee biology teacher, John T. Scopes, as 'one of the decisive battles in that warfare of science and theology which Andrew Dickson White deplored a generation earlier'. Echoing Commager, Richard Hofstadter, who was an impressionable youth in the twenties, states that 'the persistence of fundamentalism into the twentieth century is a token of the incompleteness of the Darwinian conquest'.[64]

'"When I use a word", Humpty Dumpty said, in a rather scornful tone, "it means just what I choose it to mean – neither more nor less"'. Thus the fate of 'Fundamentalism', which has become a term of disapprobation in every intellectual discipline. Sociologists and psychologists, who are accustomed to dealing in 'true believers', 'authoritarian personalities', and other convenient labels, can be forgiven for removing it from its original context. Even theologians, afflicted with a peculiar *rabies*, may be allowed occasionally to designate as fundamentalists those religiously more conservative than themselves. But surely historians, of all people, should be expected to employ a word in its historical sense, explaining carefully how it applies outside its proper context when a clear and common understanding does not do so. Especially is this required when a word comes freighted with 'local colour' and distasteful implications. Alice showed herself to be more astute in this regard than those whom we have just cited when she replied to Humpty Dumpty: 'The question is whether you *can* make words mean so many different things.'

The answer to Alice's question is, of course, No. 'Fundamentalism' has a particular historical referent and that referent lies in the twentieth century, not the nineteenth. It is now well known by historians that the word 'Fundamentalist' was coined by Curtis Lee Laws, editor of the Baptist *Watchman-Examiner*, on page 834 of the issue for 1 July 1920. Laws intended the word to denote believers who 'cling to the great fundamentals and who mean to do battle royal' for them. 'Fundamentalism' therefore must refer primarily to the movement of aggressive advocates of 'fundamental' Christianity which appeared in the United States about the year 1920. By way of confirmation, historians have pointed out that before 1920 Fundamentalism lacked much of the coherence which later made it so threatening as a movement within the Protestant churches. Despite the great emphasis on 'fundamentals', the reputed Fundamentalist creed consisting of 'five fundamentals' (which derived from a five-point doctrinal deliverance adopted by the General Assembly of the Presbyterian Church in 1910) was never recognised as normative by any large proportion of Fundamentalists. Nor was the series of twelve booklets published between 1910 and 1915 under the title *The Fundamentals* ever held to be the official repository of Fundamentalist beliefs. No creed and no book or series of books – not even the *Scofield Reference Bible* (1909) – can be considered the exclusive representative of Fundamentalism, certainly not before the twenties. In the years before its emergence as an aggressive force in American Protestantism, Fundamentalism subsisted as a disunified movement comprising diverse conservative theological viewpoints.[65]

The distinction between Fundamentalism before and Fundamentalism after 1920 – the caesura might be placed in 1918 at the end of the World War – is strikingly illustrated by comparing the attitudes towards evolution of proto-Fundamentalism, as expressed in *The Fundamentals*, with the attitudes that obtained in the controversies a decade later. In so doing we shall see that it has been Fundamentalism proper, the militant movement of the twenties, and not earlier expressions of theological conservatism, which has lent substance to the military metaphor and has thus been discerned anachronistically in the post-Darwinian controversies.

Among those who early allied themselves with the movement of conservative thought which became militant Fundamentalism were numerous outstanding Christian thinkers who had made their peace with evolution. Augustus Hopkins Strong (1836–1921), the Baptists' leading theologian, identified with and supported proto-Fundamentalism in the early years of the century, at the same time that his *Systematic Theology* was advertising a synthesis of evolution and the philosophy of personal idealism with every major Christian doctrine.[66] However, Strong was not as closely linked with the movement as was the Princeton theologian Benjamin B. Warfield (1851–1921), who contributed an article to the first volume of *The Fundamentals* on the deity of Christ. And yet Warfield accepted evolution as well. No sooner had his article in *The Fundamentals* appeared than he placed on record an old conviction, that evolution might supply a tenable 'theory of the method of divine providence' in the creation of mankind. Four years later he seemed to take particular pleasure in showing that Calvin's doctrine of the creation, 'including the origination of all forms of life, vegetable and animal alike, inclusive doubtless of the bodily form of man', was a 'very pure evolutionary scheme'.[67] Fundamentalists, notorious for denouncing and separating themselves from heterodoxy, had not yet learnt, it seems, to perform this amputation on their theologically diseased members.

That Warfield was no case of editorial malpractice is evident from the fact that he was not the only evolutionist contributor tolerated, if not welcomed, in the pages of *The Fundamentals*.[68] James Orr (1844–1913), professor of apologetics and systematic theology in the Glasgow college of the United Free Church of Scotland, contributed four articles, though as early as 1893 he had written in a well-known book that 'the general hypothesis of evolution, as applied to the organic world,...seems to me extremely probable, and supported by a large body of evidence'. In 1905, moreover, Orr affirmed that within certain limits 'no religious interest...is imperilled by a theory of evolution, viewed simply as a method of creation'. Thus it comes as no surprise to find Orr arguing in *The Fundamentals* that the Bible is not a textbook of science, that its intent is not to disclose scientific truth but to reveal the will and purpose of God, that the world is 'immensely older than 6,000 years', that the first chapter of Genesis is a 'sublime proëm' which science 'does nothing to subvert', and that although evolution is 'not yet *proved*, there seems to be a growing appreciation of the strength of the evidence for the fact of some form of evolutionary origin of species'. ' "Evolution", in short', said Orr, is coming to be recognised as but a new name for "creation" '.[69]

The most notable evolutionist contributor to *The Fundamentals* was George Frederick Wright (1838–1921), a renowned glacial geologist and professor of the harmony of science and revelation in Oberlin College. Wright had been a Darwinian for more than forty years when *The Fundamentals* appeared. In the mid-1870s he joined with Darwin's most prominent American supporter, Asa Gray, in publishing a collection of Gray's essays on Darwinism and natural theology. In 1882 he brought out a collection of his own essays, *Studies in Science and Religion*, which defended Darwinism and design with unique and compelling clarity. And in 1896 he stated in his *Scientific Aspects of Christian Evidences* that 'Darwinism really raises no new questions in the philosophy of Christianity'; it presents no greater challenge to faith than 'any other conception of the orderly processes of nature'. Fifteen years later, in the second of his three contributions to *The Fundamentals*, Wright maintained this very point of view. His essay, 'The passing of evolution', was only a refutation of those atheistic and agnostic versions of evolutionary speculation which exclude teleology *a priori*. Darwinism it defended from such metaphysical interpretations, declaring in no uncertain terms that Darwin did not eliminate the Designer behind variation and natural selection. 'Indeed', said Wright, sounding slightly more conservative than had been his custom, 'if it should be proved that species have developed from others of a lower order, as varieties are supposed to have done, it would strengthen rather than weaken the standard argument from design'.[70]

In all fairness to the editors of *The Fundamentals* their inclusion of two anti-evolutionary articles should be mentioned. But then one should not reason hastily from these articles to some generalisation about the uniformity of Fundamentalist attitudes towards evolution. Both articles appeared in volume eight, one of the volumes in the latter half of the series which, it was hoped, would be 'adapted to the more ordinary preacher and teacher', and both, accordingly, like four out of every five articles in volumes eight through twelve, were prepared by non-specialists, one by a clergyman, the other by a layman.[71] Even if the articles are regarded as tokens of hostility, however, it is none the less clear that evolution for the editors was hardly the preoccupation it became in the twenties. Only five articles – the ones we have mentioned – of the ninety which *The Fundamentals* comprise deal at any length with the subject. Therefore we shall have to look to the decade after the First World War to find a movement militantly opposed to evolution, a Fundamentalism that supplied the imagery to reinforce the metaphor in which the post-Darwinian controversies had been cast.[72]

The anti-evolution crusade of the 1920s was at least partly an instance of intellectual and cultural lag. By all rights, it seems, this popular movement should have occurred at the end of the nineteenth century.[73] Fundamentalists were suddenly overcome by the fear that American education was falling into the hands of evil men, God-denying evolutionists who were bound to subvert the faith and morals of the next generation by teaching that Genesis was not so and that mankind is akin to the apes. As proof of the insidious work of modernism and evolution Fundamentalists cited a statistical study of the belief in God and immortality among educated persons – college students and scientists, many of whom were educators – that was published in 1916 by James H. Leuba, professor of psychology in Bryn Mawr College. Leuba revealed that only fifty-six per cent of the college men surveyed and eighty-two per cent of the women believed in a personal God, and that the students' belief in human immortality suffered a ten per cent overall attrition between their freshman and senior years. For scientists the figures were even more startling: less than half of those surveyed believed either in a personal God or in human immortality, and in each case the believers among biological scientists were thirteen per cent fewer than the believing physical scientists.[74] Little did those who brandished Leuba's statistics realise that the evolutionary fate of American education was sealed before 1880; that fifty years earlier evolution had penetrated the colleges once and for all and its religious opponents had had to accommodate themselves to the doctrine. Fifty years it had taken for the teaching of evolution to filter into the high schools, for the high schools to begin to reach the people, and for the people – those, at any rate, who became militant Fundamentalists – to belong to a generation who could not remember the evangelical evolutionists among their ancestors.[75]

And here, in a reaction delayed for half a century, is an additional explanation of the Fundamentalist crusade against evolution. By 1922

Strong, Warfield, Orr, and Wright had died. The last great links with the past were gone. Bereft of intellectual leadership and convinced they were living in the midst of a major intellectual crisis, Fundamentalists did only what was natural: they panicked. 'The feeling that rationalism and modernism could no longer be answered in debate led to frantic efforts to overwhelm them by sheer violence of rhetoric and finally by efforts at suppression and intimidation.'[76]

There may also be an explanation of Fundamentalist militancy in the reactionary bitterness of the post-war 'return to normalcy', an attitude which flourished in the Ku Klux Klan, anti-Communism, and the Prohibition movement.[77] 'Having learned well that intolerance was justified when the nation was combatting foreign enemies', writes Norman Furniss, 'fundamentalists in the subsequent years of peace found themselves no longer able to meet domestic crises, especially a serious challenge to their faith, with Galilean charity'. It was no accident, moreover, that their indictments of modernism and evolution 'closely resembled Allied propaganda'. Americans had been taught to hate Germany, that barbaric nation which, to the Fundamentalist way of thinking, had uniquely fostered critical and evolutionary thought. Now that the land of Strauss and Wellhausen, Nietzsche and Haeckel, had committed political aggressions and received a just retribution, it remained to avenge its theological atrocities. Needless to say, 'the symbol of war. . .was an appealing one to the fundamentalists'.[78]

Seldom if ever since the time of Cromwell has there been such a concerted outburst of Christian military rhetoric. 'It is a battle royal. . . .There is no discharge in this war', proclaimed one aroused Fundamentalist. 'The conflict is raging. The call to arms is being heard from sea to sea', echoed another. The Moody Bible Institute announced that it was 'preparing for the greatest battle, or rather war, known to ecclesiastical history'. The largest and most prominent Fundamentalist organisation, the World's Christian Fundamentals Association, declared 'a truceless war on the worst and most destructive forms of infidelity', a conflict in which 'fundamentalists draw the weapon of their warfare from the arsenal of God's word; modernists draw theirs from the evolutionary philosophy'.[79] The repositories of such sentiments were given appropriate titles. *The Conflict* was published by the Anti-Evolution League of America and *The Crusaders' Champion* by 'General'George Washburn's Bible Crusaders of America.*Dynamite* was the official organ of Edgar Y. Clarke's 'Supreme Kingdom' and *The Defender* that of Gerald Winrod's Defenders of the Christian Faith.

There were also polemical books and tracts galore, documents such as *The Great Conflict, The Bible versus Evolution* (1923) by D. Grant Christman, which attacked evolution as a 'slimy philosophy', *The Battlefield of Faith* (1940) by S. J. Bole, a later and more temperate volume, and a British contribution to the controversy, *War on Modernism* (1931) by Avary H. Forbes, with its four chapters devoted to 'Fort evolution' and a concluding chapter proclaiming 'Victory'![80]

The campaign against evolution in education was Fundamentalism's finest hour. Between 1921 and 1929, thirty-seven anti-evolution resolutions were introduced in one form or another into the legislatures of twenty states and the District of Columbia, and in seven states official rulings were passed.[81] In fact, the political tactics of the World's Christian Fundamentals Association and the Anti-Evolution League of America, and the monkey-business of many smaller organisations such as the Flying Fundamentalists and the American Anti-False-Science League and Home–Church–State Protective Association, became so threatening that Maynard Shipley, a scientist, set out to combat them in 1925 by founding the Science League of America. Under the circumstances he is perhaps to be excused for not always writing dispassionately and for omitting sufficient documentation in his 'short history of the Fundamentalist attacks on evolution and modernism', published in 1927 as *The War on Modern Science*.[82] But the tide of battle began to turn against the zealous defenders of biblical literalism in the summer of 1925, when a thrice-unsuccessful presidential candidate and champion of the Bible champions was humiliated before a watching world.

Unlike the Huxley–Wilberforce confrontation, the court-room encounter between the populist politician, William Jennings Bryan, and the agnostic lawyer, Clarence Darrow, has been recorded for posterity in excruciating detail.[83] There is no doubt about what either protagonist said nor is there a doubt as to who carried the day. The defence for John T. Scopes, a Dayton, Tennessee high school science instructor who was accused of teaching the descent of mankind from lower animals in violation of a state statute, called Mr Bryan, attorney for the prosecution, to the witness stand as an expert on the Bible. Darrow determined to conduct the cross-examination. In full realisation that the issue for the defence lay with the law's constitutionality, Darrow nevertheless swung with the spirit of the moment, taking advantage of the popular impression that the Bible and evolution were on trial to land a crushing blow on the premier representative of the

Fundamentalist opposition. He made Bryan talk nonsense, confess ignorance, and, most important of all, admit that he did in fact 'interpret' the Bible. 'At times Darrow and Bryan rose and glowered at each other, shaking their fists', but went on at the insistence of Bryan, who stated that he was simply trying 'to protect the word of God against the greatest atheist or agnostic in the United States'.[54] Although Scopes was finally convicted, a reversal was obtained before the Tennessee Supreme Court on the grounds, not that the law was unconstitutional, but that the fine had been wrongly imposed. The Fundamentalists were reversed as well. Although their precious law had been upheld, the world could not stop laughing at their ignorance – except to feel a moment's pity at the death of Bryan within a few days of the trial's end. Never again would representatives of the Bryan Bible League, or of the Defenders of Science vs. Speculation of California, make front-page news across the nation.

'The tendency to read the previous history in light of the latter', writes John Dillenberger with reference to Fundamentalism, 'obscures the underlying problem as well as some of the more informed reactions, positive and negative, of an earlier period'.[85] The truth of this statement, which has perhaps just begun to emerge, will become increasingly apparent as we look in depth at the post-Darwinian controversies. For the present it must suffice to be reminded that if an understanding of Christian responses to Darwin in the nineteenth century is not advanced by reading them in light of later Fundamentalist militancy, neither is it advanced by conceiving them in terms borrowed from the earlier political arena or from the verbal arsenal of an outspoken contemporary. Colourful and clever though it may be, and popular though it may have been within the post-Darwinian milieu, the military metaphor must be abandoned by those who wish to achieve historical understanding.

3

<hr>

WARFARE'S TOLL IN HISTORICAL INTERPRETATION

Ideals may well be theoretically divided into good and bad, into superior and inferior, but men – and the actual battle is one of men against men – cannot be thus divided and set off against one another. . . . Each one of them contains within himself in varying degree the true and the false, the high and the low, spirit and matter.

Benedetto Croce[1]

Metaphors are indispensable figures of speech. By them the abstract is made concrete, the dull and dead is recreated with interest and vivacity, and the complex is cast in recollectable form. What would literature be without the elaborate metaphors of Dante's *Commedia* and Bunyan's *Pilgrim's Progress*, the enchanting allegories of George MacDonald and C. S. Lewis, and the evocative modern parables in Abbott's *Flatland* and Orwell's *Animal Farm*? Likewise metaphors, or 'models', are vital to the scientific enterprise. Without the model of ocean waves Huygens might not have formulated the wave theory of light; without the 'billiard ball' model of molecular collisions, the kinetic theory of gases would be difficult to explain; and apart from Bohr's planetary model of the atom, the development of atomic theory could have been greatly retarded.

Yet particular metaphors are always dispensable, if not in literature, whose themes are eternal, then in science and the philosophy of science, where discovery and understanding are often contingent on the model or metaphor employed. Thus Darwin acknowledged that the term 'natural selection', which he conceived by analogy with the 'artificial selection' of breeders, was less satisfactory than Herbert Spencer's phrase, 'the survival of the fittest', after Alfred Russel Wallace pointed out that this was 'the plain expression of the fact'. ' "Natural Selection" ', he said, 'is a metaphorical expression of it, and to a certain degree indirect and incorrect, since, even personifying Nature, she does not so much select special variations as exterminate the most

unfavourable ones'.[2] Or consider the philosophical misconception
which has been fostered by the term 'natural law'. Thinking of the
juridical sphere from which the metaphor is drawn, religious writers
sometimes claim that a 'lower' law may be overruled by a 'higher',
that the Deity can 'suspend' a law of nature in the interest of a miracle
if he so chooses, or that the very 'existence' of natural laws implies the
existence of a Divine Lawgiver.

Now John Passmore argues, and rightly so, that 'if we mean by
"science" the attempt to find out what really happens, then history is
a science. It demands the same kind of dedication, the same ruthless-
ness, the same passion for exactness, as physics.'[3] Historians, therefore,
like scientists, may employ metaphors for the purpose of elucidating
novel or complex circumstances. A judiciously chosen metaphor,
whether as a heuristic device in research, as an explanatory scheme in
teaching, or as an embellishment in writing, can be as useful to the
historian as a well-constructed model is to the physicist.

But in history no less than in natural science, particular metaphors
are always dispensable, subject to alteration or liquidation under
pressure from the facts. Thus there is a proper context for the terms
'Puritan' and 'Victorian', and there is also a common metaphorical
usage of them which has precipitated numerous corrective monographs. 'Fundamentalism', as we have seen, makes a deceptive and
confusing historical metaphor as well. Some metaphors, according to
David Hackett Fischer, have unfortunate repercussions not only in
history but, by extension, in international affairs. 'Door' metaphors in
histories of relations between Asia and the West, such as Commodore
Perry and the closed door of Nippon, and the American Open Door
Policy in China, suggest that 'Asia is all structure and the West is all
function. They communicate a sense of clear and active purpose in the
latter and of mindless passivity in the former.' In the historiography
of Poland, the traditional idea that 'Poland is the "Christ among
nations", a noble, transcendent being which has suffered for the sins of
all humanity', has encouraged the Polish people 'to develop a self-
righteous sense of persecution with few equals in the modern world'.[4]
Or, to take a case in which a political metaphor has unfortunate
historical consequences: while it may be helpful in some ways to conceive the post-Civil War 'academic revolution' in the United States in
terms of a political revolution – the 'dismissals and harassments of
teachers of evolution' as the 'inflammatory events' and the 'attack
upon religious authority in science and education' as the 'ideology of

resentment' – one wonders whether the sociology of institutional change was in fact so simple or by implication so violent, and whether the metaphor may not have been an unfortunate choice in view of the historiography of the relations of science and religion at the time.[5]

These examples demonstrate the absolute tendency of clever but ill-chosen historical metaphors to perpetuate false conceptualisation. However, to explain this tendency, and to elucidate the working of the military metaphor in particular, requires an analysis of the figure of speech itself. In a lucid essay the philosopher Max Black distinguishes carefully among three kinds of metaphor, settling finally on the 'interaction metaphor' as the only kind of importance in philosophy. Among the features of the interaction type Black specifies the following:

(1) A metaphorical statement has two distinct subjects – a 'principal' subject and a 'subsidiary' one.
(2) These subjects are often best regarded as 'systems of things', rather than 'things.'
(3) The metaphor works by applying to the principal subject a system of 'associated implications' characteristic of the subsidiary subject.
(4) These implications usually consist of 'commonplaces' about the subsidiary subject. . . .
(5) The metaphor selects, emphasizes, suppresses, and organizes features of the principal subject by implying statements about it that normally apply to the subsidiary subject.[6]

Black points out that some types of metaphor can be replaced by literal translations at the loss only of colour, vivacity, and wit, whereas interaction metaphors cannot be rendered literally without the loss of both style and insight. Yet he suggests that an elaboration of the grounds of an interaction metaphor may be extremely valuable, provided that it is not regarded as an adequate cognitive substitute for the original. 'A powerful metaphor', says Black, 'will no more be harmed by such probing than a musical masterpiece by an analysis of its harmonic and melodic structure'.[7]

The military metaphor is clearly of the interaction type. It has a 'subsidiary subject' consisting of an elaborate 'system' of military and martial concepts which are certainly 'commonplaces' for mankind. Its 'principal subject' (in the present instance) is the 'system' of personal, social, institutional, and ideological facts connected with the introduction of evolutionary thought into the late-Victorian world. And there can be no doubt that the metaphor 'selects, emphasizes, suppresses, and organizes features of the principal subject [the post-Darwinian controversies] by implying statements about it that normally apply to the subsidiary subject [warfare]'.[8] How, then, does it withstand the 'probing' of its historical basis? That it is a powerful metaphor in more ways than one is abundantly clear, but do the 'associated implications' of warfare which have been drawn out in detail by a century of historical writing adequately account for the facts of the post-Darwinian controversies? We shall answer by examining the three most important general 'commonplaces' implicit in the military metaphor: the ideas of sharp polarisation, distinct organisation, and violent antagonism.

Versus polarisation

A typical war has two sides. The 'warfare of science with theology' or the 'conflict between religion and science' is no exception. Science, the 'conqueror' with its 'forts of skepticism', stands over against the 'citadels' of Christianity, theology, and creationism. The 'despotism of science' advances before the 'retreat of theology'. Indeed, the French philosopher Emile Boutroux goes so far as to hypostatise the conflicting powers. To him religion and science are not 'concepts' but 'two actual existing things – each of them, according to the Spinozistic definition of existence, tending to persevere in its being'. As such they 'have always been on the war-path, ...they have never left off struggling, not only for the mastery, but for the destruction of one another'.[9] But of course ideas do not strive, no matter how incompatible with one another they may be, nor do nebulous entities such as 'science' and 'religion'. People who originate or support the ideas and call themselves scientists and theologians make up the contending parties. Without this distinction it is impossible to conduct an historical investigation of the relations of science and faith. If science and religion are held to be in 'conflict', irrespective of particular persons, then an evaluation of the military metaphor must be made first on philosophical, not historical, grounds.

Fortunately, most historians have not defined the post-Darwinian controversies in *a priori* philosophical categories. Beneath their vague polarising terminology one can usually discover an empirical basis for their scenes of battle. Yet it seldom seems to be recognised, even when those mysterious entities 'science' and 'religion' are resolved into their personal constituents, that the polarising terminology itself, in the words of one careful student of the subject, still 'obscures the real nature of the encounter'. What conflicts there were took place only between '*some* scientifically oriented men and *some* religiously oriented men'.[10] Moreover, even when it is acknowledged that 'some' individuals exhibited hostile attitudes, so long as a military metaphor governs the examples cited, the illusion of polarity remains. By focusing attention on the verbal pyrotechnics of T. H. Huxley and John Tyndall, 'war' historians have created the impression that these individuals commanded an army of science. By featuring the fulminations of Bishop Wilberforce in England or of clergymen such as T. De Witt Talmage and Randolph S. Foster in America, the same writers have strengthened the opinion that theirs was the battle-cry of religious leaders in general. This is thoroughly misleading. A dispute is not comprehended merely when one has heard from its loudest or its least learned partisans, entertaining though they may be.[11]

There is yet another refinement obscured by a polarising metaphor. If 'science' and 'religion' were not entities but individuals and if these individuals were not every one embroiled in hostile encounters, then it should also be pointed out that neither do they divide neatly into 'scientists' and 'theologians'. Those who reproach their forebears for failing to adhere to this hypothetical division of labour stumble badly here. 'If the theologian and the scientist had been careful to stick to their respective duties, and to learn carefully the other side when they spoke of it', says Bernard Ramm, 'there would have been no disharmony between them'.[12] The truth is better perceived, for once at least, by writers who conceive the post-Darwinian history as a record of the triumphant march of reason. The 'great inertia' of scientists under impact of Darwin's theory, writes J. M. Robertson, was 'at bottom. . .probably in most cases, of religious origin'.[13] This is quite understandable, for – to offer but one reason – a considerable portion of the scientific establishment in Great Britain was centred in universities where, until religious tests were abolished in 1871 (1889 in Scotland), students and faculty were required to subscribe to the doctrines of the established Church. And in the United States, where many colleges were controlled or deeply influenced by denominational interests well into the latter half of the century, scientists surely were not free to be scientists without being at least amateur theologians.

The union of scientific and theological interests in the post-Darwinian controversies is well illustrated by an incident at the Cambridge Philosophical Society on 7 May 1860. It was there that the

Reverend Adam Sedgwick, canon of Norwich and Woodwardian Professor of Geology in the University, emerged from the shelter of anonymity from which he had recently attacked Darwin in the *Spectator*, and nailed to the mast his profoundly anti-evolutionary colours. Darwin himself knew from correspondence with Sedgwick that his old professor 'admired greatly' certain parts of the *Origin of Species*, that parts he had 'laughed at' till his sides were sore, and that other parts – to quote Sedgwick himself – 'I read with absolute sorrow, because I think them utterly false and grievously mischievous'.[14] Now Cambridge and the world were to know. In his address Sedgwick cast a slur on all who substitute hypotheses for strict inductions and referred to the implications of some of Darwin's suggestions as 'revolting' to his sense of right and wrong. He had not, however, reckoned with the presence of his colleague, the Reverend John Stevens Henslow, rector of Hitcham in Suffolk and professor of botany. 'His moral qualities were in every way admirable', said Darwin of his former teacher, who had been instrumental in securing for him a post on the *Beagle* and whose friendship had influenced his career 'more than any other'. 'His temper was imperturbably good, with the most winning and courteous manners; yet. . .he could be roused by any bad action to the warmest indignation and prompt action.'[15] For this reason Henslow's own reservations about natural selection did not deter him from rising to his friend's defence. The origin of species is a legitimate subject of inquiry, he declared, and Darwin's work is guided by the highest motives. Like the Oxford confrontation two months later, it was an encounter charged with the issues of freedom and morality. But as the clergyman–scientists involved cannot readily be made into combatants, the incident is seldom featured as a battle. The military metaphor is unable to encompass it.[16]

'Surely the depth of the problem emerges only when the man of science and the man of faith are the same man', the Edinburgh theologian John Baillie told the British Association in 1951. 'Surely also that is the normal case. . . . Science and faith represent not so much the outlooks of two different kinds of men as two elements that are together present, though in varying degrees, in the minds of most of us.'[17] Illustrations from the later nineteenth century abound, and are by no means confined to clergymen who became professors of natural science or to scientists who took holy orders. Among men of science Christian commitment was not the exception but the rule. James Challis, Michael

Faraday, John Herschel, James Prescott Joule, James Clerk Maxwell, Charles Pritchard, G. G. Stokes, and William Thomson (Lord Kelvin) were all devout, if not always conventionally orthodox, representatives of physical science. The leading geologists were likewise strong believers; besides Sedgwick there were T. G. Bonney, James Dwight Dana, J. W. Dawson, Joseph Le Conte, Charles Lyell, Roderick Murchison, William North Rice, Alexander Winchell, and George Frederick Wright. In the biological sciences Henslow was joined by William Henry Dallinger, Philip Henry Gosse, Asa Gray, St George Mivart, and a covey of clergymen–naturalists – J. C. Atkinson, M. J. Berkeley, C. A. Johns, R. T. Lowe, W. W. Newbould, J. G. Wood, and H. B. Tristram among them.[18]

Perhaps the most telling evidence of the religious commitments of Victorian scientists appears in a survey conducted by Darwin's step-cousin, Francis Galton, and published in 1874. Having sent detailed questionnaires to 189 leading Fellows of the Royal Society and to three other eminent men as well, Galton reported that of the 104 individuals who responded, seventy per cent called themselves members of the established Churches of England and Scotland, and the recently dis-established Church of Ireland. Those remaining expressed their religious affiliations, in order of frequency, as: (1) 'none whatever'; (2) established Church with qualification; (3) Unitarian; (4) Non-conformist; (5) Wesleyan; (6) Catholic; and (7) Bible Christian. When asked whether the religion taught them in youth had had a 'deterrent effect on the freedom of your researches', the scientists replied overwhelmingly to the contrary: almost ninety per cent said 'none at all'. Although the question was poorly put, it seems clear that 'Galton's scientists did not in fact *perceive* a great conflict between their science and religion'.[19]

Furthermore, there is evidence to show that a good number of scientists were not only religious but theologically conservative. When controversy over biblical criticism had reached a peak after the publication of *Essays and Reviews*, and while believers in biblical inerrancy were being taunted by the mathematical discrepancies of Bishop Colenso and by Huxley mocking 'the cosmogony of the semi-barbarous Hebrew', several young men associated with the Royal College of Chemistry suddenly saw the value of a 'Declaration', signed by hundreds of scientists, which would reaffirm the harmony of science and scripture with all their combined authority. In 1864 a statement was drawn up and widely circulated, asserting that 'it is impossible for the Word of God as written in the book of Nature, and God's Word written in Holy Scripture, to contradict one another, however much they may appear to differ'. Although many who signed the Declaration were in fact students of the Royal College of Chemistry and others were men of no scientific achievement whatever, it is certainly note-worthy that of the 717 persons who eventually attached their names to the document, 420 were Fellows of recognised scientific or medical societies, including sixty-six Fellows of the Royal Society – roughly ten per cent of the British membership.[20] Even more individuals might have lent their names had not scientists themselves publicised their moral and theological objections to the Declaration. Charles Daubeny, the Oxford botanist, complained to *The Times* that it made scientists out to be 'peculiarly liable to the charge of infidelity'. John Herschel, the great astronomer, speaking no doubt for many Christian people, charged that the Declaration tended to add 'a fresh element of discord to the already too discordant relations of the Christian world'.[21]

If there was any polarisation over evolution such as the military metaphor implies it was not between science and theology or science and religion. Most scientists were religious men and theologians and clerics could be found among the scientists. Rather, as Alvar Ellegård suggests, 'The Darwinian controversy can probably be best character-ised as one engaging religious science against irreligious science' – individuals, that is to say, who defended to one degree or another the prevailing harmony of method, fact, and scripture versus individuals who lacked such a prepossession.[22] No better evidence of this exists than the formation in the mid-1860s of two societies which represented these very points of view.

The 'X Club', consisting of Huxley, Tyndall, Spencer, the botanist J. D. Hooker, the anthropologist John Lubbock, the anatomist George Busk, the chemist Edward Frankland, and the mathematician Thomas Archer Hirst, was rooted in old friendships and in a common concern to advance the cause of science. Huxley proposed regular meetings in January 1864 and in November of that year the group first gathered, avowing no purpose beyond 'the periodic assembling of friends' who might otherwise drift apart. However, according to Hirst, the members were also devoted to 'science pure and free, untrammeled by religious dogmas'. They often discussed theological subjects and did not hesitate to entertain theologians – Bishop Colenso and William Robertson Smith – who could reinforce their convictions.[23] By contrast, 'The Victoria Institute or Philosophical Society of Great Britain', founded in 1865, was rooted in fear of intellectual developments since the appearance of the *Origin of Species* and *Essays and Reviews*, and also in a deep concern to uphold the ideals expressed in the recent Declaration on science and scripture. The express purpose of the Institute was to defend 'the great truths revealed in Holy Scripture. . .against the oppositions of science, falsely so called', by giving 'greater force and influence to proofs and arguments which might be regarded as comparatively weak and valueless, or be little known, if put forward merely by individuals'. Among its members were the inevitable evangelical clergymen, scriptural geologists, and middle-class professionals, the ubiquitous Earl of Shaftesbury, who served until his death as president of the Institute, and, by the time of the first annual general meeting in 1867, no less than forty-two members of recognised scientific societies, including twelve Fellows of the Royal Society.[24]

Here, then, was a polarisation of sorts, but one based on questions of freedom and authority which had long divided people, regardless of their vocations. The Victoria Institute was no more a clerical reaction to the X Club than the X Club was an anti-clerical response to the Declaration on science and scripture. Both societies arose quite independently and served to bring together scientists who held particular religious viewpoints. This, however, recalls the second major implication of the military metaphor, the idea of distinct organisation.

Versus organisation

Armies do not exist without a division of rank and a chain of command. The 'Darwinian legions', therefore, are depicted as under the 'generalship' of Huxley and Asa Gray, and marshalled by various 'lieutenants'. Their religious opponents are not often credited with such efficient organisation but they nevertheless seem to have had sufficient 'strategy' to advance 'all along the line'. Indeed, one historian suggests that, although

there were as many conflicting interpretations and private constructions among the entrenched opposition as among Darwin's attacking forces, . . . the defenders acted in concert more successfully than the invaders. Perhaps this was because the latter lost their natural leader when Darwin retired from the struggle, and because even the belligerent Huxley was too subtle a tactician, too readily carried away by his own wit and rhetoric, to be an effective commander. Thus it was the opposition that enjoyed the more militant and aggressive spirit, having something resembling a chain of command and a coordinated strategy of action.[25]

The picture is thoroughly deceptive, for it connotes a polarised science and religion. More accurately does it depict – with all the limitations and banality of an out-worn metaphor – a full-scale mutiny among the 'troops' of science.

The history of science is not (as Draper supposed) the 'narrative of the conflict between two contending powers' but the record of divergent opinions within and among the sciences themselves. Naturalists in the seventeenth and eighteenth centuries could not agree on the question of the spontaneous generation of life. Geologists at the beginning of the nineteenth century argued the merits of fire and water as vulcanists and neptunists, and chemists divided sharply over the phlogiston theory of Ernst Stahl. Meanwhile, for more than two hundred years, a large number of researchers in the disciplines of pure mechanics, engineering, and physical chemistry, experienced a 'basic conflict' in regard to the conservation of energy.[26] Thus, to find scientists hopelessly at odds over Darwinism and evolution in the mid-nineteenth century should come as no surprise. As Huxley recalled in 1887, 'The supporters of Mr Darwin's views in 1860 were numerically extremely insignificant.' 'There is not the slightest doubt', he added, 'that if a general council of the Church scientific had been held at that time, we should have been condemned by an overwhelming majority'.[27]

Huxley, Hooker, and Lyell were privy to the heresy. To their judgement Darwin deferred with all his intense sensitivity.[28] Huxley long resisted transmutation, playing the devil's advocate for the better part of a decade before becoming a convert when at last an adequate causal explanation was given. 'Many and prolonged were the battles we fought on this topic', said Huxley of his early discussions with the philosophical evolutionist Herbert Spencer.[29] Hooker was converted more quickly, but Lyell, to Darwin's consternation, waited until 1868 to make public his acceptance of evolution in the tenth edition of his *Principles of Geology*.[30] Around them gathered a handful of scientists, not all well known but each welcomed by the master into his little circle. 'It is a great thing to have got a great physiologist on our side', wrote Darwin to W. B. Carpenter, who had recently pledged a measure of support. 'I say "our" for we are now a good and compact body of really good men, and mostly not old men. In the long run we shall conquer. I do not like being abused, but I feel that I can now bear it.'[31]

There was much to bear. Darwin's theory cut straight across the established way of looking at the world. Those who had given their lives to strengthening the blinkers of organic fixity with final causes were not about to let their labours go for naught. Long ago Lyell had rendered Lamarck's theory of evolution untenable and for fifteen years *Vestiges* had been a laugh. Could there be another challenge to scientific orthodoxy? 'Am I a dog, that you come to me with sticks?' demanded the leaders of contemporary science. Goliaths loomed everywhere: Roderick Murchison, second only to Lyell among British geologists; David Brewster, the Scottish natural philosopher, principal of the University of Edinburgh and founder of the British Association; William Clark, professor of anatomy at Cambridge;[32] William Henry Harvey, professor of botany in the University of Dublin; Louis Agassiz, the Harvard zoologist, whose massive intellect had been nurtured by Cuvier; Friedrich Max Müller, the leading philologist in Britain; William Whewell, the Cambridge historian and philosopher of science; William Thomson, professor of natural philosophy in the University of Glasgow; and of course Sedgwick and Henslow.[33] However, it was not cogency of argument, nor even his imposing reputation as a vertebrate anatomist and paleontologist, that made Richard Owen, superintendent of the natural history department of the British Museum, Darwin's most formidable opponent. It was above all his arrogant and underhand manner.

Huxley knew the man and his ways as well as anyone. He was duly repelled by Owen's condescension in granting a request for a testimonial in 1852, and later had cause to fear that for reasons of petty pride Owen would impede the publication of his Royal Society memoir, 'On the morphology of the cephalous mollusca'. Injury was compounded by insult when Owen arrogated to himself the title Professor of Paleontology while lecturing at the School of Mines, where Huxley taught, thereby making himself, in effect, Huxley's superior. Finally, when Owen went on record in a pontifical judgement, inspired by the idealistic morphology he had learnt from Oken, that the human skullbones are only modified vertebrae, Huxley could contain himself no longer. In his 1858 Croonian Lecture, delivered at the Royal Society in Owen's very presence, he demolished the theory with premeditated cruelty, offering sounder views based on embryological evidence which his colleague had thoroughly ignored. Owen might have had his revenge in briefing Wilberforce for his confrontation with Huxley if the bishop had made a better showing. As it was, his argument that the brain of the gorilla differs more from the human brain than it differs from the brains of the lowest Quadrumana, thinly veiled by the bishop's smooth rhetoric, simply occasioned another Huxleyan demolition.[34] Thus Owen had to content himself with penning an anonymous review of the *Origin* which was atrociously severe on Huxley, scandalously unfair to Darwin, deliberately flattering to himself, and immensely influential upon the educated public.[35] 'It is painful to be hated in the intense degree with which [Owen] hates me', lamented gentle Darwin after reading the article.[36] But Owen reached the end of his tether when, in 1866, after contradictions, qualifications, and retreat from some of his earlier statements, he quietly claimed priority over Darwin in formulating the theory of natural selection. His desperate attempts to regain the prestige snatched from him by Huxley and Darwin had failed, and 'his name to many became a measure of silent contempt'.[37]

Perhaps none was better qualified to measure the divisions of science than its up-and-coming students. Archibald Geikie, who was twenty-four years old when the *Origin of Species* appeared, recalled his impatience with his Scottish professors for failing to see the book as a 'new revelation of the manner in which geological history must be studied'. In America Nathaniel Shaler, also in his twenties at the time, spoke of his secret debates on Darwinism: 'To be caught at it', he said, 'was as it is for the faithful to be detected in a careful study of a heresy'. August Weismann, twenty-five years old when the *Origin* was published, told how in Germany the book 'excited in the minds of the younger students delight and enthusiasm' but 'aroused among the older naturalists anything from cool aversion to violent opposition'. Indeed, according to Ernst Haeckel, who read the *Origin* in his twenty-sixth year, '*all* the Berlin magnates (with the exception of Alexander Braun) were against it'. At Oxford, meanwhile, the opposition took a political turn. J. O. Westwood, professor of zoology, proposed to the University Commission 'the permanent endowment of a Reader to combat the errors of Darwinism'. And at Cambridge it was said that for years Whewell refused to allow a copy of the *Origin of Species* to be placed in the library of Trinity College.[38]

What, then, can be said for the so-called 'theological side'? If most scientists at the outset stood opposed to Darwin's theory, where stood the theologians? One writer states flatly that 'many theologians and a few scientists rejected the hypothesis outright as "the latest form of scientific infidelity"'.[39] The truth is nearer to the exact opposite: it was a few theologians and many scientists who dismissed Darwinism and evolution. On the assumption that a fair assessment is made only by comparing the more enlightened representatives of science and theology – a point consistently overlooked by authors addicted to counterposing Huxley's tirades with the outcries of ignorant clergymen – one is forced to conclude that Christian men untrained in science showed themselves on the whole considerably more open-minded than Christian men of science.[40]

To begin with we should set straight the historical record which has been unbalanced by writers preoccupied with the Huxley–Wilberforce confrontation. On Sunday, 1 July 1860, the day after that momentous encounter, the Reverend Frederick Temple, headmaster of Rugby School, preached before the University of Oxford and delegates gathered for the annual meeting of the British Association a sermon on 'the present relations of science to religion'. Although Temple did not mention Darwin's book, nor the confrontation of the previous day, his generous and incisive treatment of the points at issue between science and faith must be seen in relation to both, and especially in contrast to the late action of the bishop of Oxford. He granted scientists all the laws in the universe they could discover and promised to find 'the finger of God' in them; for the Bible he did not demand 'confirmation of minute details' but recognition of a deep identity of 'tone, character, and spirit' between the Book of God and the Book of Nature.[41]

In the same year the Reverend Baden Powell, Savilian Professor of Geometry at Oxford, a Fellow of the Royal Society, and the author of treatises in mathematics and natural philosophy, announced in his contribution to *Essays and Reviews* that 'a work has now appeared by a naturalist of the most acknowledged authority, Mr Darwin's masterly volume on *The Origin of Species* by the law of "natural selection", – which now substantiates on undeniable grounds the very principle so long denounced by the first naturalists, – *the origination of new species by natural causes*: a work which must soon bring about an entire revolution of opinion in favour of the grand principle of the self-evolving powers of nature'. So impressed was Powell that he wrote to Darwin, telling him that he had 'never read anything so conclusive' as his statement about the eye.[42]

The Reverend F. D. Maurice, professor of moral philosophy at Cambridge and a Broad Churchman who despised the label, held that 'every discovery made by Mr Darwin or Mr Huxley was a discovery of a truth which had been true in itself ages before it was discovered'. He believed 'the thing itself to be, when discovered, just in so far as it was true, a revelation to man by God whether the discoverer accepted it in that sense or not'. Indeed, Maurice was so persuaded of the efficacy of the scientific method in ascertaining truth that 'he. . .never tired of quoting the spirit of Mr Darwin's investigations as a lesson and a model for churchmen'.[48]

Broad Churchmen might have been expected to greet Darwin with tolerance, if not outright acceptance. What seems more remarkable, at a century's distance at least, is that as early as 1871 H. P. Liddon, a canon of St Paul's and a very orthodox Anglo-Catholic, could provide for belief in an original act of creation and the recognition of design in nature 'even if a doctrine of evolution should in time be accepted as scientifically, and so as theologically certain'. Evolution, he stated, 'from a Theistic point of view, is merely our way of describing what we can observe of God's continuous action upon the physical world'. Elsewhere in the same year, in a sermon preached at St Mary's, Oxford, Liddon went out of his way to cite approvingly a passage from Peter Lombard's *Sententiae* respecting the creative activity of God, a passage which, said Liddon, 'employs terms which almost read like a tentative anticipation of Dr Darwin's doctrine of the origin of species'. Even Liddon's master, the aged leader of the High Church party which came to bear his name, the Reverend E. B. Pusey, eventually allowed for the truth of evolution, provided that it did not entail 'belief in our apedom'.[44]

Some clergymen and theologians were favoured with advance preparation for Darwinism. The Reverend Henry Baker Tristram, canon of Durham, was a distinguished naturalist specialising in the fauna of Palestine and North Africa. In 1858, when the theory of natural selection was first promulgated by Darwin and Wallace in the *Proceedings* of the Linnean Society, Tristram was busy preparing a series of ornithological studies for *The Ibis*. In the third number of the series, published in October 1859, he not only took note of the theory – an occurrence rare enough in itself – but became the first naturalist publicly to accept it before the appearance of the *Origin of Species*. 'Writing with a series of about 100 Larks of various species from the Sahara before me', he said, 'I cannot help feeling convinced of the views set forth by Messrs Darwin and Wallace in their communications to the Linnean Society'. He went on to point out the 'perfectly natural causes' that 'serve to *create* as it were a new species from an old one', adding his belief that such causes 'must have occurred, and are possibly occurring still'.[45]

F. J. A. Hort, the Cambridge New Testament scholar, though a younger and less experienced naturalist than Tristram, was still sufficiently apprised of scientific subjects in 1860 to give the *Origin of Species* a warm welcome. His first six published papers were concerned with botanical topics and eventually he made four more contributions to the literature of that field. Soon after the *Origin* appeared Hort wrote to B. F. Westcott (later his collaborator in a monumental revision of the New Testament text) that, 'in spite of difficulties, I am inclined to think it unanswerable'. In their subsequent correspondence Hort sought to meet his friend's misgivings by explaining and illustrating the Darwinian theory.[46]

Charles Kingsley, poet, novelist, and rector of Eversley in Hampshire, also nourished an abiding interest in science. He was made a Fellow of the Geological Society in 1863, an honour recognising his achievements in his favourite subject of study. The great popularity of his children's books, *Water-Babies* and *Madam How and Lady Why*, likewise testified to a notable achievement, to a singular success in popularising the facts of natural history. When the time came to distribute pre-publication copies of the *Origin* Kingsley was not overlooked. In return he sent Darwin his humble thanks, confessing: 'All I have seen of it *awes* me; both with the heap of facts and the prestige of your name, and also with the clear intuition, that if you be right, I must give up much that I have believed and written.' Give up he certainly did, for to H. W. Bates he wrote in 1863 that God's 'greatness, goodness, and perpetual care I never understood as I have since I became a convert to Mr Darwin's views'.[47]

The Anglo-Catholic temper of R. W. Church was quite unlike that of Kingsley. He deplored the liberality of F. D. Maurice, Kingsley's mentor, and held in high regard those most traditional Bampton Lectures, *On Miracles*, delivered by J. B. Mozley in 1865. But in matters scientific Church was hardly less receptive. If the small chemistry laboratory in the rectory at Whatley opened his mind to science it was his long and intimate friendship with the Harvard botanist Asa Gray which opened his eyes to evolution. He 'took off a great deal of the theological edge, which was its danger', said Church, referring to Darwin's theory. In his correspondence Gray often enclosed scientific papers in support of his opinions and thus Church, the future dean of St Paul's, 'had almost acclimatized his mind to "evolution" before most clergymen had even heard of it'. The respectful review of the *Origin* which appeared in the Anglican *Guardian* for 8 February 1860,

indicating a willingness to bide time before accepting the book's implications for the human species, was almost certainly penned by him.[48]

With but few exceptions the leading Christian thinkers in Great
Britain and America came to terms quite readily with Darwinism and
evolution. In England the movement culminated when a galaxy of
younger theological lights collaborated in the essays of *Lux Mundi*
(1889) for the purpose of restating the Anglo-Catholic faith in developmental terms. Among them Charles Gore, Aubrey L. Moore, and
J. R. Illingworth stand out as particularly thoroughgoing Christian
evolutionists.[49] Scotland provided evolution with theological advocates
in John Tulloch, professor of systematic theology and apologetics at
St Andrew's, Robert Flint, professor of divinity at Edinburgh, John
Caird, professor of divinity at Glasgow, Henry Cotterill, the Episcopal
bishop of Edinburgh, Henry Drummond, the well-known evangelist
and professor in the Free Church College, Glasgow, and George
Matheson, the famous blind minister of St Bernard's, Edinburgh. The
foremost figure in Scottish church life in the last decades of the nineteenth century, Robert Rainy, principal of New College, declared himself an evolutionist at the outset of his public career in his inaugural
address, 'Evolution and theology', delivered in October 1874. He
allied his renowned piety and orthodoxy with an application of evolution even to human descent, reassuring all who were troubled by
Darwin that on this point the theologian 'may be perfectly at ease'.[50]

In the United States Darwin early received a warm reception from
the philosopher and Presbyterian president of the College of New
Jersey, James McCosh, and from younger men such as the Congregational minister George Frederick Wright. Indeed, except for
Unitarians – most notably Francis Ellingwood Abbot – Congregationalists showed the greatest affinity for evolution. Two missionaries
of the denomination, Minot Judson Savage in California and John
Thomas Gulick in the Hawaiian Islands, became evolutionists in the
sixties. Soon they were joined by the theologians Egbert and Newman
Smyth, Theodore Thornton Munger, Washington Gladden, Francis
Howe Johnson, and Lyman Abbott, and by the prince of contemporary
preachers, Henry Ward Beecher.[51] Of them all, Beecher was the most
influential and enthusiastic. As early as the winter of 1870 he led a
group of twenty-five Brooklyn ministers who invited E. L. Youmans,
the prophet of Spencerian science, to meet with them weekly (but in
secret) for the express purpose of helping them to understand evolution.
Later Beecher took the lead in encouraging the geologist Joseph Le

Conte to write his important work on evolution and religious thought.[52]

Of course many lesser voices were raised in protest – a host of clergy-men and not a few professors and American college presidents – with the result that Christians were for many years deeply divided over the question of evolution. But consider: first, that as Ellegård points out, 'only a small proportion of those who took sides one way or the other had even a rudimentary knowledge of the facts on which the theory was based';[53] second, that the issues, even from an informed scientific point of view, were comparatively complex; third, that advance preparation for the Darwinian theory, either by one's own scientific investigations or by acquaintance with one who was within the inner sanctum of theoretical opinion, was an important advantage for those such as Tristram, Hort, Kingsley, and Church, who received Darwin-ism with openness, but an advantage only for the few; and therefore, finally, that 'no blame. . .can reasonably be attached to churchmen, from Bishop Wilberforce downwards, if they accepted the prevailing judgement of men of science and joined with them in the condemnation of a novel and doubtful theory'.[54] When these points are duly weighed, the assimilation of evolution by a large proportion of leading Christian thinkers seems remarkable indeed, and its rejection by those whose formal education had ended and by others in positions of institutional authority seems both understandable and to a certain extent excusable. One must never forget that although the discord created by evolution among Christians had important theological roots, it uniquely be-fell those who were untrained in science to gauge their opinions for some time by a learned consensus that was hostile to Darwin.

Versus antagonism

The last major implication of the military metaphor is the idea of violent antagonism. Warfare has not commenced until the accoutre-ments of battle are deployed in hostile clashes of opposing armies. Thus one reads of science, dressed in the 'Darwinian armour', wheel-ing out its 'new artillery' and training the 'guns of speculation' on the fixity of species. Some say Darwin fired the first 'shot' but more often than not theology in the person of Bishop Wilberforce is blamed for instigating the hostilities. 'The immediate reaction was hostile', writes one commentator. 'Everything was ready for a battle.' Another writer somewhat romanticises the polemical ammunition: 'The controversial literature of this interesting epoch. . .reposes on the higher shelves of libraries, accumulating the peaceful dust of oblivion. These projectiles have, in fact, done their work, and if they have proved less fatal than was hoped by those who launched them, they were dispatched with good intentions, and their explosion cleared the air.'[55]

This colourful picture is shattered by historical investigation no less completely than the images of polarisation and organisation on which it depends. In a monumental study, based on an examination of 115 British periodicals (including 45 religious serials) published between 1859 and 1872, Ellegård declares the widespread belief that 'the first reaction to Darwin's theories was uniformly hostile' to be 'hardly correct'. He reports that a large number of newspaper and magazine reviews ranged from fair to favourable in their evaluations of the *Origin of Species* and that the literary reviews were for the most part cautious and non-committal. Gertrude Himmelfarb, who has produced the most substantial interpretation of Darwin's impact on the nineteenth-century intellectual world, adds that the negative response of laymen and clerics 'tended to be more tolerant and amiable than those of the professional scientists'. James Bryce, whose credentials as an observer of history can hardly be gainsaid, recalled from his student days at Oxford that 'the alarm was not quite as great as some have since represented'. Even Huxley himself, writing to the Reverend C. H. Middleton with reference to his own *Man's Place in Nature* (1863), felt obliged to confess that he had been 'pleasantly disappointed' by churchmen. 'There has been far less virulence and much more just appreciation of the weight of scientific evidence than I expected', he said, '– and that satisfactory state of things is due, I doubt not, to the much wider dispersion than I imagined of such liberal thought as is manifest in your letter'.[56]

Evidence drawn from the personal affairs of Darwin and his followers is especially effective in disappointing the expectations of violent antagonism raised by the military metaphor. 'Several clergymen go far with me', wrote Darwin, who was as much the non-combatant as Huxley was the warrior. He thanked his old friend, the Reverend Leonard Jenyns, curate of Woolley and Langridge near Swanwick, for the 'kind things' which he had to say, adding, 'You go with me much further than I expected. . . . Your going some way with me gives me great confidence that I am not very wrong.' The Reverend J. Brodie Innes, vicar of Down, where the Darwins had retired, was perhaps less receptive, yet he was pleased to report that he and Darwin 'never attacked each other'. He recalled that on his last visit to the

Darwin home (in the company of the atheists Edward Aveling and Ludwig Büchner) his host had said at dinner: 'B[rodie] I[nnes] & I have been fast friends for 30 years. We never thoroughly agreed on any subject but once and then we looked hard at each other and thought one of us must be very ill.'[57] Nor did the Reverend F. W. Farrar accept the Darwinian theory and yet Darwin wrote to him in 1867, commending him warmly for a lecture on the defects of public school education. Indeed, Darwin had been so impressed with an essay by Farrar on the origin of language (for Farrar was an evolutionist in philology if not in biology) that he had proposed him for the Fellowship of the Royal Society, to which he was duly elected in 1866. At the time of Darwin's death in 1882 Farrar was rector of St Margaret's, Westminster and a canon of Westminster itself. He expedited the Abbey burial on his own initiative, served as a pall-bearer with Huxley, Wallace, and Lubbock, and on the Sunday evening preached the funeral sermon at the Nave Service.[58]

As for Huxley, it should be pointed out that he was not altogether the man of war his essays reveal. A most amiable and philosophic correspondence – one which 'unlocked Huxley's most sacred interiors'[59] – passed between Kingsley and himself during the sixties, beginning with a letter of condolence from Kingsley after the death of his son Noel. Clerical members of the London School Board, to which Huxley was elected in 1870, were generous with appreciation for their colleague. 'Towering as was his intellectual strength', wrote the Reverend Benjamin Waugh, '...he did not condescend to me.... There were no tricks in his talk. He did not seem to be trying to persuade you of something. What convinced him, that he transferred to others.' Farrar, who knew Huxley for many years and often had 'very earnest and delightful conversations' with him on religious subjects, found him 'perfectly open-minded, reverent, and candid', though he held certain unfortunate stereotypes of the clergy and the Christian creed.[60]

The Metaphysical Society (of which we shall speak again shortly) provided Huxley with both an opponent and a comrade in candour in W. G. Ward, the ultramontane Catholic editor of the *Dublin Review*. When one of the speakers at an early meeting insisted 'on the necessity of avoiding anything like moral disapprobation in the debates' Ward responded in his genial and light-hearted manner: 'While acquiescing in this condition as a general rule, I think it cannot be expected that Christian thinkers shall give no sign of the horror with which they would view the spread of such extreme opinions as those advocated by Mr Huxley.' After an appropriate silence, allowing Ward's words to settle fully upon his hearers, Huxley took up the challenge: 'As Dr Ward has spoken, I must in fairness say that it will be very difficult for me to conceal my feeling as to the intellectual degradation which would come of the general acceptance of such views as Dr Ward holds.' In fact the feelings referred to remained quite perfectly concealed. Ward drew Huxley aside after a meeting and informed him with a twinkle – in the latter's words – that 'we Catholics hold that so and so and so and so (naming certain of our colleagues who were of less deep hue than mine) are not guilty of unpardonable error; but your case is different, and I feel it is unfair not to tell you so'. Huxley responded with a hearty handshake, 'My dear Dr Ward, if you don't mind, I don't.' On another occasion, Ward's son recalled, the two held a peripatetic conversation until the dawn. And in yet another friendly encounter Huxley, who was passing the time before a meal by gazing out into the garden, was asked by Ward what he was doing. He replied, 'I was looking in your garden for the *stake*, Dr Ward, which I suppose you have got ready for us after dinner.'[61]

Nor did that other hammer of orthodoxy, John Tyndall, experience anything like the hostile relations pictured in the military metaphor. He thankfully acknowledged the greatness of Michael Faraday's character, having found him to be 'above all littleness and proof to all egotism' in their differences over scientific matters. Faraday, who was a life-long member of an obscure and isolated Christian sect, the Sandemanians, esteemed Tyndall as 'a true philosopher and friend'. During the fourteen years they were colleagues at the Royal Institution they consulted together 'on all matters of doubt and difficulty, and though they differed profoundly on matters of religion, and had joined issue on scientific interpretation, there does not seem ever to have been a quarrel'.[62]

Among those who may properly be called theologians Tyndall enjoyed a special relationship with the Reverend A. P. Stanley, a relationship so durable that even Tyndall's presidential address to the British Association at Belfast in 1874, the spirit of which is summed up in the words, 'We claim, and we shall wrest from theology, the entire domain of cosmological theory', was unable to break it.[63] In 1866, when the sceptical physicist made public his desire for a form of prayer which would express the heart without embarrassing the head,

Stanley responded with an invitation to attend a service at Westminster Abbey in the day appointed by the bishop of London to pray for the cattle plague of that year. Tyndall declined to come, expressing distaste for prayers requesting material good and approval, on the other hand, of petitions for 'strength of heart and clearness of mind to meet it manfully and fight against it intelligently'. Stanley sent him in return an eloquent prayer of his own composition, embodying these very words, and Tyndall recorded it in full in his journal.[64] Ten years later Stanley's wife lay dying in agonies after a distressing and protracted illness, yet she insisted that he go to perform Tyndall's marriage ceremony. Stanley, the dean of Westminster, 'looked wretched' and a helper was there to take his place. But he met the wedding party in Henry VII's Chapel – Tyndall, Huxley, Hooker, Thomas Carlyle, and the bride and her family. Long afterwards Mrs Tyndall wrote: 'The intensity of feeling pervading the whole group was manifested in his voice and handgrasp: the treasure of the happiness which was so fast ebbing out of his own life.' Stanley himself died in 1881 and Tyndall attended the funeral in Westminster Abbey. Once again his wife recorded their heart-felt sentiments: 'As long as he lived and to the last year of his life Stanley never failed to visit us on our wedding day. The little difficulty of our leap year wedding was no hindrance to him. If it was not on our anniversary he came, it was on his own that he brought us the handgrasp of friendship and the kiss of peace. He frequently dined with us and of these little gatherings I have the most delightful reminiscences.'[65]

Stanley, Tyndall, Huxley, and fifty-nine other prominent British intellectuals were members of an organisation which was as much unlike a battlefield as the General Assembly of the United Nations. In November 1868 the idea came to James Knowles, later the editor of *The Nineteenth Century*, to form a 'Theological Society' in which representatives of all schools of religious thought would come together 'in an effort to counteract scientific materialism and unite warring theological factions in a common cause'. Knowles soon discovered that the aura of pugnacity surrounding his proposal repelled the theologians. James Martineau, a Unitarian, 'refused to join a society of believers to fight unbelievers'. Dean Stanley, too, felt that the Society could only 'widen the breach between the religious and scientific points of view'. Broad Churchman that he was, however, Stanley suggested that *rapprochement* should characterise the organisation.

Knowles saw the wisdom in this counter-proposal. With a flourish of the fair play for which Englishmen are justly famous, he changed its name to the 'Metaphysical Society' and made plans to invite the leaders of science and theology, literature and politics, regardless of their religious views. The first meeting was held in the Deanery at Westminster on 2 July 1869. Thereafter at the Grosvenor Hotel, once a month, except when Parliament was in recess, and until the organisation ceased to exist on 16 November 1880, from five to twenty members of the Metaphysical Society met together for dinner and an evening of discussion.[66]

The startling dissimilarity of the Society's members was described with lasting freshness by W. C. Magee, the bishop of Peterborough, after his first introduction to them following his own election:

I went to dinner duly at the Grosvenor Hotel. The dinner was certainly a strangely interesting one. Had the dishes been as various we should have had severe dyspepsia, all of us. Archbishop Manning in the chair was flanked by two Protestant bishops right and left – Gloucester and Bristol [C. J. Ellicott] and myself – on my right was Hutton, Editor of the *Spectator* – an Arian; then came Father Dalgairns, a very able Roman Catholic priest; opposite him, Lord A. Russell, a Deist; then two Scotch metaphysical writers – Freethinkers [probably Sir Alexander Grant and Prof. A. C. Fraser]; then Knowles, the *very* broad Editor of the *Contemporary*; then, dressed as a layman and looking like a country squire, was Ward, formerly Rev. Ward, and earliest of the perverts to Rome; then Greg, author of 'The Creed of Christendom', a Deist; then Froude the historian, once a deacon in our Church, now a Deist; then Roden Noel, an actual Atheist and red republican, and looking very like one! Lastly Ruskin who read after dinner a paper on miracles! ...which we discussed for an hour and a half! Nothing could be calmer, fairer, or even, on the whole, more reverent than the discussion. Nothing flippant or scoffing or bitter was said on either side, and very great ability, both of speech and thought, was shown by most speakers. In my opinion, we, the Christians, had much the best of it. Dalgairns, the priest, was very masterly; Manning, clever and precise and weighty; Froude, very acute, and so was Greg; while Ruskin declared himself delighted 'with the exquisite accuracy and logical power of the Bishop of Peterborough'. There is the story of the dinner. Altogether a remarkable and most interesting scene, and a greater gathering of remarkable men than could easily be met elsewhere. We only wanted a Jew and a Mahometan to make our Religious Museum complete.[67]

Other notable members, who are often supposed to have been waging war on behalf of science or theology, were: the Darwinian John Lubbock, the Positivist Frederic Harrison, J. Fitzjames Stephen and his *littérateur* brother Leslie, W. K. Clifford, and John Morley on the one hand; and on the other, the bishop of St David's, Connop Thirwall; the dean of Canterbury, Henry Alford; the archbishop of York, William Thomson; J. B. Mozley, Regius Professor of Divinity at Oxford; the Duke of Argyll; St George Mivart; W. E. Gladstone; and A. J. Balfour.[68]

Not every member attended every meeting, it is true, but in all the various volatile combinations of personalities that appeared there never was, evidently, an explosive reaction. Some of the early natural politeness did eventually give way to a more formal politeness and theistic members, after a time, tended to withdraw from open controversy with the feeling that basic differences had been exhaustively explored and could not be resolved. But this, surely, is a far cry from the violent exchanges that obtain under conditions of 'war'. In fact, none but those who joined the Society in its later, declining years – Morley, Leslie Stephen, and Frederick Pollock – ever spoke of it 'with anything but enthusiasm or admiration'. Henry Sidgwick, who conceived the Society's aim to be 'a diminution of mutual misunderstanding', must not have been disappointed.[69]

Therefore we conclude that the military metaphor has taken a dreadful toll in historical interpretation. Our probing of its basis in the facts connected with the post-Darwinian controversies has shown that each of its three major 'associated implications' is entirely misleading if not utterly false. There was not a polarisation of 'science' and 'religion' as the idea of opposed armies implies but a large number of learned men, some scientists, some theologians, some indistinguishable, and almost all of them very religious, who experienced various differences among themselves. There was no organisation apparent on either 'side' as the idea of rank and command implies but deep divisions among men of science, the majority of whom were at first hostile to Darwin's theory, and a corresponding and derivative division among Christians who were scientifically untrained, with a large proportion of leading theologians quite prepared to come to terms peacefully with Darwin. Nor, finally, was there the kind of antagonism pictured in the discharge of weaponry but rather a much more subdued overall reaction to the *Origin of Species* than is generally supposed and a genuine amiability in the relations of those who are customarily believed to have been at battle. In each of its major implications the military metaphor perverts historical understanding with violence and inhumanity, by teaching one to think of polarity where there was confusing plurality, to see monolithic solidarity where there was division and uncertainty, to expect hostility where there was conciliation and concord. Henceforth interpretations of the post-Darwinian controversies must be non-violent and humane.

4

∞∞∞∞∞∞∞∞∞∞∞∞∞∞∞∞∞∞∞∞∞∞∞∞∞∞∞∞∞∞

TOWARDS A NON-VIOLENT HISTORY

The many issues which Darwinism brought into focus. . .were the grounds
of the spiritual struggle through which innumerable Victorians passed. . . .
The popular name of the struggle was 'science vs. religion', but it was
much more complicated than that crude simplification would suggest.
                                                        Richard D. Altick[1]

The elaboration of metaphors is an important way in which human
beings disclose their innermost feelings and beliefs. Metaphors, to use
Max Black's terminology, present a 'principal subject' in terms of a
'subsidiary' one. A subsidiary subject consists of a 'system of things'
which implies certain 'commonplaces' about itself. The system of
things may therefore express some fundamental notions about the
metaphor-maker and his world. The system is but a symbol and, as
Paul Tillich, Colin Turbayne, and Mary Douglas have argued from
their different points of view, the choice of symbols in religion reflects
basic attitudes and assumptions about God and nature, society and
mankind.[2]

What then does the military metaphor reflect? What are the attitudes and assumptions which 'conflict' has expressed for historians of
the post-Darwinian controversies? Most obviously, perhaps, it reveals
the absence of any deep moral aversion from war. Historians would
not have elected to portray the interaction of scientists and theologians
in a figure which features violence and inhumanity had they been
convinced that these things should be deplored. To ask why historians
have not been so convinced, however, is to suggest a second level of
analysis. The military metaphor has proved a congenial figure of
speech because, as the foregoing chapters repeatedly illustrate, historians have been children of their times. They have employed the
vocabulary of a milieu in which they themselves were partisans and
participants. Political conflicts in the nineteenth and early twentieth
centuries gave currency to military terminology and a fillip to Christian

militarism; Christians employed military metaphors with a vengeance throughout the period, thereby depicting their intellectual and moral stance towards the world. Thus historians, being creatures of politics and religion, have merely expressed their own predilections in familiar terms. Draper, a freethinker of Methodist stock, did not disregard the political pretensions of the Roman Catholic Church. Nor was White, a Broad Churchman, oblivious to the political influence on higher education of the smaller Protestant denominations. Huxley and the Fundamentalists had definite political and religious aims, and those who have subsequently adopted their bellicose imagery have usually done so to congratulate or to oppose. 'Conflict' may fail as a model of the post-Darwinian controversies but it remains a potent symbol of the social and intellectual ferment of the last one hundred years.

There is yet another level at which the military metaphor reveals basic attitudes and beliefs – those less of latter-day historians than of the *dramatis personae* of the post-Darwinian controversies. How the phenomena at this level relate to the political correlations of the previous one is at present of little account, for whether effects or causes, mediations or originations, the phenomena are significant in themselves. They testify poignantly of the human casualties of 'war' and point at the same time towards a non-violent historiography. We refer to the abundant evidence of a different kind of 'conflict', the Victorian crisis of faith.[3]

The crisis of faith

It requires no new insight to establish that Christians in the later nine-teenth century were beset with spiritual disorders and intellectual strife. Indeed, the collective crisis is acknowledged explicitly in the titles of some recently published books: Appleman's *1859: Entering an Age of Crisis* and Symondson's *Victorian Crisis of Faith*, composite works dealing with developments in Great Britain; Weisenburger's *Ordeal of Faith: The Crisis of Church-going America, 1865–1900* and Carter's *Spiritual Crisis of the Gilded Age*, monographs which present a comple-mentary picture of American culture.[4] However, what does demand fresh insight, or at least a renewed emphasis, is the contribution made by Darwinism to the turmoil of Victorian minds. Though obscured by sounds of battle, this plaintive theme has not always gone unheard. At the turn of the century, for example, the geologist William North Rice reminded a younger generation of the 'agonies of terror' with which the Darwinian theory was regarded in his student days. Forty years later Bert James Loewenberg denied that the struggle which Darwin precipitated took place simply between reason and emotion. 'It was a struggle among complex psychological states contingent upon both', he observed, and it had a 'shattering impact on the human spirit'. More recently Walter E. Houghton has published a sympathetic analysis of 'the Victorian mind', showing how Darwinism and evolution contributed to 'anxiety' by raising the fear of atheism and its corollaries, immorality and revolution, and by inflicting the pain of doubt that came from successive retrenchments before scientific explanation. 'The new conception of man and nature', he said, 'was to drive sensitive minds into the mood of ennui and frustration'.[5]

These statements are not exaggerations. They can be substantiated time and time again. Of course not all who were unsettled by science were affected by Darwinism;[6] nor, on the other hand, did those specifically influenced by Darwin always say so. The factors were many and complex which gave rise to spiritual crises and it was not exactly a mark of distinction to expose them before a religiously self-righteous society. Yet there were many Victorians, eminent and otherwise, whose manuscripts and memoirs reveal to posterity a common struggle, the conflict of minds steeped in Christian tradition with the ideas and implications of Darwinism.

Clergymen, more than others, were affected by tensions in their beliefs. One cannot long minister to the spiritually needy when the grounds of one's own faith is uncertain. In the United States Minot Judson Savage (1841–1918) laboured as a Congregational missionary and pastor until 1873, when, under the impact of Darwin and Spencer, he became a Unitarian. Among the liberal and radical clergy he joined, others, too, were undergoing the trauma of conversion: John White Chadwick (1840–1904) and Octavius B. Frothingham (1822–1895) renounced their transcendentalist faith in 1876 and embraced the empirical science of Darwin. In England Savage's move was paralleled by the Reverend Stopford Brooke (1832–1916), chaplain in ordinary to the Queen. Brought up as an evangelical, Brooke enlarged his religious sympathies according to an 'original intuition of Love as the master-principle of life' until at last, in 1880, he burst from the confines of the Established Church to officiate as a Unitarian minister. Wide reading in science and literature, of which books by Darwin and Huxley formed a part, was instrumental in his controversial and widely publicised defection.[7]

Leslie Stephen (1832–1904) moved farther and faster. He was an evangelical as well, the grand-nephew of William Wilberforce and the grandson of a leading member of the Clapham Sect. Dutifully he studied for holy orders at Cambridge and was ordained a priest at Ely in 1859. But sometime between 1862 and 1867 his faith flickered out. 'I became convinced, among other things, that Noah's flood was a fiction...and that it was wrong for me to read the story [before a congregation] as if it were a sacred truth', he said. Without the verbal inspiration of scripture there was 'no real stopping-place' and, in time, under the influence of Darwin, Mill, Comte, and Kant, Stephen decided that he had 'never really believed'.[8] Yet there was a price to pay for infidelity, not the least part of which was the loss of his tutorship and fellowship at Trinity Hall. Even more costly was the 'misery' which led up to his decision, the 'doubt as to the truth of revealed religion according to the orthodox view'. Although Stephen in later life maintained that the process had been painless – a burden removed, not a crumbling of ground from beneath his feet – one who knew him during his crisis years said that 'the pain he suffered was very acute...and was made doubly so because he knew what grief his determination would cause to some of his family who were nearest and dearest to him'.[9] In 1875 Stephen finally renounced his orders under the provisions of the Clerical Disabilities Act.

Not all who abandoned the Christian faith took up the genteel agnosticism of Stephen or the imperious naturalism of Huxley, Tyndall, and Spencer. William James (1842–1910), the son of a Swedenborgian theologian, entered Harvard in 1861 and attended the lectures of Louis Agassiz. Beset with his father's monistic spirituality and the idealistic anti-Darwinism which Agassiz espoused, James developed a 'brooding preoccupation with philosophy'. 'The first philosophical problem to which he devoted himself systematically was the problem of evolution.'[10] He read Darwin and Spencer, Comte and Mill, Buckle and Lecky, and French and German philosophers by the score. By 1868 his beliefs were in disarray and depression had invaded his mind. By mid-1870 he had passed through a 'spiritual crisis' which involved the loss of the desire to live. The problem for James was to find a philosophy which was scientifically respectable yet propitious as a creed for life. There had to be a moral order; there had to be freedom of the will.[11] Spencer's evolution and the Darwinism of Huxley could not honour these demands, nor indeed could the naturalism, agnosticism, and materialism of other leading writers. Although 'his sense of conflict [was] exacerbated by his devotion to Darwin's biology', James also found in Darwin's work the *indicia* of a new philosophy. Spontaneous variations of life and mind were serviceable as a phenomenal basis for the pluralism and pragmatism which he would later elaborate for thirty years as a Harvard professor of philosophy and psychology.[12]

Among others who came to occupy the Victorian never-never land between science and Christian faith Samuel Butler (1835–1902) is especially notable. The son of a canon and the grandson of a bishop, Butler was intended for the Church. But between 1859 and 1862 he gave up baptismal regeneration, left Cambridge for the antipodes, read Gibbon on the voyage, and embraced Strauss' theory of the Resurrection while tending New Zealand sheep. He also read the *Origin of Species*, which he credited with destroying his faith in a personal God. When, however, it came to expounding an alternative religious vision based on an alternative doctrine of evolution, Butler found himself alienated from the 'Church scientific' as well. Evidently a layman was not entitled to a hearing when he disputed the historical development of evolutionary theory and the pre-eminence of natural selection. Orthodox scientists could not countenance a doctrine of evolution based on the conscious will and unconscious memory of an omnipresent psychic energy. Again Darwin administered faith's *coup de grâce* when in 1879 he snubbed Butler's attempts to repristinate the theories of Erasmus Darwin and Lamarck, and carefully ignored his accusations of personal treachery. Butler could only conclude that naturalistic science had become another oppressive religion, with Darwin and Huxley as its priests. Soon men such as he would be hunted as heretics for claiming that professional scientists were not omnicompetent in the acquisition of truth. 'From this time onwards Butler felt himself to be a confirmed malcontent, in sworn opposition to. . .all Establishments whether ecclesiastical or scientific.'[13]

In some ways the most instructive of spiritual crises were experienced by Darwin's scientific friends. And without doubt the most revealing of these crises was that of Charles Lyell (1797–1875). For thirty years, beginning with the publication in 1830 of the first volume of his *Principles of Geology*, Lyell maintained the idiosyncratic view that the geological record testifies, not to the appearance of successively more complex and diversified forms of life, but to a steady-state in which plants and animals have at all times been continually and specially created in response to a perpetually fluctuating environment. The motivation for this belief was neither biblical nor highly theological. It was rather a complex religious longing – psychological, aesthetic, and social – to preserve a high genealogy for mankind. Lyell saw clearly that geological progression could become the correlate of theories of transmutation; that transmutation, once admitted, could not stop short of degrading 'time's noblest offspring' to the status of a beast.[14] Thus progression had to be resisted – resisted, that is, until the mid-1850s, when Lyell first encountered the arguments of Wallace and Darwin. In November 1855 Lyell began a series of notebooks on transmutation which for six years would record his struggle with the idea. Development and stasis, law and miracle, chance and providence and design – the issues led inevitably through painful twists and turns to the question of the origin and destiny of mankind. At one point Lyell seems to have wondered whether he might not have been happier had he never begun his researches. Indeed, it was only a few months later, early in 1859, that he sought consolation in the Unitarian theology of James Martineau and in the counsel of the Reverend J. J. Tayler, minister of Little Portland Street Unitarian Chapel, where the Lyells had attended for some time.[15]

Darwin, for his part, was counting on a swift and favourable verdict from his old respected mentor. Since, however, he could not enter sympathetically into Lyell's intellectual struggle, he consistently over-estimated the progress of his deliberations. In the *Antiquity of Man*, published in 1863, Lyell abandoned non-progression and endorsed transmutation; but in referring to a higher 'law of development', which rendered natural selection a 'subordinate agency', and to the 'intervention' of 'new and powerful causes' associated with the 'moral and intellectual faculties of the human race', he merely demonstrated how far apart were Darwin's views from his. Although he should have known better, Darwin felt betrayed. Lyell's response was a plea of personal incapacity: 'I have spoken out to the utmost extent of my tether, so far as my reason goes, and farther than my imagination and sentiment can follow.'[16] By 1868, when the tenth edition of the *Principles of Geology* had appeared, neither reason nor imagination nor sentiment had advanced. The book which Lyell conceived four decades earlier as a bulwark against transmutation now testified in unmistakably autobiographic phrases to the 'disquiet' and 'alarm' which had arisen at the prospect of human evolution.[17]

If Lyell was the outstanding representative of an older generation of naturalists for whom Darwin created unavoidable spiritual conflicts, then George John Romanes (1848–1894) was the foremost among the fledgling scientists who first grappled with Darwinism during the 1870s. Yet unlike many who left their faith once and for all in a mighty wrench of intellectual pain, Romanes had the shattering experience of both losing and rediscovering his Christian commitment. He was the son of a clergyman and professor of Greek at the University of Kingston in Canada, and in 1867 he entered Gonville and Caius College, Cambridge, to read mathematics and prepare for holy orders. In 1870, owing to pressure from family and friends, Romanes abandoned these plans and took the B.A. in natural science. Thereafter he remained four years in Cambridge to study medicine and physiology, reflecting all the while on the connexion of his researches with his lingering theological interests. During this time the physical efficacy of prayer was being severely questioned in the press by the advocates of scientific naturalism.[18] Romanes, whose studies were of a subject much discussed on bended knee, felt constrained to take a position in the debate, particularly as it was the topic prescribed in 1873 for the Burney Prize Essay competition at Cambridge. In fact Romanes gained the prize with a lengthy and closely reasoned composition which he published in 1874 as *Christian Prayer and General Laws*. The book gave evidence of a firm belief in the efficacy of prayer according to biblical promises and a commitment no less tenacious to the ubiquity of natural laws. Theism for Romanes was the basic postulate of all physical investigation. Rightly understood, it allowed that particular providences might occur in response to prayer without requiring science to notice them as miraculous violations of the lawful order of nature.[19]

But within two years all had changed. Romanes moved to University College in 1874 and began researches *inter alia* on the physiology of inheritance. In July of that year he began corresponding on the subject with Darwin and Spencer; by December he had met Darwin in person and given him a copy of his book. Spencer he met in 1875, as well as Huxley and Hooker, who joined with Darwin in proposing him for a Fellowship of the Linnean Society.[20] The effect of these events on a philosophically minded young man, not thirty years old, who stood at the threshold of his career, can easily be imagined. Romanes was flattered and awed. But he was also deeply troubled, for the doubts which had been kindled in writing *Christian Prayer* were inflamed by the enormous authority and intellectual prestige of his new-found agnostic friends. The result was a 'conflict between faith and scepticism which grew more and more strenuous'.[21] By 1876 Romanes had written another essay to set forth his beliefs. He published it two years later under the pseudonym 'Physicus' as *A Candid Examination of Theism*. Drawing heavily on Darwin and Spencer, on natural selection and 'persistence of force', the book disposed of theistic arguments and their advocates with an air of certainty that defied identification of its author with his earlier work. An air of certainty, yes, but not an air of triumph. 'It is...with the utmost sorrow that I find myself compelled to accept the conclusions here worked out', Romanes confessed.

So far as the ruination of individual happiness is concerned, no one can have a more lively perception than myself of the possibly disastrous tendency of my work. . . . I am not ashamed to confess that with this virtual negation of God the universe to me has lost its soul of loveliness; and although from henceforth the precept to 'work while it is day' will doubtless but gain an intensified force from the terribly intensified meaning of the words that 'the night cometh when no man can work', yet when at times I think, as think at times I must, of the appalling contrast between the hallowed glory of that creed which once was mine, and the lonely mystery of existence as now I find it, – at such times I shall ever feel it impossible to avoid the sharpest pang of which my nature is susceptible.[22]

'No one felt the strain, the positive agony of soul, in greater degree than did George Romanes', wrote his wife. He was deeply religious by nature and the 'hallowed glory' of the creed which had once bestowed on the universe a 'soul of loveliness' was yet unforgettably radiant in his mind. In time, under the influence of Oxford churchmen such as Charles Gore, P. N. Waggett, and Aubrey L. Moore, Romanes gradually came to doubt his disbelief.[23] A strict determinism became untenable; materialism no longer gave a satisfactory account of the origin of mind; a 'metaphysical teleology', in which the general order of nature rather than particular phenomena gives evidence of design, seemed a more compelling witness to God; the historical evidences of Christianity, relieved of *a priori* objections, took on greater strength; and religious needs and intuitions assumed a larger role in the quest for philosophical contentment. In 1893 or 1894, as Romanes saw the night closing upon him 'when no man can work', he began to make notes for a volume to be published under the name 'Metaphysicus' as 'A Candid Examination of Religion'. Regrettably, death overtook him on 23 May 1894, before the work could be finished. In 1895 Canon Gore published the notes as *Thoughts on Religion*, a book which none the less reveals the unmistakable predisposition towards faith expressed by Romanes in the stanzas of a poem written on his last Easter Day.

> Amen, now lettest Thou Thy servant, Lord,
> Depart in peace, according to Thy Word:
> Although mine eyes may not have fully seen
> Thy great salvation, surely there have been
> Enough of sorrow and enough of sight
> To show the way from darkness into light;
> And Thou hast brought me, through a wilderness of pain,
> To love the sorest paths if soonest they attain.
>
> . . . .
>
> As Thou hast found me ready to Thy call,
> Which stationed me to watch the outer wall,
> And, quitting joys and hopes that once were mine,
> To pace with patient steps this narrow line,
> Oh! may it be that, coming soon or late,
> Thou still shalt find Thy soldier at the gate,
> Who then may follow Thee till sight needs not to prove,
> And faith will be dissolved in knowledge of Thy love.

On Thursday of Whit week Romanes took Holy Communion and during the day remarked, 'I have now come to see that faith is intellectually justifiable'. Later, according to his wife, he added, 'It is Christianity or nothing'. His pilgrimage from faith to faith was complete.[24]

Romanes, Lyell, Butler, James, and Stephen afford prominent but by no means isolated examples of the spiritual crises which were instigated and catalysed by post-Darwinian evolutionary thought. To them might be added Darwin's own religious difficulties as he moved from his early orthodoxy through a liberal theism to the uneasy agnosticism that dominated his later years;[25] the *angst* which accompanied Huxley's importunate need to retain value and purpose in a universe where only the fittest survive, as revealed in his lecture on evolution and ethics, delivered in 1893 at the end of his career; Jeffries Wyman's 'deep distress, emotional as well as rational', over the question of the Pithecoid origin of mankind; the 'long soul-searching struggle' which James Dwight Dana underwent as he sought to reconcile his faith with the Darwinian theory; Henry Sidgwick's protracted 'interior debate' between irrepressible religious sentiments and irresistible demands for a 'scientific study of Human Nature'; the 'troubles and doubts' through which Alfred Marshall passed as he abandoned evangelicalism under the influence of Darwin, Spencer, and Mill; W. K. Clifford's 'intellectual and moral struggle' in exchanging the theological authorities of Anglo-Catholicism for the philosophical liberties of Darwinism; the painful 'inner conflict' which seized H. C. G. Moule as a Cambridge undergraduate, beset by 'the continual droppings of the controversies and questions' of the early 1860s; E. Ray Lankester's earnest but unsuccessful struggle for a 'great religious belief' such as the Anglican creed from which Darwinism claimed him; the 'spiritual crisis' experienced by many of the leading New England writers between 1868 and 1872; and the countless unrecorded crises of unknown individuals who found evolutionary doctrines in tension with their beliefs.[26] One would seem less than human if one did not at once feel a deep sympathy with this late-Victorian generation. A 'shaking of the foundations' must come to every person, to be sure, but it was no fault of theirs that a compelling argument for biological evolution should be thrust on the world at a time of particular religious rigidity and in the midst of intellectual currents which in themselves were sufficient to overwhelm all but the strongest faith.

How inappropriate, then, is the military metaphor! If, as one historian puts it, the judgement of Andrew Dickson White on the contribution of science to the Christian life 'documents rather the religious insensitivity of America in the 1890's than the insight of the apologist of science',[27] how much more does the habitual use of warfare to characterise the post-Darwinian controversies demonstrate the thoughtless and unimaginative outlook of later interpreters. If critical historiography presupposes sympathetic understanding, can it be wondered that these interpreters have failed to shed much light on Christian responses to Darwin?[28] An appreciation of Darwin's contribution to the Victorian crisis of faith, on the other hand, suggests an approach to the post-Darwinian controversies which is both empathic and constructive.

Darwinism and Darwinisticism

Since Christians experienced painful conflicts between Darwinism and their beliefs, it seems reasonable to expect that some general understanding of the manner in which these conflicts were resolved would assist in interpreting the most tangible evidence of the post-Darwinian controversies: namely, the bewildering variety of books on evolution and religion that were published in the last third of the nineteenth century. For, clearly, each book had an author or authors; each author expressed in writing a discernible response to Darwin; and each response *ex hypothesi* was the product of at least a minimal crisis of faith. (Some evidence of this crisis–reality exists in the frequent occurrence of its symbol – the military metaphor – within the primary texts.) Presumably, therefore, if some general statements can be made about the resolution of such conflicts as each crisis was likely to entail, then on the basis of these statements the literature in which the conflicts were resolved could more readily be interpreted.

The difficulty of course consists in making general statements about the interior life of historical individuals. Even with the fullest of literary remains at one's disposal, such statements would at best be an inconclusive mixture of the vague, the obvious, and the untestable. However, at a theoretical level, where generalisations have been checked empirically and thus retain an heuristic value, there are various interpretations of conflict-resolution which might prove useful for our ends. We propose to elucidate the post-Darwinian controversies by an application of the most important of these interpretations, the one which has had 'by far the greatest impact on the study of attitude change' as well as on 'the entire field of social psychology': Leon Festinger's theory of cognitive dissonance.[29]

According to Festinger, who published the theory in 1957, the human response to opinion-making information proceeds in four stages. At first there is a *conflict* as new knowledge challenges old. Incompatible alternatives are established in the mind and until the discrepancy between them is eliminated one lives in a state of tension which may lead in time to frustration, anger, and aggression. The normal way to eliminate this tension and regain one's intellectual composure is by making a *decision*. Conflict is eliminated by committing oneself to one or another of the incompatible alternatives established by the new information. Following a decision, however, there is usually another kind of conflict which Festinger calls *dissonance*. In this case the discrepancy is not between incompatible alternatives towards which one feels pushed simultaneously, requiring that a choice be made, but between what one chose and what one might have chosen – between what Festinger calls 'cognitive elements'. A cognitive element is 'any knowledge, opinion, or belief about the environment, about oneself, or about one's behavior'. 'Two elements are in a dissonant relation', says Festinger, 'if, considering these two alone, the obverse of one element would follow from the other'.[30]

If, for example, a member of the 'Flat Earth Society' is shown photographs of his favourite planet taken during the several moon voyages, he will experience a definite conflict. The new evidence compels a decision on the configuration of the earth. If he remains a member in good standing of the Flat Earth Society and goes on evangelising in behalf of his cock-eyed cosmology, his cognition of the photographs will be dissonant with his behaviour and his beliefs. If, on the other hand, he gives up the flat earth theory, his new belief in the earth's sphericity will be dissonant with all his arguments for the old position, with all his past pronouncements in its favour, and with his continuing membership in the Flat Earth Society.

The last stage in the human response to opinion-making information is *dissonance reduction.* A normal individual finds dissonance, like conflict, to be an unpleasant state of mind and therefore automatically experiences pressure to harmonise his cognitions. He will seek to reduce his dissonance 'by thinking about, considering and reconsidering, and re-evaluating these dissonant cognitions until adequate reinterpretations are invented or discovered'. A reinterpretation which is adequate to reduce dissonance, says Festinger, must involve a change in the dissonant cognitions. Either one or more of the dissonant elements must be changed or the relationship of the dissonant elements must be changed by adding new cognitions which dilute their dissonance or reconcile them.[31]

Thus a flat-earther might be convinced by extra-terrestrial photographs that the earth is a globe and yet increase his commitment to the Flat Earth Society, attending more meetings, taking on new responsibilities, soliciting new members, all in an effort to compensate for his departure from flat-earth orthodoxy. Alternatively, he might retain the flat-earth theory by choosing to believe that the photographs were artificially contrived for the purpose of inducing doubt in flat-earthers. He might dilute the dissonance between his cognition of the photographs and the flat-earth theory by contemplating other obscure and mysterious phenomena – flying saucers, extra-sensory perception, paradoxes in theoretical physics, etc. – and classifying the enigmatic photographs among them. Or he might reconcile his cognitions by a belief in the strange refractive powers of trans-lunar space – a belief expressed in December 1968 by a prominent British flat-earther who is supposed to have said, 'Isn't it strange that even from so great a distance the earth still *appears* to be a globe!'

Admittedly, in the encounter between Darwinism and Christian faith neither the evidence nor the issues were so well defined, at least not at the outset. Flat-earth theories have been out of fashion for more than a millennium whereas in Darwin's day special creation was virtually the only acceptable theory of origins. But the point of the illustration is to explicate the process of conflict resolution and in this respect it accomplishes its purpose. As our putative flat-earther experienced an intellectual crisis which may be described in the terms of Festinger's theory, so in the lives and writings of the participants in the post-Darwinian controversies we can expect to find evidence of conflicts, decisions, dissonance, and dissonance reduction. Moreover, if history is not to be reduced to retrospective psychology, we should expect to make some connexions by means of the theory of cognitive dissonance between Darwinism, on the one hand, and the welter of religious responses it elicited on the other.[32]

The conflict between Darwinism and Christian beliefs is already familiar. So compelling was the new view of nature that some abandoned the faith to which, in a few cases, they owed their entire livelihood. For their part Darwin seemed to annihilate singlehandedly the entire body of Christian doctrine. If creation did not proceed according to the record in the first chapters of Genesis, they asked, can the Bible be trusted to provide inerrant knowledge in other matters? If human beings were not specially created, as Genesis teaches, but descended from lower animals in a bloodthirsty struggle for existence, then what becomes of their fall into sin and whence their need for redemption? If things 'make themselves', if the manifold forms of life, with their wonderfully intricate structures and their consummate adaptations to the environments they inhabit, have developed quite without reference to any plan or purpose, then what need is there of a wise and beneficent Creator to superintend the course of nature? Indeed, is it conceivable that such a Creator should exist in view of the world's pain and suffering, the enormous waste of life as countless millions of one variety are extinguished in favour of countless millions of another and so on as each new species emerges? And if this be the true course of things, if neither revelation nor redemption nor a wise and loving Creator are to be found, but instead struggle, suffering, and death, then how can one speak of Christian morality, the very cement of society, and of the afterlife which is its chief reward?[33]

Others drew the same inferences from Darwinism and for that

reason rejected it. 'Acceptance of the Darwinian theory seemed to them to necessitate a complete spiritual revolution, a total change of outlook towards life.' Only the most worthy and compelling cause could make such an upheaval worthwhile. To the vast majority, obviously, 'the possible truth or falsehold of an abstract scientific theory, which few were in a position to judge on its merits, was not at all a strong incentive.'[34] There were of course a number of clergymen, scientists, and men of letters who experienced Darwin's theory as a challenge to which intellectual integrity demanded a response. Many of these individuals were equipped to appreciate the theory, some to judge it on its merits; and almost all of them went beyond the popular attitude by responding with philosophical and scientific objections. But like the majority they felt no compulsion to undergo a 'spiritual revolution'. There was no incentive in Darwinism strong enough to overcome the appeal of traditionary interests.

Out of cognitive conflict, then, came two contrary responses to Darwin. Some individuals embraced Darwin's theory and abandoned Christianity. Others turned their backs on Darwin and clung the more to conventional Christian beliefs. Whatever their differences, on one point all were agreed: Christianity, rightly understood, is incompatible with Darwinism, rightly understood.

On one point also historians have been generally, if tacitly, agreed: neither response *per se* is very interesting. In science as in war (and here there *is* a legitimate parallel), history is written by the victors. Those who first embraced a new science are styled as precursors of the latest orthodoxy. Those who stubbornly clung to the old are featured as historical curiosities. One group is absorbed, the other is absurd. The post-Darwinian controversies are particularly susceptible of this interpretation, for the Darwinians have been largely vindicated in their Darwinism and Christians who opposed evolution did so frequently in the *genre* of comic relief. Too easily, therefore, historians preoccupy themselves with the pageantry of vulgar anti-Darwinians or the vulgar remarks of leading anti-Darwinians while almost entirely overlooking the intellectual background of their attacks, a subject less entertaining, perhaps, but certainly much more interesting and instructive. We shall seek to correct this misplaced emphasis by reviewing some of the philosophical and scientific objections raised by the leading representatives of *Christian Anti-Darwinism*. Of those who gave up Christianity in favour of Darwinian evolution we shall say no more save to add that in most cases they experienced a sharp and painful dissonance. They alone abandoned what once was cherished. The consequences of their loss and their efforts at dissonance reduction are perhaps best illustrated in the life of Romanes.

Dissonance, says Festinger, is the 'inevitable consequence' of a decision; but its magnitude may vary considerably, depending on the 'relative attractiveness' of the chosen and unchosen alternatives. In general, decisions involving alternatives of equal or nearly equal appeal result in a greater dissonance than obtains from decisions which involve alternatives of decidedly unequal appeal.[35] Thus the dissonance of Christian anti-Darwinians was negligible because Darwinism to them was unattractive, a poor incentive to undergo a spiritual revolution. The dissonance of Christians who wished to be evolutionists, however, was very large indeed. To some Darwinism promised release from an oppressive orthodoxy. To most it promised a new and larger truth. But to all, regardless of creed, it brought the intellectual distress involved in choices between attractive alternatives. Ellegård has pointed out that the idea of evolution, the Darwinian theory of natural selection, and the extension of evolution to mankind were the main points at issue, broadly speaking, before 1872; and there is every reason to think that the issues persisted for the rest of the century.[36] In reckoning with Darwinism Christians therefore had to decide between transmutation and the special creation of immutable species, between natural selection and the venerable argument from design, between human descent from animals and mankind's unique creation in the image of God. These decisions (and others as well) had to be undertaken while conserving truth wherever it was found. A few concessions might be made – the literal interpretation of Genesis and the absolute fixity of species, for example – without jeopardising the faith. But sooner or later a naturalistic doctrine of evolution would be seen to conflict with the doctrines of a supernaturalistic religion, a truth chosen with a truth neglected. Under these circumstances only some kind of dissonance reduction could put the mind at ease.[37]

Dissonance reduction, according to Festinger's theory, must involve a change in the dissonant cognitions themselves or a change in their relationship by the addition of cognitions that dilute their dissonance or reconcile them. Thus we may assume that Christians who wished to be evolutionists were compelled to alter Darwinism, Christian beliefs, or both, or to draw on ideas that reduced their dissonance or reconciled the dissonant doctrines. As in our illustration, a flat-earther confronted with extra-terrestrial photographs of the earth might compensate a decision in favour of conventional geography with a renewed behavioural commitment to the unchosen alternative; he might deny or misunderstand one of his cognitions; he might absorb both cognitions in a larger outlook that makes his dissonance bearable; or he might contemplate a *tertium quid*, some notion that explains and reconciles the dissonant cognitions. So in the confrontation of Christian beliefs and Darwinism we can expect to find similar *modi vivendi*. The number of possible responses employing the various methods of dissonance reduction, singly or in combination, are of course legion. We could not begin to describe them before examining the particular responses themselves. But since it will be instructive to understand the terms on which Christians were willing to accept *Darwinian* evolution in particular, we shall observe one critical distinction.

With Morse Peckham, editor of the variorum text of the *Origin of Species*, we shall refer to those propositions and implied assumptions which may be properly ascribed to a source in Darwin's publications as 'Darwinian', and to those propositions and derived assumptions which are not properly so ascribed as 'Darwinistic'. 'Darwinism', Peckham explains,

is a scientific theory about the origin of biological species from pre-existent species, the mechanism of that process being an extraordinarily complex ecology.... Darwinisticism can be an evolutionary metaphysic about the nature of reality and the universe. It can be a metaphysical and simplistic notion of natural law. It can be an economic theory, or a moral theory, or an aesthetic theory, or a psychological theory. It can be anything which claims to have support from the *Origin*, or conversely anything which claims to have really understood what Darwin inadequately and partially presented.

In observing this distinction (and refining it as well) we can bisect the varieties of dissonance reduction.[38] On the one hand there are the Darwinian responses – what we shall denominate *Christian Darwinism*. Its representatives understood Darwin's theory and left it substantially intact, neither emasculating it nor adulterating it with foreign ideas in the interests of dissonance reduction. Christian Darwinians may have made adjustments to their religious beliefs or drawn on cognitions that mitigated their dissonance, but they did not do so in the name of or at the expense of Darwin. On the other hand there are the Darwinistic responses – what we shall denominate *Christian Darwinisticism*. Its representatives either misunderstood, misinterpreted, or modified

Darwin's theory, adulterating it as they had need with non-Darwinian ideas. Christian Darwinists may also have altered their religious beliefs or entertained some cognitions that were compatible with Darwinism in order to mitigate their dissonance, but their acceptance of evolution none the less involved a departure from Darwin. As more than one historian has felt uneasy over the fact that 'books on evolution and theology spend more time on the opposition to Darwin and on ways of undermining or getting round [his] theory than on absorbing and interpreting it', we shall devote separate chapters to considering these two types of dissonance reduction.[39]

For the present, however, nothing will clarify the distinction between Christian Darwinism and Christian Darwinisticism better than an historical example, though in giving it we anticipate a later discussion.

The case of St George Mivart

St George Jackson Mivart (1827–1900) was Darwin's most influential Christian critic in Great Britain. Twice in his life he underwent a conversion, once in religion and once in science. In religion Mivart was raised as an evangelical. But evangelicalism could not satisfy his romantic longing for an absolute faith which expressed itself aesthetically in historic ritual and architecture. He found that faith only in the spring of 1844 when, under the impact of the Anglo-Catholic revival, he entered the Church of Rome. In science Mivart was influenced first by his father, a Fellow of the Zoological Society, who allowed him to copy illustrations from the family's folio edition of Buffon's *Histoire naturelle*, and later, after his religious conversion, by Richard Owen, the renowned anatomist.[40] But neither Buffon nor Owen was the match of Huxley, whom Mivart came to know in 1861. Charmed by his friendship, his ruthless honesty in exposing pretence (Owen's pretence doubtless included), and his brilliant grasp of anatomical subjects, Mivart became Huxley's admiring student and thus – in his own words – a 'hearty and thoroughgoing disciple of Mr Darwin' who 'accepted from him the view that Natural Selection was "*the* origin of species"'.[41]

During the sixties Mivart published a series of authoritative studies of Primate osteology. The subject was timely considering that Huxley and others were then much exercised by the question of mankind's place in nature. Mivart's treatment was also timely, for he supported the new theory which was coming under attack from every side. But his support did not last. By 1868 he had begun to question the efficacy of natural selection, particularly as a cause of human evolution. Natural selection might explain the development of the human body, he conceded, but it can hardly account for mankind's unique psychological nature. The longer Mivart reflected the greater was his conflict, the wider became the chasm between human beings and their fellow Primates. At last he placed himself squarely 'on the side of the angels': the human soul with its intellectual and moral attributes is a supernatural infusion, he maintained, and natural selection is not, in fact, '*the* origin of species'.

It was a grave decision. Mivart had been a convinced Darwinian, a trusted and respected member of a rising school of biology. Now his faith (coupled, no doubt, with a lingering commitment to Owen's transcendental philosophy) had caused him to fail what was, in effect, a loyalty test, and this fact could not long be kept from his colleagues. What then should he do? Towards the end of his life he recalled his course of action.

For the rest of that year [1868] and the first half of the next I was perplexed and distressed as to what line I should take in a matter so important, and which more and more appeared to me one I was bound to enter upon controversially.

After many painful days and much meditation and discussion my mind was made up, and I felt it my duty first of all to go straight to Professor Huxley and tell him all my thoughts, feelings, and intentions in the matter without the slightest reserve, including what it seemed to me I must do as regarded the theological aspect of the question. Never before or since have I had a more painful experience than fell to my lot in his room at the School of Mines on the 15th of June, 1869. As soon as I had made my meaning clear, his countenance became transformed as I had never seen it. Yet he looked more sad and surprised than anything else. He was kind and gentle as he said regretfully, but most firmly, that nothing so united or severed men as questions such as those I had spoken of.[42]

Unlike the experience of most Christians who assimilated evolution in the decade after 1859, Mivart's dissonance derived from a prior adherence to Darwinism. Neither the intention nor the character of the Church in England was so carefully defined in his Darwinian years that Mivart should have experienced a conflict between his faith and his science during that time. After 1864, however, and particularly after 1870, there were compelling reasons for one who would be a scientist as well as a loyal son of the Church to examine his conscience. The message of the 'Syllabus of errors' and *Pastor Aeternus* was inimical to modern science in so far as it did not accord with the Roman congregations, the ancient scholastic doctors, and the infallible pontiff himself. Science, if it were to be conducted at all, would have to demonstrate its compatibility with all that opposed it. Mivart felt equal to the task. Through such a demonstration he hoped to persuade critics of the Church's genuine liberality and, conversely, to convince Catholics of the respectability of science. Above all, a synthesis of Catholic faith and Darwinian science would make bearable that painful inner conflict which accompanied his departure from the Darwinian circle.

Confronted by the intellectual impasse produced by the apparently contradictory views of the universe provided by Darwinism on the one hand and Catholicism on the other, Mivart could resolve his own conflict only by a complete rejection of one position or a reconciliation of the two. Dedicated to both he could reject neither. As the phenomena observable in nature, the subject matter of science, were true, so too were the data derived from revelation, the subject matter of religion, true. Since both represented truth, neither could be rejected without danger to that harmonious world of truth whose ultimate description and understanding were the goal of both scientist and theologian. Since both represented truth, albeit of different orders, the goal of the scientist as well as the religious philosopher, lay in the harmonious reconciliation of the two sets of data.[43]

On the Genesis of Species, Mivart's first and most important controversial work, was published in January 1871, a month before Darwin's Descent of Man. Its objects were to show that the Darwinian theory was untenable because natural selection is not the origin of species and that nothing in the 'general theory of evolution' was necessarily incompatible with Christianity. In rejecting natural selection as the principal factor in evolution Mivart did not deny its efficacy altogether, but made it subordinate to 'special powers and tendencies existing in each organism', which, under the influence of the environment, give rise to new forms. These powers and tendencies were God's special endowment and the environment was the divine instrument employed in directing them to produce just those forms which God had preconceived. The human body was derived from this evolutionary process but the soul, the source of mankind's rational and ethical nature, appeared de novo by creative fiat. In the book's concluding chapter Mivart argued that although the Fathers of the Church may not have taught these doctrines, neither did they condemn them. On the contrary, there was evidence in his judgement that the Fathers actually allowed for transmutation as a possible explanation of specific origins.

Mivart continued to promulgate his mediating views in *Man and Apes* (1873), *Contemporary Evolution* (1876), *Lessons from Nature* (1876), *Nature and Thought* (1882), and *The Origin of Human Reason* (1889), each time setting himself against the Darwinians while synthesising the remnants of his Darwinism with Christian and Catholic dogma. In the Church his labours were well received: the Catholic press sang his praises; in 1876 Pius IX conferred on him the degree of doctor of philosophy; and in 1884, at the invitation of the Belgian Episcopate, he was made professor of the philosophy of natural history in the University of Louvain. But among the Darwinians his reputation was irreparably damaged. Huxley devoted nearly half his review, 'Mr Darwin's critics', to an excoriation of his former student, much to the pleasure of Hooker, Tyndall, and Darwin. Chauncey Wright, the American philosopher and mathematician, published a weighty article which upheld Darwin on philosophical grounds by arguing that Mivart had failed to comprehend the problems addressed in the *Origin of Species*. Darwin himself was so impressed with Wright's article that he had it published as a pamphlet.[44] Mivart did not endear himself to his old colleagues with a stinging review of the *Descent of Man*, and when in 1874 he blundered into accusing George Darwin of encouraging profligacy by means of his eugenic ideas, his alienation from them was complete. Darwin took it as a slanderous insult on his son and through him on himself. His friends also bore the accusation vicariously, so that 'until the day of his death Mivart was haunted by the hostility, latent and overt, of the small circle which had surrounded Darwin'.[45]

Like his conversions, Mivart's excommunications were two in number. The vision of transcendental truth and the craving for intellectual autonomy which brought about his schism with the Darwinian party in time brought him into conflict with the Church. Mivart did not oppose the Church *per se* but only its static, illiberal, and ultramontane posture. He pressed for a nobler and grander institution, evolving under the revelations of science and embodying all that the human spirit knows as true Christianity. Though old doctrines and superstitions would have to be shed, Mivart was convinced that the Church so reformed would yet remain the locus of divine truth. Rome did not agree nor did any but a handful of Catholic intellectuals. Some of his essays were placed on the *Index expurgatorius* and in 1895 the 'Mivartian' theory of biology was judged untenable. Sick with bitter-

ness and desperation, Mivart made a last assault on the Church, only
to be insulted' in return, the responsibility for which he promptly
placed on Cardinal Vaughan, the archbishop of Westminster. The
prelate responded with the terms for his unconditional surrender. On
23 January 1900 Mivart replied, '*Liberavi meam animam*'. His
second excommunication was sealed. Within four months he was dead,
the tragic outcast of science and faith.[46]

The failure of the military metaphor to comprehend Mivart's sad
career is a striking instance of its historical inadequacy. Mivart was
neither a sceptical scientist nor a hostile clergyman but a deeply
religious biologist and lay-theologian. His life-long controversies were
not part of an organised campaign of religious opposition to Darwin,
but a lonely personal struggle within science itself to combat a philosophy antagonistic to his own and to reconcile the traditions he held
dear with the truths of Darwinism he could accept. The hostilities
Mivart experienced were not so much heated polemical exchanges as
distressing disagreements and misunderstandings with old friends, rifts
which sometimes (as in Huxley's case) were later bridged. Moreover,
these personal conflicts did not place Mivart either against science or
against the Church but squarely between them both, one whose autonomous quest for absolute truth led him to cast off every authority save
his own, both scientific and religious.

The theory of cognitive dissonance as applied above does not force
Mivart to be what he was not, but helps to explain the way he was. It
predicts the dissonance which drove him to confess to Huxley and it
anticipates the dissonance reduction evident in his several controversial
works. According to the theory, Mivart should have reduced the
dissonance between Darwinism and his faith by altering one or the
other and by injecting new concepts that weakened the dissonance or
brought the dissonant doctrines into harmony. In fact Mivart employed
all these means. Darwinism was made to give up the dominant role of
natural selection and the right of natural selection to account for the
human mind. Catholic theology, on the other hand, had to relinquish
the special creation of the human body. To reconcile the two Mivart
posited 'special powers and tendencies' to supplement natural selection
in bringing species into existence according to God's preconceived ideas
and the supernatural infusion of the human body by a rational and
ethical soul to account for the higher faculties of the mind. To weaken
the dissonance between Christian doctrines and Darwinism Mivart
sought to establish the compatibility of evolution with the teachings of

the Church's theologians. The synthesis expressed in these new and altered concepts we have referred to as Christian Darwinisticism, for Mivart's understanding of evolution was not consistent with Darwin's.

Other Christian responses to Darwin we have distinguished as Christian Anti-Darwinism and Christian Darwinism. These three categories are intrinsic to the post-Darwinian controversies. They are not derived from a universe of discourse which wreaks historiographic violence, reducing the controversies to its own terms in Procrustean fashion. Rather, they are chosen to represent the conflict and confluence of ideas proper to the subject itself. The categories are intrinsic in a deeper sense as well. While there are any number of schemes by which Christian responses to Darwin might be classified such that attention is drawn to the concepts they express, the theory of cognitive dissonance has enabled us to derive an interpretation directly from the widespread and well-attested crisis in which these concepts were rooted. The struggle to come to terms with Darwin in the later nineteenth century is better understood on its own terms than by means of the military metaphor, which was its popular symbol at the time.[47]

PART II

DARWINISM AND EVOLUTIONARY THOUGHT

5

DARWINISM IN TRANSITION

When one discusses Darwinism, to what does one refer? Is it a theoretical
Darwinism, . . .what Neo-Darwinism retained or adapted from Darwin?
Or is it the historical Darwinism, with its doubts, its retractions, its
concessions to critics, from the first edition of the *Origin of Species* to the
'Essay on Instinct'?. . .[Would it not be] simpler to start with Darwin
himself and what he wrote than with an imaginary Darwinism whose
existence in the realm of ideas creates unnecessary difficulties?

Jacques Roger[1]

The works of Charles Darwin (1809–1882) have been cited more
frequently than read, and read far more often than understood.
Religious reactionaries and the sycophants of scientism have alike
forced Darwin to serve their dubious ends, while even persons of
moderation have represented as Darwinism either more or less than
the primary texts allow. Under these circumstances, which have only
lately begun to change, a just understanding of Darwin's theory of
evolution by natural selection must be founded on Darwin's statement
of it. Fortunately, there is at least one passage in the *Origin of Species*
that remained substantially unaltered throughout the book's six editions
(1859–72), which distils the essence of Darwinism into less than five
hundred words.

If during the long course of ages and under varying conditions of life,
organic beings vary at all in the several parts of their organisation, and I
think this cannot be disputed; if there be, owing to the high geometrical
powers of increase of each species, at some age, season, or year, a severe
struggle for life, and this certainly cannot be disputed; then, considering
the infinite complexity of the relations of all organic beings to each other
and to their conditions of existence, causing an infinite diversity in structure, constitution, and habits, to be advantageous to them, I think it
would be a most extraordinary fact if no variation ever had occurred
useful to each being's own welfare, in the same way as so many variations
have occurred useful to man. But if variations useful to any organic being
do occur, assuredly individuals thus characterised will have the best

chance of being preserved in the struggle for life; and from the strong principle of inheritance they will tend to produce offspring similarly characterised. This principle of preservation, I have called, for the sake of brevity, Natural Selection. . . . Amongst many animals, sexual selection will give its aid to ordinary selection, by assuring to the most vigorous and best adapted males the greatest number of offspring. Sexual selection will also give characters useful to the males alone, in their struggles with other males. . . . [Natural selection] entails extinction; and how largely extinction has acted in the world's history, geology plainly declares. Natural selection, also, leads to divergence of character; for more living beings can be supported on the same area the more they diverge in structure, habits, and constitution, of which we see proof by looking at the inhabitants of any small spot or at naturalised productions. Therefore during the modification of the descendants of any one species, and during the incessant struggle of all species to increase in numbers, the more diversified these descendants become, the better will be their chance of succeeding in the battle of life. Thus the small differences distinguishing varieties of the same species, will steadily tend to increase till they come to equal the greater differences between species of the same genus, or even of distinct genera.[2]

Darwin's argument for the origin of species is a good approximation to the Newtonian, hypothetico-deductive ideal which was held high by contemporary philosophers of science. The argument rests on a series of axioms or physical laws: if these laws are true, if the conditions they describe do in fact obtain, then it follows logically that certain consequences will be observed in nature.[3] If there is some *variation* in plant and animal populations, and if, owing to the *overproduction* of offspring, each individual at some time undergoes a *struggle for existence*, then those individuals embodying variations that are advantageous in the struggle will have a better chance of surviving to produce offspring on which *inheritance* will bestow similar characteristics. This 'principle of preservation' is *natural selection*. Its corollary is *sexual selection*, for (as Darwin believed) success in leaving offspring is determined not only by variations which confer advantage in the struggle for the environment's limited means of sustenance, but also by variations of structure and habit in individual animals which assure their competitive advantage over others in the mating process. Furthermore, if some variant forms are preserved in this manner, then *extinction* of others will necessarily result; for individuals which lack advantageous variations will fail to reproduce themselves in the struggle for existence. But if some forms are continually preserved, by reason of having possessed and passed on favourable variations, while others with injurious variations are extinguished by them, form upon form, age upon age, then every variation which confers advantage in leaving offspring will tend to persist, resulting in a great *divergence* of structure and function in the natural world and leading eventually, through the ecological and geographical *isolation* of interbreeding populations, to *speciation*.[4]

No aspect of Darwin's argument was incontestable. Yet so generally admitted were its premises, so reasonable were Darwin's deductions from them, that to reject natural selection was practically to deny a great deal of otherwise undoubted truth. There could be little question that the 'artificial selection' of desired stocks by animal breeders, a well-known and venerable art, was dependent on the very heritable variations which served the mechanism of natural selection. Had not Darwin himself been at pains in the *Origin*'s first chapter to lay the basis for his theory with this analogy? Moreover, it was a commonplace of Victorian social life, if not of nature, that a struggle for the means of existence results from overpopulation. When Darwin applied the doctrine of Malthus 'with manifold force to the whole animal and vegetable kingdoms' he did not adopt a novel or unfounded idea.[5] Extinction was an undeniable feature of the geological record, and in no way better explained than by the effects of a struggle for existence and a failure to reproduce – both concepts intimately a part of the Darwinian theory. As for divergence, isolation, and speciation, there were no empirical grounds to deny them. Indeed, after 1859 it became increasingly clear that the only grounds on which the origin of species through 'descent with modification' might effectively have been opposed were *a priori*.[6]

Yet despite his cogency of argument, Darwin experienced real difficulties in maintaining his theory, even in the face of those who could accept the idea of descent with modification. The reason was that both the phenomena of natural selection and its presuppositions, as he conceived them, were vulnerable to attack. For twelve years Darwin adjusted his theory to compensate for these conceptual weaknesses. In the end the theory stood, but neither so elegantly nor so impressively as before.

Variation and inheritance

The primary phenomena of natural selection were variation and inheritance. In each instance Darwin made assumptions that became highly problematic in his own time and have subsequently been abandoned. 'Our ignorance of the laws of variation is profound', Darwin confessed in each edition of the *Origin*. 'Not in one case out of a hundred can we pretend to assign any reason why this or that part differs, more or less, from the same part in the parents.'[7] To variation he did nevertheless assign three basic characteristics: its nature is continuous, both from individual to individual and over an entire population; its extent is unlimited by the nature of a species; and its causes lie ultimately within the environment.

In the first case *Natura non facit saltum* was his byword.[8] If a variant organism is to be favourably selected it cannot be a 'saltation'. Discontinuous forms (Darwin also called them 'sports' and 'monstrosities') are extremely rare and generally sterile; their characteristics are seldom advantageous. Only organisms in close continuity with their parent forms – manifesting 'individual differences' or (somewhat larger) 'small deviations of structure', rather than gross anomalies – can serve as the raw material of natural selection. Under compulsion from Huxley, Asa Gray, and William Henry Harvey to temper this view and give more place to variations of an intermediate size, Darwin relented somewhat. Beginning in the *Origin*'s third edition (1861), for example, his 'infinitesimally small inherited modifications' became 'small inherited modifications'.[9] But in 1867, when the physicist H. C. Fleeming Jenkin (1833–1885) showed to his satisfaction that discontinuous variations of whatever size are not serviceable to natural selection, Darwin concluded that individual differences are supreme. In so doing he created for himself a new problem.[10]

No one doubted that variations appear continually in every living organism. But that these variations should never cease to appear, that they should be accumulated without limit in those individuals preserved by natural selection, was quite unthinkable for the great majority of Darwin's colleagues. Lyell, Owen, Harvey, Gray, and Louis Agassiz, each for his own reasons – sometimes very good reasons – believed that the extent of variation in every organism tends to be strictly limited by formidable specific barriers. Saltations aside, all were agreed: mere individual differences could never supply the material for organic evolution. But to Darwin this was nothing but a 'simple assumption'. Likewise, he said, 'that the process of variation should be. . .indefinitely prolonged is an assumption, the truth of which must be judged of by how far the hypothesis accords with and explains the general phenomena of nature'.[11] Since there were powerful independent reasons (such as the geographical distribution of plants and animals) for believing in the transmutation of species, as well as strong reasons for believing that the real source of evolutionary change is individual differences, not discontinuities, Darwin concluded that his assumption was the better of the two. Again, however, Jenkin's criticism had a telling effect. Jenkin, a brilliant mathematician, maintained on the basis of experimental evidence that each animal or plant is contained, as it were, in a 'sphere of variation', with the average form at its centre and a steadily decreasing number of forms towards its surface. Darwin's theory, he argued, had not only to fix a variant population near the outside of the sphere of variation, but to endow it with the power to vary beyond that sphere. This power, Darwin responded, is inherent in the unfailing ability of the environment to elicit continuous variation: 'It would be...rash to assert that characters now increased to their utmost limit, could not, after remaining fixed for many centuries, again vary under new conditions of life.'[12] Given enough time, the influence of 'new conditions of life' would produce individual differences sufficient to overcome every specific barrier. Time, as we shall see, was precisely what Darwin was denied.

The causes of variation, according to Darwin, lie ultimately within the environment. Variations arise directly or indirectly from 'new conditions of life'. Three causal factors appear consistently throughout Darwin's works, while in his *Variation of Animals and Plants under Domestication* (1868) 'correlation' and 'compensation' appear as secondary causes, subordinate to them. In the first edition of the *Origin*, for example, Darwin wrote that 'the conditions of life, from their action on the reproductive system, are so far of the highest importance as causing variability. . . . Something must be attributed to the direct action of the conditions of life. Something must be attributed to use and disuse.'[13] Now the environment causing variation indirectly, through its action on the reproductive organs, was neither a new nor a particularly troublesome concept for Darwin and his colleagues. But the other causal factors turned out to be positively contentious. Darwin maintained that the environment might sometimes cause variations directly, by action on the parent bodies, and definitely rather than randomly over a period of time. He also held that the organism itself, by change of habit in response to a changed environment and thus by use and disuse of its parts, might similarly bring about variation in its offspring. Acceptance of these two factors, however much Darwin may have qualified them, embroiled him and his followers in controversy for more than fifty years. Furthermore, the fact that through 'correlation' and 'compensation' secondary variations might occur in organic parts linked physiologically with others in which primary variations had taken place only exacerbated the dispute. To many it seemed that Darwin had opened a leak from the environment to the entire organism, through which all manner of adaptive change might directly flow, quite independently from natural selection.[14]

The assumptions Darwin made about inheritance also created diffi- culties for his defence of natural selection. Like variation, inheritance to him was an enigma. Its laws were also 'quite unknown'.

No one can say why the same peculiarity in different individuals of the same species, and in individuals of different species, is sometimes inherited and sometimes not so; why the child often reverts in certain characters to its grandfather or grandmother or other much more remote ancestor; why a peculiarity is often transmitted from one sex to both sexes, or to one sex alone, more commonly but not exclusively to the like sex.[15]

But whereas in the case of variation Darwin made controversial assumptions on the whole, in the case of inheritance he simply accepted what naturalists of his generation already believed: that inheritance means primarily the transmission of parental characters, resulting in their duplication (parents alike) or blending (parents differ) in the off- spring. In other words, the reappearance of characters or an inter- mediate mixing of them was regarded as the rule. 'Prepotency' (dominance of one parent's character over that of the other), reversion, variation, and non-inheritance were thought to be exceptions, having nothing to do with inheritance, which was simply a process of trans- mission, but with the effects of the environment and the use and disuse of parts.[16]

Neither Darwin nor his critics were slow to perceive the consequences of 'blending inheritance' for his theory: when a few variant individuals appear in a population, whatever advantageous characters they possess will rapidly disappear through intercrossing – by blending in their off- spring with the disadvantageous characters of non-variant parents. Indeed, Darwin himself had anticipated the problem long before 1859 and had carefully provided for its solution. First, by insisting that the variation on which natural selection depends consists mainly of indi- vidual differences that occur widely – not just in a few members of a population – because of the uniform effects of the environment; second, by stipulating that the changed conditions under which these variations appear result in the destruction of many members of the population that have not varied favourably; and finally, by positing a degree of isolation between the variant and non-variant members, he dealt with blending inheritance in the conceptual basis of his theory. Variation, selection, and isolation were so conceived as to increase the probability that two variant forms would reproduce together, thereby duplicating their advantageous characters in the next generation. Darwin's questionless commitment to blending inheritance was also evident in his somewhat *ad hoc* idea – mentioned in the *Origin*'s first edition, underscored in the fourth (1866), and asserted with confidence in the *Variation of Animals and Plants* – that the tendency to manifest similar variations may itself be inherited, resulting in a continual supply of favourable variations, even under conditions of blending.[17] The route to reappraisal, on the other hand, was sealed off in the latter book, where prepotency and reversion, both signposts pointing towards a different understanding of inheritance, were again misconceived as exceptions to the rule of blending and duplication, and identified as tendencies of limited duration.[18]

Thus, when Jenkin argued that discontinuous variations cannot furnish the raw material for natural selection because they are immediately swamped by blending, Darwin felt unscathed. Could he not rely entirely on individual differences, which are incomparably more plentiful? Would not the persistence of favourable differences be assured, regardless of blending inheritance, by the uniform action of the environment on the members of a population? Previously Darwin had paid but passing attention to the environmental causes of variation. After 1867, however, it became necessary to invoke them repeatedly 'in order to ensure the necessary conditions for the selective modification whose limitations were becoming more apparent'. The consequences for Darwin and Darwinism were not altogether favourable.[19]

Utility and time

The theory of natural selection presupposed that variations are 'favourable' or 'advantageous' only in so far as they are useful to the variant individual. The theory also presupposed a lapse of time sufficient for useful variations to be accumulated, through selection, into contemporary forms of life. Both assumptions, Darwin discovered, were seriously open to dispute.

'Natural selection cannot possibly produce any modification in any one species exclusively for the good of another species', wrote Darwin.

If an example could be produced to the contrary 'it would annihilate my theory'. If the doctrine that some structures 'have been created for beauty in the eyes of man, or for mere variety', were true, it would be 'absolutely fatal'. Nor will natural selection ever 'produce in a being any structure more injurious than beneficial to that being', Darwin maintained, 'for natural selection acts solely by and for the good of each'.[20] Strict individual utility is therefore the standard of evolutionary change. If a variation does not bestow advantage on its possessor in the struggle for existence then natural selection will tend to eliminate it. But H. G. Bronn and Karl Nägeli, among others, were quick to point out the existence of many trivial and, to all appearances, useless morphological features in the plant and animal kingdoms. Of what utility, for example, is the arrangement of the cells in the tissues of a plant or the distribution of the leaves on its axis? Of what survival value are the various lengths of ears in hares and mice or the complex folds of enamel in the teeth of many animals? Darwin could plead ignorance in some cases while pointing to characters formerly considered to be useless which were later found to serve important functions. In other cases he could account for inutility through the inheritance of features formerly useful but now neither useful nor injurious. To explain the large residue of useless structures, however, he had recourse to causes quite independent of selection: reversion of characters, correlation of growth, and the direct action of the environment.[21]

The critics did not relent. Darwin's problem with utility was compounded in 1871 when St George Mivart objected in his *Genesis of Species* that natural selection is incompetent to account not only for structures which are useless at present, but also for the early persistence of the incipient stages of structures which are found useful only in their present, fully developed forms. How, for example, could natural selection have produced the giraffe? A gradually lengthening neck implies an increasing mass of body and thus a growing need for food, which in times of scarcity would be a disadvantage more than compensating for the benefits of a gradually lengthening neck. Or if a lengthened neck was, on balance, beneficial, why should it not persist in other Ungulata?[22]

In answering his critics Darwin redoubled his reliance on the causes of variation as agents of structural change. To take but three examples: in the first edition of the *Origin* Darwin wrote that 'physical conditions probably have some little effect on structure, quite independently of any good thus gained'; in the fourth edition the sentence was changed to read 'some little direct effect'; in the fifth (1869) and sixth (1872) editions it was 'the definite action of changed conditions', together with other causes of modification; and in the sixth edition the words, 'probably a great effect', were added.[23] Again, in 1859 Darwin made 'some little allowance for the direct action of physical conditions' in producing useless structures; in the fifth edition it became 'due allowance for the definite action of changed conditions' and other causes of modification; and in the sixth edition 'due allowance' was transmuted into the assertion that 'it is scarcely possible to decide how much allowance should be made'.[24] Finally, in responding to Mivart, Darwin went so far as to claim that natural selection, 'combined no doubt in a most important manner with the inherited effects of the increased use and disuse of parts', accounts for the development of useful characters through their useless incipient stages. There could be no denying henceforth that natural selection needed all the help it could get.[25]

Above all, Darwin's theory of natural selection demanded a vast amount of time. Problems with variation, inheritance, and utility could be dealt with easily enough by elaborating or re-emphasising causal factors – the direct action of the environment and the use and disuse of organs – that were proposed in the first edition of the *Origin of Species*. But without sufficient geological time for the innumerable variations, selections, and extinctions required to produce the observed phenomena of life, natural selection could at best be an adjunct to the real causes of evolution. Indeed, had not Darwin come to believe that natural selection acts almost exclusively on continuous variations and that continuous variations must occur for countless generations if natural selection is to accumulate them into forms which exceed the apparent limits of variation for each species?

At the outset Darwin could assume all the time he liked. His calculation of the denudation of the Weald by marine erosion gave the assumption a veneer of respectability and gave natural selection 'in all probability a far longer period than 300 million years...since the latter part of the Secondary period'. Unfortunately, the veneer itself was soon eroded by geological critics, and in the *Origin*'s third edition the calculation disappeared.[26] Still, however, Darwin felt able to meet the objection that 'time will not have sufficed for so great an amount of organic change...through natural selection'. He waxed eloquent on the several proofs from Lyell's *Principles of Geology* that the past ages have been 'incomprehensibly vast', exclaiming at length, 'What an infinite number of generations, which the mind cannot grasp, must have succeeded each other in the long roll of years!' These words were first written in 1859. A decade later the ages were only 'vast' and the exclamation was entirely omitted.[27]

The interim had seen a revolution in geochronology. Following in the steps of Hutton, Playfair, and Lyell, many geologists at mid-century could presume that the earth's crust revealed no 'vestige of a beginning' and held out 'no prospect of an end'. Their theories knew no exact mathematical checks but only extrapolation of nature's apparent uniformity into a uniformitarianism based on implicit trust in the boundless extent of geological time.[28] William Thomson (1824–1907), however, knew otherwise. With the intent to 'insure that the results of geological speculation be made physically and philosophically sound', Thomson (later Lord Kelvin), professor of natural philosophy in the University of Glasgow and possibly the leading British physicist of the nineteenth century, published three physical arguments which placed radical constraints on the geologists' estimate of time.[29] Two appeared in articles, 'On the age of the sun's heat' and 'On the secular cooling of the earth', which were published in 1862 and 1863. Employing Joule's newly quantified principle of the conservation of energy and the sophisticated mathematics of Fourier's analytical theory of heat, Thomson calculated that the earth could not be more than about 100 million years old (or, within limits, between 20 and 400 million years). In 1868 he presented his third argument in an address before the Geological Society of Glasgow, confirming the accuracy of his earlier estimate by calculating the frictional effect of tides in retarding the period of the earth's diurnal rotation. Thomson's tenacity was great, his reputation formidable, his assumptions generally acceptable, and his mathematics as faultless as it was appealing to those whose science as yet lacked quantitative precision. In consequence his chronology remained a part of the orthodoxy of British geology for nearly fifty years.[30]

To Lyell, who was already struggling to assimilate Darwin's theory, the new time-scale must have come rather as a blow. Like most geologists of the day he could at first neither entirely comprehend nor effectively refute the arguments in its behalf. Thomson left him 'grasping for any possible mechanism which might provide the earth with a continuous and uniform supply of energy' – just the state, ironically, in which he himself had left an earlier generation of theo-

logians in their quest to sustain the Deluge. As for Huxley, his efforts
to reply to Thomson in his presidential address before the Geological
Society of London in 1869 'must be praised more for their vigor than
their strength'.[31] Indeed, although Darwin later sent his thanks,
Huxley at one point actually conceded all that, from Darwin's point
of view, was essential. 'Biology takes her time from geology', he
declared. 'The only reason we have for believing in the slow rate of the
change in living forms is the fact that they persist through a series of
deposits which, geology informs us, have taken a long while to make.
If the geological clock is wrong, all the naturalist will have to do is
modify his notions of the rapidity of change accordingly.'[32] The 'only
reason' for believing in a slow rate of evolution was not of course
geological. If there was but one important reason it was methodological:
natural selection requires a vast amount of time. 'All the naturalist will
have to do' was not to 'modify his notions of the rapidity of change'.
If the geological clock was wrong nothing less than a reassessment of
the causes of evolution would suffice. Later in the same year Thomson
made this very point in his reply to Huxley before the Geological
Society of Glasgow. 'The limitations of geological periods, imposed by
physical science, cannot, of course, disprove the hypothesis of transmutation of species', he said, 'but it does seem sufficient to disprove
the doctrine that transmutation has taken place through "descent with
modification by natural selection"'.[33]

Darwin himself took the new chronology with increasing seriousness
as time went on. Although Jenkin employed it in 1867 to bolster his
argument against natural selection, Darwin could claim later in the
same year that Hooker's case of plant migration in Fernando Po, with
all the time it implied, was 'worth ten times more than the belief of a
dozen physicists'.[34] However, by the time he began work on the
Origin's fifth edition in December 1868, his attitude had changed.
Mid-way through his revision he wrote to the geologist James Croll:
'I am greatly troubled at the short duration of the world according to
Sir W. Thomson, for I require for my theoretical views a very long
period *before* the Cambrian formation.' When the new edition
appeared it gave evidence of how deeply Darwin felt: ideas of lapsed
time were reduced or qualified; a section on sedimentation as a chrono-
meter was deleted; and Darwin seemed almost desperate to impress his
readers with the vastness of even such time as the new calculations
allowed.[35] The problem had been faced and Darwin thought his
changes would 'do fairly well'. But still the problem persisted. Alfred

Russel Wallace accepted Thomson's chronology late in 1869, though his defence of the theory which he and Darwin had discovered remained virtually intact. Mivart, on the other hand, cited Thomson, Jenkin, and Wallace in his cumulative argument against natural selection. Even young George Darwin, doing a brilliant course in mathematics at Cambridge, would not let his father forget the ravages of time.[36]

Could it be that Thomson was right after all – that his arguments were 'sufficient to disprove the doctrine that transmutation has taken place through "descent with modification by natural selection"'? Darwin would not give in. Again, for the third time, he had recourse to the action of the environment in producing evolutionary change. Assuming with Thomson that in order to account for the facts of geology 'the world at a very early period was subjected to more rapid and violent changes in its physical conditions than those now occurring', Darwin reasoned in the sixth and last edition of the *Origin* that 'such changes would have tended to induce changes at a corresponding rate in the organisms which then existed'. Natural selection, he believed, would simply have to work the faster to make up for lost time.[37]

Again Darwin had assimilated the objections of his critics. The irritating difficulties presented by the nature, extent, and causes of variation, the effects of blending inheritance, the origin of useless characters, and the limitation of geological time – the last difficulty being 'probably one of the gravest as yet advanced'[38] – each had been taken into the *Origin* and coated with layer upon editorial layer of adjustments and qualifications. The result was not an elegant theoretical structure enhanced by veritable pearls of insight. It was rather a brilliant book vitiated by striking inconsistencies.

The pre-eminence of natural selection

Was Darwinism both omnivorous and indestructible? able to withstand every adverse criticism by incorporating the appropriate changes into its founding document? In 1871 the *Descent of Man* appeared *de novo*, without the burden of revisions from five previous editions, and in it one could at last see clearly how Darwin proposed to deal with the objections that had been raised against his theory. The old points of controversy remained. Variation was continuous and unlimited. 'Slight individual differences', wrote Darwin, '...are by far the most important for the work of selection'. The causes of variation were pre-eminently 'the direct and definite action of changed conditions' and the 'effects of the increased use and disuse of parts', with 'correlated variation' again denoting changes in organs related to those in which variations were initially induced. Inheritance was a matter of transmission, and blending was the result. The individual utility of selected variations and the enormous time needed for their accumulation into new species were still the premises of natural selection, though the limitation of geological time was nowhere dealt with explicitly.[39] Furthermore, as Darwin reiterated his basic assumptions he continued to meet the objections they raised by invoking the direct action of the environment and the use and disuse of parts for the purpose of producing heritable modifications. At first, it will be recalled, Darwin regarded these two factors as primarily the causes of variation. Then, as criticism of natural selection mounted, Darwin invoked them again and again, either to produce the conditions necessary for the action of natural selection or to produce characters for which natural selection could not account. Finally, in the *Descent of Man* the two factors emerge as integral parts of the evolutionary mechanism.[40]

Referring to the non-selective origins of useless structures, for example, Darwin stated that 'each peculiarity must have had its efficient cause'.

If these causes, whatever they may be, were to act more uniformly and energetically during a lengthened period, ...the result would probably be not a mere slight individual difference, but a well-marked and constant modification, though one of no physiological importance. Changed structures, which are in no way beneficial, cannot be kept uniform through natural selection, though the injurious will be thus eliminated. Uniformity of character would, however, naturally follow from the assumed uniformity of the exciting causes, and likewise from the free intercrossing of many individuals. During successive periods, the same organism might in this manner acquire successive modifications, which would be transmitted in a nearly uniform state as long as the exciting causes remained the same and there was free intercrossing.[41]

In another context Darwin summed up the causes of adaptive change in mankind, asserting 'that the inherited effects of the long-continued use or disuse of parts will have done much in the same direction with natural selection'.[42] Previously he had only claimed disuse as a cause of rudimentary organs; now he pressed both use and disuse into service to explain why savages are long-sighted and watch-makers and engravers are short-sighted, or why the Quechuas and Aymaras, living in the rarified atmosphere of the Peruvian plateaux, have bodies of greater length and girth than those of other races.[43] Previously, as Darwin freely admitted, 'I perhaps attributed too much to the action of natural selection'.[44]

We have not represented Darwinism aright, however, if we conclude our discussion of its transitional years with this admission. Darwin was no pedestrian naturalist, driven to confusion and tergiversation by a galaxy of better-informed critics. Not only did he formulate the theory of evolution by natural selection in all its simplicity and illustrate it from the accumulated observations of twenty years' research; he also served as his own first and best critic, anticipating in the original edition of the *Origin of Species* almost every objection that would be raised against his theory in its initial decade. The numerous adjustments to that volume are not tokens of confusion, much less of dishonesty; and they were certainly not intended to be retractions of the theory which, even in his own time, transformed the study of natural history. Adam Sedgwick, Darwin's old professor of geology at Cambridge, once remarked that a man who talks about what he does not in the least understand is invulnerable.[45] Darwin was vulnerable precisely because he talked about what he understood perfectly well, and better, certainly, than most of his contemporaries. Analytical and tough-minded theoretician that he was, however, Darwin had a deep reverence for facts and for their bearing on natural selection; sensitive, retiring, and self-effacing person that he was, Darwin was quick to admit an oversight. Thus when new facts were brought to his attention or implications of his theory and of its assumptions drawn out, which he had not fully taken into account, the *Origin* soon reflected the new light its author had received. Far from being a sign of retreat, the very fact that the book passed through five painstaking revisions is an enduring witness to Darwin's basic honesty and to his 'gracious ability to compromise'.[46]

Darwin did not abandon natural selection. Rather he looked to other change-producing factors 'only in those instances where he could not see selection as a possible means'; he looked only to factors for which, in his opinion, there existed substantial evidence; and in every case these factors were subordinate to, or coordinate with, the dominant action of natural selection.[47] In the very context where Darwin admits attributing perhaps too much to it, he also declares that he is 'very far from admitting' that he 'erred in giving to natural selection great power'. Elsewhere in the *Descent of Man* Darwin tells 'how subordinate in importance is the direct action of the conditions of life, in comparison with the accumulation through selection of indefinite variations'; and in the *Origin*'s sixth edition the effects of use and disuse are coupled closely with natural selection, in one case being said to aid it, and in another, to be strengthened by it.[48] Darwin's definitive statement and a *sine qua non* of Darwinism for half a century after its first appearance in 1859 is the concluding sentence of the introduction to each edition of the *Origin of Species*: 'I am convinced that Natural Selection has been the most important, but not the exclusive, means of modification.'[49]

6

∞∞∞

THE CHALLENGE OF LAMARCKIAN EVOLUTION

Natural selection. . .replaces a transcendental explanation with a natural one. To be sure, it does not explain force, and thus leaves the whole subject shrouded in as deep fundamental mysticism as ever. But science does not hope to explain force and power, and will be satisfied to account for natural phenomena by. . .natural forces acting in accordance with natural laws. Natural selection was a great step in this direction.

H. W. Conn[1]

Evolution has not taken place by the action of 'Natural Selection' *alone*, but. . .partly, perhaps mainly, through laws which may be most conveniently spoken of as special powers and tendencies existing in each organism; and partly through influences exerted on each such organism by surrounding conditions and agencies organic and inorganic, terrestrial and cosmical, among which the 'survival of the fittest' plays a certain but subordinate part.

St George Mivart[2]

The last of Darwin's works directly concerned with organic evolution was the *Descent of Man*. From the time of its publication until his death in 1882 Darwin occupied himself with long-standing researches which served to illustrate the broader significance of his theory. The first and last fruits of his labour – *The Expression of the Emotions in Man and Animals* (1872) and *The Formation of Vegetable Mould through the Action of Worms, with Observations on Their Habits* (1881) – had a psychological interest. Other volumes reported his work on plants: *Insectivorous Plants* (1875), *The Movements and Habits of Climbing Plants* (1875), *The Effects of Cross and Self Fertilisation in the Vegetable Kingdom* (1876), and *The Different Forms of Flowers on Plants of the Same Species* (1877). In September 1872 Darwin told Wallace, 'I have taken up old botanical work and have given up all theories'.[3]

Are we therefore to conclude that the critics of natural selection had badgered the elderly Darwin into retirement from theoretical debate to the seclusion of his Kent garden, where he might do and think as he pleased? There is perhaps an element of truth in this interpretation.[4] But one must hasten to add that for three decades Darwin had been in constant ill health, that in 1858 he had been forced by the prospect of anticipation in announcing the discovery of natural selection to interrupt researches which he still wished to complete, and that, above all, he would rather have died than have had his work subjected to the lapses, the excesses, and the growing intransigence of an aged mind.[5] The biological sciences were advancing at an unprecedented rate, beginning even to exceed the grasp of the one who had placed them on a broad theoretical foundation, and specialisation was rapidly becoming the *modus vivendi* among the younger naturalists. Wisely, Darwin recognised these trends and saw his labours in perspective. 'I shall continue to work as long as I can', he told Ernst Haeckel late in 1872, 'but it does not much signify when I stop, as there are so many good men fully capable, perhaps more capable than myself of carrying on our work'.[6]

Darwin's years of controversy were over. For his followers they had just begun. 'Within ten years after the publication of the *Origin of Species* all the diversity of opinion which confronts us today', wrote J. Arthur Thomson at the turn of the century, 'was either clearly expressed or existed in rudiment'.[7] The problem for Darwinians was no longer anti-evolutionary hostility from fellow-naturalists. Transmutation had triumphed by reason of its numerous compelling arguments, leaving Louis Agassiz as its chief Anglo-American opponent, though more as a superannuated oracle than as an argumentative force to be reckoned with. The problem facing Darwinians was primarily the 'diversity of opinion' in Darwin's works themselves, which by 1870 had revived a school of evolutionary thought that was at once transmutationist and anti-Darwinian.

Darwin, it will be recalled, never faltered in his belief that natural selection had been the principal cause of the origination of species. Yet in rescuing natural selection from its critics he drew repeatedly on the heritable effects of non-selective factors – the direct action of the environment and the use and disuse of parts – which at first he thought were simply the causes of indefinite, or random, variation. To meet William Thomson's objection that geological time had been insufficient for evolution by natural selection, Darwin supposed that the rapidly changing conditions of life had induced continuous variations at a corresponding rate, thereby enabling organisms, under natural selection,

soon to exceed their apparent specific limits. Meanwhile the blending effects of inheritance – another objection – would have been
overcome as the influence of the environment on life and habit
assured that many beneficial variations had occurred within plant and
animal populations. At the same time many structures of no apparent
utility, inexplicable by means of natural selection, would have
developed under the impact of new conditions and habits. But regardless of these adjustments to his theory, Darwin continued to believe
that the heritable effects of the environment and of use and disuse
either supported or were subsidiary to natural selection. As causes of
variation they were necessary and sufficient. Only natural selection,
however, would suffice as a cause of speciation. It was at this point that
Darwin and his followers had to face the challenge of Lamarckian
evolution.

The components of Lamarckian theory

Jean Baptiste Pierre Antoine de Monet, Chevalier de Lamarck (1744–
1829), professor in the Museum of Natural History at Paris, did not
live to see his fame. In his own day, when the indomitable authority of
his paleontologist colleague, Georges Cuvier, sanctioned geological
révolutions and the fixity of species, the uniformitarian and transform-
ist doctrines of his *Philosophie zoologique* (1809) and *Histoire naturelle
des animaux sans vertèbres* (1815–22) were roundly condemned. Few
were prepared to believe that species may be related by descent and
fewer still could accept Lamarck's explanation of it. For while
Lamarck was an outstanding naturalist – the first, for example, to
make a clear distinction between vertebrates and invertebrates – he
was also a natural philosopher with a bent for excessive speculation.
His theory of transmutation developed over a period of years, but in its
most general form it embraced two causal factors: first and foremost,
an innate power (*pouvoir de la vie*), conferred on nature by God, that
tends to produce a series of plants and a series of animals, each of
which shows an orderly progression in complexity and perfection
among its major taxonomic groups, the animal series culminating in
mankind; and second, an inner disposition peculiar to living bodies
(*sentiment intérieur* in higher animals), which assures the performance
of actions sufficient to meet new needs (*besoins*) created by a changing
environment, the actions becoming habitual, the habits becoming
instinctive and producing organic change, and the instincts and
changes thus induced being inherited by the offspring. (Changes in

plants and lower animals are induced more directly by the environment.) These two factors of Lamarck's theory – an innate complicating power and an inner adapting disposition – may be regarded generally as two 'components' of phylogenetic evolution. The first component is a 'vertical' one, causing a 'rise' in organic complexity, producing continually 'higher' forms of life. The second component is 'lateral' in direction, causing heritable adaptations to particular circumstances, producing deviations (even regressions) from the ascending linear series. Theories of development that feature one or both of these components we shall refer to as Lamarckian evolution.[8]

In the decades after his death Lamarck, as in life, made more enemies than friends. To some extent this may be attributed to the kind of friends his theory kept. Etienne Geoffroy Saint Hilaire (1772–1844), professor of geology at the Museum of Natural History, adopted the second factor of Lamarck's theory in the mid-1820s, though within a few years he came to hold that the environment (*monde ambiant*) induces lasting and significant changes in embryos alone. Geoffroy also believed in the ideal structural unity of living things – a piece of metaphysics from the German *Naturphilosophie* – and in the modifying influence of Cuvier's sudden geological catastrophes. Neither belief was likely to commend Lamarck to Charles Lyell, whose *Principles of Geology* (1830–3) inoculated a generation of British naturalists against 'the development hypothesis'.[9]

Thus it was not a naturalist but a layman, the Scottish publicist and publisher Robert Chambers (1802–1871), who first commended Lamarckian evolution to the English-speaking world. In 1844, the year of Geoffroy's death, Chambers issued anonymously a sketch of universal evolutionary development, from astronomy to psychology, entitled *Vestiges of the Natural History of Creation*. The book did much to pave the way for Darwin. While absorbing a good deal of clerical choler, it advertised some of the evidence that would later form the basis of a respectable theory of descent.[10] Chambers, however, was a rank amateur, and his credulous speculations obscured *Vestiges'* undoubted virtues and further damaged the reputation of Lamarck. Although at first Chambers claimed to have little regard for Lamarck's theory, placing its second factor 'with pity among the follies of the wise', he did in fact maintain that species had evolved by means of a 'higher generative law' and through the indirect and heritable influence of the environment on embryonic life.[11] Indeed, in the much-revised tenth edition of *Vestiges*, published in 1853, Chambers plainly adopted both components of Lamarckian evolution. 'The proposition determined on after much consideration', he said,

is that the several series of animated beings, from the simplest and oldest up to the highest and more recent, are, under the providence of God, the results, *first*, of an impulse which has been imparted to the forms of life, advancing them, in definite times, by generation, through grades of organisation terminating in the highest dicotyledons and vertebrata, these grades being few in number, and generally marked by intervals of organic character, which we find to be a practical difficulty in ascertaining affinities; *second*, of another impulse connected with the vital forces, tending, in the course of generations, to modify organic structures in accordance with external circumstances, as food, the nature of the habitat, and the meteoric agencies, these being the 'adaptations' of the natural theologian.[12]

Lamarck would have said that organisms are gradually complicated and perfected, rather than advanced through 'intervals of organic character'; he would have added that habit is the medium through which 'vital forces' achieve adaptive modifications. Otherwise there was little to choose between Lamarck's two factors and the two 'impulses' of *Vestiges*.[13]

While *Vestiges* was achieving notoriety for its far-fetched views, a similar but rather more sophisticated theory was in the making. Richard Owen (1804–1892), the 'British Cuvier', whose reconstruction of fossil vertebrates at the British Museum of Natural History took inspiration from the *Naturphilosophie*, would have been the last to admit – before 1859 – a belief in the derivative origin of species, though in fact his acceptance of the doctrine was scarcely veiled by his synthesis of Cuvier's anatomical method and the transcendental morphology of Oken. Owen's basic theory was set forth in 1848 in his *Archetype and Homologies of the Vertebrate Skeleton*. Like the theories of Lamarck and Chambers, it consisted of two distinct elements: a 'Platonic *idea*', 'organizing principle', 'vital property', or 'force', which produces diversity of form; and a 'polarizing force' in counter-operation to the *idea*, which produces similarity of form, repetition of parts, and unity of organisation. In his *Nature of Limbs*, published in 1849, Owen confessed ignorance of the 'natural laws or secondary causes' which have brought about 'the orderly succession and progression of...organic phenomena', but he maintained none the less that these causes, 'guided by the archetypal light', had advanced with 'slow and stately steps...from the first embodiment of the Vertebrate idea under its old Ichthyic vestment, until it became arrayed in the glorious garb of the Human form'. In his notorious review of the *Origin of Species* in 1860 Owen referred this evolution to the 'continuous operation of the ordained becoming of living things', to a 'constantly operating secondary creational law' or a 'pre-ordained law or secondary cause'. Moreover, he explicitly identified one element of this law, the *idea* or archetype, with the second 'impulse' of Chambers' theory.[14] That the law itself functioned in a manner resembling Chambers' first, or vertical, 'impulse' became clear in 1868, when Owen officially stated his theory of derivation. Minimising the effects of environment and habit, he explained that evolution takes place through large discontinuous variations, by 'departures from parental type, probably sudden and seemingly monstrous, but adapting the progeny inheriting such modifications to higher purposes'. This was strongly reminiscent of *Vestiges*, with its advance 'through grades of organization. . .marked by intervals of organic character'.[15] As a causal explanation Owen's archetype did not differ in principle from Lamarck's innate complicating power. Thus, although Lamarck offered consistently uniformitarian and materialistic accounts of life, his affinity with the transcendentalists was in some ways 'a close one'.[16]

If Owen stressed the vertical component of Lamarckian theory, the 'slow and stately steps' which the archetype had inspired, it was the lateral component that seemed to hold the greatest promise in the decade after 1859. Lamarck offered clearly and simply what (some believed) Darwin had obfuscated by his repeatedly revised insistence on the supremacy of natural selection: evolution by means of the inherited effects of environment and habit. Clarity and simplicity were much in demand in the years that saw the *Origin of Species* modified in a bewildering number of ways, and so just in proportion to Darwin's growing reliance on environment and habit to prop up natural selection there was a movement away from natural selection towards environment and habit as the principal factors of organic development. In England the philosopher Herbert Spencer, who had been a Lamarckian since he first read Lyell's refutation of the theory, blended natural selection with use-inheritance in the first volume of his *Principles of Biology* (1864) and assumed the latter doctrine for more than thirty years as the basis of his massive *Synthetic Philosophy*.[17] Meanwhile in the United States there emerged an influential group of evolutionists headed by a triumvirate of self-proclaimed 'Neo-Lamarckians': Alpheus Hyatt (1838–1902) of the Boston Society of

Natural History, Alpheus S. Packard, Jr (1839–1905), professor of zoology and geology in Brown University, and Edward Drinker Cope (1840–1897), paleontologist *par excellence* and professor of geology and mineralogy in the University of Pennsylvania.

Cope and Co.

Hyatt, Packard, and Cope were young and ambitious members of the first generation of naturalists to reckon with Darwinism. Each became a Lamarckian before he was thirty years old and each lived out his most productive years in the last third of the century. In 1866 Hyatt published a memoir on the fossil cephalopods, which formulated the characteristic Neo-Lamarckian interpretation of embryonic growth. Cope published a similar interpretation quite independently within the same year and described it in 1868 as the 'law of acceleration and retardation'. Thereafter he advanced the Lamarckian cause through an enormous literary output, consisting largely of descriptive memoirs that reported his extensive researches in vertebrate paleontology. Packard abandoned Darwinism by 1870 and devoted himself to publicising Lamarckian evolution and applying its theory to the peculiar characteristics of cave animals. Together, in 1867, Hyatt and Packard became founding editors of *The American Naturalist*, a journal occupied for more than thirty years with the 'American' theory of evolution they were setting forth. Cope, a man of means, joined Packard as an editor in 1878 when he purchased a share in the journal. In time, through publications, publicity, and their periodical, Hyatt, Packard, and Cope obtained the support of many prominent naturalists. Some, such as the botanist Thomas Meehan and the geologist Joseph Le Conte, came from an earlier generation. The majority, however, were their students and peers: William H. Dall and Henry Fairfield Osborn, both paleontologists; John A. Ryder and William Keith Brooks, zoologists; Joel A. Allen, an ornithologist; and Clarence King, a geologist. By 1877 Packard could write of 'a large and increasing school of American naturalists who believe that natural selection is but the last term in a series of factors, which together make up a true evolution theory'.[18]

The intellectual force of Neo-Lamarckism was largely concentrated in Cope. (Judging from the indefatigable zeal with which he pursued his fossil-hunting expeditions in the West, he must have possessed much of the movement's kinetic energy as well.) *The Origin of the*

Fittest, a collection of his essays published in 1886, and *The Primary Factors of Organic Evolution*, the definitive statements of his views, published ten years later, contained perhaps the most substantial and sophisticated expositions of Lamarckian evolution that appeared in the English language during the nineteenth century. Cope was not at first a conscious disciple of Lamarck. He was simply impressed with the inability of natural selection to account for the origin of useless characters and with the fact that Darwin himself did not attempt to account for them by this means.[19] Like Darwin, he turned to environment and habit to solve the problem. Unlike Darwin, however, he saw no reason why these should not become the 'primary factors' of evolution. For him as for Mivart and others, it was obvious that while 'these factors worked admirably in accounting for the origins of useless and/or incipient structures, they could nevertheless be applied with equal facility to account for structures which Darwin saw as strictly the objects and results of natural selection'.[20]

According to Cope, neither non-adaptive *nor* adaptive characters can originate by natural selection. Natural selection accounts for the survival of the fittest but it cannot originate the variations from which the characters of the fittest are derived.[21] The 'origin of the fittest' or, in other words, the origin of species, takes place through the inherited effects of environment and habit, and according to a law quite unlike natural selection, the law of acceleration and retardation. Cope explained it as follows:

The superposition of characters which constitutes evolution, means that more numerous characters are possessed by the higher than the lower types. This involves a greater number of changes during the ontogenetic growth of each individual of the higher type. In other words, characters acquired during the phylogenetic history are continually assumed by the progressive form at earlier and earlier periods of life. This process has been metaphorically termed by Professor Alpheus Hyatt and myself 'acceleration'. All progressive organic evolution is by acceleration, as here described. Retrogressive evolution may be accomplished by a retardation in the rate of growth of the taxonomic characters, so that instead of adding, and accumulating them, those already possessed are gradually dropped; the adults repeating in a reversed order the progressive series, and approaching more and more the primitive embryonic stages. This process I have termed 'retardation'.[22]

Evolution by acceleration and retardation (or 'anagenesis' and 'catagenesis' in Cope's specialised vocabulary) is evolution directed through embryonic growth. This recalls the theories of Geoffroy,

Chambers, and Owen, where the embryo is also prominent. In fact, Cope's acceleration describes precisely the outcome of Chambers' first impulse, or 'higher generative law'. Taking the development of sex in bees as an example, Chambers stated that all the changes 'may be produced by a mere modification of the embryotic process. . . . All that is done is merely to *accelerate* the period of the insect's perfection.' 'Thus', he concluded, 'the production of new forms, as shewn in the pages of the geological record, has never been anything more than a new stage of progress in gestation'.[23] For Chambers, however, a law of generation could be referred to as a cause; for Cope the law of acceleration and retardation merely summarised the effects of environment and habit. These causes, in turn, depended for their efficacy on an elaborate metaphysical synthesis of the components of Lamarckian evolution.[24]

The metaphysics of Neo-Lamarckism

Cope believed that 'every variation in the characteristics of organic beings, however slight, has a direct efficient cause'. The cause in question can be of two kinds: physico-chemical, resulting from the direct action of the environment; and mechanical, resulting from use and disuse. Accordingly, he termed the appearance of variations due to a physico-chemical cause, which are found mainly in the plant kingdom, 'physiogenesis', and the appearance of variations brought about by a mechanical cause, which are found mainly in animals, 'kinetogenesis'.[25] The variations thus acquired are inherited by means of an underlying growth-energy which Cope called 'bathmism'. 'Bathmogenesis' takes place under the impact of environment and habit, in each instance of physiogenesis and kinetogenesis, resulting, accordingly, in incremental energies called 'physiobathmism' and 'kinetobathmism'. When these energy levels impress themselves through the nervous structure on the reproductive elements, where they are recorded, then the variations acquired in physiogenesis and kinetogenesis become heritable. The variations reappear as part of embryonic growth, thereby accelerating or retarding the rate of phylogenetic evolution. Cope referred to this encoding as 'mnemogenesis'.[26] 'It is evident', he explained, 'that [the reproductive elements] and the other organic units of which the organism is composed

possess a memory-structure which determines their destiny in the building of the embryo. This is indicated by the recapitulation of the phylogenetic history of its ancestors displayed in embryonic growth. This memory has perhaps the same molecular basis as the conscious memory, but for reasons unknown to us, consciousness does not preside over its activities. The energy which follows its guidance has become automatic.... It is incapable of a new design, except as an addition to its record.[27]

The design to which Cope refers, like all the designs or intentional acts apparent throughout the world, originates in consciousness. Indeed, consciousness has been 'the *primum mobile* in the creation of organic structure', according to Cope's hypothesis of 'archaesthetism', first stated in 1882.[28] Whereas 'natural selection includes no *actively* progressive principle whatever', bathmism itself is a conscious energy (in Cope's later formulations); all kinetobathmisms therefore result from the 'efforts' of animal consciousness to meet particular needs (a metaphysical advance on Lamarck, as Cope pointed out); these efforts and their effects become habitual, their bathmisms become embryonic and hence automatic; and, finally, through the survival of the 'most intelligent' – those whose designed acts have enabled them to leave the most offspring – there emerges an ascending, or anagenetic, scale of evolution.[29] The process of catagenesis naturally accompanies this ascent, its organic and inorganic energies – neurism, myism, chemism, etc. – deriving from anagenetic (i.e. conscious) energy through a universal 'retrograde metamorphosis'. Catagenesis can be seen first of all in the physiobathmisms of plants, where energy has become automatic and thus unconscious; at length catagenesis appears in the degeneration and extinction of species and the reversion of individuals to mineral elements. 'Why', asked Cope, 'should evolution be progressive in the face of universal catagenesis? No other ground seems discoverable but the presence of sensation or consciousness.' Without continual intelligent direction evolution would not have occurred at all.[30]

Cope's jargon lends an occult atmosphere to much of his writing. This seems quite fitting in view of the religious and mystical background of his thought. Born and raised in strict Quaker orthodoxy, Cope was narrow-minded and dogmatic in his early years. Later, as he tried to reconcile his scientific beliefs with his father's religion, he became 'volubly pious, even less tolerant, and a little too fervently attached to his own opinions for the liking of his Quaker neighbours'. Finally, after his father's death in 1875, Cope resigned from the Society of Friends, though 'not without a pang', and maintained broader and more tolerant views in keeping with the Unitarian faith he professed.[31] His *Theology of Evolution*, a lecture published in 1887, was written from this standpoint, and in it the metaphysics of archaesthetism became even more explicit. Uniting realism and a spiritual monism, Cope argued that 'mind was one at the start', that 'mind has had one or more physical bases prior to the origin of that material which we now find in living beings', and therefore that 'a great mind from which lesser minds may have been derived, has a material basis' and experiences the 'limits' of time and power which constrain 'mind of any kind or sort whatever'. This conclusion held happy implications for the problems of theodicy and free will, and Cope believed he had accomplished 'what Job said could not be done'. 'The Neo-Lamarckian philosophy', he declared, is 'entirely subversive of atheism'. The primitive conscious energy, the first principle of Cope's cosmology, was but the emanation of a material and finite god.[32]

The discovery of divinity at the headwaters of the evolutionary continuum, and the 'fatherhood of mind' behind each eddy of kine- tobathmic energy, was by no means the exception among American Lamarckians. The metaphysics of Neo-Lamarckism (though not necessarily Cope's metaphysics) were nearly as conspicuous as the movement's subordination of natural selection to a vital energy and the inherited effects of environment and habit. In a statement prefixed to the first number of the *American Naturalist*, the editors projected that 'the value of our Magazine will depend more on its power to awaken the absorbing interest invariably excited by the contemplation of nature, and of illustrating the wisdom and goodness of the Creator, than on any adornment of style, or cunning devices of the artist'.[33] Clarence King, the founding director of the U.S. Geological Survey, who would later receive Lord Kelvin's commendation by allowing Darwinism only 24 million years of geological time, inferred from the equine fossils of O. C. Marsh that

the evolution of environment has been the major cause of the evolution of life; that a mere Malthusian struggle was not the author and finisher of evolution; but that He who brought to bear that mysterious energy we call life upon primeval matter bestowed at the same time a power of development by change, arranging that the interaction of energy and matter which make up environment should, from time to time, burst in upon the current of life and sweep it onward and upward to ever higher and better manifestations.[34]

Evolution's 'ever higher and better manifestations' were central to the social metaphysics of Neo-Lamarckism. Packard, who looked for a 'second Paley' to write a 'new natural theology' in the light of 'the law of evolution', found evidence everywhere of progress 'from the simple to the complex, from the lower to the higher, from evil to good'. 'The whole outcome of evolution', he believed, 'is from the imperfect to the perfect; a constant improvement of the world and its inhabitants'.[35] Cope fortified this conviction, the idea that 'evolution implies optimism', with deductions from his panpsychist philosophy. The 'profitable direction of human energy' has accelerated the development of mental and moral attributes, thereby advancing 'the whole race' at 'an increasing rate of progress'. This advance, however, was immanently predetermined, for 'the Creator of all things has set agencies at work which will slowly develop a perfect humanity out of his lower creation, and nothing can thwart the process or alter the result'. 'Every revolution of a wheel', Cope declared, 'is moving the car of progress, and the timed stroke of the crank and the rhythmic throw of the shuttle are but the music the spheres have sung since time began'.[36] In later years Cope was somewhat less sanguine about the likelihood of moral progress, and this due perhaps to the growing visibility and influence of those whom he regarded as less 'intelligent': Negroes, women, immigrants, and the urban poor.[37]

On the whole, Darwin was sharply critical of Lamarckian evolution. So little regard had he for Lamarck's theory, so anxious was he to dissociate his views from it, that he seems never to have understood it well. *Vestiges* he found more compelling, if only because its arguments popularised transmutation and did not encroach upon his own. Yet the book's geology was 'bad', its zoology 'far worse', and its 'two supposed "impulses"' could not 'account in a scientific sense for the numerous and beautiful co-adaptations which we see throughout nature'. For Owen *qua* comparative anatomist Darwin had the utmost respect, but his metaphysical meanderings, coupled with jealous personal affronts, were utterly contemptible. Such 'miserable inconsistencies and rubbish' Darwin exposed, in a rare public attack, in the 'Historical sketch' prefixed to the third and later editions of the *Origin of Species.*[38]

The Neo-Lamarckians came to Darwin's notice through the memoirs published by Hyatt and Cope in 1866 and 1868. These, evidently, he read with some difficulty, judging from the manner in which acceleration and retardation are misconceived in the sixth edition of the *Origin.* Hyatt wrote him at once, hardly a month after Darwin had forsworn 'all theories', and pointed out that the law in question had to do with the acquisition of taxonomic characters, not the 'period of reproduction'. Darwin replied with an apology and a confession that he was unable 'to grasp fully what you wish to show'. The law of acceleration and retardation did not seem to be an explanation of anything, but a mere 'statement of facts'. Whereupon Hyatt went on writing as if the law could 'explain'; Darwin, still uncomprehending, suggested that variations in ontogeny could be explained by 'laws of growth', the inheritance of acquired characters, and natural selection. 'After long reflection', he concluded, 'I cannot avoid the conviction that no innate tendency to progressive development exists, as is now held by so many able naturalists, and perhaps by yourself. It is curious how seldom writers define what they mean by progressive development.'[39]

The admonition had no effect. Acceleration had caught hold of the younger American naturalists and nothing in the *Origin*'s sixth edition, or from its venerable author, could retard them. Darwin thought Cope wrote 'very obscurely'; he also knew that Cope was 'an *excellent* Naturalist'. The new generation was full of naturalists as excellent, if not so obscure, who seemed willing to be enthralled by the law of acceleration and retardation, though his own attempt to grasp its meaning had been 'given up in despair'. When Hyatt sent a monograph in 1881 there was more than pique in Darwin's thanks: 'It is all the kinder in you to send me this book, as I am aware that you think that I have done nothing to advance the good cause of the Descent-theory.'[40] No sooner had the *Origin of Species* 'come of age' than its central doctrine was being abandoned. No sooner had the life sciences begun to be purged of gratuitous explanatory concepts, from catastrophes to archetypes and tendencies, than their practitioners were rushing back headlong to the weak and beggarly elements in an effort to overcome the difficulties of the argument for natural selection.

7

THE VOGUE OF HERBERT SPENCER

The biological theorist took for his central problem the question of the mutability of organic species and the conditions of their origin. The philosopher of science, on the other hand, proposed to cover a broader field, seeking to trace out not simply the course of biological development, but the evolution of the entire phenomenal universe from star-dust up to mind and social life. The aim of the one is a theory of species, of the other a doctrine of cosmical progress. . . . The theory of Darwin accounts for the genesis of natural kinds through adaptation to environment in virtue of natural selection under the conditions of the struggle for existence. Spencer's 'synthetic system' explains the world and life on the basis of 'the continuous redistribution of matter and motion'. Darwinism acquires a bearing on fundamental problems because of its relations, for in itself it is no more than the first principle of a special department of science. The Spencerian philosophy. . .is so inclusive in its scope that the synthesis undertaken involves from time to time the transcending of the limits of phenomenal inquiry.

A. C. Armstrong[1]

Britain's leading Lamarckian was Herbert Spencer (1820–1903). His *System of Synthetic Philosophy* (1860–96) set forth a physical law which determined the evolution of 'definite, coherent heterogeneity' throughout the universe; his life-work proceeded from beginning to end on the assumption that biological adaptation occurs primarily through the inherited effects of environment and habit. Unlike the American Neo-Lamarckians, however, Spencer was neither a naturalist nor an avowed opponent of Darwinism. He was a self-styled philosopher and a member – in so far, that is, as his conspicuous individuality would permit – of Darwin's inner circle of friends. Had his doctrines been an overt challenge to the primacy of natural selection, it is doubtful whether he would have remained in that privileged position; had they appealed deliberately to religious sentiments, it seems likely that Mivart's lot would soon have become his own. The bearing of Spencer's philosophy on Darwinism was different from the bearing of other theories of Lamarckian evolution. Yet if anything, ironically, it was even more challenging. For besides depreciating the influence of natural selection and admitting, in fact, of certain theological interpretations, the *Synthetic Philosophy* attained a vogue which far surpassed the popularity enjoyed by other Lamarckian works. To account for this vogue, and thus to understand the nature of Spencer's challenge to Darwinism, it is necessary to begin by considering Darwin's method of reasoning in science and its application to the evolution of mankind.

Mankind and the Darwinian method

Rightly or wrongly, Darwin thought of himself as a 'poor critic', as one whose ability to follow a 'long and purely abstract train of thought' was limited and whose memory was 'extensive, yet hazy'. But of one thing he was confident: that in natural history he had the ability to think clearly and validly. In his autobiography he remarked, 'The *Origin of Species* is one long argument from the beginning to the end, and it has convinced not a few able men. No one could have written it without having some power of reasoning.'[2] Darwin's self-assessment was basically correct. His reasoning, though not without its shortcomings, conformed closely to accepted canons of scientific explanation. On occasion, of course, Darwin mistook himself for a practitioner of old-fashioned Baconian induction. In the very context where he defended his ability to reason, for example, he recalled that from early youth he 'had the strongest desire to understand or explain' whatever he observed – 'to group all facts under some general laws', he added like a veritable Baconian. But then he went on to state that this desire, together with his love of natural science, had given him 'the patience to reflect or ponder for any number of years over any unexplained problem'. 'As far as I can judge', Darwin said,

I am not apt to follow blindly the lead of other men. I have steadily endeavoured to keep my mind free, so as to give up any hypothesis, however much beloved (and I cannot resist forming one on every subject), as soon as facts are shown to be opposed to it. Indeed I have had no choice but to act in this manner, for with the exception of the Coral Reefs, I cannot remember a single first-formed hypothesis which had not after a time to be given up or greatly modified. This has naturally led me to distrust greatly deductive reasoning in the mixed sciences.[3]

Darwin was not a mere collector and sorter of facts. From his earliest scientific education he was a framer and tester of bold hypotheses in the sciences of natural history.[4]

As we have seen, Darwin's argument for evolution by natural selection is a good example of hypothetico-deductive reasoning. It consists of a system of hypothetical premises concerning plants and animals, from which follow logically certain evolutionary consequences. In formulating his argument Darwin did not seek to build up a cosmic generalisation from numerous diverse inductions; nor, on the other hand, did he endeavour to construct a metaphysical generalisation with which any and all inductions would be compatible. He simply posed hypotheses from which deductions could be made that were susceptible of testing by a particular group of phenomena. Thus his theory had nothing to do with the formation of the solar system, the derivation of the chemical elements, or the origin and nature of life. It pertained only to 'the origin of species by means of natural selection'. Biological phenomena were altogether ample for many lifetimes of investigation without adding to them the phenomena of physical, chemical, or biochemical nature. 'In the distant future', Darwin remarked in the concluding paragraphs of the *Origin*, 'I see open fields for far more important researches' – researches in which, he added cryptically, 'Light will be thrown on the origin of man and his history.'[5]

The statement concealed an old conviction. In 1838 Darwin wrote in a private notebook, 'I will never allow that because there is a chasm between man...and animals that man has different origin.' For more than thirty years he kept his vow and kept it to himself. Premature disclosure would certainly have meant persecution – witness the vengeance with which the anonymous author of *Vestiges* was pursued – and the notion would have been prejudged for lack of an acceptable theory of transmutation. Only after his reputation was well established and his theory of natural selection widely respected, if not altogether accepted, did Darwin think it advisable to present his views on the origin of mankind.[6] The burden of the *Descent of Man*, which appeared early in 1871, is that human existence has not been specially exempt from the process of evolutionary change. In the first chapter Darwin drew attention to evidences in the human body – homologous structures, embryonic development, and rudimentary parts – that place mankind's ancestry among the animals. Chapter two shows that variations occur in the physical features of mankind, that these variations may be inherited, and that overproduction of offspring and a struggle for existence have occurred equally in human and animal populations – in short, that natural selection has been the primary mode of human phylogenetic development. The third to fifth chapters then take up those features which distinguish mankind most sharply from other forms of life: the intellectual powers and moral sense of the species.

Here above all was the opportunity for unchecked speculation. Surely at this point it was inevitable that some deep-seated prejudice, perhaps some overwhelming desire not to transgress the bounds of propriety, should have caused Darwin to flout the limitations of his hypothetico-deductive argument. On the contrary, however, he dealt with the higher human faculties much as he did the other biological phenomena which were within his province to treat. By numerous comparisons of ape and 'savage' – comparisons of their instincts, emotions, curiosity, imitation, attention, memory, imagination, and reason; comparisons of their use of tools and weapons, their self-consciousness, their use of language, their sense of beauty, and their superstitions – Darwin reasoned that mankind's intellectual powers, like the bodily attributes of the species, must have been slowly and gradually acquired. This was not, of course, to minimise the 'enormous' differences that exist between even the 'lowest savages' and the 'higher apes'. It was only to deny the presence of an 'insuperable barrier', an irreducible essence which separates mankind from the beasts. Darwin wished to establish that 'the difference in mind between man and the higher animals, great as it is, certainly is one of degree and not of kind'.[7]

Similarly, though 'the moral sense or conscience' affords 'the best and highest distinction' between mankind and other animals, Darwin determined to approach the subject 'exclusively from the side of natural history'. In the behaviour of many species he found evidence of social instincts, acquired through natural selection, which would lead to a deepened feeling of sympathy and a growing sense of conscience in proportion as the reflective intellectual faculties were evolved. Sympathy and conscience, in turn, would be encouraged by the wishes of the community, as soon as the power of language was acquired, and the social utility of these attributes in the struggle for existence would cause them to prevail. Finally, as human beings themselves regarded more and more the welfare and happiness of their fellows, as 'from habit, following on beneficial experience, instruction, and example', their sentiments gradually extended, first to society, then to all the species, then to the lower animals, 'so would the standard of [human] morality rise higher and higher'.[8] The culmination of moral evolution, Darwin believed, is the Golden Rule: 'As ye would that men should do to you, do ye to them likewise.' This has become 'the foundation-stone of morality' and could hardly have evolved apart from reason, instruction, and 'the love or fear of God'. Indeed, 'the grand idea of God hating sin and loving righteousness' is not only a powerful moral sanction; it also represents 'the highest form of religion'.[9]

Darwin was evidently not without his own moral and metaphysical beliefs – a point on which we shall presently enlarge. But while these beliefs were part of the anthropological data that evolution had to explain, they did not intrude upon the natural and mechanical explanations furnished by his argument. The proof of this lies in the analogy by which Darwin commended his analysis of the mental evolution of mankind. Repeatedly he urged that the difficulties presented by his views are no different from the difficulties of conceiving how the intellectual and moral faculties develop within the life of the human individual. 'At what age', he asked, 'does the new-born infant possess the power of abstraction, or become self-conscious, and reflect on its own existence? We cannot answer; nor can we answer in regard to the ascending organic scale.' Or again, in concluding the *Descent of Man*, Darwin observed, 'Few persons feel any anxiety from the impossibility of determining at what precise period in the development of the individual, from the first trace of a minute germinal vesicle, man becomes an immortal being; and there is no greater cause for anxiety because the period cannot possibly be determined in the gradually ascending organic scale.' For Darwin the evolution of mankind, mind and body, was no less natural than individual growth; phylogeny was no more 'irreligious' than ontogeny.[10]

If Darwin observed the constraints of his hypothetico-deductive argument in offering an evolutionary account of human ontology, how did he fare in his treatment of the sociology of human evolution? Here there was not so much a temptation to indulge in metaphysical speculation as an opportunity to extrapolate his theory carelessly or simplistically, heedless of the acknowledged fact that mankind's intellectual powers and moral consciousness, and hence the social existence of the species, differ vastly from comparable phenomena in the rest of the animal kingdom. Once more, however, Darwin proceeded with caution and with a due regard for the peculiar and limited evidence at his disposal. The bodily evolution of mankind, he maintained, was largely superseded as soon as the intellectual and moral faculties could confer some special advantage in procuring means of subsistence. Thereafter evolution affected mankind's mind and morals, and this primarily by means of natural selection in 'the rudest state of society'. Tribes with many inventive, clever, and sagacious people; tribes whose members inculcate sympathy, fidelity, patriotism, and courage; tribes in which these attributes have become habitual and therefore heritable – such tribes predominate in the 'never-ceasing wars of savages.' 'The standard of morality and the number of well-endowed men', said Darwin, 'will thus everywhere tend to rise and increase'. Yet this by no means implies that all tribes tend to rise in the 'scale of civilisation'. History rebukes the inclination 'to look at progress as normal in human society. . . . Progress seems to depend on many concurrent favourable conditions, far too complex to be followed out.' Even 'the problem. . . of the first advance of savages towards civilisation' was, in Darwin's view, 'at present much too difficult to be solved'.[11]

With the advent of civilisation, according to Darwin, natural selection becomes a 'subordinate' factor of evolution. The advanced morality of 'highly civilised nations' forbids that they should 'supplant and exterminate one another as do savage tribes'. Moreover, unlike primitive societies, such nations care for the physically weak and the mentally deficient, allowing them to propagate their kind. To do otherwise, 'even at the urging of hard reason', Darwin declared, would be to choose the greater of evils. It would cause a 'deterioration in the noblest part of our nature', which itself is the highest product of moral evolution. Under civilised conditions natural selection may continue to have some influence on the development of the human body because of the better nourishment it receives, on the human mind because of the greater propensity of able people to rear healthy children and educate them, and on morals because intemperance, profligacy, and criminal behaviour normally interfere with the ability to leave off-spring. But again, neither natural selection nor the inherited effects of environment and habit – the principal cause of evolution in civilised nations – can ensure the direction or inevitability of social development. Degeneration may well occur and 'progress is no invariable rule'. The 'assumption, so often made with respect to corporeal structures, that there is some innate tendency towards continued development in mind and body', was wholly unacceptable to Darwin. 'Development of all kinds', he cautioned, 'depends on many con-

current favourable circumstances. Natural selection acts only tenta-
tively. Individuals and races may have acquired certain indisputable
advantages, and yet have perished from failing in other characters.'
At least, however, it could be seen that 'a nation which produced
during a lengthened period the greatest number of highly intellectual,
energetic, brave, patriotic, and benevolent men, would generally
prevail over less favoured nations'.[12]

Such temperate statements about progress, morality, and natural
selection show how carefully Darwin applied his theory and how he
dealt 'as a naturalist...with the questions of sociology'. Other statements, however, reveal the extent to which, in treating of these subjects, Darwin remained a man of his time.[13] Progress to him was a fact
of nature and history; natural selection, aided by the inherited effects
of environment and habit, was an explanation of progress which could
not ensure its past or future inevitability. Yet it was 'a truer and more
cheerful view' to believe that 'progress has been much more general
than retrogression', for this could give mankind 'hope for a still higher
destiny in the future'.[14] Morality to Darwin was a product of evolution; the Golden Rule was the foundation of morality in civilised
nations, its virtues – loving one's enemies, returning good for evil, and
refusing to 'violate the dignity of humanity' – would always and
ultimately triumph in the struggle for existence. Yet society was divided
into 'grades', the 'poor' were among society's 'inferior members', the
'rich' among 'the better class of men', and the advantage of rich over
poor in the 'race for success' was not to be reduced by hindering 'the
moderate accumulation of wealth'.[15] Natural selection, in Darwin's
view, was a 'subordinate' cause of evolution among civilised nations,
for such nations fulfil the highest morality, keeping the peace outside
their borders, preserving the weak within. Yet in order that mankind
should 'advance still higher', it was 'to be feared that [the species]
must remain subject to a severe struggle', 'barbarians' with 'civilised
nations', 'lower races' with the 'civilised races' which are 'everywhere
extending their range', and the 'less gifted' with the 'more gifted' in
the 'battle of life' in advanced societies.[16]

Clearly, Darwin's world-view may be variously regarded as
Christian, Victorian, Anglo-Saxon, capitalist, and middle-class. It
forms the setting apart from which the theory of natural selection, and
Darwin's theory of human evolution in particular, can hardly be
understood. At the same time, however, in so far as the theory and its
setting may be dissociated and compared, it is useful to do so in order

to reveal their countervailing demands. From the foregoing paragraphs it may be inferred that Victorian popular optimism, and a belief in progress derived ultimately from a Christian view of history, encouraged Darwin on occasion to mitigate the demands of natural selection as an evolutionary mechanism.[17] Likewise a strong humanitarianism, both cultural and religious, restrained Darwin from drawing most of the harsher implications of his theory. The ambiguity of his position is well summarised in the statement, 'It is impossible not to regret bitterly, but whether wisely is another question, the rate at which man tends to increase.'[18]

Natural selection and sexual selection, on the other hand, lent credence to the unity of the human races, thereby preserving Darwin from some (but not all) of the excesses associated with the imperial spirit.[19] Darwin was also kept from much of the racism and elitism of domestic eugenics by the fact that the laws of inheritance were not 'thoroughly known'. Indeed, the notion that the inherited effects of environment and habit predominate over struggle and selection in the evolution of civilised nations is implicit in Darwin's harshest eugenic words: 'There should be open competition for all men; and the most able should not be prevented by laws or customs from succeeding best and rearing the largest number of offspring.'[20] According to the theory of natural selection, the 'most able' are those who do, in fact, succeed in leaving the largest number of offspring, not some group which 'should not be prevented by laws or customs' from doing so. Darwin vacillates here because for him the 'most able' among civilised *Homo sapiens* are not necessarily defined by their reproductive success in the struggle for existence. They are rather those who are 'highly intellectual, energetic, brave, patriotic, and benevolent', those, that is, whose attributes may be ascribed less to natural selection than to the inherited effects of environment and habit. Had not Darwin earlier established that the 'more efficient causes' of intellectual progress consist of cultural endowments, and that 'the moral qualities are advanced, either directly or indirectly, much more through the effects of habit, the reasoning powers, instruction, religion, &c., than through natural selection'?[21]

What, therefore, becomes of Darwin's much-alleged connexion with 'Social Darwinism'? With a host of his twentieth-century detractors, shall we say that, in extending the theory of natural selection to mankind, Darwin became the progenitor of all those ideologies of force and conflict which have claimed him as their own – racism and imperialism, whether that of Karl Pearson, Theodore Roosevelt, or Alfred Rosenberg; the militarism of Moltke and Treitschke; and the cut-throat capitalism of Carnegie and Rockefeller, to name only the more prominent?[22] By now it should be obvious that Darwin was not a social Darwinist in any straightforward or unambiguous sense. Elements of his theory were useful to ideologies of human conflict and his own prejudices coincided with such ideologies at several points. But theory and prejudice were tempered with that caution which caused Darwin's scientific reputation to endure and with those noble virtues, comprised in the Golden Rule, which endeared his character to every race and class and nation.[23]

It is not hard to understand why Darwin's name has become attached to movements for which he bore little direct responsibility. Many have taken the name as a symbol of prestige, few have read carefully what Darwin said, and in some cases, as we have seen, the manner in which he said it may have 'led those, who were looking for scientific support for opinions already held, to infer that he meant what they already believed'.[24] Another reason why Darwin has been confused with those who carelessly extended his doctrines lies in the popularity of Spencer's evolutionary philosophy. Lest the confusion be perpetuated, we shall do well in considering Spencer's approach to evolution to begin with the judgement of a cautious and eminently successful practitioner of hypothetico-deductive argument, one who had learnt from experience to 'distrust greatly deductive reasoning in the mixed sciences'.

Darwin and the Spencerian synthesis

Darwin can be read to compliment Spencer in the highest terms. He first corresponded with him in 1858, sending thanks for the gift of a volume of essays and expressing admiration for Spencer's remarks 'on the general argument of the so-called development theory'. 'I treat the subject simply as a naturalist, and not from a general point of view', he said, alluding to his *Origin of Species*, then in preparation, 'otherwise, in my opinion, your argument could not have been improved on, and might have been quoted by me with great advantage'. Later Darwin wrote again, informing Spencer that his 'excellent essay' on the development hypothesis was duly mentioned in the 'Historical sketch' prefixed to the third edition of the *Origin* (1861). In 1867, after Spencer's *Principles of Biology* had appeared, Darwin told the author, 'I was fairly astonished at the prodigality of your views. Most of the chapters furnished suggestions for whole volumes of future researches.' Twice thereafter, in 1871 and 1872, Darwin exposed his sentiments for the general public to read as compliments, referring to Spencer as 'our great philosopher' in the *Descent of Man* and as 'the great expounder of the principle of Evolution' in *The Expression of the Emotions in Man and Animals*.[25]

And this, clearly, is how Spencer read Darwin. One soon learns to expect from Spencer the kind of reticence which tells the readers of his *Autobiography* how he was about to reproduce Darwin's commendation of his essay on the development hypothesis, but has instead decided to omit it as the compliments contained in the letter would be 'out of taste'.[26] It is a pity that he could not have profited from passages in other letters – Darwin complaining about his 'detestable style', his '*a priori* conclusions', and his aversion for research[27] – and from a passage which Francis Darwin judiciously excised from the published version of his father's autobiography. After admitting dislike for Spencer's 'extremely egotistical' personality, admiration for his 'transcendent talents', and very little knowledge of the great men such as Descartes and Leibniz with whom Spencer had often been ranked, Darwin pronounced his final judgement, thereby setting all his previous statements in their proper perspective:

I am not conscious of having profited in my own work by Spencer's writings. His deductive manner of treating every subject is wholly opposed to my frame of mind. His conclusions never convince me: and over and over again I have said to myself, after reading one of his discussions, – 'Here would be a fine subject for half-a-dozen years' work'. His fundamental generalisations (which have been compared in importance by some persons with Newton's laws!) – which I daresay may be very valuable under a philosophical point of view, are of such a nature that they do not seem to me to be of any strictly scientific use. They partake more of the nature of definitions than of laws of nature. They do not aid one in predicting what will happen in any particular case. Anyhow they have not been of any use to me.[28]

Nor have the ten ponderous tomes of Spencer's *Synthetic Philosophy* been of much use to scientists since Darwin's time. Their contrast with the *Origin of Species* could not be more complete. One is the *Novum Organum* of modern biology; the other is a period-piece of metaphysical speculation. One expounds a fertile theory which has stimulated research; the other enshrines a sterile prolixity which has repelled its readers. Darwin's work remains scientifically important because of its originality, its unique relevance to the problems of specific origins, and the tenacity with which its author held to natural selection as 'the most important, but not the exclusive, means of modification'. The philosophy of Spencer is historically interesting because of 'the traditions on which it drew, the areas in which he applied it, and the tenacity with which he clung to belief in the inheritance of acquired characteristics'.[29]

With a rich heritage of Nonconformity, Spencer must have incarnated the worst fears of every High Churchman. Like his father before him, who moved from the Wesleyans to the Quakers to agnosticism, he possessed a restless reactive mind, fiercely individualistic and indifferent to almost any authority save his own. At the tender age of eleven he imbibed Gall's phrenology from an itinerant lecturer and thereafter, for nearly two decades, was under the influence of its heady dogmas. At twenty he read Lyell's *Principles of Geology* and promptly converted to Lamarckism, the up-and-coming heresy – *Vestiges* was yet four years in store – which Lyell was at pains to refute. In the meantime Spencer played the boy-scientist, dabbling in geology, entomology, and electromagnetism. He became a civil engineer and worked for the railways, applying his knowledge of geometry and mechanics to various practical problems, but chiefly to the invention of sundry labour-saving devices. For, according to Spencer's recollection later in life, his 'physical constitution did not yield such overflow of energy as prompts some natures to spontaneous activity. In many directions action was entered upon rather reluctantly; while thinking was a pleasure'. Although there are perhaps more concise expressions for a disinclination to engage in manual labour, it must be remembered that Spencer believed this 'predominant tendency to contemplation' to have been an important factor in his career.[30]

Spencer was most certainly right. His contemplation did not involve the intellectual effort required to master the essential books in his chosen fields of research, and this had a grave effect on his career. 'Except for his own works', remarked J. Arthur Thomson, 'he did not set great store on the invention of printing'.[31] Spencer could believe for thirty years that he had originated the doctrine of 'the right of every man to do what he wills so long as he does not trench upon the similar rights of any other man'. Only after that lapse of time did he discover (from a reference in *Mind*) that Kant had preceded him. As for writings with which he disagreed, he was not ashamed to say that, 'being an impatient reader, . . .I did not go far'.[32] Consequently, when at mid-century Spencer came to formulate his views on evolution, his fund of information was at a rather low ebb. Scattered about in his mind were the 'tendency to individuation' of Coleridge and Schelling, Milne-Edwards' 'physiological division of labour', Von Baer's formula of 'the development of every organism as a change from homogeneity to heterogeneity', and so much of Lewes' *Biographical History of Philosophy* as had proved stimulating to his thoughts. But these were all the data he needed. In Spencer's words, 'Various ideas, forming components of a theory of evolution, were lying ready for organisa-tion.'[33] Organisation, like geometry, mechanics, and the invention of labour-saving devices, was his speciality.

Spencer began to pull things together. His declaration for evolution came in 1852 in 'The development hypothesis', the polemical essay against 'special creations' which earned him Darwin's praise. The first edition of his *Principles of Psychology* – in some ways his most important book – appeared in 1855, uniting the concept of develop-ment and some old phrenological lessons with the theory of associa-tionism. Then in 1857 Spencer published 'Progress: its law and cause', a lengthy article given to illustrating the proposition – henceforth central to all his thought – that the universal transformation of the homogeneous into the heterogeneous, or Progress, is the necessary consequence of the universal multiplication of effects from causes.[34] Now all that he lacked was a moment of inspiration in order to com-plete a truly grand organisation of the universe on evolutionary lines. So revealing is Spencer's account of that moment, and so typical of his general pattern of thought (not to say his literary style), that we repro-duce it in full.

During a walk one fine Sunday morning (or perhaps it may have been New Year's Day), in the Christmas of 1857–8 I happened to stand by the side of a pool along which a gentle breeze was bringing small waves to the shore at my feet. While watching these undulations I was led to think of other undulations – other rhythms; and probably, as my manner was, remembered extreme cases – the undulations of the ether, and the rises and falls in the prices of money, shares, and commodities. In the course of the walk arose the inquiry – Is not the rhythm of motion universal? and the answer soon reached was – Yes. Presently – either forthwith or in the course of the next few days – came a much more important result. This generalization concerning the rhythm of motion recalled the general-ization. . .that motion universally takes place along the line of least resistance. Moreover there had become familiar to me the doctrine of the Conservation of Force, as it was then called – in those days a novelty; and with this was joined in my mind Sir William Grove's doctrine of the correlation of the physical forces. Of course these universal principles ranged themselves alongside the two universal principles I had been recently illustrating – the instability of the homogeneous and the multiplication of effects. As, during the preceding year, I had been showing how throughout all orders of phenomena, from nebular genesis to the genesis of language, science, art, there ever goes on a change of the simple into the complex, of the uniform into the multiform, there naturally arose the thought – these various universal truths are manifestly aspects of one universal transformation. Surely, then, the proper course is thus to exhibit them – to treat astronomy, geology, biology, psychology, sociology and social products, in successive order from the evolution point of view. Evidently these universal laws of force to which conforms this unceasing redistribution of matter and motion, constitute the *nexus* of these concrete sciences – express a community of nature which binds them together as parts of a whole. And then came the idea of trying thus to present them. Some such thoughts they were which gave rise to my project, and which, a few days later, led to the writing out of the original programme.[35]

Following five years' observations on a global voyage Darwin spent more than two decades forming hypotheses, amassing experimental data, forming more hypotheses, and gathering still more data from correspondents throughout the world, and would no doubt have continued this process for several more years before completing his 'larger book' had not his hand been forced by Wallace.[36] Spencer, having travelled but a little in Europe, and without the intent of making scientific observations, spent less than ten years in eclectic reading, formed no hypotheses,[37] and thus early in 1858, nearly two years in advance of the *Origin of Species* – a fact to which he frequently found occasion to advert[38] – conceived the main idea of his *Synthetic Philosophy*, the law of evolution: 'Evolution is an integration of matter and concomitant dissipation of motion; during which the matter passes from an indefinite, incoherent homogeneity to a definite, coherent heterogeneity; and during which the retained motion undergoes a parallel transformation.'[39] The difference is not simply that Darwin's method was inductive and Spencer's deductive; it is that in Darwin's work induction, deduction, and what C. S. Pierce called 'abduction' – the imaginative process of hypothesis formation – were fairly in balance, whereas in Spencer's philosophy abduction was absent, induction was inadequate, and deduction determined the whole. Spencer rightly, albeit rather defensively, maintained that his 'law of evolution' was not arrived at deductively; yet he seems never to have appreciated that the law explained nothing, but was merely a comprehensive definition, and that its factual basis – essentially the facts reviewed in his essay on progress – was miserably weak. For Spencer, 'brief inspection made it manifest that the law held in the inorganic world, as in the organic and super-organic'. Thus it remained simply to apply the cosmic generalisation to all the disorganised facts of the universe. 'Bearing the generalization in mind', said the philosopher, 'it needed only to turn from this side to that side, and from one class of facts to another, to find everywhere exemplifications'.[40]

The pattern was set for his life's work: collect some data; make some inductions; apply the law of evolution; draw manifold deductions; and collect more data to confirm them. Of course there had to be some labour-saving devices as well. Various amanuenses were secured and Spencer dictated to them the entire *Synthetic Philosophy* – much of it, one gathers, in the intervals between bouts of rowing, games of racquets and the like, before 'cerebral congestion' set in. When the time came to collect and abstract the enormous data of sociology, Spencer also obtained the assistance of several compilers. Forty years of desultory dictation and scavenging for facts began in 1860, and in the same year *First Principles* began to appear. Both volumes of the *Principles of Biology* were published by 1867 and a two-volume revision of the *Principles of Psychology* was completed in 1872. Publication of the *Principles of Sociology* extended over two decades, its three thick tomes emerging in instalments, like the others, between 1876 and 1896. And while these were in preparation Spencer produced yet another double-decked addition to his philosophy, the *Principles of Ethics*, which appeared between 1879 and 1893.[41]

The bases of Spencer's appeal

Why, then, in view of this intimidating mass of deductive philosophising – over 6000 large octavo pages – have we proposed to account for the *vogue* of Herbert Spencer? Is it possible that an author capable of turgidity on such a scale should have enjoyed anything like popular acclaim? The answer to this question is Yes – with one important qualification. Spencer, like so many visionaries, lacked honour in his native land. His veneration by R. A. Armstrong as 'the supreme teacher of the doctrine of evolution' and Edward Clodd's proud proclamation that 'the Theory of Evolution dealing with the universe *as a whole*, from gas to genius, was formulated some months before the publication of the Darwin–Wallace paper' were not altogether typical English responses to the *Synthetic Philosophy*.[42] Spencer was assaulted from left and right, by rationalists and by churchmen, and even within the friendly agnostic confines of the Darwinian circle dissent with mollification seems to have been the prevailing attitude towards many of his views.[43] Furthermore, the *Synthetic Philosophy*, with its doctrines of individualism and *laissez-faire*, appeared just as the working class and social reformers were rejecting these doctrines as the political basis of a burgeoning industrial society. By the time Spencer's definitive statement of social and political conservatism, the *Principles of Ethics*, was in print, cooperative, labour, and trade union movements had coalesced in support of the liberal socialism which is the special characteristic of modern Britain, and the Labour Party had been born. 'He saw his political advice disregarded, and on all sides an exuberant growth of the socialistic organisations which he had spent himself in criticising', wrote W. H. Hudson, who knew Spencer well in the years before his death in 1903, and these developments 'cast a very black shadow over his declining path'.[44]

It was in the United States, that century-old adolescent nation, that awkward, impetuous, free-wheeling society then going through its 'Gilded Age' – it was there that Spencer found his fame. Within forty years, from the mid-sixties until 1903, the sales of his works came to nearly 370 000 volumes in their authorised editions alone. The *Synthetic Philosophy* became the *pièce de résistance* of philosophers, its author 'the metaphysician of the homemade intellectual and the prophet of the cracker-barrel agnostic'.[45] However, it was not simply American marketing know-how, from the unstinting promotion of E. L. Youmans, to the financial backing of a group of eminent subscribers to the *Synthetic Philosophy*, to the presses of D. Appleton and Company, which convinced common people and intellectuals alike that Spencer in all his garrulity was worth reading. There were other incentives without which no amount of salesmanship would have availed.

In the first place, his philosophy could be made compatible with religion. Considering that Spencer was heralded as an evolutionist and widely believed to be the *ex officio* philosopher of the British Darwinians, this was an important point in its favour – a selling point in a day when science and religion were supposed to be in conflict. Chapter one of *First Principles* is entitled 'Religion and science' and chapter five 'The reconciliation'. The intervening pages draw on the Reverend H. L. Mansel's 1858 Bampton Lectures, *The Limits of Religious Thought Examined*, and through them on Sir William Hamilton's 'philosophy of the unconditioned', to show that neither the Ultimate Reality of religion nor the ultimate reality of science is knowable, and thus that science and religion together 'may keep alive the consciousness that it is alike our highest wisdom and our highest duty to regard that through which all things exist as The Unknowable'. To many who preferred to use the word 'God' – among them John Fiske and Henry Ward Beecher, each of whom had an enormous following – this appeared tantamount to a theistic proof, and so it was noised abroad that the *Synthetic Philosophy* had laid the groundwork for a new theology. To Count Goblet d'Alviella, Spencer's Belgian admirer, writing in 1886, the phenomenon seemed 'characteristic of the American mind, which, when it adopts the philosophy of the old world, immediately transforms it into religion'.[46]

Another incentive for Americans to read and revere Spencer was the plain fact that his philosophy, though not in every detail, yet in its fundamental assumptions and affirmations, provided a thoroughgoing rationale for the American way of life. From his earliest political essays to the last volume of his *magnum opus*, Spencer championed the cause of liberty and individual rights. Free trade, free enterprise, voluntary association, *laissez-faire*, disestablishment – all were at the heart of his thought. Being at the heart of Spencer's thought, these doctrines came under the law of evolution, according to which all heterogeneity, all individuality, is the inevitable product of natural forces and a manifestation of universal progress. Thus where markets are freely competitive, where government is decentralised, and where religious liberty expresses itself in a proliferation of denominations and sects, there, one could be sure, human beings are cooperating with the forces that mould their hopeful destiny.[47] And where else were these conditions more fully realised than in the United States? Post-bellum America was apparently undergoing the metamorphosis from Spencer's 'militant' type of society to his 'industrial' type. The country manifested 'a growing independence, a less-marked loyalty, a smaller faith in governments, and a more qualified patriotism'.[48] Business was booming, untrammelled by federal regulation, and the fittest were proudly surviving in a ruthlessly competitive marketplace. Immigrants diversified the population and railroads dispersed its members across the continent. 'Rugged individualism' had never been more rugged. Religious individualism had never been so variously religious. Bigness was goodness, newness was trueness, and the *Synthetic Philosophy*, which was both big and new, baptised it all in the name of Science and Progress. 'The peculiar condition of American society', wrote Beecher to Spencer, 'has made your writings far more fruitful and quickening here than in Europe'. 'In America', commented the *Atlantic Monthly*, 'we may...confess our obligations to the writings of Mr Spencer, for here sooner than elsewhere the mass feel as utility what a few recognize as truth'.[49]

Spencer came to see his philosophy at work in 1882. For nearly three months he travelled about the country, fraternising with the *élite* of science and industry, inspecting the products and respectable by-products of American civilisation. The visit was exhausting for one whose predominant tendency was towards contemplation. How much more exhausting life must be, he reasoned, for those men of wealth, power, and prestige who honoured him with their presence at a farewell banquet in New York City on the ninth of November. To them on that occasion Spencer therefore preached a sermonette on the 'gospel of relaxation'. His appeal was to the serious physical effects of overwork.

Damaged constitutions reappear in children, and entail on them far more of ill than great fortunes yield of good. When life has been duly rationalized by science, it will be seen that among a man's duties care of the body is imperative, not only out of regard for personal welfare, but also out of regard for descendants. His constitution will be considered as an entailed estate, which he ought to pass on uninjured if not improved to those who follow; and it will be held that millions bequeathed by him will not compensate for feeble health and decreased ability to enjoy life.[50]

By thus addressing his audience in the only terms many of them were able to understand, Spencer revealed as clearly as possible the biological mechanism through which his law of evolution was expressed and the importance of that mechanism to social progress. For without the inheritance of characters acquired under the influence of the universal movement towards 'definite, coherent heterogeneity' – the 'improved' constitution of a life well spent, for example – human progress would have to depend on the protracted and unpredictable process of natural selection.

The place of functional adaptation

In apportioning the factors of evolution in civilised populations Darwin and Spencer did not in fact greatly differ. Darwin, it will be recalled, believed that the development of civilised nations 'depends in a subordinate degree on natural selection' and that the inheritance of characters acquired through habit, thought, instruction, and religious faith is more important for the development of 'the highest part of man's nature' than the struggle for existence.[51] In his *Factors of Organic Evolution*, published in 1887, Spencer was pleased to underscore Darwin's reliance on the Lamarckian factor, which he preferred to call the inheritance of 'functionally-produced modifications'. 'Eventually, among creatures of high organization', he wrote, 'this factor became an important one; and I think there is reason to conclude that, in the case of the highest of creatures, civilized men, among whom the kinds of variations which affect survival are too multitudinous to permit easy selection of any one, and among whom survival of the fittest is greatly interfered with, it has become the chief factor: such aid as survival of the fittest gives, being usually limited to the preservation of those in whom the totality of the faculties has been most favourably moulded by functional changes'.[52]

But beneath this relatively superficial agreement was a vast difference in approach. In the years before 1859, during which he conceived and charted the *Synthetic Philosophy*, Spencer held that the inheritance of functionally-produced modifications was the sole cause of organic evolution and thus a factor ultimately reducible to the terms of his general law. '*The Origin of Species*', he later admitted, 'made it clear to me that I was wrong; and that the larger part of the facts cannot be due to any such cause'.[53] Now there were two factors contributing to organic evolution, of which only one, the less extensive of the two, had been brought under his law. The obvious conclusion to be drawn was that 'the changes produced by functional adaptation...and the changes produced by "natural selection" had both to be exhibited as resulting from the redistribution of matter and motion everywhere and always going on'. No sooner said than done.[54] By mid-1864 Spencer had defined life as a dynamic equilibrium of forces, with natural selection as a mode of 'indirect equilibration' and functional adaptation as 'direct equilibration'. Then in the *Principles of Biology* he proceeded to deduce the interaction of these factors in the later stages of evolution:

As fast as essential faculties multiply, and as fast as the number of organs that co-operate in any given function increases, indirect equilibration through natural selection, becomes less and less capable of producing specific adaptations; and remains fully capable only of maintaining the general fitness of constitution to conditions. Simultaneously, the production of adaptations by direct equilibration, takes the first place – indirect equilibration serving to facilitate it. Until at length, among the civilized human races, the equilibration becomes mainly direct: the action of natural selection being restricted to the destruction of those who are constitutionally too feeble to live, even with external aid.[55]

Darwin objected to Spencer's reasoning not merely because he had evidence that artificial selection, and hence natural selection, could do much to modify the higher animals;[56] he objected fundamentally because Spencer's deductive approach to evolution was 'wholly opposed' to his own method of reasoning in science. The problem of explaining 'the first advance of savages towards civilisation' – indeed, the problem of explaining the progress of any civilisation – was not unlike the difficulty of accounting for 'the first steps in advancement or in the differentiation and specialisation of parts' among the simplest organisms. 'Mr Herbert Spencer', wrote Darwin in the *Origin*, 'would probably answer that. . .his law would come into action, namely, "that homologous units of any order become differentiated in proportion as their relations to incident forces become different"'. However, having just dispensed with any 'necessary and universal law of advancement or development' – the law in question was Lamarck's – Darwin could only register his dissent: 'As we have no facts to guide us, all speculation on the subject is useless.' His own theory, on the other hand, was not a Spencerian speculation, for even at the dawn of life, as he immediately pointed out, natural selection might have occurred through the preservation of favourable variations in a struggle for existence.[57]

For Darwin it was an observed fact that natural selection plays only a 'subordinate' role in the evolution of civilised nations, and a reasoned assumption that, since progress by means of the survival of the fittest is 'no invariable rule', whatever beneficial changes do occur may be attributed mainly to the inheritance of acquired characters. For Spencer it was a universal law of evolution which ultimately determined the factors of human development, a law originally formulated in keeping with a single Lamarckian factor, a law to which natural selection was later assimilated, and a law which, by subordinating natural selection to the inheritance of functionally-produced modifications in the higher forms of life, was able to guarantee the inexorable progress of human society.[58]

Spencer's deductive edifice rested on slender premises. None was so vital and none therefore was so vulnerable as the Lamarckian doctrine of the inheritance of acquired characters.[59] Twice Spencer felt obliged to descend from the Olympian heights at which he was labouring on the *Synthetic Philosophy* in order to shore up its shaky support. On neither occasion did he attempt to conceal his fears. In his preface to *Factors of Organic Evolution* he mentioned the 'indirect bearings upon Psychology, Ethics, and Sociology' of the argument contained in the book. 'My belief in the profound importance of these indirect bearings', he added, 'was originally a chief promoter to set forth the argument'.[60] The situation became more desperate when he tangled over the inheritance of acquired characters with the German zoologist August Weismann in an exchange of articles published in the *Contemporary Review* between 1893 and 1895 – perhaps 'the longest and most detailed and characteristic controversy on the subject'.[61] Weismann could find no convincing evidence of the inheritance of 'functionally-produced modifications' and had challenged naturalists to show him otherwise. Typically Spencer retorted, 'Either there has been inheritance of acquired characters or there has been no evolution.' In a later article he revealed why his reaction was so intense: 'I...have been led to suspend for a short time my proper work only by consciousness of the transcendent importance of the question at issue. . . . A right answer to the question of whether acquired characters are or are not inherited, underlies right beliefs, not only in Biology and Psychology, but also in Education, Ethics, and Politics.'[62] So threatened did Spencer feel that William Thistleton-Dyer, according to Wallace, thought him 'dreadfully disturbed'. 'He fears that acquired characters may not be inherited, in which case the foundation of his whole philosophy is undermined.' The *Synthetic Philosophy* had indeed begun to crumble but, mercifully, its author did not live to see the ultimate collapse. Not until 1909 did one venture to say on the Oxford lectureship endowed in Spencer's name that 'one of the most important links in the chain of argument used in the synthetic philosophy is broken, and the sociological conclusions founded upon the biological principles set forth in that system are vitiated'.[63]

It would be a mistake, however, to allow the obscurity and reproach into which Spencer's work has fallen to condition an assessment of its importance in the late-Victorian world. The *Synthetic Philosophy* was a monumentally impressive achievement. So significant was its impact on life and thought that one historian, in a felicitous figure, has called it 'a fossil specimen from which the intellectual body of the period may be reconstructed'.[64] By offering a scientific rationale for human progress, which was susceptible of a religious interpretation, Spencer seduced the multitudes with his specious pan-evolutionism and lent plausibility to the Lamarckian assumptions on which it was based. Moreover, in achieving this vogue the Spencerian philosophy provoked – and has continued to provoke – confusion of itself with Darwinism, the theory to which it bore so very little resemblance. 'What brought him rapid victory and prolonged sway over his age', writes Jacques Barzun of Darwin, 'was...the ability of the age to recognize itself in him'.[65] The more appropriate referent for this statement is, of course, Herbert Spencer.

8

∞∞∞

DARWINISM AND NEO-DARWINISM

Species have been modified, during a long course of descent, . . .chiefly
through the natural selection of numerous successive, slight, favourable
variations; aided in an important manner by the inherited effects of the
use and disuse of parts; and in an unimportant manner, that is in relation
to adaptive structures, whether past or present, by the direct action of
external conditions, and by variations which seem to us in our ignorance
to arise spontaneously. It appears that I formerly underrated the frequency and value of these latter forms of variation, as leading to
permanent modifications of structure independently of natural selection.
But as my conclusions have lately been much misrepresented, and it has
been stated that I attribute the modification of species exclusively to
natural selection, I may be permitted to remark that in the first edition of
this work, and subsequently, I placed in a most conspicuous position –
namely, at the close of the Introduction – the following words: 'I am
convinced that natural selection has been the main but not the exclusive
means of modification.' This has been of no avail. Great is the power of
steady misrepresentation; but the history of science shows that fortunately
this power does not long endure.

Charles Darwin[1]

If there is but one passage in all Darwin's publications that set the
stage for the discussion of evolutionary theory in the closing decades
of the nineteenth century, it is this paragraph from the conclusion to
the last edition of the *Origin of Species* (1872). The first half of the
paragraph recalls the tension that Darwin created within the *Origin*
and the *Descent of Man* as he came more and more to supplement
natural selection with the inherited effects of environment and habit.
It brings to mind those who relieved the tension in favour of Lamarckian evolution. The latter half of the passage brings to mind those who
relieved the tension in the opposite manner: by emphasising, even to
the point of misrepresentation, the all-sufficiency of natural selection.[2]

The tension among the causes of evolution in Darwin's works did
not originate by accident. Nor did it result from ignorance or confusion, but from Darwin's characteristic caution, as manifested in his

desire to uphold natural selection as 'the most important, but not the exclusive, means of modification' while retaining every workable adjunct explanation in the face of troublesome evidence. Many naturalists, however, found Darwin's tension unbearable, and thus in time a polarisation over the factors of evolution occurred. The logic of the situation is not difficult to understand. If one granted that acquired characters are inherited, as Darwin did, there was no reason in principle why environment and habit could not usurp most, if not all, of the power attributed by Darwin to natural selection. But if, on the other hand, one severely restricted the effects of environment and habit, as Darwin was inclined to do, then there arose the temptation to conclude that acquired characters could not be inherited and that natural selection is the sole causal factor in evolution.[3] Those who minimised the role of natural selection called themselves Neo-Lamarckians and found their greatest strength in the United States. Those who made natural selection all-powerful were called Neo-Darwinians and were known, with at least one notable exception, as the compatriots of Darwin.[4]

We must make distinctions carefully however. The Lamarckian evolution of Spencer was certainly not identical with the Neo-Lamarckism of Cope. While Spencer claimed a much larger role for environment and habit than did Darwin, and believed that his comparison of statements in successive editions of the *Origin* proved his claim to be a legitimate extrapolation of Darwin's own views, he nevertheless allowed that natural selection may account for a large part of the phenomena of evolution. Cope, on the contrary, maintained that natural selection is altogether subordinate to the law of acceleration and retardation throughout the organic world, and that Spencer, therefore, no less than Darwin, erred in attributing to it any constructive function in the evolution of life.[5] Now this distinction among those who enlarged the limited role to which Darwin assigned the inheritance of acquired characters is paralleled by a distinction among those who sought to restrict the efficacy of this Lamarckian factor. 'Neo-Darwinism', properly speaking, refers to the views of Alfred Russel Wallace, the co-discoverer of the theory of natural selection, and August Weismann, professor of zoology in the University of Freiburg in Breisgau, both of whom believed that natural selection alone is sufficient to account for the origin of species – for Wallace, as we shall see, every species except mankind. 'Darwinism', in turn, may be understood to denote an outlook more nearly like that of Darwin, an outlook represented, appropriately enough, by those who were closest to him: his 'bulldog' T. H. Huxley (1825–1895), his stepcousin Francis Galton (1822–1911), and his devoted disciple George John Romanes (1848–1894).

Huxley, Galton, and Romanes

Huxley and Galton were the most prominent members of an older generation of naturalists to embrace the Darwinian theory. Both greatly advanced its cause, Huxley by winning for it a public hearing and a place in scientific education, Galton by applying to it statistical methods which would later bear fruit in the theoretical study of population genetics. But despite their strong allegiance to natural selection as the primary factor in evolution, neither Huxley nor Galton was entirely happy with Darwin's exposition of the theory. In their view it was unnecessary to assume that only continuous variations can serve as the raw material of natural selection. Huxley seems to have maintained for many years the opinion expressed in his 1860 review of the *Origin of Species*: that 'Mr Darwin's position might. . .have been even stronger than it is if he had not embarrassed himself with the aphorism, "*Natura non facit saltum.*" ' 'We believe', he added, '. . .that Nature does make jumps now and then, and a recognition of the fact is of no small importance in disposing of many minor objections to the doctrine of transmutation'.[6] Galton, in *Hereditary Genius* (1869), the first of his series of remarkable studies in anthropometry, conceded the validity of Jenkin's argument that continuous variation occurs within definite limits, and stated the case for evolution by 'a series of changes in jerks', which he was to enlarge and defend for the rest of the century.[7]

The advantages of this position were several, but chiefly that it relieved some of the tension between the competing factors of natural selection and the inheritance of acquired characters. Blending inheritance, it will be recalled, demanded a multitude of continuous variations to overcome its effects; continuous variations required much time to accumulate through natural selection into forms which exceed the apparent limits of variability for each species; and physicists required that the geological time allotted for this process be reduced. Darwin met the demand for continuous variations and resolved the dilemma over geological time by relying increasingly on the direct action of the environment and the use and disuse of organic parts to produce heritable modifications. Huxley and Galton, on the other hand, dealt with blending inheritance, not by providing for a greater supply of continuous variations, but by showing that discontinuous forms are often stable and fertile, and therefore able to counteract the effects of blending. The exigencies of time could then be met by supposing that these saltations had hastened the accumulation of specific differences by means of natural selection. Thus there was no need for Huxley and Galton to insist on a leading role for the inheritance of acquired characters. And since neither of them could find convincing evidence of the phenomenon, their position was a coherent one indeed.[8]

Huxley held that 'there is really nothing to prevent the most tenacious adherent to the theory of natural selection from taking any view he pleases as to the direct influences of conditions and the hereditary transmissibility of the modifications which they produce'. Personally, in fact, he had 'no a priori objection to the transmission of functional modifications whatever', and could even confess, 'I should rather like it to be true.' Putting proof before preference, however, he reminded Spencer, 'I argued against the assumption (with Darwin as I do with you) of the operation of a factor which, if you will forgive me for saying so, seems as far off sufficiently trustworthy evidence now as ever it was.'[9] Whether natural selection alone sufficed for the production of species remained to be seen. But 'few can doubt', Huxley declared, 'that if not the whole cause, it is a very important factor in that operation'. His residual misgivings were those of an anatomist and physiologist with limited experience of species in a state of nature. The proof of natural selection lay for him, not in its predictive and explanatory power, as it did for Darwin and other working naturalists; it lay rather in the production of mutually infertile breeds from a single stock by means of artificial selection. This proof was lacking and thus Huxley's old *thätige Skepsis* never entirely disappeared.[10]

Galton, for his part, was led to minimise the inheritance of acquired characters through testing the theory of heredity proposed in Darwin's *Variation of Animals and Plants under Domestication*. According to the 'provisional hypothesis of pangenesis' (which Darwin regarded merely as a 'thread of connexion' for the 'many large classes of facts' that were cluttering his mind), the cells of living bodies produce minute particles – 'gemmules' – that disperse throughout the system, multiply by self-division, collect by mutual affinity in the sexual elements, and finally, through generation, either pass dormant to the offspring or develop into units like those from which they were derived.[11] The effect of pangenesis was to provide a physiological mechanism for the inheritance of characters acquired by the different parts of an organism. To test the mechanism Galton sought to transfer gemmules from common to pure-bred rabbits through a partial trans-fusion of blood. His results were negative: the transfused rabbits con-tinued to produce pure-bred offspring. Pangenesis had failed as a theory of heredity, though conceptually it might still prove useful, as Galton revealed in his own theory, published in 1875. He retained Darwin's information-bearing gemmules but confined them to the 'stirp', the individual genotype (to use a modern term) contained in the reproductive germ. Galton did not deny that the stirp could be influenced by somatic changes: that is, he did not reject a priori the inheritance of acquired characters. However, he did maintain that 'it is indeed hard to find evidence...that is not open to serious objec-tion'.[12]

Darwin and Galton, the grandsons of the very first 'Lamarckian', Erasmus Darwin, were the intellectual forebears of George John Romanes, the most prominent second-generation Darwinian. Romanes learnt from Darwin, whom he revered, a deep respect for evidence and a cautious and critical attitude towards scientific explanation. From Galton, whose experiments he repeated, he learnt a theory of heredity which minimised the inheritance of acquired characters.[13] Thus scientifically equipped, and possessing a trenchant literary style, Romanes cut a wide swathe through the tangled debates between Neo-Lamarckians and Neo-Darwinians in the decade before his un-timely death in 1894. With the faith of a convert and disciple, he rebuked all who sought to eliminate the tension that Darwin had created within the Origin of Species and the Descent of Man, declar-ing, 'I myself believe that Darwin's judgement with regard to all these points will eventually prove more sound and accurate than that of any of the recent would-be improvers upon his system.'[14]

With Darwin, Romanes acknowledged the inability of natural selection by itself to account for three classes of phenomena: the origin and persistence of useless characters; the occurrence of divergent evo-lution under conditions of free intercrossing and blending inheritance; and the origin of cross-infertility between allied species occupying common or closely contiguous areas. Objections to natural selection based on these limitations he considered valid and formidable, but ultimately irrelevant to a doctrine of evolution in which natural selection figures as 'the most important, but not the exclusive, means of modification'.[15] Such was Romanes' doctrine, for besides stressing, as did Darwin, that environment and habit give rise to useless characters and that environment and isolation help to counteract the effects of blending, Romanes proposed another causal factor, subordinate to natural selection, which could originate that most useless of useless characters, cross-infertility. As was his custom, Romanes found precedent for this factor in his master's views.

Darwin had always allowed that a small amount of adaptive change might result from 'variations which seem to us in our ignorance to arise spontaneously'. He had also emphasised – increasingly in his later years – that 'some degree of separation', either ecological or geographical, would interfere with free intercrossing and thereby facilitate speciation under conditions of blending inheritance.[16] In 1886 Romanes merged these two ideas in his 'supplementary hypothesis' of 'physiological selection': if the reproductive system, which in most organisms is highly variable, were to vary spontaneously and simultaneously in some individuals of an interbreeding population; if the variant individuals thus became sterile in any degree with respect to non-variant individuals, but remained fertile *inter se*; then the sterility would tend inevitably to be preserved and augmented; the variant individuals thus isolated, though on the same terrain, would form an incipient species; and the incipient species would begin an independent and divergent course of evolution.[17]

The response to Romanes' 'supplementary hypothesis' was about as negative as the response to Darwin's 'provisional hypothesis' of pangenesis.[18] In one sense this seems appropriate, for disciple and master could thus also be identified by the manner in which their speculations were received. Yet in another sense the response to physiological selection was unfitting and largely unfair. Romanes assuredly had not tried to undermine natural selection, as his Neo-Lamarckian and Neo-Darwinian critics alleged. Nor did his supplementary hypothesis necessarily have that effect – unless, of course, one disagreed with Darwin and regarded natural selection as the sole and sufficient cause of evolution. For this response, however, Romanes had a ready reply. For those who preached the all-sufficiency of natural selection the objections based on useless characters, blending inheritance, and cross-infertility were 'not merely...valid and formidable, but...logically insurmountable'.[19] Such, Romanes believed, were the obstacles confronting the Neo-Darwinians.

Wallace and Weismann

In 1889, seven discreet years after Darwin's death, Alfred Russel Wallace (1823–1913) published a book entitled *Darwinism* in which he announced that 'Natural Selection is supreme, to an extent which even Darwin himself hesitated to claim for it'. This he regarded as a profession of 'pure Darwinism', though 'Neo-Darwinism', as others called it, was surely a more apposite term.[20] Already in fact, twenty-five years earlier, Wallace had diverged from purely Darwinian doctrines. Not only, as we shall see, did he reject Darwin's account of human evolution; he also argued, almost by way of compensation, that natural selection could account for much more than Darwin was willing to admit. Thus, rather than explain colouration and other secondary sexual characters in volitional terms, by means of sexual selection, Wallace subjected the phenomena to natural selection under a theory of mimicry and protective resemblances, which rendered sexually dimorphic characters immediately useful in the struggle for existence.[21]

The strict utilitarian viewpoint required of those who took natural selection as the all-sufficient factor of evolution was prominent throughout Wallace's evolutionary writings, and not untypically in the pages of *Darwinism* which discuss Romanes' three crucial objections. Free intercrossing and blending inheritance furnish no obstacle to speciation, given a multitude of useful continuous variations and a measure of geographical and ecological isolation.[22] Cross-infertility, the *sine qua non* of speciation, cannot, however, be the result of physiological selection, for 'no form of infertility or sterility between individuals of a species, can be increased by natural selection unless correlated with some useful variation, while all infertility not so correlated has a constant tendency to effect its own elimination'. As for useless characters – the real crux of the matter – Wallace maintained on the one hand that their existence had 'not been proved', and on the other that it could not be: 'The assertion of "inutility" in the case of any organ or peculiarity which is not a rudiment or a correlation [i.e. neither previously useful nor presently correlated with a useful character], is not, and can never be, the statement of a fact, but merely an expression of our ignorance of its purpose or origin'. Although there was always the abstract possibility that use and disuse, the direct action of the environment, and the laws of growth might produce some inherited change, 'no modification thus initiated', Wallace insisted, 'could have advanced a single step, unless it were, on the whole, a useful modification'.[23]

In these bold assertions there was nothing particularly new. Wallace had expressed similar views on natural selection and utility as early as 1867.

> Perhaps no principle has ever been announced so fertile in results as that which Mr Darwin so earnestly impresses upon us, and which is indeed a necessary deduction from the theory of Natural Selection, namely – that none of the definite facts o[f] organic nature, no special organ, no characteristic form of marking, no peculiarities of instinct or of habit, no relations between species or between groups of species – can exist, but which must now be or once have been *useful* to the individuals or the races which possess them.[24]

Romanes might well have been frustrated with Wallace's 'necessary deduction' merely on account of its persistence. But to Darwin's arch-admirer the persistence was less troubling than the perversion, which of course had persisted as well. In fact, Wallace's pan-utilitarian deduction had as little to do with the *Origin of Species* as the views expressed in *Darwinism* had to do with 'Darwin's earlier position'.[25] Still more frustrating, however, was the specious and irrefutable truth of Wallace's argument. 'It is a question of theory', Romanes complained in his *Darwin, and after Darwin* (1892–7), 'not a question of fact; our difference of opinion is logical, not biological: it depends on our interpretation of principles, not on our observation of species'. 'Wallace's deductive argument', he continued, 'is a clear case of circular reasoning. We set out by inferring that natural selection is a cause from numberless cases of observed utility as an effect: yet, when "in a large proportional number" of cases we fail to perceive any imaginable utility, it is argued that nevertheless utility must be there, since otherwise natural selection could not have been the cause.' In a score of closely reasoned pages Romanes proceeded to defend Darwin and to adduce the evidence which Wallace found necessary to reject *a priori*. 'Darwin', he said, 'would have had an immeasurable advantage in this imaginary debate'. It was he, not Wallace, who took seriously the theoretical problems arising from the many cases of apparently useless structures. It was he who saw fit to make natural selection 'the most important, but not the exclusive, means of modification'.[26]

If Romanes defended Darwin's understanding of natural selection against the deductive Neo-Darwinism of Wallace, it was Galton's theory of heredity which he sought to protect from the deductive Neo-Darwinism of August Weismann (1834–1914). In an 1883 essay, 'On heredity', the founding document of the Neo-Darwinian school, Weismann set forth the principle and the premise on which his subsequent work was based. The fundamental principle was an application of the so-called Law of Parsimony, which states that new forces or occult agencies should not be invoked to account for phenomena which can be readily explained by known causes. Weismann believed that the theories of heredity then current – Darwin's pangenesis in particular – violated this principle by appealing to some unknown factor in order to secure the transmission of parental characters, whether congenital or acquired, by representing them in the reproductive germ. Whereupon he reasoned that 'an explanation can. . .be reached by an appeal to known forces, if we suppose that characters acquired. . .by the parent cannot appear in the course of the development of the offspring, but that all the characters exhibited by the latter are due to primary changes in the germ'.[27] This, in turn, became his premise – the non-inheritance of acquired characters – and Weismann proceeded to deduce its consequences, first, for the substance of heredity, the 'germ plasm', and second, for the theory of evolution. From the non-inheritance of acquired characters it followed that the germ plasm is uninfluenced by changes in the body, or 'soma plasm'. Weismann therefore asserted the 'perpetual continuity' of the germ plasm 'from the first origin of life'.[28] From the perpetual continuity of the germ plasm, together with the additional postulate of its absolute stability, or invariance under the impact of external forces, since the origin of sexual reproduction, it followed that the raw materials of evolutionary change consist solely of the congenital variations that arise randomly in the offspring from the mixture of parental germ plasms. Thus Weismann concluded that the only causal factor in evolution was the one which assures the survival and accumulation of the most beneficial of these variations: namely, natural selection.[29]

Weismann may have thought that he proposed his theory of the continuity of the germ plasm in the 'provisional' spirit of Darwin's pangenesis, but in fact there was a strong atmosphere of dogmatism about it. 'Every other theory of heredity is founded on hypotheses which cannot be proved', he declared. And again: 'It has never been proved that acquired characters are transmitted, and it has never been demonstrated that, without the aid of such transmission, the evolution of the organic world becomes unintelligible.'[30] Romanes was frankly repelled. For if the case for the inheritance of acquired characters had yet to be convincingly set forth, the case for natural selection as the all-sufficient factor in evolution also lacked a firm empirical basis. The Darwinian knot was very tightly tied; he who would undo it had to deal with two bodies of evidence, not only by discounting the one, but also by meeting the formidable objections to the other. 'It is forgotten', Romanes replied to Weismann's application of the Law of Parsimony, 'that the question in debate is whether causes of the Lamarckian order *are* necessary to explain all the phenomena of organic nature. Of course if it could be proved that the theory of natural selection alone is competent to explain all these phenomena, appeal to the logical principle in question would be justifiable. But this is precisely the point which the followers of Darwin refuse to accept; and so long as it remains the very point at issue, it is a mere begging the question to represent that a class of causes which have hitherto been regarded as necessary are, in fact, unnecessary.'[31]

Since Weismann's rejection of the inheritance of acquired characters was a *petitio principii*, the deductions from that premise could not be upheld. Romanes devoted an entire volume, *An Examination of Weismannism* (1893), to dismantling the unwieldy superstructure which Weismann had erected in the decade since his essay, 'On heredity', had appeared: a full-blown theory of organic evolution, with natural selection as its sole factor in all multi-cellular forms of life, and an elaborate theory of the composition of the germ plasm, replete with unknown factors and fictions. It was a relatively simple task, for all depended on the perpetual continuity and absolute stability of the germ plasm. Remove the words 'perpetual' and 'absolute' by reopening the question of the inheritance of acquired characters, and nothing remained of Weismann's speculations save a theory of heredity which was, Romanes announced, 'indistinguishable from that of Galton'. And as Darwin's views obtained an 'immeasurable advantage' over Wallace's deductive elimination of the problems connected with the existence of apparently useless structures, so Galton's theory of the stirp, a modification of Darwin's hypothesis of pangenesis, turned out to be the corrective to Weismann's deductive elimination of the Lamarckian factor in evolution. 'The truly scientific attitude of mind with regard to the problem of heredity', Romanes concluded, 'is to say, as Galton says, "that we might almost reserve our belief that the structural [i.e. somatic] cells can react on the sexual elements at all, and we may be confident that at most they do so in a very faint degree; in other words, that acquired modifications are barely, if at all, *inherited*, in the correct sense of the word" '.[32]

The problem of human evolution

The conflicts and agreements among Darwin, his followers, and his Neo-Darwinian successors were still more various than those we have reviewed. Galton became increasingly enamoured of eugenic ideas, advocating selective social policies which Darwin, who knew 'far too little about the laws of inheritance to argue about them', regarded as somewhat dubious and 'utopian'. Romanes eventually stated 'the whole theory of organic evolution', including natural selection, in terms of the causes and conditions of isolation; this led Huxley, for one, to doubt whether he thoroughly understood the *Origin of Species*. Weismann, on the other hand, before his latter divagations, received Darwin's praise for supporting natural selection over against an 'innate tendency to perfectibility' in his *Studies in the Theory of Descent* (1882). In 1893, after Weismann had fully set forth his speculations, his critic Romanes nevertheless honoured him with an invitation, which he accepted (Spencer having declined), to deliver the third annual Romanes Lecture at Oxford.[33]

Of all the anomalies in post-Darwinian evolutionary thought, whether anomalies of dissent or accord, the most outstanding by far was Wallace's divergence from Darwin over the question of human evolution. The co-discoverer of the theory of natural selection, the one who admitted to being 'more Darwinian than Darwin himself', could not subscribe to one of Darwin's cardinal beliefs: that 'the difference in mind between man and the higher animals, great as it is, certainly is one of degree and not of kind'.[34]

In 1864 Wallace became the first publicly to apply the theory of natural selection to human evolution. Lyell's *Antiquity of Man* and Huxley's *Man's Place in Nature* had touched on the subject in the previous year, but these works dealt more with fossils and physiology than with the causes of organic change. Wallace, in a paper read before the Anthropological Society of London, described how the causes comprised in natural selection bear upon 'the origin of human races and the antiquity of man'. His leading idea was that the physical differences separating mankind from other animals, together with the external characters which distinguish the several human races, must have been produced before the appearance of mankind's distinctive mental faculties. For by means of a 'superior intellect' and 'superior sympathetic and moral feelings' mankind overcame the physical effects of natural selection at 'a comparatively remote geological epoch'; and since that time, as other animals have undergone generic and higher physical modifications, mankind has undergone an equivalent evolution of the head and brain. From this, Wallace concluded, 'it must inevitably follow that the higher – the more intellectual and moral – must displace the lower and more degraded races; and the power of "natural selection", still acting on his mental organisation, must ever lead to the more perfect adaptation of man's higher faculties to the conditions of surrounding nature, and to the exigencies of the social state,...till the world is again inhabited by a single nearly homogeneous race, no individual of which will be inferior to the noblest specimens of existing humanity'.[35]

Darwin thought highly of Wallace's paper and drew on it in writing the *Descent of Man.* However, he did not refer to the text which Wallace reprinted with alterations six years later in his *Contributions to the Theory of Natural Selection.* There the initial advance of human mental development is ascribed to 'some unknown cause', while the present and future advances result from 'the inherent progressive power of those glorious qualities which raise us so immeasurably above our fellow animals, and...afford us the surest proof that there are other and higher intelligences than ourselves, from whom these qualities may have been derived, and towards whom we may ever be tending'.[36] Between 1864 and 1870 Wallace had been 'beaten' by the 'facts' of spiritualism. Rappings, table-turnings, and the like had caused him to doubt whether natural selection could originate the part of mankind that made these things possible, and whether, in fact, it could secure the future progress of the human race. Furthermore, the phenomena had set him to refurbishing the universe with a hierarchy of purposeful spirits, conceived as efficient causes, grading upwards from the material, and all presided over by a 'Great Mind' or 'Supreme Intelligence'. Since neither the phenomena nor his philosophy of nature, as it turned out, seemed compelling to fellow-naturalists, Wallace sought other grounds for showing that natural selection could not produce the distinctive features of mankind.[37]

In 1869 Wallace first disclosed his evidence in a review of new editions of Lyell's *Principles* and *Elements of Geology*, a review which first caused Darwin to realise how 'grievously' he differed from his colleague.[38] Wallace pointed out that primitive people possessed attributes that were either useless or plainly harmful, yet prerequisites for life in a civilised state: an over-sized brain, an erect posture, a well-formed hand, exquisite physical characters, naked and vulnerable skin, and highly developed organs of speech. A year later, in the concluding essay of his *Contributions*, Wallace augmented this evidence by detailing the significance of the human brain. The brains of primitive people were enlarged by the presence of intellectual and moral faculties for which they had little or no need: powers of 'abstract reasoning' and 'ideal conception', capacities for 'pure morality' and 'refined emotion', all of which, however, are necessary for the 'full development of human nature' and the 'rapid progress of civilisation'.[39] Wallace by this time had been a convert to phrenology for over twenty years. The faculties to which he referred were irreducible units, correlated with the size and shape of the brain as manifested by the external form of the cranium. They could not have been acquired by accretion, but only suddenly, perhaps through 'some unknown cause'.[40] Yet Wallace chose rather to stress that these faculties, together with the other distinctive features of mankind, being at first useless or harmful, were acquired prospectively, and that some 'unknown higher law' had therefore determined the course of human evolution. 'A superior intelligence', he inferred, 'has guided the development of man in a definite direction, and for a special purpose, just as man guides the development of many vegetable and animal forms'. The 'intervention of some distinct individual intelligence' has aided in 'the production of ...the ultimate aim and outcome of all organised existence – intellectual, ever-advancing, spiritual man'.[41]

Although Wallace never ceased contending for spiritualism and the spiritualistic origin of the human mind, he did wait until after Darwin's death to contrast his own views with those of the *Descent of Man*. This occurred in 1889, in the last and most anti-Darwinian chapter of his *Darwinism*. Denying that Darwin had proved natural selection to be the cause of a continuous evolution of mental faculties, Wallace summarised his earlier arguments and invoked 'some other influence, law, or agency', some 'unknown cause or power' of a 'higher order than those of the material universe', which 'superadded to the animal nature of man' a 'spiritual essence or nature' capable of 'indefinite life and perfectibility'. Three times such a cause had 'come into action' in the history of the world, first bringing forth organic life, then sensation and consciousness, and finally the distinctive human faculties – all, according to Wallace, probably through 'different degrees of spiritual influx', all securing the progressive realisation of a spiritual ideal. To those who accept the existence of 'the unseen universe of Spirit', he explained, 'the whole purpose, the only *raison d'être* of the world, . . .was the development of the human spirit in association with the human body'.[42]

In treating of human evolution *Darwinism* disclaimed its title. As Wallace later frankly admitted, 'My view. . .was, and is, that there is a difference in kind, intellectually and morally, between man and other animals'.[43] Thus it hardly seems surprising that the spiritualist pretender to 'pure Darwinism' was soon reproached by the heir apparent to Darwin's unpublished notes and manuscripts on psychological subjects. That heir was of course Romanes, who stated in his review of *Darwinism* that its concluding chapter seemed 'sadly like the feet of clay in a figure of iron'. Yet he refrained from discussing Wallace's views on human evolution because, for one reason, as he subsequently explained, he had elsewhere dealt with the subject at considerable length. And so indeed he had, for in *Animal Intelligence* (1882), *Mental Evolution in Animals* (1883), and *Mental Evolution in Man* (1888) Romanes had ably carried forth his master's work, pioneering the study of comparative and developmental psychology on the basis of the materials which Darwin had placed at his disposal. The books were unified by a single over-arching purpose: to vindicate Darwin's judgement that the minds of animals and mankind, like their bodies, had a common origin, and thus that mankind's 'own living nature is identical in kind with the nature of all other life'.[44]

In the first two volumes Romanes presented a mass of evidence, largely anecdotal, to show that mental traits, including those thought to be distinctively human, have developed by accretion. All mankind's emotional faculties except the ones related to 'religion, moral sense, and perception of the sublime' – the subjects of another volume on mental evolution which Romanes did not complete – appear elsewhere in the animal kingdom. Likewise the same instincts are found in animals and mankind, with instinct shading off into reason among the higher animals. Volition is everywhere identical in kind, though human volitions surpass all others in complexity, refinement, and foresight. The intellectual faculties exhibit enormous differences as well, yet, according to Romanes, there is a correspondence between psychogeny in the human individual and the mental evolution of beasts. The correspondence is a close one up to the point where human intelligence, building on the foundation of inherited animal faculties, begins to outstrip animal intelligence by virtue of its unique capacity for conceptual thought.

The origin of human 'faculty' in this larger sense, as beginning with distinctively human ideation, is the concern of *Mental Evolution in Man*. Romanes did not linger over Wallace's views on the subject save at the outset of the book, pointing out somewhat cheaply that these views, though coincident with Mivart's, were held on exactly opposite grounds – primitive people being slightly, rather than incomparably, more intelligent than the higher apes – and therefore that the views of the 'two leading dissenters' from Darwin were mutually cancelling.[45] However, in dealing with the familiar objection that the human consciousness alone possesses a superadded immortality, Romanes furnished a revealing rejoinder to Wallace's spiritualistic philosophy. Concerning immortality, he declared, psychology has nothing to say.

From the nature of the case, any information of a positive kind relating to these matters can only be expected to come by way of a Revelation; and, therefore, however widely dogma and science may differ on other points, they are at least agreed upon this one – namely, if the conscious life of man differs thus from the conscious life of the brutes, Christianity and Philosophy alike proclaim that only by a Gospel could its endowment of immortality have been brought to light.[46]

Darwin, as we have seen, dealt with the same objection by recalling that an analogous problem exists in accounting for the origin of immortality within the life of the nascent individual. This was but one instance in which he used the analogy of ontogenetic growth to illustrate the continuity of phylogenetic history. Romanes, for his part, took the analogy as his leading idea in *Mental Evolution in Man*, thereby opposing all who, like Wallace, sought the origin of human faculty in some additional, preternatural cause.

'In the childhood of the world, no less than in that of the man', Romanes maintained, 'we may see the fundamental change from sense to thought'. Even as historical and comparative philology traces the history of intelligence in the evolution of the human race, marking the gradations to that self-conscious and conceptual thought which constitutes 'the whole distinction between man and brute', so linguistic psychology discovers a parallel history in the mind of the developing child. And if, in the latter case, there can be no question of a difference

of origin, or 'kind', for the distinctively human faculties, neither can
the question arise in the case of mankind's mental evolution from
animals.[47] 'In the growing intelligence of a child', Romanes concluded,

we have. . .as complete a history of 'ontogeny', in its relation to 'phylogeny', as that upon which the embryologist is accustomed to rely when he
reads the morphological history of a species in the epitome which is
furnished by the development of an individual. . . . Those are without
excuse who, elsewhere adopting the principles of evolution, have gratui-
tously ignored the direct evidence of psychological transmutation which
is thus furnished by the life-history of every individual human being.[48]

Romanes thought Wallace was also 'without excuse' for his
'deplorable weakness as a "philosopher"', and for this reason as well
he refrained from discussing human evolution in his review of
Darwinism. His respect for Wallace as a naturalist, according to a
subsequent article, had prevented him from dealing with 'the Wallace
of spiritualism and astrology, the Wallace of vaccination and the land
question, the Wallace of incapacity and absurdity'.[49] In thus referring
abusively to some of Wallace's recent preoccupations (and falsely to
his belief in astrology), Romanes somewhat compromised his claim to
represent the mind of Darwin. For although Darwin was sceptical
about spiritualism and fearful that Wallace might 'turn renegade to
natural history', he would never have expressed himself so harshly.[50]
Romanes, moreover, not only failed to treat his master's old friend
with respect; he also failed to be entirely candid about his own misgivings towards spiritualism. In 1876, almost coincident with his loss of
faith, Romanes began to investigate psychical phenomena. He shared
his first experiences in detailed, confidential letters to Darwin, obtain-
ing sceptical encouragement in reply.[51] Later, in 1880, after publishing
an anonymous query in *Nature*, Romanes had a brief correspondence
and two guarded interviews with Wallace, whom he informed that
'any definite evidence of mind unassociated with any observable
organization. . .would be to me nothing less than a revelation – "life
and immortality brought to light"'. Little in fact was revealed to
Romanes except Wallace's overdeveloped 'faculty of deglutition', as
exemplified in his refusal to dismiss astrology out-of-hand.[52]
    There matters stood for a decade, Romanes writing about mental
evolution as if spiritualism were untrue, yet unable – as he was never
able – to 'assure himself that there was absolutely nothing' in the
phenomena.[53] Meanwhile, however, Wallace happened on copies of

the original Romanes–Darwin correspondence. He took careful note of all that Romanes had concealed from him at the time of their own exchanges, thinking that some day the information might prove of use. When Romanes ventured to accuse him of 'incapacity and absurdity' the day of reckoning had arrived. Deeply incensed, Wallace told his critic in a private letter that 'there is a Romanes "*of incapacity and absurdity*!!" but he keeps it secret. He thinks no one knows it. He is ashamed to confess it to his fellow-naturalists; but he is *not* ashamed to make use of the ignorant prejudice against belief in [spiritualistic] phenomena, in a scientific discussion with one who has the courage of his opinions, which he himself has not.'[54]

A brief and argumentative correspondence ensued. The result was a stalemate so far as spiritualism was concerned. But for Romanes and Wallace personally the outcome was a waymark in the history of post-Darwinian evolutionary thought. A year later, when Romanes requested that most Victorian courtesy, a photograph, for use in his *Darwin, and after Darwin*, Wallace declined with thanks.[55] Like Darwin and the Neo-Lamarckian Hyatt ten years before, like Spencer and the Neo-Darwinian Weismann a few years thence, the controversialists went their separate ways.

PART III

THEOLOGY AND EVOLUTION

9

CHRISTIAN ANTI-DARWINISM:
THE REALM OF CERTAINTY AND FIXITY

The vivid and popular features of the anti-Darwinian row tended to leave the impression that the issue was between science on one side and theology on the other. Such was not the case. . . . Although the ideas that rose up like armed men against Darwinism owed their intensity to religious associations, their origin and meaning are to be sought in science and philosophy, not in religion.

John Dewey[1]

The military metaphor has taken its heaviest toll among Christian anti-Darwinians. With visions of polarisation, organisation, and antagonism filling their minds, commentators on the post-Darwinian controversies have made Darwin's religious opposition appear as hostile as possible, sometimes by emphasising the incoherent polemical utterances of vulgar writers while neglecting the responses of more competent critics, more often by overlooking the philosophical and scientific objections implicit in the vulgar responses and explicit in the competent ones. To take but a single example: in a recent study of the impact of evolutionary naturalism on American thought there is a chapter entitled 'The warfare of science and religion'. This chapter surveys the rhetoric of twenty-one anti-Darwinian writers, among them three who figure in the following pages, Enoch Fitch Burr, Charles Hodge, and John William Dawson. And what becomes of these prominent individuals? Burr is twice enlisted for a colourful quotation to match the lurid remarks of the popular preacher T. De Witt Talmage (who is cited three times), thereby illuminating the chapter's title. Hodge appears, not as the author of a trenchant theological analysis of Darwin's theory, but as an obscurantistic bibliolater who simply equated Darwinism with atheism. Dawson is mistaken for his son George Mercer, who was also a geologist, and neither his scientific stature nor his discerning approach to Darwinism is mentioned. The author does point out that natural selection was not always understood and that the 'factual gaps' in Darwin's theory were much exploited, but this is no substitute for dealing sympathetically with the beliefs that brought Christians into conflict with Darwin.[2]

In this chapter we accord Darwin's opponents such consideration. If, as Andrew Dickson White observed, theologians 'swarmed forth angry and confused',[3] our concern will be, not to escalate their inflamed attitudes and incoherent arguments into a 'war', but to excavate the anger and confusion which have served to conceal the philosophical foundations of Christian Anti-Darwinism.

The quest for certainty

In 1859 Francis Bacon had been the patron saint of students of nature for more than two centuries, his induction the method to be followed nominally, if not sedulously, in every investigation. Bacon taught in his *Novum Organum* (1620) that it would be possible to control and modify the forces of nature by gaining true and certain knowledge of the basic physical properties of natural objects. Such knowledge, he believed, could be gradually acquired by collecting data widely and systematically, without preconceptions; by classifying the data according to their similarities and differences; by eliminating from the classification those data with attributes which are accidentally correlated; and, finally, by employing the data with essential correlations as the basis of further inductive generalisation. The outcome of repeated inductions would be a series of propositions, decreasing in number, increasing in generality, and culminating in 'those laws and determinations of absolute actuality' which can be known to be certainly true.[4]

Bacon's inductive method was enlarged and improved in the first half of the nineteenth century by Britain's leading philosophers of science. John Herschel in his *Preliminary Discourse on the Study of Natural Philosophy* (1830), William Whewell in his *Philosophy of the Inductive Sciences* (1840), and John Stuart Mill in his *System of Logic* (1843) argued a greater role for deduction. They held that scientists, while practising a rigorous inductive method, may also devise explanatory generalisations, or hypotheses, from which particular consequences can be deduced and then tested against the facts. But none of these philosophers abandoned the quest which Bacon had taken up from Aristotle and classical philosophy: the quest for ultimate certainty in scientific inferences. Herschel and Mill looked for proof of hypotheses in increasingly thorough inductions. Mere probability was of little account, only a half-way house on the path to complete factual confirmation. 'What, in the actual state of science, is...important for us to know', said Herschel, 'is whether our theory truly represent *all* the facts, and include *all* the laws to which observation and induction lead'. Verification of laws, he believed, would come through 'comparison with *all* the particulars'. For Mill hypotheses were legitimate tools of discovery 'on one supposition, namely, if the nature of the case be such that the final step, the verification, shall amount to, and fulfil, the conditions of, a complete induction'. Whewell, who belonged to a different philosophical tradition, also sought proof through induction: that is, through the superinduction upon relevant facts of an ideal 'conception'. If the conception is 'clear and appropriate' and embraces all the facts then, according to Whewell, it can be regarded as a law of nature and a necessary truth. Fortified thus by the most influential contemporary philosophers of science, the standard of proof in scientific explanation remained, in Darwin's day, one of full and final certainty.[5]

The *Origin of Species* represented a new departure in scientific explanation. Facts swarmed its pages in orthodox Baconian proportions. Hypotheses such as its author could not resist forming 'on every subject' governed the choice of facts. Yet the book set forth natural selection, not as a theory for which absolute proof had been obtained, or even might be obtained, but merely as the most probable explanation of the greatest number of facts relating to the origin of species. 'Any one whose disposition leads him to attach more weight to unexplained difficulties than to the explanation of a certain number of facts', wrote Darwin at the end of the book, 'will certainly reject my theory'.[6]

The lesson was a hard one but Darwin had to learn it: a thorough induction could yield at best only probable results and, at worst, egregious error. Following the procedure for induction by elimination as adopted from Bacon by Herschel and Whewell, Darwin argued in his first full-length scientific paper, published in 1839, that the 'parallel roads' of Glen Roy must have been produced by the action of the sea at different stages in the elevation of the Highland region. For he reasoned that if the enigmatic terraces had been created by the action of lake water at different elevations, it would be impossible to account for the erection and removal at the inlets to the Glen of the great rock barriers required to change the level of the lake. But to his chagrin he was forced to admit in 1861 that the ice of Louis Agassiz's glacier theory, as improved by James Thomson and Thomas Jamieson, provided the very dams required by the lake hypothesis he had eliminated. The lesson was plain: induction, no matter how rigorous, could never rule out the possibility of alternative explanations.[7] An acceptable explanation in natural history is not the one shown by induction to be fully and finally certain – that is inconceivable – but merely the one which accounts for the greatest number of facts at a given time. 'The line of argument often pursued throughout my theory', Darwin noted in 1838, 'is to establish a point as a probability by induction, & to apply it as hypothesis to other points, & see whether it will solve them'. More than twenty years later he commended F. W. Hutton as 'one of the very few who see that the change of species cannot be directly proved, and that the doctrine must sink or swim according as it groups and explains phenomena'. 'It is really curious', Darwin added, 'how few judge it in this way, which is clearly the right way'.[8]

For Darwin's opponents it was clearly the wrong way to judge his theory. Ellegård has shown that before 1872 Darwin's idea of scientific explanation was one of the most offensive aspects of his work. To the chant of Newton's *hypotheses non fingo* popular writers invoked the safe, sure, traditional, and inductive spirit of Bacon. 'Induction was the good positive ideal which Darwin had deserted for its bad negative counterpart, variously named hypothesis, theory, speculation, assumption, imagination, fancy, guess, and the like.' More competent critics reflected the opinions of contemporary philosophers of science. They 'fully admitted that hypotheses were necessary tools. But hypotheses had to be proved: and it was very common to insist that as long as Darwin had not proved his theory, the traditional view [of creation] should stand.'[9] After 1872 the same objections prevailed, though often in a state of confusion and seldom in a work of scientific stature.

Among the many clerical anti-Darwinians in Great Britain 'there was probably not one who wrote at greater length, more outspokenly, vehemently, and decidedly' than the ornithologist and rector of Nunburnholme in the East Riding of Yorkshire, FRANCIS ORPEN MORRIS (1810–1893).[10] Beginning with papers read before the British Association at Norwich in 1868 and Exeter in 1869, Morris crusaded uninhibitedly for more than twenty years to eradicate Darwinism from English intellectual life. His pamphlets – with titles such as *Difficulties of Darwinism* (1869), *All the Articles of the Darwin Faith* (1875), and *The Demands of Darwinism on Credulity* (1890) –

appeared on the railway bookstalls and they appealed with increasing rhetorical violence to 'the common sense of the people of England'. In phrases which provoked no less than Darwin himself, Morris expressed his sentiments with wonted vigour:

'Ineffable contempt and indignation' is the only feeling which any person of common sense and of a right mind must feel at the astounding puerilities of Darwinism, its ten thousand times worse than childish absurdities; contempt for them in themselves, and indignation at the criminal injury the miserable infidelity of the wretched system has done to the minds of too many. If the whole of the English language could be condensed into one word, it would not suffice to express the utter contempt those invite who are so deluded as to be disciples of such an imposture as Darwinism.[11]

Although Morris' reverence for the Bible (on which he declined to base his arguments) and his fervid anti-vivisectionism may have accounted for the depth and duration of his opposition, its 'main exciting cause', according to his son, was the Darwinian 'process of reasoning'.[12] Darwin's 'one great mistake' was his acceptance of the *non sequitur* that because species vary they may also descend one from another. The *Origin of Species* was a 'valuable collection of interesting facts', Morris conceded, but he had never read 'a more inconclusive, illogical book' until he took up *The Variation of Animals and Plants under Domestication*. By supposing that variation, which can be observed, may lead to transmutation, which has never been proved, Darwin had 'pointed out a path for others to wander on in into the most hopeless chaos of thought, in a "confusion worse confounded" than even his own'.[13]

Another clergyman–scientist, ENOCH FITCH BURR (1818–1907), was 'one of the most influential of all...writers against evolution' in the United States.[14] While serving as minister of the Congregational church in Hamburg, Connecticut, Burr distinguished himself in physical astronomy and higher mathematics, thereby earning an appointment in 1868 as lecturer in the scientific evidences of religion in Amherst College. His lectures before college audiences at Amherst and elsewhere were published in two series as *Pater Mundi* (1869, 1873) and in 1883 Burr brought out *Ecce Terra* to complete the natural theology he had begun sixteen years earlier in *Ecce Coelum*. By this time, however, his appointment at Amherst had long ceased. In 1870 the chair of natural theology and geology, lately occupied by the anti-Darwinian geologist Edward Hitchcock, was filled by Benjamin

K. Emerson, who had recently completed doctoral studies in Germany with Ernst Haeckel.[15] Emerson, on being consulted by the president of the College about continuing Burr's lectures, advised, 'If you want to make all the boys evolutionists, keep him on.' Thereupon, in 1874, the lectures were terminated.[16]

Burr, unlike Morris, did not conduct a vendetta against Darwinism but against the larger philosophy which he thought it implied. Evolution, he declared, teaches that 'all things we perceive, including what are called spiritual phenomena, have come from the simplest beginnings, solely by means of such forces and laws as belong to matter'. This 'law hypothesis', 'law scheme', or 'law dream' was therefore, according to Burr, 'not merely the most noted, plausible, influential, and violent enemy of Theism in our day, but what is its only possible enemy for all ages to come'.[17] The 'Theistic hypothesis', on the other hand, seemed a superior 'explanation of Nature' because it was the 'simplest' and 'surest' one possible. On the supposition of creation by 'an eternal Being with power and wisdom indefinitely greater than the human', all natural wonders are immediately and elegantly explained. On the supposition of evolution, however, the nebular hypothesis, spontaneous generation, and the transmutation of species all have to be proved.[18] Indeed, Burr maintained that

the possibility of all this needs not only to be shown, but to be shown to a demonstration; since all the assumptions of the rival hypothesis are possible to an absolute certainty. It is self-evident that there is some eternal substance, and that an eternal power is, in the nature of things, just as possible as eternal matter – self-evident that there is nothing in the nature of things to limit an eternal intelligence to a given breadth of knowledge and power – self-evident that there is nothing to prevent that intelligence from being as much greater than man in these respects as man is greater than a worm. Thus is the Theistic Hypothesis. So everything in the rival hypothesis must be put on a basis of absolute certainty.[19]

The last of the clerical anti-Darwinians in America was LUTHER TRACY TOWNSEND (1838–1922). Old enough to have mature memories of the publication of the *Origin of Species*, he was also young enough to live through the post-Darwinian controversies and on into the Fundamentalist period. Townsend graduated from Dartmouth College in 1859, studied at the Congregational seminary in Andover, Massachusetts, and served in the Union army during the Civil War. Thereafter he pastored Methodist churches for several years and, in 1868, took up an appointment as professor of biblical languages in Boston

Theological Seminary (later Boston University School of Theology). He taught there until 1893, assuming in turn the professorships of church history and practical theology, and all the while he built a reputation as a popular apologist for traditional evangelical beliefs. At college Townsend had passed through two years of intense 'personal scepticism'. His return to 'the faith of early boyhood' came about through the study of Christian evidences and the harmonising of science and scripture.[20] This was a salutary experience for a Christian educator in the later nineteenth century, and Townsend made much of it in his numerous apologetic works, among them *The Mosaic Record and Modern Science* (1881), *Bible Theology and Modern Thought* (1883), and his most substantial book, *Evolution or Creation* (1896).

In *Bible Theology* Townsend recommended that scientific theories should be held as mere 'working hypotheses'. These, in turn, he defined so narrowly as practically to exclude theoretical innovation. To 'possess weight', for example, a working hypothesis 'must receive the assent of all, or nearly all, who are capable of investigating the subject'. Tested by this rule, the 'extreme Darwinism' of Huxley and the hypotheses of spontaneous generation and 'atheistic evolution by natural selection' are of very little consequence, for the vast multitude of scientists reject them. On the other hand, 'Darwinism, as held by its author, is a working hypothesis', Townsend granted, leaving the issue of its 'weight' unresolved.[21] By the time *Evolution or Creation* came to be written, however, the consensus of scientists was hardening against Darwinism in any form. Seizing the opportunity, Townsend applied his rule. When 'scores of men eminent in the field of natural science do not see any reason for giving their assent to the hypothesis ...of evolution as announced by Mr Darwin and defended by Mr Spencer and others,...then we must conclude that for the present such hypotheses should have no weight except that accorded to other very questionable speculations'. The evolutionary hypothesis was, after all, merely 'a revival and enlargement of views entertained by philosophers and church fathers, skeptics and scientists, during the last twenty centuries'. Now as ever, according to Townsend, it is 'not supported as a whole or in any of its parts by a single well-established fact in the whole domain of science and philosophy'.[22]

The physician CHARLES ROBERT BREE (1811–1886) was one of Darwin's more philosophically-minded critics in Great Britain. No sooner did the *Origin of Species* and the *Descent of Man* appear than Bree subjected each in turn to a book-length refutation. *Species Not Transmutable nor the Result of Secondary Causes* was published in 1860, and in 1872, while Bree served as senior physician to the Essex and Colchester Hospital, he delivered himself of *An Exposition of Fallacies in the Hypothesis of Mr Darwin*. Darwin told J. D. Hooker that he 'need not attempt' the first book and complained to his old professor, Henslow, that its author 'had not the soul of a gentleman in him'.[23] Wallace reviewed the second one in *Nature*, eliciting from Bree a reply which accused him of 'blundering' in his rendition of Darwin's views on mankind's early pedigree. Darwin came to Wallace's defence in the next issue of the magazine. He stated that Bree's letter was 'unintelligible' to him and suggested that no one who had read an earlier work by Bree on the same subject 'will be surprised at any amount of misunderstanding on his part'.[24]

Darwin might as well have been referring to Bree's *Exposition of Fallacies*. Most of the fallacies Bree undertook to expose were only reflexions of his own fractured understanding of Darwinism. Concerning the nature of proof, however, there was certainly the consistent discrepancy between his science and Darwin's which he sought. The proofs of fixity and design were, to Bree, as numerous and absolute as his present observations of stable organic forms. The proof of Darwinism, on the other hand, could never be obtained. Reasons for this appear *ad nauseam* throughout his book, not as biblically based objections – Bree thought that 'the scriptures are not scientific authorities, nor ever were intended to be' – but chiefly as instances of Darwin's 'false mode of reasoning', his assumption of unproven facts, his attempt to see how much his theory could explain before arraying all the facts, present and prehistoric, that are required to support it. Again and again Bree rebuked Darwin for 'imagination', for believing something 'without a shadow of evidence'. 'Where. . .are Mr Darwin's variations? Where his causes? Where his proof of transmission?' he demanded in a chapter on mental evolution. Pangenesis is an 'enormous guess', he objected, for it is insusceptible of proof in the present state of science. Human descent from animals is no less a speculation, for all the intermediate forms are missing. 'Mr Darwin's mind. . .is warped by the necessity of considering everything in human structure as the product of a theory which has never been proved', a theory which is but 'a cold, unsound, unphilosophic, degrading system of assumed probabilities'.[25]

Britain's foremost evangelical anti-Darwinian was THOMAS RAWSON BIRKS (1810–1883). Raised a Nonconformist and educated at Cambridge, Birks took holy orders on graduation and assumed the curacy at Watton in Hertfordshire under the 'most colourful and godly of the evangelical clergy', Edward Bickersteth.[26] He married the vicar's daughter in 1844, joined with his father-in-law in founding the Evangelical Alliance two years later, and in 1866, after more than twenty years as rector of Kelshall in Hertfordshire, sealed his identity with evangelicalism by accepting the charge of Trinity Church, Cambridge, where Charles Simeon had held forth for more than half a century. In 1872, with his appointment to the Knightbridge Professorship of Moral Philosophy at Cambridge, Birks found his churchmanship at the centre of a small controversy. The followers of F. D. Maurice, who had occupied the chair until his death in the same year, protested that the appointment of an out-and-out evangelical was a setback for the cause of liberal religion. For his part, Birks was pleased to oblige the opposition by publishing two exceedingly illiberal books on evolution during his tenure, *The Scripture Doctrine of Creation* (1872) and *Modern Physical Fatalism and the Doctrine of Evolution* (1876).[27]

As befitted a professor of philosophy, Birks preoccupied himself in both books with Herbert Spencer. As befitted an evangelical professor of philosophy, Birks stressed that the belief in special creation is the product of the testimony of scripture together with '*a priori* reasonings'. Spencer and Darwin had not only contravened the plain teaching of Genesis, but had ignored, evaded, or deliberately transgressed these reasonings, and so were doubly liable to criticism.[28] In his first book Birks devoted a chapter to exposing the evolutionists' violation of the laws of inductive and deductive inference, principally, by neglect of 'direct evidence' – the Bible, its claim to be the Word of God, and its revelation of prehistory – and by failing to hold 'the most comprehensive lesson from the great body of evidence', the 'one main inference' of natural history, and the induction than which none is more firmly established: that like produces like, with only minor variations. 'The larger our knowledge of existing species, and the wider the range of time over which our discoveries extend,' he declared, 'the more complete is the refutation' of the theory of transmutation.[29] In *Modern Physical Fatalism* Birks indulged in the conceptual realism of Whewell, his predecessor in the chair of moral philosophy. He distinguished sharply between physical and vital phenomena, though allowing that in living matter the laws of physical nature are not 'suspended' because a 'different and higher law mingles with them'.[30] When Darwin submitted natural selection as a *vera causa*, the 'law' it expressed as nothing more than 'the sequence of events as ascertained by us', the opponent of 'uncaused, uncreated' laws of evolution objected that natural selection is not a 'proved result of induction from all the known facts', that it 'consists in the free invention of conjectural ante- cedents, through millions of years or ages, before man was born, or experiment and observations could be made' – hardly a 'clear and appropriate' conception. 'The excuse offered for building a theory of the universe on a metaphor, that it is hard to avoid personifying, is very worthless', Birks stated, for it only proves that 'a deep instinct of the mind, if it be violently repressed by falsely pretended science, will assert itself in some other way'.[31]

In the United States it was neither a physician nor a philosopher but an eminent lawyer, GEORGE TICKNOR CURTIS (1812–1894), who undertook an extended philosophical critique of evolution. Indeed, of all the legal anti-Darwinians in the century past, none is more illus- trious and none more variously accomplished.[32] As a lawyer Curtis often practised before the United States' Supreme Court, where in 1858 he argued the defence in the famous case of the fugitive slave Dred Scott. As a patent attorney he obtained a high reputation, serving clients such as Goodyear, Morse, and Cyrus McCormick. Even as a man of letters Curtis was also well known. Three biographies and a novel, in addition to his extensive legal writings, won him a large and appreciative audience in the later nineteenth century. In 1887, at the end of his professional career, Curtis appealed to this popular jury – 'that great mass of people of average intelligence' – to settle a notorious and outstanding case. He published his arguments as a 'philosophical inquiry' entitled *Creation or Evolution?*[33]

The book was intended to be a 'philosophical inquiry' in the strictest sense. Issues of dogmatic theology would not be allowed to intrude. Curtis declined to discuss Genesis and its interpretations, claiming in fact to be unconscious that his work had been influenced by his belief in revealed religion. The question of God's existence as deducible from the phenomena of nature, and its relation to the hypo- theses of evolution and special creation, could be answered in accord- ance with 'the principles of belief which we apply in the ordinary affairs of life'.[34] Accordingly, Curtis began his enquiry by setting forth its rules of evidence: first, that every fact in a valid argument must be independently proved, not inferred from another; and second, that all the facts must be placed in the proper relation to each other if a just inference is to be drawn.[35] In the latter two-thirds of the book he would apply these criteria to Spencer's evolutionary philosophy, but the primary task was to employ them in testing the distinctive theory of Darwin. Having shown that the theory violates the second rule of evidence – the arrangement of facts in support of mental evolution is equally compatible with the Platonic theory of the transmigration of souls as set forth in the *Timaeus* – Curtis applied his first rule to the facts of paleontology. If Darwin's is a 'true account' of human evolution, he stated, 'we ought practically to find no missing links in the chain' of life which culminates in mankind. The chain 'forms a chain of evidence; and, according to the rational rules of evidence, each distinct fact ought to be proved to have existed at some time before our belief in the main hypothesis can be challenged'. Of course the proofs did not exist, as Darwin himself acknowledged, and Curtis was much impressed with 'the inconsequential character of the reasoning, and the amount of assumption which marks the whole argument' for human evolution by natural selection.[36] 'I need not say', he summed up, 'that this kind of argument will not do in the common affairs of life, . . .and no good reason can be shown why our beliefs in matters of science should be made to depend upon it'.[37]

Among Darwin's more discerning critics on both sides of the Atlantic was the great and learned CHARLES HODGE (1797–1878). For half a century, from the chairs of exegetical, didactic, and polemical theology in Princeton Theological Seminary, Hodge propagated the most powerful forces in behalf of conservative Christianity which then prevailed among the American churches. Three thousand former students carried forth his 'Princeton Theology', the Calvinism of the Westminster divines and François Turretin, grounded in the Scottish philosophy of common sense and the verbal inspiration and inerrancy of Holy Scripture. The *Biblical Repertory and Princeton Review*, founded by Hodge in 1825 and edited by him for over forty years, argued the case for the Princeton Theology and the cause of Old School Presbyterianism in scores of polemical articles, many from the pen of the redoubtable editor himself. Meanwhile, adding to his reputation as an authoritative teacher and a fearsome controversialist, Hodge published numerous scholarly books, among them his greatest work, a three-volume *Systematic Theology* (1872–3). Grounded in thorough biblical inductions and constructed by rigorous Calvinistic deductions, the book was a masterful attempt to 'adapt theology to the methodology of Newtonian science'. Its 2200 large octavo pages remain a fitting monument to the last great representative of Calvinist orthodoxy before the spread of the modern historical consciousness. In 1874, not long after the final volume of the *Systematic Theology* appeared, Hodge made his last book-length contribution to theological discussion under the title *What Is Darwinism?*[38]

Hodge's notorious answer to the question his title posed – 'It is atheism' – should not obscure the fact that this conclusion was reached only after a perceptive and even-tempered (albeit logic-chopping) analysis of the Darwinian theory and its theological implications. Darwin, according to Hodge, was not a speculative philosopher like Spencer but 'simply a naturalist, a careful and laborious observer; skillful in his descriptions and singularly candid in dealing with the difficulties in the way of his peculiar doctrine'.[39] However, where Darwin had failed, and failed irremediably, was in reasoning 'as to matters of theory' and not 'as to matters of fact'. To Hodge and his colleagues at Princeton the acquisition of truth in science and theology depended on strict adherence to the Baconian method. 'The Bible is to the theologian what nature is to the man of science', wrote Hodge in the introductory pages of his *Systematic Theology*. Each is the repository of sacred facts that reveal God's truth. 'The theologian must be guided by the same rules in the collection of facts, as govern the man of science.' One arranges the facts of scripture, the other the facts of nature, so that mankind may know the will of God. Darwin had argued in a manner which would be presumptuous in theology – 'if we suppose this and that, then it may have happened thus and so' – and he had made no pretence that his theory was either proved or provable. Therefore the atheism of Darwinism could be readily and rapidly dismissed.[40]

After the death of Louis Agassiz in 1873, the geologist JOHN WILLIAM DAWSON (1820–1899), principal of McGill University in Montreal, became the most distinguished anti-Darwinian naturalist in the English-speaking world. As a young man Dawson cut his teeth on Lyell's *Principles of Geology*, acquired an expert knowledge of the geology of Nova Scotia, his home province (where in 1842 and 1852 he geologised with Lyell himself), and studied at Edinburgh under Robert Jameson in the midst of the *Vestiges* controversy. He was a 'modified uniformitarian', averse to the hypothesis of development.[41]

In religion Dawson was a Scottish Presbyterian, well instructed at a denominational academy in the rigours of Hebrew exegesis. Between geology and Genesis stood Hugh Miller. Miller, a self-taught 'uniformitarian' geologist who replied so eloquently to *Vestiges* in his *Footprints of the Creator* (1847) and a Free Church Presbyterian who abetted the Disruption of 1843 in the Church of Scotland, figured largely in all Dawson's dozen or more works on the relations of religion and science, from *Archaia* (1860) through *The Story of the Earth and Man* (1873) to his definitive critique of Darwinism, *Modern Ideas of Evolution* (1890). One might even say that Dawson was Miller's true successor, a geologist in the same 'spirit and power'. Like Miller, he accepted it as a scientist's highest privilege and obligation to harmonise the works of God in creation with his words in scripture. 'If there was any possible way to preserve a literal reading, Dawson was determined to follow it; only if this were clearly impossible did he search for other means of accommodation.'[42]

Yet for Dawson, as for the other anti-Darwinians we have met, it was not primarily the Bible against which Darwin had offended, but against the methods and truths of established science. Darwin had reasoned 'as to possibilities, not by facts'; his theory was '*not a result of scientific induction* but a mere *hypothesis*, to account for facts not otherwise explicable except by the doctrine of creation'. In the *Story of the Earth and Man* Dawson declared that the Darwinian account of human evolution is an 'arbitrary arrangement of facts in accordance with a number of unproved hypotheses', a 'system destitute of any shadow of proof, and supported merely by vague analogies and figures of speech'. Similarly, in *Modern Ideas of Evolution* he insisted that the true be sifted from the hypothetical, that science be separated from 'speculation', that 'supposition' regarding the paleontological record have no part in 'scientific certainty'. Darwinism and other such 'crude and simple hypotheses', Dawson believed, could never achieve the certainty which attached to the doctrine, both scientific and scriptural, of the immutability of species.[43]

The belief in fixity

The conviction that ultimate certainty is the desirable and attainable product of inductive inference – that to be acceptable a theory has to be proved, and to be proved it has to explain all the facts – this conviction forms one of the philosophical premises that underlay the anger, confusion, and theological pettifogging of Christian Anti-Darwinism. Though of ancient and pre-Christian lineage, it was perfectly respectable in its day and supported by eminent philosophers of science. Another premise is closely related to it. Induction can produce ultimate certainty only in a world which contains a finite number of fixed natural 'kinds'. Neither a complete elimination of non-essential data from an enquiry nor a full enumeration of relevant data is possible if the data are liable to change. Once admit unrestricted or indefinite change in the facts of nature and their relations can only be expressed in theories which have more or less probability. Darwin submitted his theory as a probable explanation of organic diversity because its material basis was the unlimited variation of plants and animals. Anti-Darwinians could demand that the theory be made absolutely certain because they believed in the fixity of biological species.[44]

From Plato and Aristotle until Darwin the mainstream of western philosophers explained the orderliness and stability of the biological world by positing an immutable 'nature', 'form', or 'essence' for every organism that naturally breeds true. Platonic interpreters located each essence in the world of ideas, the fixed eternal realm on which the ephemera of the material world are modelled. Aristotle and his followers placed each essence within the individuals it characterised. From a scientific standpoint the Aristotelian doctrine could be regarded as an advance on the Platonic. For if essences do not exist apart from individuals, then, since individuals are understood by defining what is 'essential' to them, a comprehensive understanding of nature can be achieved by a complete classification of essences – 'species', as they came to be called – on the basis of the perceptible differences of their individual embodiments. The high-water mark of this Aristotelian philosophy of nature (though without Aristotle's metaphysical sophistication) was reached in the taxonomic labours of John Ray in the seventeenth century and Carolus Linnaeus in the eighteenth. Their theological legacy to the nineteenth century was a synthesis of the static species concept and a literal interpretation of the creation account in Genesis.[45]

With the nineteenth century, however, came a resurgence of Platonic essentialism. Earlier, it is true, the French naturalist Buffon suggested that Aristotelian essences might give way to a Neo-Platonism based on vertebrate homologies, but he stopped short of affirming unequivocally either the fixity or the mutability of species.[46] No such hesitation characterised the founder of vertebrate paleontology, Georges Cuvier

(1769–1832). In 1795 Cuvier started work at the Museum of Natural History in Paris, and, as David Hull has observed, 'there is nothing so well calculated to turn a man into a neoplatonist than to put him in a storeroom full of fossil remains and set him the task of reconstructing the original organisms. The idea that there are a series of basic plans of organization with numerous variations emerges quite forcibly.'[47] Cuvier divided the animal kingdom into four branches – vertebrates, articulates, molluscs, and radiates – each distinguished by its own general plan of structure. Within each branch every animal that breeds true is a distinct species and every species was specially created according to its own fixed structural plan, a teleological adaptation of the branch's general plan.[48] The strict conditions of existence peculiar to each species determine its persistence in nature: if an individual organism varies beyond the limits imposed by its sensitive and harmonious internal organisation, or if the environment exceeds these limits, then the organism ceases to exist. Cuvier regarded these generalisations as purely descriptive, the clear and incontrovertible results of the most scrupulous induction. He would give no quarter to the metaphysical speculations of his colleagues, Jean Baptiste Lamarck and Etienne Geoffroy Saint Hilaire, and of the German *Naturphilosophen*, Lorenz Oken and F. W. J. von Schelling. Thus it remained for Cuvier's most celebrated follower, a former student of Oken and Schelling, to make explicit the Neo-Platonism on which his master's generalisations were based.[49]

In the mind of Louis Agassiz (1807–1873) 'it was possible to formulate a synthesis of Cuvier's empiricism and metaphysics, supported by the intellectual impulses that had...made Oken and Schelling seem such attractive philosophers'. The great Swiss naturalist, who came to the United States in 1846, assumed the professorship of geology and zoology at Harvard in the following year, and thereafter became in himself a virtual institution of American science, never did 'discard a single item of his education: no matter how contradictory Oken and Geoffroy might seem in contrast to Cuvier, in Agassiz's mind the motivation of *Naturphilosophie* to view nature as a whole was substantiated by the techniques and presuppositions learned from the French naturalist'.[50] For Agassiz, as for Cuvier, the world was constructed by God on rational plans that can be discovered by strict and diligent inductions. As the whole manifests design, so every factual part is sacred; the scientist may not adulterate God's truth with his own speculations.[51] The structure of animal life derives from the general plans of Cuvier's four branches. These plans have been variously adapted to form the plans of different species, each specific plan being as changeless as the eternal Creator himself. Yet Agassiz recognised that the individuals of each species do change within the limits of their plan. Human perceptions of species change as well. The unity and stability of the creation, he concluded, can lie neither within itself nor, ultimately, within the human mind, but only within the mind of the Creator, who willed it into existence.

Here was the pantheistic idealism of Oken and Schelling made personal, the *nous* of Plotinus made Christian. For forty years Agassiz taught that the entire animal kingdom, living and extinct, manifests an overall 'plan' of creation which 'has been preconceived, has been laid out in the course of time, and executed with the definite object of introducing man upon the earth'.[52] Animal species therefore have no more material existence than the transcendental plan they manifest, nor do any of the higher taxa. Each is a discrete act of the divine intellect and, as such, none can be related to another by physical descent. Radiated animals, according to Agassiz, afford an excellent example.

The more we penetrate into the differences among these animals, the more do we see that between all there is an intellectual link which brings them into close relation and shows them to be but variations of an idea, and not the result of diverse circumstances and influences operating on them. They were made what they are by an intellectual process which connects them all and combines them under one original plan. They are not the product of accident or chance; and the evidence of the fact that they are the work of intellect may be derived from the facility with which our mind can grasp the idea which lies at the foundation of their structure, and generalize it.[53]

Mankind, the end and epitome of the ascending vertebrate series, the object of creation's master plan – mankind has thus become 'instinctively and...unconsciously the translators of the thoughts of God' because, through creation, the human mind has obtained an affinity with the divine. 'It is surely not amiss', Agassiz declared in his 'Essay on classification', 'for the philosopher to endeavor, by the study of his own mental operations, to approximate the workings of the Divine Reason, learning from the nature of his own mind better to understand the Infinite Intellect from which it is derived'.[54]

This was hardly the faith that Agassiz had learnt from his father, a Swiss Protestant pastor, though it did have unmistakable, if somewhat disturbing, theological implications. Agassiz took over the catastrophes with which Cuvier had punctuated the geological record and improved on them, asserting that in each instance all life had been wiped out and then recreated on new plans. Had the 'Infinite Intellect' changed his mind or was he merely callous and capricious? Agassiz's belief that the human races were specially created on separate plans which had been adapted to different environments – a logical extension of his doctrine of species – was likewise fraught with theological significance. Polygenism contradicted the Bible no less thoroughly than successive worldwide extinctions.[55] For Agassiz, however, 'the scientific study of nature ...supplied the materials for an idealistic world view that transcended mere recorded assumptions about the powers of the Deity. The essence of the Creative Power was to be discovered in the book of nature itself, not in the Bible.'[56]

Agassiz's transcendental paraphernalia – 'categories of thought', 'prophetic types', and so forth – were to Darwin 'merely empty sounds'. When Agassiz asked in his review of the *Origin of Species*, 'If species do not exist at all...how can they vary? And if individuals alone exist, how can differences which may be observed among them prove the variability of species?' he believed he had impaled every theory of transmutation on a dilemma. In fact, as Darwin immediately perceived, Agassiz had simply demonstrated his inability to conceive of living things apart from an essentialistic notion of species. 'How absurd that logical quibble – "if species do not exist how can they vary?"' wrote Darwin to Asa Gray. 'As if anyone doubted their temporary existence. How coolly he assumes that there is some clearly defined distinction between individual differences and varieties.'[57] Darwin conceived of species not as eternal and distinct but as temporary and indefinite. As temporary, they are the historic products of a world in which variability is unlimited and extinction the outcome of unfavourable variations. 'The only distinction between species and well-marked varieties is, that the latter are known, or believed, to be connected at the present day by intermediate gradations, whereas species were formerly thus connected.' As indefinite, species consist of forms 'sufficiently constant and distinct from other forms, to be capable of definition': forms, that is, which manifest differences 'sufficiently important to deserve a specific name'. These differences do not constitute the diagnostic marks of transcendent or immanent essences; they are 'merely artificial combinations made for convenience'. 'This may not be a cheering prospect', said Darwin, 'but we shall at least be freed from the vain search for the undiscovered and undiscoverable essence of the term species'.[58]

Darwin's opponents certainly did not regard the new definition of species as a cheering prospect. Christian anti-Darwinians to a man embraced an essentialistic philosophy of nature, justifying themselves from scripture and science. Their scriptural authority was a literal interpretation of Genesis, the legacy of Ray and Linnaeus, and science they usually made synonymous with the teaching of Agassiz. Among the anti-Darwinians we have considered, Morris and Birks alone omitted to invoke the name of the great naturalist, though neither left a doubt as to his devotion to the Bible and his steadfast opposition to any change in species. Birks in particular was uncompromising at this point. Human individuals, like the members of every species, he declared, are endowed with their own 'fixed types', in accordance with the 'precise and clear' account of creation given in Holy Scripture. But this doctrine also accords with a theory of vitality which is 'at once in harmony with the facts of science, and with the natural and instinctive feelings of mankind'. Drawing on the researches of the physiologist Lionel S. Beale, Birks specified the characteristics of life as 'individuality of being', 'active vital power' in each individual, and the 'presence' of some 'specific type or form', either 'in the structure of the plant or animal, or in the products or direction of its activity', towards which the vital power 'tends continually'.[59]

Some anti-Darwinians who appealed to Agassiz had a high regard for the Bible but seldom if ever brought it into their arguments. Burr maintained in *Ecce Terra* that, while each species possesses 'its own measure of structural flexibility', observation proves that 'the different species never come to overlap'; they remain 'as far apart to day as they were at the dawn of history'. Their *termini* of structure' are evidence of the unity, order, and design of the divine creative 'plans' which Agassiz discerned throughout the living world. Structural variation is thus limited 'by no necessity of nature or construction' but by 'the current choice and agency of God'.[60] Bree devoted an entire chapter in his *Exposition of Fallacies* to the 'immortal' Agassiz, who had shown to his satisfaction that structural similarities are 'the expressions of the thoughts of the Deity'; that the successive embodiments of the vertebrate plan were created 'with special reference to the ultimate structure of man himself'.[61] Curtis had known Agassiz well for 'a long period of years' before 1862, when his legal practice moved from

Boston to New York City. Of all the scientists within his acquaintance 'Agassiz always seemed...the broadest as well as the most exact and logical reasoner'. Having written of the 'ideal plan' that unites mankind and other vertebrates, and of the Creator's adapting his 'plans of construction' to particular ends, it was therefore perhaps less than candid of Curtis to claim that the 'opinions and reasoning' contained in *Creation or Evolution?* were 'adopted independently of any influence' from his friend.[62]

Other anti-Darwinians – those who were arguably the most influential – implicitly endorsed Agassiz's philosophy of nature and took it explicitly as support for a traditional reading of the early chapters of the Bible. Townsend could only be selective in his aversion for theories, and in his anonymous *Credo*, published in 1869, he adumbrated a 'scriptural theory' of creation which he later expanded and embellished from numerous scientific authorities in *The Mosaic Record and Modern Science*. The theory was based on 'the law of type and antitype', according to which 'a prophetic or typical principle' runs throughout the universe. Applied to Genesis and geology, the law yields the theory that the six 'cosmical days of the original creation', as revealed in the six major geological epochs which occurred within the time-span of the first verse of Genesis 1, are typical of the six ordinary days of Genesis 1 and 2, when the earth, having first been laid waste, was repopulated by divine fiat with perfectly adapted creatures. Applied again, the law of type and antitype teaches that the creatures thus produced, whether of the second creation or the first, were typical of those which would succeed them in time. The fins of fishes, the wings and feet of birds, the fore and hind feet of other beasts – all, wrote Townsend, garnering support from Agassiz, Owen, and Mivart, are 'typical or prophetic of the arms and feet of man'. Even 'prehistoric human remains, should they be discovered, might be those of irrational creatures, not from which man has been developed, but which are simply a type, in accordance with which God designed to make rational and existing man when the ordained period had arrived'.[63]

Hodge did not obtrude the Bible in *What Is Darwinism?* for he reasoned that since Darwin's theory could not be reconciled with 'the declarations of the Scriptures', there would be many other reasons why it should be rejected. One such reason was of course that true science, practised according to Baconian principles, contradicted Darwinism as fully and inevitably as it agreed with the inductions of true theology. For the biblical inductions Hodge was his own best authority. In science his authority was Agassiz. 'Religious men believe with Agassiz that facts are sacred', he declared. Sticking to the sacred facts, biblical theology and biological science alike teach that the 'primordial forms' of every species were specially created, the incarnations of 'type ideas' which together constitute an 'ascending series' that culminates in mankind. 'It has in the progress of science been discovered', Hodge announced in his *Systematic Theology*, 'that the whole vegetable and animal world has been constructed on one comprehensive plan. . . . So obviously is this the case that Professor Agassiz's "Essay on classification" is, to say the least, as strong an argument for the being of God as any of the "Bridgewater Treatises."'[64]

Dawson may have referred on occasion to some 'law of creation' or 'theistic form of evolution', but in fact the only such theory he could accept was the discontinuous transcendental progressionism, based on the works of Miller and Agassiz, which he embraced in *Archaia* and enlarged on seventeen years later in its successor, *The Origin of the World according to Revelation and Science*.[65] In these volumes Dawson made an elaborate attempt to show how the geological and biblical records 'exhibit the progressive character of creation', how the two 'are agreed not only as to the fact and order of progress, but also as to its manner and use'. Both records, he argued, testify that life appeared on earth, periodically and abruptly, in great waves of species, each wave with species more advanced than those of the previous, each species 'immutable, except within narrow limits', and appearing 'at once in [its] most perfect state'. Both records testify that mankind is the 'culminating-point of the whole creation', the species towards which all matter and life has progressed.[66] In *Modern Ideas of Evolution* Dawson remained abreast of the major developments in evolutionary theory but he deviated not an inch from his progressionistic point of view. The successive introduction of new species into prepared environments, coupled with the extinction or degradation of lower forms, he stated, 'indicates not a mere spontaneous evolution, but a progressive plan carried on by a great variety of causes'. Weismann's theory of heredity holds promise because isolation of the germ plasm would encourage 'constancy to the ideal plan of the species' and would help to account for the 'wonderful permanence of types in geological time'. On the other hand, it is 'entirely gratuitous' to assume that structural similarities among species are evidence of common ancestry, for these similarities may simply 'represent a planning mind following the same ideas in different works'. Again Dawson honoured *Homo sapiens* as the pinnacle of organic progress. Mankind's body and spirit are united to the whole creation, not by descent, but through the mind of the Creator. The human spirit is a special endowment which enables mankind alone of all the creatures to know the Creator's mind. For what other animal has been able to discover the progressive character of creation, the immutability of species, and its own place in nature? The very fact that human beings perceive 'a theistic principle of development...in all nature' and attribute it to 'the plan and methods of creation' gives 'inexpressible dignity' to them and to their science. 'It shows that the human reason must be after the model of the infinite Divine reason, that in scientific inquiry we are studying God's laws and revelation of Himself in nature.'[87]

Ellegård has pointed out that 'a direct synthesis of idealistic philosophy and Biblical orthodoxy...was evidently very widespread among the early opponents of Darwin's theory'.[68] The same synthesis, it is safe to say, was not uncommon at a later period as well.

The problem of faith and philosophy

It would appear that Christian Anti-Darwinism was neither so anti-Darwinian nor so Christian as might be thought. That most of Darwin's critics were less opposed to what he wrote than to their misconceptions of it should hardly require elaboration. Indeed, if prominent and influential critics such as Morris, Burr, Townsend, Bree, and Birks were as imperceptive as their books reveal, what must be said for the mass of lay and clerical anti-Darwinians who read and profited from them? In these books evolution is usually conceived in terms of maximum generality, often with allusions to the cosmic developmentalism of *Vestiges* or to Spencer's universal 'law'. Questions of materialism and the origin of life frequently intrude because evolutionists such as Huxley, Tyndall, and Spencer, who were supposed to represent Darwinian views, could be understood to favour these doctrines. Human evolution, on the other hand, is so far from a question of serious debate that ridicule on occasion displaces reason. Natural selection is neglected or hopelessly confused, typically by assimilation to discredited theories like Lamarck's, though the reiteration of scientific criticisms obtained at second-hand sometimes leaves the impression that Darwin has been understood. It is sad and ironical to contemplate that Darwin's epoch-making discovery, the one theory that made biological evolution for the first time scientifically cogent and theologically challenging, was the very doctrine that most Christian anti-Darwinians were unable to comprehend.[69]

That these strictures do not entirely apply to Darwin's few discerning critics – Hodge, Dawson, and perhaps also Curtis – should be obvious as well. Each distinguished between Darwinian and other versions of evolution, particularly that of the *Synthetic Philosophy*. Each gave evidence of having correctly understood the theory of natural selection, both in its explanatory role and in its crucial bearings on received notions of creation, providence, and design. Each expressed himself, for the most part, with composure and restraint, even when dealing with that most provocative question of human ancestry. Dawson, a professional geologist, stayed abreast of the debate *für-und-gegen* Darwin and made his criticisms in its light. Yet, while these critics may be regarded as more discerning than the others because they were more expressly anti-Darwinian, their opposition was based none the less on philosophical premises to which the name 'Christian' cannot distinctively apply.

'Although the ideas that rose up like armed men against Darwinism owed their intensity to religious associations, their origin and meaning are to be found in science and philosophy, not in religion.'[70] Thus wrote John Dewey on the centennial of Darwin's birth, and his words sum up admirably the conclusions we have reached. The military metaphor is a crude but telling by-product of conflicts variously experienced by persons scientific and religious in the post-Darwinian period. At one level these conflicts occurred between Darwinian and Christian beliefs: that is, between concepts expressed or entailed by Darwinism as a scientific theory and concepts deemed essential to the Christian religion. Among the latter were many 'ideas that rose up like armed men', some of which no doubt were genuine matters of faith. Two such ideas, however, had little or nothing to do with Christianity – indeed, may well have embarrassed Christian doctrines – and each alone was sufficient to produce a conflict leading to the complete rejection of Darwinism.

The quest for certainty was a chimerical undertaking, even in a world populated by a finite number of fixed natural kinds. Not only therefore was it a perennial quest, far antedating the Christian era; it might also have been regarded, from a Christian standpoint, as a presumptuous one, for its success depended on a degree of human competence which took no account of creatureliness, finitude, and sin. A religion that stressed faith as well as works, believing as well as seeing, should hardly have been thought compatible with the godlike pretensions to omniscience implied by a search for full and final verification of inductive inferences. Nor, for that matter, should the implication that Providence is in any respect ultimately predictable have set lightly with those who believed in the sovereignty and transcendence of God. But this implication had already been codified in the doctrine of the fixity of biological species.[71]

The belief in fixity, likewise of pre-Christian origin, persisted in the post-Darwinian period largely as an amalgam of biblical literalism and Neo-Platonism, the latter deriving from German romantic philosophy through the idiosyncratic and widely influential teaching of Louis Agassiz. Far from representing the Christian or biblical view of nature that figured in the scientific renascence of the sixteenth and seventeenth centuries, the *Naturphilosophie* revealed Kant's 'Copernican revolution' of the mind as the Ptolemaic counter-revolution it could become. To Oken and Schelling the meaning of all creation was divinely resident in the human spirit; the 'romantic chaos' which their philosophies so greatly inspired was 'ruled and ordered by the simple wisdom that man is the measure of all things'.[72] Agassiz did not escape this anthropocentrism, nor did his anti-Darwinian followers, who imbibed it with their biblical exegesis and imbedded it in their natural theologies. Ideal types, creative plans, and a progressive plan of creation that culminates in a being whose thoughts on the matter are supposed to be like unto God's – none of this was less presumptuously anthropocentric, none less discordant with doctrines otherwise professed, than the theological implications of a quest for ultimate certainty in inductive inferences. The anti-Darwinian element in Christian Anti-Darwinism may thus in fact have had little to do with Christian doctrines. Perhaps, after all, what conflicted with Darwinism were the philosophical assumptions with which the Christian faith had been allied.

Faith has never been able to seal itself off from philosophy and few believe it should. But if faith's philosophy goes unacknowledged or if faith thinks it has no philosophy whatever, then, insensibly and inevitably, a prevailing world-view seeps in, colouring whatever pretends to be a pure apprehension of Christian truth.[73] Doing philosophy thus by default is a risky business and anti-Darwinians illustrate the result. In the name of Christian and biblical teaching they set the static world of antiquity over against a theory that helped to resolve the enigmas of natural history which the old world had merely enshrined. The fixity and certainty banished from the heavens by Christian philosophers, from Galileo and Newton to contemporary interpreters of the nebular hypothesis, they domesticated on the earth, where Darwin found naught but process and probability.[74] Thus, while Darwin won the best minds of the next generation, the faith that had attached itself to the old philosophy and the old science was quietly abandoned.

'The tendency of theology to conform itself to the philosophical and scientific hypotheses which are ever cropping up and disappearing', wrote one eminent anti-Darwinian, 'for a time. . .carries all before it, but it incurs the danger that when the false and partial hypotheses have been discarded the higher truths imprudently connected with them may be discarded also'.[75] He wrote better than he knew.

IO

CHRISTIAN DARWINISTICISM:
THE ROLE OF PROVIDENCE AND
PROGRESS

Pseudo-science. . .has grown and flourished until, nowadays, it is becoming
somewhat rampant. It has. . .an army of 'reconcilers', enlisted in its
service, whose business seems to be to mix the black of dogma and the
white of science into the neutral tint of what they call liberal theology.

T. H. Huxley, 1887[1]

Only those who could inject spiritual dimensions into Darwinism could
directly come to terms with it. For others, Darwinism produced conflicts
in which the real issue was frequently obscured. In essence, one can say
that Darwinism could be reinterpreted or transformed.

John Dillenberger[2]

In the polemical world of T. H. Huxley liberal reconcilers of
Christianity and evolution could be nothing but an 'army' bent on
blending scientific truth with theological error. Their exploits were
simply 'pseudo-science', the neutralising of issues as plain as black and
white. Huxley was of course mistaken, though his caricature has
persisted in a military metaphor. The reconcilers did not always con-
front clear-cut issues, much less did they blur them. One might well
argue that they 'engaged the advance lines of the realistic modern
mind', but to chide them for a 'strange insensitiveness to all the implications of science' and for a delusive belief that they had 'made
contributory to their faith the grand army of scientific inquiry' seems
altogether unjust.[3] Scientists themselves, most of whom were overtly
religious, could hardly agree on the theological 'implications' of their
theories. The 'grand army of scientific inquiry' never did exist. Evolutionary science in particular was a stricken field in the closing decades
of the nineteenth century; Christians who wished to be evolutionists,
even the small minority who were philosophically and scientifically
astute, had no clear consensus to follow. The 'neutral tint of liberal
theology' in their reconciliations was not simply a blend of 'the black

of dogma and the white of science' but an expression of the meta-physical diversity of post-Darwinian evolutionary thought.

'Christian Darwinisticism' is the term we have applied to recon-ciliations of Darwinism and Christian doctrine that embodied non-Darwinian evolutionary theories. Among such theories are those we have denominated 'Lamarckian evolution'. In this chapter we explain how Christian Darwinists, with two exceptions, found in Lamarckian evolution the 'spiritual dimensions' by which Darwinism could be 'reinterpreted or transformed'.

The theological conflict

The travails of Darwinism are often attributed to two related sources, according to Charles Coulston Gillispie: 'a belief, first, in both the inspiration and the literal sense of the Biblical texts; and secondly, a requirement that man be a unique species, specially created as a vessel for the immortal soul which absolutely distinguishes him from all other forms of life, existent or extinct'. Gillispie adds,

If it were simply a question of the reaction of clerical opponents of science, who produced a literature of denunciation, reconciliation, and exegesis, ...this explanation would, on the whole, suffice. But it does not account for the religious difficulties experienced by scientists themselves, by popularizers and theologians of science, and by the more open-minded individuals who made up the greater part of the educated public.[4]

This judgement has proved quite correct. Ellegård points out in his detailed study of the British periodical press from 1859 to 1872 that questions of the bearing of evolution on the biblical cosmogony and on mankind's creation dominated only the more popular and less sophisticated literature of the period.[5] And in the last three decades of the century, as we may now observe, there was a similar pattern of response. Biblically related objections to evolution were the stock-in-trade of only two of the anti-Darwinians we have discussed (Townsend and Dawson), and of the Darwinists discussed in the present chapter only Joseph S. Van Dyke was significantly concerned with reconciling evolution and scripture. The special creation of the human body and soul was defended by all the anti-Darwinians and by at least two Darwinists (Van Dyke and the Duke of Argyll), but Curtis alone seems to have been preoccupied with the problem. There were in fact more urgent and underlying 'religious difficulties' experienced by most of the post-Darwinian controversialists.[6]

Christian anti-Darwinians came into conflict with Darwinism because they sought certainty through inductive inferences and because they believed in the fixity of biological species. Their difficulties, we have argued, were primarily philosophical in character. The difficulties of Christian Darwinists were primarily and properly theological. Having abandoned, for the most part, a philosophy of certain inferences and fixed essences, and having accepted as a revelation of divine truth the discoveries of a science based on process and probability, Darwinists had to reconcile these discoveries with the God whose facts they were. God created everything, God designed what he created, and God created mankind in his own image. These were received and accepted doctrines, none of which was optional for Christians. Thus, it seemed, the only conclusion to be drawn was that God created everything by evolution, executed his designs through evolution, and brought his noblest creature into existence by means of evolution. If, however, evolution were simply equated with Darwinism, then the inference could not be so simple and straightforward. God created everything but Darwin spoke almost exclusively of the biological world. God designed what he created but Darwin substituted for design the natural selection of numerous minute and apparently purposeless variations. God created mankind in his own image, a rational and immortal species, but Darwin held that mankind had evolved – body, mind, and morals – from lower animals in an uninterrupted line of descent. Moreover, God created everything, including mankind, in accordance with his character – he 'saw that it was good' – but Darwin made central to evolution, from the lowest life to the highest, a brutal struggle for existence in which merely the fittest, not the best, survive. In short, Darwinism seemed to exclude God from the world by explaining away the evidence of his design in the creation of life and mankind; Darwinism also seemed to reflect on God's character – his omnipotence and beneficence – by treating evolution as a mere biological phenomenon and by casting the creative process as a gladiatorial contest.

Christian Darwinists therefore came into conflict with Darwinism because they believed that God's purposes are manifested in the world and that these purposes disclose God's omnipotent and beneficent character: because, more precisely, they believed in a God whose purposes could not have been realised through evolution as Darwin conceived it. Their dissonance stemmed from the fact that, while they felt compelled to preserve theological truth, Darwinism obviously contained much undeniable truth as well. And as for the attempts made by Darwinists to reduce the dissonance involved in arbitrating between Darwinian and theological truths, we can now understand how they adopted theories of evolution which, by altering and adulterating Darwinism, were congenial to the purposes and character of God.[7]

Teleology and providence

The main hindrance to the accomplishment of God's purposes in nature was natural selection. Darwin called it 'the most important, but not the exclusive, means of modification'. This statement needed alteration if some form of teleology were to be preserved. Thus to Darwinists natural selection became a less important, but generally not excluded, means of modification. Although a few overlooked or misunderstood it, the great majority of Darwinists, regardless of their perceptions, made it subordinate to other causes of change.

Some interposed a divine agency as the cause of evolution.[8] The Broad Churchman FREDERICK TEMPLE (1821–1902), whose liberality was much in evidence, from his contribution to *Essays and Reviews* in 1860 until his consecration as archbishop of Canterbury in 1896, believed that Darwinism did not 'affect the substance of Paley's argument' in view of the 'limitations and modifications' which 'necessarily attached' to an acceptance of the theory. Natural selection, according to Temple's 1884 Bampton Lectures, *The Relations between Religion and Science* (1885), is simply one partial expression of the 'original properties' – physical, chemical, and teleological – with which matter was created. The Creator 'impressed on certain particles of matter which, either at the beginning or at some point in the history of His creation He endowed with life, such inherent powers that in the ordinary course of time living creatures such as the present were evolved'.[9]

HENRY WARD BEECHER (1818–1887), America's foremost pulpiteer of the later nineteenth century, sold his Calvinistic birthright for the mess of philosophical pottage contained in the *Synthetic Philosophy*. Yet, combining 'Romanticism, religion, and science – the epistemology of Kant, the Gospel of Jesus, the teleology of Spencer' – Beecher remained a Congregational clergyman and preached mightily to the throbbing human heart.[10] In sermons and lectures published as *Evolution and Religion* (1885) he offered a crude, though eloquent, explanation of natural selection, then exulted in the creation's 'moving onward and upward in determinate lines and directions' through 'the mediation of natural laws'.

If single acts [of creation] would evince design, how much more a vast universe, that by inherent laws gradually builded itself, and then created its own plants and animals, a universe so adjusted that it left by the way the poorest things, and steadily wrought toward more complex, ingenious, and beautiful results! Who designed this mighty machine, created matter, gave to it its laws, and impressed upon it that tendency which has brought forth the almost infinite results on the globe, and wrought them into a perfect system? Design by wholesale is grander than design by retail.[11]

The botanist and Anglican clergyman GEORGE HENSLOW (1835–1925), believing that 'Natural Selection Plays no Part in the Origin of Species', sought through his scientific works, *The Origin of Floral Structures* (1888), *The Origin of Plant Structures* (1895), and *The Heredity of Acquired Characters in Plants* (1908), 'to revive the "Monde ambiant" of Geoffroy Saint Hilaire, as the primal cause of change'.[12] If the wisdom and goodness of God were to be secured in the face of Darwinism there had to be a 'natural law' in the process of adaptation, some assurance that in nature there are only definite, never indefinite, variations. Henslow found this assurance in the power of protoplasm to respond to external influences. 'Directivity', he called it, a force superadded to physical and chemical forces, immeasurable in itself, by which the Creator endowed the first blob of protoplasm with 'practical omnipotence'. From this blob all life has descended, the protoplasm of plants and animals alike adapting each structure to the environment in which it is placed. 'If the Argument from Design be not restored', wrote Henslow in *Present-day Rationalism Critically Examined* (1904), 'that of *Adaptation* under *Directivity* takes its place; and Paley's argument, readapted to Evolution, becomes as sound as before; and, indeed, far strengthened as being strictly in accordance with facts'.[13]

The eighth DUKE OF ARGYLL, George Douglas Campbell (1823–1900), a distinguished statesman of the Liberal Party, was a lay-geologist and amateur ornithologist who accepted Darwinism only in so far as it could be reconciled with the transcendental morphology of his mentor, Richard Owen.[14] 'The analogies which the disciplined intellect sees in external nature', the Duke explained, though 'at least substantially representative of the truth', may nevertheless be formed by higher or lower faculties of the mind. Natural selection is formed by faculties which are 'simple and almost infantile'. It might dimly express the truth that God plans and chooses what he creates, but it does so by elevating the superficial distinction between the organism and its environment and ignoring the truly profound and fundamental fact of the 'self-containedness' of every organism. The same truth is better expressed when, with Owen and others, the individual organism is made the centre of attention, especially its subtle and intricate correlations of growth. For only then can one begin to understand 'that inexhaustible wealth of primordial inception, of subsequent development, and of continuous adjustment, upon which alone selection can begin to operate'. Supposing that several creations were as likely as one, the Duke proposed in *Organic Evolution Cross-Examined* (1898) that the major orders originated in separate germs, each germ, like an embryo, possessing its own 'internal directing agency or force', a 'Plan' with a prevision of all the complexities of adaptation that would be expressed, step by step, through 'ordinary generation'. The sudden emergence of new species in this manner, he stated, would 'harmonise with those intellectual instincts and conceptions of our mental nature to which the idea of chance is abhorrent, and which demand for an orderly progression in events some regulating cause as continuous and as intelligible as itself'.[15]

ST GEORGE MIVART (1827–1900), the twice-converted and twice-excommunicated critic of natural selection whom we have met in earlier chapters, embraced a theory of evolution which, he claimed, 'accepts, distributes, and harmonizes' three conceptions of the organic universe: 'the teleological, the typical, and the transmutationist'. A student of Owen's and a thoroughgoing rationalist, Mivart found purpose in nature, not as an inference from material contrivances, but 'as a necessary truth of reason deduced from. . .primary intuitions'.[16] It followed from this assumption – mind, not matter, being the arbiter of the actual – that an organic teleology could be expressed in a typology which did not preclude transmutation as the divine method of creation. The belief in transmutation and the belief 'which represents all organic forms as having been created according to certain fixed ideal types', wrote Mivart in *Lessons from Nature* (1876), '. . .can and do co-exist in perfect harmony in one and the same individual mind' and therefore must also be present in the same manner within the mind of God. 'Let the idea of God be once accepted, and then it becomes simply a truism to say that the mind of the Deity contains all that exists in the human mind, and infinitely more.'[17] In uniting teleology and typology with a theory of transmutation Mivart first argued in *On the Genesis of Species* (1871) that Darwin's non-teleological and non-typological theory of the natural selection of numerous 'minute, indefinite, and fortuitous' variations could not explain, on its own terms, the phenomena of organic nature; and second, he set forth a Christian cosmogeny which explained these phenomena by subsuming natural selection under teleological forces directed towards typological ends. 'Cosmical forces of all kinds', according to *Lessons from Nature*, 'unite and "transform" themselves in each living creature into a single force which, regarded abstractly, may be said to be the dynamical side of [each] such creature'. This composite force, under divine guidance, accounts for the origin and persistence of organic novelty, the emergence of new species as macro-mutations, occasioned by the action of the environment on embryonic life, and all the phenomena of homology for which reason demands an intelligible explanation.[18]

The American philosopher JOHN BASCOM (1827–1911), whose struggle for spiritual liberty took him from the Congregational ministry, via the philosophy of Fichte, Schelling, and Hegel, to a career in higher education at Williams College and the University of Wisconsin, was initially a reluctant evolutionist.[19] However, in his *Natural Theology* (1880), having enlisted all Mivart's objections to natural selection, especially the fortuitous character of variation, Bascom accepted a theory in which material and mechanical causes are transformed by 'the presence of a spiritual agency.' 'By a spiritual evolution', he wrote in *Evolution and Religion* (1897), 'we understand one of distinct increments and of an over-ruling purpose, which in its entire process contains and expresses personal, spiritual power in the means employed, in their combination, and in their evolution.... The physical forces are, in every stage of their development, permeated and borne forward by intellectual ones.'[20] Whereas in Darwinism, with its 'purely accidental variations', there is 'no law...under which fitting material for an organic kingdom would be furnished', in a spiritual evolution the 'anticipatory work' is assigned to 'life', an inscrutable 'plastic power' which, superintended by the divine mind, has given rise to certain non-selective factors, or 'laws', of development: direct adaptation, heredity, correlation of growth, and use-inheritance. Direct adaptation, said Bascom, 'may be passed, as offering explanations of no magnitude or interest'. Heredity, however, 'first makes its appearance among organic powers, and is part of that endowment which belongs to them as a distinct increment'; correlation of growth is 'not included in matter and motion, and does disclose design'; and use-inheritance is 'one of those primary principles which make up the regimen of life and disclose it as a new thing in the world'. Indeed, 'the response of life to its conditions, and the transmission of the organic changes so induced, remain peculiar and inscrutable kinds of causation, and call, in the new forms they take on and in the new directions they assume, for an informing and over-ruling idea'. 'By far, then', Bascom concluded, 'the most simple and sufficient theory of life is the spiritual one'.[21]

Other Darwinists preserved God's purposes in nature by absorbing natural selection in immanent divine power, though this power could also sometimes be referred to realistically as an agent of evolution. HENRY DRUMMOND) (1851–1897), the Scottish naturalist and Free Churchman, abandoned experiments in mesmerism for the 'inquiry room' of D. L. Moody's revivals, where he obtained the data for his best-selling *Natural Law in the Spiritual World* (1883), a collection of sermons which supplied the basic doctrines of the spiritual life as interpreted by evangelical Christianity with proof texts from the *Synthetic Philosophy*.[22] In his *Ascent of Man* (1894) Drummond remained Spencer's ardent disciple, finding evidence of natural law in the spiritual world, not merely of the individual, but of the entire human race. The 'Struggle for Life' had gradually given way to the 'Struggle for the Life of Others'; Darwin's 'final clue' to organic development was not 'the sole or even the main agent in the process'. The prevalence of altruism as a factor of evolution, Drummond believed, has resulted from the action of the environment, an environment which is 'Nature, the world, the cosmos – and something far more, some One more, an Infinite Intelligence and Eternal Will'. The physical forces which produce evolution, not by an internal 'innate tendency to progress' nor by 'the energies inherent in the protoplasmic cell), but 'by a continuous feeding and reinforcement from without', are in reality spiritual and moral. Evolution therefore is not merely 'progress in matter'. 'It is a progress in spirit', Drummond declared. 'Evolution is Advolution; better, it is Revelation – the phenomenal expression of the Divine, the progressive realization of the Ideal, the Ascent of Love. . . . The aspiration in the human mind and heart is but the evolutionary tendency of the universe becoming self-conscious.'[23]

America's senior Neo-Lamarckian naturalist, JOSEPH LE CONTE

(1823–1901), professor of geology and natural history in the University of California, defined evolution as '(1) continuous *progressive change*, (2) *according to certain laws*, (3) and by means of *resident forces*'. The 'progressive change' from primitive simplicity to mature complexity and the 'certain laws' relating to the divergence and diversification of types in geological time Le Conte learnt from his teacher, Louis Agassiz. The 'resident forces' he interpreted in accordance with a theology learnt from Kant, Fichte, Hegel, and Berkeley.[24] According to Le Conte's *Evolution and Its Relation to Religious Thought* (1888), an 'external progressive force' resulting from the several factors of adaptive change – particularly use-inheritance and the direct action of the environment – comes into conflict with an 'internal conservative force' resulting from the 'law of heredity' – like produces like. The tension between these forces within an organism is relieved periodically in outbreaks of mutations which accelerate the evolution of new forms. However, as Le Conte went on to explain, the phenomena in question are 'naught else than objectified modes of divine thought, the forces...naught else than different forms of one omnipresent divine energy', the 'law of evolution' itself 'naught else than the mode of operation of the...divine energy in originating and developing the cosmos'. 'Science', he declared, 'is...a rational system of natural theology.... There is no real efficient force but spirit, and no real *independent* existence but God.'[25]

Le Conte's authoritative work was widely influential on 'clergymen of every denomination' and also on numerous scientifically minded young men who wrote to the author, thanking him for a book which, they said, 'saved them from blank materialism'.[26] One of these young men was THOMAS HOWARD MACQUEARY (1861–1930), a priest of the Protestant Episcopal Church in Canton, Ohio, whose belligerent and iconoclastic volume, *The Evolution of Man and Christianity* (1890), brought about his deposition from the ministry in 'the first outcome of a trial for heresy in the history of the American Church'.[27] Mac-Queary dedicated his book to Le Conte as an expression – in his own words – of 'my heart-felt appreciation of the unfailing sympathy and invaluable assistance I received from you during the long, dark period of mental and spiritual struggle which resulted in my emancipation from the thralldom of a crude and irrational Traditionalism'. The dedication may also be interpreted as a disclaimer of plagiarism. A 'popular summary of scientifico-theological opinions from the evolution point of view' that was intended to 'suggest lines of thought which every one may follow out for himself' need not have been open to such a charge, but MacQueary's work, his literary first-born, made itself vulnerable. For example, in calling for a 'reconstruction of Christian theology' in the light of evolution, MacQueary took his cue from the closing chapters of Le Conte's book and began the task with the doctrine of God. Aided by Spencer and John Fiske, he claimed to have found in Berkeleyan idealism a basis for reconciling God's providence in evolution with his transcendence over nature. 'Spirit', he wrote, '. . .is the only Eternal Absolute Substance. Nature is an outward and visible sign of this inward-underlying-Energy or Being. Its phenomena are naught else than *objectified modes* of the Eternal I Am; the forces of Nature are naught else than *different manifestations* of one Divine Will; the laws of Nature, naught else than the regular *modes of operation* of that will, unchangeable because He is unchangeable.'[28]

Among the numerous clergymen of an older generation who were grateful to Le Conte for his synthesis of evolution and religion, none was more prominent than LYMAN ABBOTT (1835–1922), Beecher's successor at Plymouth Congregational Church in Brooklyn, New York and America's outstanding representative of evangelical liberalism at the turn of the century.[29] Adopting Le Conte's definition of evolution, Abbott maintained that it referred to 'the history of a process, not the explanation of a cause'. Yet with one cause he did concern himself, and that in order to minimise it. 'Evolution is not to be identified with Darwinism', he declared in *The Theology of an Evolutionist* (1897). 'It is not the doctrine of struggle for existence and survival of the fittest.' Biologists with hardly an exception were evolutionists, but they were not necessarily Darwinians, regarding natural selection as an adequate causal explanation of organic development. Indeed, Darwin himself did not regard natural selection as a 'complete summary of the process'.[30] In *The Evolution of Christianity* (1892) Abbott had adopted the Darwinian explanation as 'the most common, if not the most accurate, formula of evolution' in order to show that as all life is a perpetual battle against forces hostile to progress, so the Christian life is a constant 'warfare' against evil. In his later book the emphasis was different. He made no attempt 'to show that Christianity can be harmonized with Darwinism'. The view that 'all animate nature is wrestling, every fellow with his fellow, and that every life depends on the destruction of some other life, slain in the struggle by the selfishness of the victor', was repulsive to him and, moreover, was certainly not, in his opinion, 'the view of the great evolutionists'. Huxley's Romanes

Lecture on evolution and ethics, and Drummond's *Ascent of Man*, showed to his satisfaction that evolution involves not only a struggle for self but a 'struggle for others', altruism as well as competition.[31] All particular causes of evolution, however, Abbott simply attributed to Le Conte's 'resident forces' – or, rather, to a 'Resident Force'. God 'transcends all phenomena', he said, 'and yet is the creative, controlling, directing force in all phenomena'. The theistic evolutionist holds that 'God is the one Resident Force;...that His method of work in His world is the method of growth; and that the history of the world ...is the history of a growth in accordance with the great law interpreted and uttered in that one word evolution'. Teleology thus becomes a manifest certainty. 'Science', Abbott believed, 'perceives in nature a real thoughtfulness' and in its investigations 'follows along a path which preëxisting thought has marked out for it'.[32]

FRANCIS HOWE JOHNSON (1835–1920), a retired Massachusetts Congregational minister and the 'most scholarly' of the liberal divines who brought notoriety to Andover Theological Seminary after 1880, rejected the 'subjective analysis' of Hegel and Fichte and the 'objective analysis' of Spencer's physical realism for an evolutionary theology based on the panpsychism of Rudolf Lotze and the personal idealism of Andrew Seth.[33] The problem with both the Lamarckian and the Darwinian theories of evolution, according to Johnson's *What Is Reality?* (1891), is that they 'are exclusive of the idea of a supreme constantly working creative intelligence'. Lamarck posited a meagre animal disposition which exercises organs as the environment elicits their use, resulting in heritable modifications. Darwin introduced a causal agent which, though 'selective', has no intelligence whatever to commend it. 'Nature' does not select; only animal breeders do. Yet, Johnson observed, 'the selective act of man is the very element in the analogy of which no use is made'. In Darwinian evolution the creative element is missing. Natural selection therefore can only play an 'exceedingly subordinate' role, intensifying the separateness of forms but never serving as the 'form producer'. Its 'radical defect' is the omission of 'intelligent guidance'.[34] The 'intelligent guidance' in which Johnson believed was a teleology founded on a 'new and totally unknown principle', inaccessible to science. 'The fact that we can nowhere detect the points at which intelligence exerts its shaping influence', he explained, 'is no argument against the reality of such influence'.

We cannot detect the points at which the modifying power of the human mind is brought to bear upon the apparently closed circle of physical causes. And, if foresight and intellectual guidance are properly regarded as the cause of human adaptations, they are just as properly regarded as the cause of the adaptations in nature that so closely resemble them. The chain of mechanical events is just as accommodating in the one case as in the other, and not one whit more so.[35]

Johnson here speaks of two worlds: the 'whole protoplasmic world', which is the 'specialized part of the greater organism upon which the Creator impresses his thought', and mankind's cerebral world. The evolutionary events in the former, the macrocosm, are produced analogously to the physiological events in the latter, the microcosm. 'We have...within the realm of human activities', he said, 'a true instance of the creation, by intelligence, of specialized organisms' – modified cells, each possessing its own intelligence – which, 'subject to further modifications from the *ego*, reproduce their kind like the different species of animals'. Thus, as the *ego* does not have to exert its creative or sustaining power independently from the subordinate intellects within its body, so God accomplishes his will in the world, 'making modifications at innumerable points, while leaving most of the details of the great conflict to be determined by those whose lesser intelligence has been given them for that very purpose'.[36] 'I hold this to be the truth', Johnson declared, 'that, wherever we come upon factors in evolution that defy analysis, it is our privilege and duty to recognize God as acting immediately'. 'This is not unscientific', he added. 'It is not to assume that no further analysis can be made, but that no amount of analysis ever has or can arrive at an originating cause other than intelligent will.'[37]

Still other Darwinists established a teleological interpretation of nature by absorbing natural selection in the divine immanence itself, though generally not without some equivocation about God's relation to the world. GEORGE MATHESON (1842–1906), an outstanding preacher of the Church of Scotland, studied under John Caird at the Divinity Hall of the University of Glasgow, learning from him a christianised Hegelianism which he later found to be validated by 'the most advanced representative' of the 'modern scientific spirit'.[38] 'We have no hesitation in saying', wrote Matheson in *Can the Old Faith Live with the New?* (1885), 'that the modern doctrine of evolution, and especially the modern doctrine as expanded by Mr Spencer, is more favourable to the existence of an analogy between the human and the Divine than any previous system of nature with which we are acquainted'. Whereas evolution aims 'to abolish ultimately all plurality of existence, to reduce the many to the one', to unite the entire cosmos by the omnipresence of force, Spencer goes a step farther, unifying all things in The Unknowable, that great inscrutable substratum of which nature and its forces are but 'manifestations'. The Unknowable, or 'God' in Spencer's universe, 'supports and manifests' all phenomena while yet transcending them. Such a doctrine, Matheson went on to say, has been allied with the Darwinian theory – a transcendent force with a 'material chain of evolution' – in order to explain the laws of variation and heredity. Mivart cannot countenance natural selection apart from some innate tendency for variations to form organic homologies; Spencer holds that the hereditary transmission of characters is contingent on the innate power of 'physiological units' to reproduce a likeness of the parent form. Spencer, for whom all force is a manifestation of The Unknowable, is therefore 'more than a Darwinian; he is a Darwinian *plus* a transcendentalist'. Darwinism has required the support of 'a system...which in all essential particulars is identical with that ancient creed which declared the heavens and the earth to be the work of an Almighty God'.[39]

MINOT JUDSON SAVAGE (1841–1918), perhaps the first clergyman in America to accept evolution from the pulpit and attempt to 'reconstruct religious and theologic thinking' in its light, resigned from the Congregational ministry after a 'severe mental struggle' and for more than twenty years proclaimed the Spencerian gospel from Boston's Church of the Unity.[40] In sermons and lectures published as *The Religion of Evolution* (1876), *The Morals of Evolution* (1880), *The Evolution of Christianity* (1892), and *The Irrepressible Conflict between Two World-Theories* (1892), Savage held evolution responsible for teaching that 'the worlds came into being by processes of continuous variation, change, growth', that 'infinite, eternal spirit and life is in and through the universe and that matter is...only a manifestation of the eternal life and spirit', a reliable and adequate revelation of 'the Unknowable One'. Spencer himself had assured Savage in personal conversation that he regarded 'the existence of [the] "Infinite and Eternal Energy", that religion calls God, as the *one most certain object of all our knowledge*'. Savage therefore could claim with absolute authority that God's existence had been placed on a 'secure and scientific foundation' and that 'the noblest religion and the noblest morality' derive from 'a grand reason in the nature of things'.[41] Natural selection, or 'creation after the Darwinian idea', did not contradict such religion and morality if it were conceived as 'a struggle on the part of all things and creatures to fit themselves to the conditions of their existence'. In any event, the Darwinian struggle, though 'central to the whole philosophy of Evolution', was merely a partial expression of the larger Spencerian 'law' to which Savage paid poetic tribute.[42]

> Hear me, O jarring peoples! I am one,
>   In deep abysses or in heavens high:
> One law swings the long circuit of the sun,
>   And by one law the new-fledged birdlings fly.
>
> Religion binds thee to my law divine,
>   And this law binds thee to thy fellow-man.
> 'Tis one law in the market, at the shrine:
>   Earth, heaven, – see! they're built upon one plan.[43]

Spencer's foremost American interpreter, the lay-philosopher and historian JOHN FISKE (1842–1901), discovered in the mid-1860s that the *Synthetic Philosophy* honoured the religious impulses which had sent him reeling into rebellion against the New England Congregationalism of his youth. For the rest of his career, in more than thirty volumes – among them *Outlines of Cosmic Philosophy* (1874), *Darwinism* (1879), *The Destiny of Man* (1884), *Excursions of an Evolutionist* (1884), *The Idea of God* (1885), *A Century of Science* (1899), *Through Nature to God* (1899), and *Life Everlasting* (1901) – Fiske bore testimony that Spencer's philosophy confirmed belief in 'a Power to which no limit in time and space is conceivable, and that all the phenomena of the universe, whether they be what we call material or what we call spiritual phenomena, are manifestations of this infinite and eternal Power'.[44] Darwin's theory may have won a 'complete and overwhelming victory', gaining a sway in biology 'hardly less complete than that of gravitation in astronomy', but Fiske agreed with Mivart and Spencer in assigning large roles to use-inheritance and direct adaptation to the environment and with Spencer in particular in attributing the efficacy of both factors to an organism's 'inherent capacity for adaptive changes' through the direct and indirect 'equilibration' of internal and external forces. The theory of natural selection was, after all, only 'the most conspicuous portion of that doctrine of evolution...which Herbert Spencer had already begun to set forth in its main outlines before the...theory had been made known to the world'. And Spencer's 'law of universal evolution' was but the description of how 'all the myriad phenomena of the universe. . .are the manifestations of a single animating principle that is both infinite and eternal' and therefore ultimately Unknowable.[45]

Whether by a divine agency, immanent divine power, or the divine immanence itself, whether by innate tendencies, resident divine forces, or a panpsychic or hylotheistic interpretation of Spencer's law, each of the Darwinists we have considered subjected natural selection to an orthogenetic impulse and assumed that the adaptive characters acquired under the influence of this impulse and the impact of environment and habit are inherited. By thus conceiving the components of Lamarckian evolution in terms of the providential metaphysics of post-Kantian liberal theology, they sought to preserve a teleological interpretation of nature in the face of the Darwinian challenge.[46] Natural selection, however, was not the only hindrance to the accomplishment of the divine purposes. There was also the doctrine that mankind as a psychosomatic whole had descended in an unbroken succession from lower forms of life. God, it was generally believed, had created mankind uniquely in his own image, a rational species destined to live forever. But in Darwin's view only evolution could account for the striking physical resemblances between mankind and the higher animals; and on the basis of evolution there was every reason to think that 'the difference in mind between man and the higher animals, great as it is, . . .is one of degree and not of kind'. To many this outlook on human origins seemed too narrow. It needed augmentation if God's purposes in creating mankind were to be preserved.

Some Darwinists therefore maintained that an act of special creation introduced an essential difference between mankind and the higher animals. The Duke of Argyll, holding it as a philosophical necessity that the human mind, though intellectually limited and morally debased, 'is really the type, and the only type, of that which men call the Supernatural', agreed with Owen and Cuvier that *Homo sapiens* occupies a separate class, that the method of mankind's creation was entirely unique. 'There is no ground whatever for supposing that ordinary generation has been the agency employed, seeing that no effects similar in kind are ever produced by that agency, so far as is known to us', he declared in *Primeval Man* (1869). The Duke also argued that paleontological evidence is conclusively against 'any change whatever in the specific characters of Man since the oldest Human Being yet known was born' and that, contrariwise, 'all scientific evidence is in favour' of mankind's origin from a 'single pair'. Mankind had not risen from anthropoid apes but had degenerated from the pristine state in which its two primeval ancestors were created – according to the Duke's *Unity of Nature* (1884) – 'by operations as exceptional as their result'.[47]

Unlike Argyll, Mivart did not suppose 'that any action different in kind took place in the production of man's body, from that which took place in the production of the bodies of other animals, and of the whole material universe'. But to account for human self-consciousness, rationality, and moral intuitions, which together constitute 'not a difference of degree but of *kind*' between mankind and the higher animals, according to *Man and Apes* (1873), he postulated the 'direct and immediate creation' of a 'rational soul' in every human being. In the early 1870s this was necessary both as a means of explaining the phenomena which Wallace had urged against the mental evolution of mankind and as a means of preserving the teachings of the Roman Catholic Church. As time went on, however, and particularly after 1889, when Mivart published *The Origin of Human Reason* in an attempt to refute the work of Romanes on mental evolution, the issues at stake had less to do with science and the Catholic faith and more to do with Mivart's rationalistic philosophy of nature. The evolution which he conceived as unifying teleology, typology, and transmutation depended entirely on the ability of human reason to penetrate to the reality of things behind the flow of phenomena and to discern the mind of God. Such an ability, Mivart and Argyll agreed, could only be guaranteed if the human mind had a purely transcendental origin.[48]

Temple and Henslow were at first inclined to attribute mankind's essential features to an act of creative intervention, though both eventually made other provisions. In view of Darwin's uncertainty regarding the causes of variation, Temple suggested that the 'unity of the plan' among vertebrates might be due, 'not to absolute unity of ancestry, but to unity of external conditions at a particular epoch in the descent of life'. Mankind therefore may well be 'regularly descended, without branching off at all', from one original form. This 'exceedingly early difference of origin', Temple believed, is corroborated by the 'enormous gap' which 'separates [human] nature from that of all other creatures known'.[49] The chasm is an ethical one; it involves not the emotions, the 'subordinate powers of the soul', which 'it is perfectly reasonable to find subject to laws of Evolution', but the 'spiritual faculty', mankind's 'inmost essence', which proclaims the 'Moral Law'. 'Science', wrote Temple, 'cannot yet assert, and it is tolerably certain will never assert, that the higher and added life, the spiritual faculty, which is man's characteristic prerogative, was not given to man by a direct creative act as soon as the body which was to be the seat and the instrument of that spiritual faculty had been sufficiently developed to receive it'. Yet, preferring not to rest in the unknowable past, Temple was careful to make provision for the soul in the unknown present. 'We cannot tell, and the Bible never professes to tell', he said, 'what powers or gifts are wrapped up in matter itself, or in that living matter of which we are made'. Perhaps the soul lies dormant from the beginning, awakens at conception, and springs into self-expression during the course of generation. One thing, in any event, seems certain: that 'it is impossible to say that anything which Science has yet proved, or ever has any chance of proving, is inconsistent with the place given to man in Creation by the teaching of the Bible'.[50]

In the light of the conviction with which Henslow wrote of an inscrutable force in nature, it is little wonder that he took up spiritualism at the turn of the century. Perhaps he thought that the adaptive power of protoplasm was correlated with some pneumatic power which bears human personality beyond the grave. *The Spiritual Teaching of Christ's Life* (1906), *The Proofs of Spiritualism* (1919), and *The Religion of the Spirit-World Written by the Spirits Themselves* (1920) may not answer this suggestion decisively, but the books do bear witness that their author had learnt more about the additional 'impulse' or 'power' which formerly he had regarded as the distinguishing mark of mankind. In his early works, *Genesis and Geology* (1871) and *The Theory of Evolution of Living Things* (1873), Henslow agreed with Wallace that mankind's peculiar physical, intellectual, and moral attributes prove that the species 'cannot have been evolved solely by Natural laws'; that 'some special interference of the Deity' was required to bestow on human beings the moral and religious nature which separates them 'essentially' from other animals. Later, however, Henslow read Romanes' books on mental evolution and adopted the ethereal hypothesis as an explanation of psychic phenomena. He came to hold that evolution not only had equipped mankind with its one 'essential' feature, its 'self-consciousness based on the power of realising abstract ideas'; evolution also gave promise that the human soul, vibrating in the ether, may be just beginning a career to be pursued hereafter.[51]

Other Darwinists maintained that evolution itself introduced (perhaps abruptly) an essential difference between mankind and the higher animals, though this was not always clearly stated. The conservatism with which Beecher beclouded the subject of human evolution did not fail to have its silver lining. It is 'by no means proved' that mankind was 'evolved from the inferior mammalian world', he stated, '. . .nor do I see yet how any bridge can be constructed over the abyss between man and his ancestors, if such there were, . . .which shall lead us to an absolute certainty'. Hence there is no necessity to 'push the origin of the race below the line of mammals'. Primitive people, it is 'safe to say', entered life 'in the lowest savage or barbaric conditions'. Similarly, it is 'very difficult' to explain how mankind's unique moral qualities might have originated, for 'between man even in his lower animal forms and the animal from which many suppose he sprang' there is a 'long and deep gulf'.[52] 'But', expanded Beecher,

if the whole theory of evolution is but a slow decree of God, and if He is behind it and under it, then the solution not only becomes natural and easy, but it becomes sublime, that in that waiting experiment which was to run through the ages of the world, God had a plan by which the race should steadily ascend, and the weakest become the strongest and the invisible become more and more visible, and the finer and nobler at last transcend and absolutely control its controllers, and the good in men become mightier than the animal in them.[53]

With God all things were possible. One of them was the 'ascending scale by which men, beginning at organized matter and steadily going up from animal to the social, to the moral, to the intellectual, and to the spiritual, at last find out. . .their birthright, and by a successive series of unfoldings learn that God is their God and that they are his children'.[54]

In Bascom's 'spiritual' evolution 'thought-power' is added to 'life-power', resulting in a 'new grade of beings' which is mankind. 'The rational element', Bascom explained, 'is superinduced on the vital element as wholly above and beyond it'. Yet since evolution remains a spiritual process, since life and mind are present and at work every-where, variably and spontaneously, in and through all physical forms, the higher human nature, no less than the human body, may be regarded as an outgrowth of organic development. It is a 'distinct increment' achieved without intervention, an 'unfolding' which con-verts the highest product of animal life into the first term of a spiritual life which itself must undergo a further evolution.[55]

From his *Outlines of Cosmic Philosophy* to his posthumously published *Life Everlasting*, Fiske took his stand with Wallace and Mivart: mankind, he believed, is essentially different from other animals and the distinguishing features of the species cannot be explained by natural selection. Wallace held that there came a critical point in human evolution beyond which mental variations were selected in preference to physical variations. Seizing on this 'luminous suggestion', Fiske drew the conclusion – an original contribution to evolutionary theory, he believed – that

the increase of cerebral surface, due to the working of natural selection in this direction alone, has entailed a vast increase in the amount of cerebral organization that must be left to be completed after birth, and thus has prolonged the period of infancy. And conversely the prolonging of the plastic period of infancy, entailing a vast increase in teachableness and versatility, has contributed to the further enlargement of the cerebral surface.[56]

In the mutually reinforcing processes of cerebral development and the prolongation of infancy Fiske discovered the origin of the family and of human society, with their accompanying moral sentiments. And through this Lamarckian evolution of mind, morals, and society 'the heaping up of minute differences of degree', he maintained, 'has ended in bringing forth a difference in kind' between mankind and other animals. There came another 'critical point' when mankind 'suddenly' put on immortality, when the omnipresent divine energy acquired in the species 'sufficient concentration and steadiness to survive the wreck of material forms and endure forever'. 'Mr Mivart has truly said', Fiske remarked, 'that, with regard to their total value in nature, the difference between man and ape transcends the difference between ape and blade of grass'.[57]

Darwin did not isolate mankind in this fashion, nor indeed did Lamarck, Spencer, and Cope. Thus many Darwinists were not even thoroughgoing Lamarckians. Those who were – those, that is, who refrained from adulterating Darwin's anthropology with the creation, by a special providence or by evolution, of an essential difference between mankind and the higher animals – were Drummond, Le Conte, MacQueary, Johnson, Matheson, and possibly also Savage. But this was just because each had embraced a theology of divine immanence in which natural selection could be absorbed. If a spiritual essence were always present everywhere in nature then no differences in kind need ever arise; God's purposes in creation could be continually and universally secured.[58]

Theodicy and universal progress

As a naturalist who knew his limitations, Darwin concerned himself first and foremost with evolution in the biological world. That there were hypotheses unifying physical, social, and spiritual development he had no doubt. That such hypotheses had been taken up by naturalists and natural philosophers, beginning in the ill-famed *Vestiges*, extending through Spencer's synthetic speculations, and culminating in the conceptual turgidity of Cope's essays, he was well aware. And for this reason he was the more determined to conduct his theorising in the realm he knew best. Now at first sight there seems little cause for Christians to have found Darwin's limitation objectionable. It might well have been argued, for example, that pan-evolutionary speculation was a metaphysical Tower of Babel. But in fact many Christian Darwinists assumed a quite different point of view. Having abandoned the old doctrines of special creation and the fixity of species, they looked to evolution as a description of the divine method of creation, not merely in the biological world, but throughout the universe. To them – if we may take a statement made by the German liberal theologian Otto Pfleiderer as representative – there was 'only the one choice: either the evolutionary mode of thought is right, in which case it must be uniform in all fields of investigation; . . .or it is wrong, in which case the views of nature acquired by means of it are not justified, and we have no right to prefer them to the traditions of faith'.[59] To these Christians, who accepted the universal validity of the evolutionary approach as a matter of theological conviction, any limit placed on evolution was *ipso facto* a virtual limit on the presence and power of God. And this, certainly, was not in keeping with a right conception of the divine character.

In this light it is possible to understand more fully why most of the Darwinists we have discussed assimilated Darwinism to a system of universal evolution. It was not just because a cosmic metaphysic provided for a Lamarckian expression of divine providence with which to overrule natural selection, though this in fact was its primary function. Another reason, which runs throughout the Darwinistic literature as an underlying assumption, was that any evolution less than universal evolution cannot be a method worthy of the universal and omnipotent Creator. This assumption is most evident, appropriately enough, among those who embraced the grandest of all the schemes of evolution, the *Synthetic Philosophy*. A regular feature of their writings is an emphatic commitment to universal creation by evolution coupled with an invidious comparison of Darwin and Spencer.

Beecher heralded the 'great cosmic doctrine' of evolution as a description of 'God's methods in universal creation' and its author as 'the ablest man that has appeared for centuries'; Darwin he did not so much as mention in *Evolution and Religion*, despite the fact of his death within the year before the second series of sermons contained in the book was commenced. Drummond made creation coextensive with the divine presence in the environment while casting Darwin and his Struggle for Life as the villains and Spencer and his altruistic law as the heroes of the evolutionary love-story of the world. Matheson stressed the agreement of Spencer's system with the opening words of the Apostles' Creed while demeaning Darwinism for giving 'a very limited view of the real scope of the evolutionist'. 'The doctrine of the unity of species would at present be better indicated by the name of Spencerism than by that of Darwinism', he observed, 'for it is in the hands of Mr Spencer that it has assumed its largest proportions and sought to embrace in its law all existing things'. Savage held that the entire creation, from star-dust to civilisations, was the outcome of God's immanent evolution. 'Long before Darwin had published his theories', he exaggerated, '. . .Spencer had conceived the idea and the outlines of a philosophy of which the work of Mr Darwin is only a subordinate part'. Fiske likewise regarded the cosmos as an evolution of the divine immanence while noting carefully that Spencer formulated his doctrine both earlier and on a larger scale than the theory of natural selection.[60]

Other Darwinists were no less insistent that creation is a universal process and evolution is its method. Temple accepted that, according to the 'abstract doctrine' of evolution, all things, organic and inorganic, are the outgrowth by 'regular law' of previously existing things, and that their growth is governed by a 'great design'. Human knowledge, he claimed, is also an orderly growth – the doctrine of evolution has been applied to it with 'perfect justice' – and so, too, is religion, which under the impulse of divine revelation has undergone 'in some respects a development parallel to that of Science'. Henslow welcomed the 'general principle of evolution' as 'one of the grandest and most comprehensive laws in the universe', embracing creation, redemption, and the hereafter. Mivart held that 'the whole phenomena of the universe – physical, biological, political, moral, and religious – may be explained and understood as a *continuous* evolution towards a *pre-ordained* end'. Supplemented by the postulates of a rational theism, he believed, Spencer's account of the process becomes the 'material expression' of 'the formal law of Cosmical Evolution': namely, 'the continuous progress of the material universe by the unfolding of latent potentialities through the action of incident forces. . .in harmony with a preordained end, such unfolding exhibiting a succession of changes from indefinite, incoherent homogeneity to definite coherent heterogeneity'.[61]

Bascom interpreted evolution as 'the development of each successive stage of the Universe and of the World as part of it, from the previous one, without external addition or modification'. To him the doctrine expressed 'the universality and continuity of intelligible relations, of creative processes'. Le Conte, followed by MacQueary, maintained that evolution 'pervades the whole universe, and the doctrine concerns alike every department of science – yea, every department of human thought'. As a 'law of continuity' and a 'universal law of becoming' it describes the 'progressive movement' of any system to 'higher and higher conditions'. Abbott regarded evolution as God's 'method of growth', whether in 'the history of creation, of providence, or of redemption in the race or of redemption in the individual soul'. 'All life, including the religious life', he declared, 'proceeds by a regular and orderly sequence from simple and lower forms to more complex and higher forms, in institutions, in thought, in practical conduct, and in spiritual experience'. Johnson understood evolution to be the 'universal method' of God's free and purposeful creativity, a creativity manifested throughout the macrocosm and notably in the progress of divine revelation, in the history of the church, and in the development of society.[62]

Even more important than universal evolution, which secured God's omnipotence, was some means by which God's beneficence could be preserved. Darwin not only conceived evolution on a limited scale; he made central to the entire course of organic development a seemingly cruel and wasteful process of natural selection which, however one might try to minimise its role in the actual creation of new species, did indubitably play an important part in determining the species that would survive. Darwinists, therefore, compelled by tooth and claw to shoulder the burden of theodicy, took up a notion that was at once Christian in origin and harmonious with the spirit of the age: they adulterated Darwinism with the concept of inevitable material, social, and spiritual progress. Now historians seem generally agreed that the belief in progress during the later nineteenth century, especially in its Christian manifestations, was usually 'a way of avoiding the unpleasantly relativist implications of a world in which many of the old certainties were disappearing'.[65] There might be more truth in this observation if it read 'reinterpreting' for 'avoiding'. For among Christian Darwinists, at any rate, the conviction that the world was moving on to bigger and better and more desirable states of affairs had its origin less in an effort simply to avoid the unpleasant implications of natural selection than in a shrewd attempt at reinterpreting natural evils as preconditions for progressively greater goods.

This certainly was the main rationale offered by Spencer's henchmen. Beecher followed his paean to the divine plan 'by which the race should steadily ascend' with the assurance that if this progressive plan 'be added to evolution', then the entire process becomes a contribution to faith in God's existence and 'gives revelation and rest to many a doubt as to why the world was left as it was'. Drummond admitted that the Struggle for Life 'appears irreconcilable with ethical ends, a prodigious anomaly in a moral world', but argued that since it is the 'efficient instrument of progress', and then merely 'half the instrument' (the other half being the Struggle for the Life of Others), there can be no question of vindicating the spiritual forces of evolution. Even at its 'very terrible price', he believed, evolution was 'none too dear', for the thing purchased was 'nothing less than the present progress of the world'. Matheson announced that the difficulty of death in the struggle for existence had been 'dispelled by the system of evolution'. 'The history of evolution', he wrote, 'is the history of a progressive vitality – we might almost say, of a progressive immortality; it is the record of that process by which life gradually obtained such a mastery over death as to enable it to become a living soul'. 'So progressive' is the scheme of evolution in realising its 'underlying purpose', the creation of the immortal human spirit, that it 'cannot be accidental, so orderly that it cannot be unintelligent'. Savage regarded temporal evil as 'not a real thing at all' but only 'temporary maladjustment' which is swallowed up in 'vistas of eternal progress, star-lighted pathways that lead on and on in light, in truth, in joy, in peace, in service, forever and ever'. Less glibly, Fiske discerned in moral progress the sanction for natural selection. He believed that 'the very same forces. . . which through countless ages of struggle and death have cherished the life that could live more perfectly and destroyed the life that could only live less perfectly, until humanity. . .has come into being as the crown of all this stupendous work – . . .these very same subtle and exquisite forces have wrought into the very fibres of the universe those principles of right living which it is man's highest function to put into practice'.[64]

Other Darwinists developed the same rationale in closely similar ways. Temple maintained that nature's 'imperfections' are a 'necessary part' of the 'great design' which is gradually being realised, a beneficent plan of 'perpetual progress'. Henslow interpreted the world's 'inideality' as 'the correlative result of the processes of development and evolution', the inideal providing the conditions of mankind's temporal probation, the 'law' of evolution promising temporal progress and eternal reward. Bascom viewed the Kingdom of Heaven, towards which all things inexorably press, as the 'adequate goal' that compensates for the transient sufferings of organic development. In a spiritual evolution, he explained, 'all conflicting tendencies are partial and temporary, mere eddies in the stream, a retreat simply to gain fresh vantage for progress'. Le Conte held that evil 'has its roots in the necessary law of evolution'; that evolution, however, dictates the 'progressive movement' of the world to 'higher and higher conditions', a movement which has culminated gloriously in the ascent of mankind; and that therefore evil is a 'necessary condition of all progress, and pre-eminently so of moral progress'. Johnson perceived in nature a balance between conflict and cooperation which reflects on the Creator neither good nor ill, but in *God in Evolution* (1911) he reasoned that since 'evolution *is* a progressive continuity, a unified process of ever-increasing complexity', and since this process has produced the highest morality by repeated victories of cooperation over conflict, God's method in creation can be positively vindicated.[65]

Thus, in sum, it would appear that liberal Christians who preserved God's purposes in nature by subjecting natural selection to the components of Lamarckian evolution had, in so doing, embraced concepts of a universal progressive providence through which the divine omnipotence and beneficence could also be preserved. For these Darwinists human evolution was also a problem but, given the right metaphysics, not an insuperable one. For others, however, the reverse was true. While maintaining their teleology and theodicy intact, they could approximate to Darwin's views on the nature and extent of evolution, including the role of natural selection, and yet draw back from the Darwinian doctrine of the descent of mankind. Two such Darwinists came from the College at Princeton, New Jersey, which was otherwise noted for its Old School Presbyterianism and the oracular anti-Darwinism of Charles Hodge.

The Darwinists nearest Darwin

After Hodge's death in 1878 his students and colleagues could safely entertain an evolutionary account of creation. There is no better evidence that they did, and did so publicly, than the volume published in 1886 under the title *Theism and Evolution*. Its author was JOSEPH S. VAN DYKE (1832–1915), a graduate of the College of New Jersey and, in 1861, of Princeton Theological Seminary, a pastor of Presbyterian churches, and a sometime tutor at the College.[66] The author of the book's introduction was Archibald Alexander Hodge, the son of Charles and his successor as professor of theology in the Seminary. Neither Hodge nor Van Dyke regarded evolution with much enthusiasm. Their acceptance was half-hearted at best. But so far do the statements of *Theism and Evolution* differ from those of *What Is Darwinism?* that both men seem to be conspicuous evolutionists.

Hodge's introduction is a dogmatic statement of the limitations of evolutionary science. Evolution cannot account for the 'origin, causes, and ends of all things'. Much less can it explain 'the plain facts of man's spiritual nature, his reason, conscience, and free-will'. Restricted thus to a 'scientific account of phenomena and their laws of co-existence and of succession', Hodge declared, '...evolution is not antagonistic to our faith as either theists or Christians'. 'Evolution considered as the plan of an infinitely wise Person and executed under the control of His everywhere present energies can never be irreligious; can never exclude design, providence, grace, or miracles.' There is 'nothing to fear from the ultimate results' of such a doctrine as a factor in science.[67] But, he warned, if a theory of evolution 'assumes to be a philosophy, or becomes associated with a philosophy supplying the ideas, the causes, and the final ends which give a rational account of the facts collected', then Christians have everything to fear. If, for example,

progress along the entire line of biological advance is explained wholly on the hypothesis of an all-directioned variation, and the selection of special forms by an accidental environment (the precise position of Darwin), then certainly the universe and its order is referred to Chance, teleology is impossible, theism stripped of its most effective evidence, and therefore Dr Charles Hodge was abundantly justified in indicating this phase of evolution as atheistic.[68]

The younger Hodge was, after all, very conservative. He discarded few of his father's beliefs. Although he knew which versions of evolution could not be true, he had no intention of expressing 'any opinion as to the truth of evolution in any of its forms'. Yet he must be credited for placing his *imprimatur*, the honoured name of Hodge, and a marginally favourable statement on evolution in a book which was intended, by its very title, to counteract the teaching of *What Is Darwinism?* Surely this, coming in the last year of his life, was a turning point for the acceptance of evolution among American Protestants.[69]

Theism and Evolution was a timid book, preponderantly negative in its approach. Van Dyke occupied himself more with defending the faith from evolution than with scrutinising the faith in its light. While 'conceding that evolution may give a new impulse to embodied christianity, relieving it of some objectionable features, furnishing attractive arguments in its favour, and teaching the church to employ new agencies in the elevation of humanity', he was concerned primarily 'to present an argument against those forms of the evolutional theory which seem to tend towards atheism'.[70] Accordingly, most of the book is given to showing that any theory which presumes to account for the origins of matter, energy, life, mind, and freedom of the will apart from the creative and superintending intelligence of a personal God can have no claim to credibility. What little Van Dyke had to say constructively about evolution is confined to a few initial chapters.

Whether evolution as a theory to account for 'the successive appearances of new species' from pre-existing species is true, is a question of fact. It 'must be settled by induction', Van Dyke declared, 'not by *a priori* arguments'. Induction may yield divergent results. Biologists may disagree concerning the agencies by which new species have emerged. Yet it is possible that 'they should succeed in establishing a law of evolution without being able to specify the causes which produce the ever changing series of effects'. And any law thus conceived would be entirely reasonable. To regard evolution 'as absurd, especially in its more modest pretensions, is to acknowledge ignorance of facts, or to confess oneself under the influence of strong prejudice'.[71] However, it would bespeak an equal prejudice to regard evolution in any of its forms as fully established. Thus, Van Dyke temporised, it would seem to be 'the dictate of prudence to concede that at present it is difficult, practically impossible, to fix the limits of species; more difficult still to fix those of genera; simple folly to attempt to determine those of tribes and families'. To those who found even this concession objectionable, he explained that

a system of faith which outlived the scientific dictum of the fixity of the earth can easily display vitality sufficient, if necessary, to survive even after the doctrine of the immutability of species has been reverently laid away in the roomy receptacle of perished beliefs. We shall be forced to acknowledge that the permanence of species is a doctrine which is in no sense needed for the defence of Scripture.[72]

Van Dyke was receptive to evolution primarily as a biological theory. Spencer and his 'law' gave him pause for jest but Darwin and natural selection were worthy of serious consideration. Darwin, he said, occupies the 'foremost rank among evolutionists'. According to his hypothesis of natural selection, 'the inherent predisposition in plants and animals to vary has sufficed, in conjunction with the causes originating in the intense struggle for existence, to modify all species and to produce the present diversity'. Van Dyke admitted that 'this hypothesis, though embodying much and explaining not a few of the facts, is not viewed, even by a majority of evolutionists, as an adequate explanation'. Indeed, if Mivart regards his work 'as an agency', it alone 'furnishes theologians with an unanswerable argument against this particular form of evolution'.[73] But Van Dyke was not about to dismiss Darwinism on the authority of contemporary science. He could anticipate that one day the theory might be proved, and against that time, as well as against the elder Hodge, he wished to vindicate it from the charge of atheism.

In a chapter entitled 'Is it atheism?' Van Dyke's first task was to prove that 'progressive development is not necessarily hostile to theism, nor to any statement contained in Scripture'. Having thus established that organic evolution in general raises no new issue between theist and atheist, he undertook to show that the Darwinian theory in particular is not atheistic. 'Its author rejects spontaneous generation', he said.

His hypothesis has to do, not with the cause of the phenomena, but with the mode of their manifestation, thus leaving the question of design untouched. Are we to conclude that the diversification of organic forms, consequent upon the struggle for existence or upon other secondary causes, excludes the possibility of design? Certainly not. Unless an evolutionist affirms that the causes to which he refers changes are self-sufficient, he is not open to the charge of atheism.[74]

'It is true', Van Dyke admitted, 'that Darwinism by its apparent substitution of Natural Selection for design seems to have taken a step towards atheism'. But suppose that the selected variations 'are of the nature of origination and occur under the guidance of Divine Omniscience, secondary causes being employed to produce them'. Then, he declared, far from being undermined, the teleological argument would be strengthened. Or, rather, a new teleology would emerge, one which, unlike the old, would find in over-production, the struggle for existence, and the survival of the fittest, a larger rationale for much of the world's seeming purposelessness. 'We are at liberty to assert', he concluded, 'that if Darwinism should become an established theory, . . .there is no just cause for fear'.[75]

Van Dyke did not differ from the elder Hodge so obviously on the subject of human evolution. Certainly he could countenance the possibility that mankind's descent from some ape-like creature might be proved, but in his opinion the proof would probably not be forthcoming. Surrounding Darwinism with all the standard obstacles and a few less common ones – the fittest survive as a degenerate class because they have been weakened by their struggle, human infants are weak and helpless whereas young monkeys are sprightly and independent – he thought it merely 'conceivable' that these could be surmounted in the evolution of the human body. At this point Darwinism was as yet a 'conspicuous failure'.[76] However, Van Dyke found it 'almost impossible to conceive' that the human mind had developed from the instincts of lower animals. There was, he thought, a tremendous presumption in favour of its 'immediate creation' by a Supreme Being. Indeed, because mind and body are one, there was a large presumption in favour of the special creation of the human body as well. Together these presumptions grew in the light of the moral consciousness of mankind, 'the clearest evidence' of being created in the divine image, and at last became fully certain in the light of mankind's religious nature. Van Dyke therefore expressed 'firm faith in the final adoption, even by scientific men, of the Scriptural account of man's origin'.[77]

In the last result Van Dyke also was very conservative. So well had he learnt his lessons at the feet of Charles Hodge that Darwinism seemed hardly a challenge. A small adjustment here, a minor conces-

sion there – if the theory should be proved theology need do no more
to accommodate it. Thus Van Dyke concluded his work with two
chapters entitled 'Science and the Bible; no conflict'. Effortlessly he
argued for 'perfect harmony', throwing the burden of disproof on
unproven theories of evolution. But, after all, the depth of *Theism and
Evolution* was not the measure of its importance. Its significance lay
rather in its deliberate openness to Darwin and in its calculated
departure from the anti-Darwinism of the elder Hodge.

The first American Protestant theologian and religious leader to express
publicly his sympathy for Darwinism was JAMES MCCOSH (1811–
1894). Born in Ayrshire and educated at the Universities of Glasgow
and Edinburgh, McCosh served at Arbroath and Brechin as a minister
of the Church of Scotland, becoming while in the latter pastorate a
'trusted member of the inner circle which carried the Free Church
movement to a successful conclusion' in 1843.[78] His first book, *The
Method of the Divine Government, Physical and Moral* (1850), earned
him the praise of William Hamilton, whose intuitionist philosophy he
had adopted, and of Hugh Miller, the lay-geologist and publicist of
the Disruption. On the strength of the volume and the commendations
it received McCosh was appointed to the chair of logic and metaphysics in Queen's College, Belfast. He remained there from 1851
until 1868, when he crossed the Atlantic and assumed the presidency
of the ailing College of New Jersey – later Princeton University.

McCosh was not a progressive educator. He defended a classical,
prescribed curriculum and thereby embroiled himself in controversy
with Harvard's innovative President Eliot.[79] However, the College of
New Jersey did see some substantial changes during the twenty years of
his administration. Its enrolment more than doubled, an extensive
building programme was undertaken, and a school of science begun.
Partly for religious reasons, science at Princeton had never achieved
the prominence it enjoyed at Harvard and Yale. McCosh, who, unlike
most Americans, was thoroughly conversant at an early date with the
Darwinian debates, did what he could to rectify the situation. In fact,
his attitude towards the religious bearings of science was as liberal as
that of Eliot. 'We give to science the things that belong to science, and
to God the things that are God's', he declared. 'When a scientific
theory is brought before us, our first inquiry is not whether it is consistent with religion, but whether it is true.'[80] McCosh did not occupy
his new post for a week before expressing to the upper classes of the

College that he was fully in favour of evolution, provided that it was 'properly limited and explained'.[81]

The Method of the Divine Government not only established McCosh in an academic career; by its teaching that God governs the world both by law and through a 'complication' of laws which produces 'fortuities', it allowed him to assimilate Darwinism with minimal difficulty. Such was not the case with *Typical Forms and Special Ends in Creation*. Published in 1855 under the joint authorship of McCosh and his colleague George Dickie, professor of natural history in Queen's College, the book represented the last and greatest strength of natural theology in the pre-Darwinian period. In a detailed historical review of teleological reasoning in science and theology L. E. Hicks called it 'the most admirable design argument ever produced'.[82] McCosh and Dickie, inspired by Owen's transcendental morphology, endeavoured to show that principles of unity and order run throughout the universe and that these principles are subject to designed adjustments which adapt them to various conditions. In the organic world the forms and colours of plants, the typical forms of articulates, molluscs, and radiates, and the archetype and homologies of the vertebrate skeleton; in the inorganic world the forms of crystals and the movements of the heavenly bodies – all testify by their combination of typical order and special adaptations to the wisdom, power, and goodness of God. The fixity of biological species, needless to say, was everywhere assumed.[83] 'What ruined the run of this book', remarked George Macloskie, professor of biology at Princeton and a student at Queen's College during McCosh's tenure there, 'was the appearance ...of Darwin's "Origin of Species", which carried the whole controversy into new regions'. Although Dickie was not about to inhabit the new regions, 'in this juncture...McCosh showed his characteristic readiness to learn, his honesty in discarding his published opinions, and his courage'. Coupled with an outspoken acceptance of a chastened doctrine of evolution, this attitude was perhaps the best service McCosh rendered the cause of Christianity in his day.[84]

Throughout his numerous publications on subjects concerned with science and Christian faith, including *Christianity and Positivism* (1871), *Development* (1883), *Herbert Spencer's Philosophy* (1885), and *The Religious Aspect of Evolution* (1888), McCosh employed the terms 'evolution' and 'development' interchangeably, sometimes to refer to Darwinism and sometimes to refer to Spencer's philosophy. By employing both terms he defused the explosive potential of 'evolution' and taught his readers that the word simply stands for the divinely governed succession of phenomena everywhere in the world. At the same time, however, in so generalising or simplifying the matter, he was neither unfair to Darwin nor uncritical of Spencer. In *Christianity and Positivism* McCosh referred to Darwin's 'extensive and accurate acquaintance with all departments of Natural History, the pains taken by him in the collection of facts, and the simple and ingenious way in which he stated them'. Of Spencer, on the other hand, he wrote disparagingly, 'Give him a set of facts, and he at once proceeds to generalize them, and devise a theory to account for them. He evidently regards it as his function to unify the metaphysics of the day and the grand discoveries lately made in physical science.' Spencer was the philosopher of the universe, Darwin the historian of organic nature, and although McCosh aligned himself with a philosophy to which Spencer was also indebted, he left little doubt as to whose account of 'development' he preferred.[85]

McCosh also left little doubt as to the mechanism of evolution he preferred. At first he was cautious. Either special creation or evolution by natural selection might have been the divine method of originating species, he wrote in *Christianity and Positivism*. But since Darwin had 'copiously illustrated' the latter method he was 'inclined to think the theory contains a large body of important truths'. To be sure, he added that natural selection 'does not contain the whole truth, and. . .it overlooks more than it perceives'. The nature of heredity was unknown, and ill-explained by Darwin's 'exceedingly vague and confused and complicated hypothesis of pangenesis'; the geological record was flagrantly imperfect; and the critics of natural selection were of no mean stature, whether Agassiz and Thomson or those such as Mivart, who had initially been taken with the theory. 'I am far from saying that some of these formidable objections, supported as they are by an array of facts, . . .may not be answered', said McCosh. 'But this is certain, that for years, perhaps for ages to come, it will be an unsettled question whether Natural Selection can account for all the ordinary phenomena of the modification of organisms.'[86]

A decade later the question was still of course unsettled. In his pamphlet entitled *Development* McCosh took a cue from Darwin's admission that he had perhaps attributed too much power to natural selection and, following the lead of Owen, Mivart, and Cope, explained that the origin of species was due to both external and internal causes, in some cases the one and in some cases the other being more important. But in later years the ambiguity in this position was largely eliminated. In the *Religious Aspect of Evolution* McCosh specified the causes of transmutation as: first, natural selection, an agency 'undoubtedly operating in all organic nature' and having 'mighty influences'; second, use and disuse, an agency 'for physiologists to explain'; and third, the direct influence of the environment, 'an agency much dwelt upon by Cope and Hyatt, and...undoubtedly acting everywhere in nature'. This seems to have been McCosh's final statement on the factors of evolution and one cannot but doubt whether Darwin would have put it much differently. Was it not after all a Neo-Lamarckian, Clarence King, who twitted McCosh as 'a spinner of Darwinian theology'?[87]

As McCosh approached the subject of human evolution, however, he became less and less Darwinian. There was a 'profounder set of facts' connected with life, sensation, instinct, intelligence, and morality which lay beyond the pale of natural selection. Indeed, these phenomena probably could not even have been produced by the 'ordinary powers of nature'. At this point McCosh nearly abandoned the doctrine of a perpetual providence in favour of supernatural interventions. He kept hold of his old belief only by allowing that science may possibly explain the 'profounder' facts. The presence of a vital principle in nature, he said, is 'for science to settle'; if it is discovered it will certainly be found to have its own laws. Sensation and instinct are 'new powers', but 'whether introduced by natural or supernatural causes we may not be able to determine'. All these powers, including intelligence and morality, may well have been 'evolved...out of other agencies'. If so, however, they retain their uniqueness in the fact that they were 'called in at an appropriate time'.[88]

McCosh believed that the uniqueness of *Homo sapiens* is not, in the last result, the providential outcome of immanent causes. It is the product of divine intervention. For this Wallace offered evidence, showing that natural selection alone could not account for the frame and constitution of mankind, and Genesis provided proof.[89] The biblical narrative did not seem to preclude the interpretation that the human body was formed 'out of existing materials' and 'by secondary causes'. Nor did it state clearly whether the 'breath of life' came by natural or supernatural means. But that the text in fact required some kind of 'interposition and addition' from God in the fashioning of the human soul was unquestionable. 'If any one asks me if I believe man's body to have come from a brute, I answer that I know not', McCosh stated in the last chapter of the *Religious Aspect of Evolution*. 'I believe in revelation, I believe in science, but neither has revealed this to me; and I restrain the weak curiosity which would tempt me to inquire into what cannot be known. Meanwhile I am sure, and I assert, that man's soul is of a higher origin and of a nobler type.'[90]

Thus, save for his views on human descent, McCosh might be called a Darwinian. He attributed much to natural selection when others found it impossible. He could do so because it never occurred to him that natural selection should diminish the force of the argument from design. Not once in more than forty years did he question that the order and adaptations of nature testify to the wisdom, power, and goodness of the Creator. Not once did he doubt that living organisms had been created according to 'common forms with adjustments to a purpose' and by means of 'common causes' instituted and collocated by God. The older naturalists held that these forms governed the original creation of fixed species. McCosh came to believe they were in error.[91] Yet he was able to retain their teleological point of view in conjunction with a belief that a sovereign God had superintended the causes by which, in time, each form in all its various adaptations had been evolved. 'The supernatural power', he declared, 'is to be recognized in the natural law. The Creator's power is executed by creature action. The design is seen in the mechanism. Chance is obliged to vanish because we see contrivance.... Supernatural design produces natural selection. Special creation is included in universal creation.'[92]

There were objections to a Darwinian natural theology and McCosh had ready answers for them. The survival of the fittest is ultimately benevolent, for it promotes the strengthening of a race. Where beauty is absent there is utility, and this also is one of God's provisions. If the suffering and calamity in nature seem great, it must be remembered that the world is not unmitigatedly evil, that there is enough goodness and wisdom revealed to assure us that behind all stands a beneficent and omnipotent Creator.[93] This Creator, moreover, has given evidence in the progressive development of the organic world that one day evil will vanish and the earth will become 'a perfect abode for a perfected man'. 'Nature is struggling', said McCosh, 'but it is in order to improvement. It is plowing and sowing, but in order to reap in due season. It is moving onward, but also upward. It is groaning, but it is to be delivered from a load. It is travailing, but it is for a birth. It is not perfect, but it is going on toward perfection.' The causes were already at work which would effect this millennial transformation.

Education was spreading, agriculture advancing, and commerce binding nations closer together. These were but the first fruits of the better era to which the scriptures continually look forward.[94]

Critics of McCosh have not always done him justice. Typically, they have tried to make him less Darwinian than he actually was. One claims that by 'development' McCosh 'did not for a moment mean Darwinism': that is, he did not mean 'explicit Darwinism'. Another says that natural selection was 'rendered superfluous' by his belief in providence. Another says that he assigned natural selection a 'subordinate role' among the causes of evolution. Another, referring to his 'irrationalism', states that he 'lacked the courage' to extend evolution to mankind, that he 'shrank with horror from the picture drawn by Darwin'. Still another critic even tries to show that, regardless of his public pronouncements on evolution, McCosh 'privately. . .seemed to *have had some doubts*'. But none of these attempted detractions is quite successful.[95] With the possible exception of certain psychological phenomena and the creation of the human soul, McCosh brought the entire natural world, in principle at least, within the sphere of causomechanical explanation. The explanation of evolution which he seems to have favoured most was natural selection. And that theory, or any other theory of evolution, he confined to its proper domain, reserving statements of universal providence and progress for a biblical theology. 'Evolution might be the key that should unlock the mysteries of nature; for McCosh, God was the key that should unlock the mysteries of Evolution.'[96]

God was of course the key that unlocked the mysteries of evolution for all the Christian Darwinists. But the God in whom one believed had everything to do with the kind of evolution whose mysteries had been unlocked. Those who could only discern God's purposes in nature if they were ascribed primarily to causes other than natural selection seemed bound to interpret evolution as the expression of a universal progressive providence through which the divine character could also be vindicated. Human evolution for them did not constitute an insuperable difficulty if the divine immanence were rightly construed. Those, on the other hand, who could see God's purposes being realised through natural selection had ways of accounting for his omnipotence and beneficence which were not necessarily incompatible with Darwinism. For them the main stumbling-block was human evolution and in this matter their outlook remained pre-Darwinian. In the first instance we find the majority of Christian Darwinists transforming Darwinism with theories of Lamarckian evolution which embodied doctrines of providence and progress characteristic of post-Kantian liberal theology. In the second instance we find a minority of two representing a theological tradition which served well as a basis for Christian Darwinism.

I I

~~~~~~

CHRISTIAN DARWINISM:
THE RELEVANCE OF ORTHODOX
THEOLOGY

PALEY. May I, without impertinence, Mr Darwin, inquire in what sort
of estimation you were held among the orthodox Christians of your time?
    DARWIN. I hardly know how to answer your question. I enjoyed the
friendship of many of them, and incurred, so far as I am aware, the ill will
of none. Why should I, indeed? I never assailed their doctrines.
    PALEY. But was not the Christian world alarmed by your speculations?
Did it not protest against them? Had your contemporaries grown wiser
than the Apostle, and did they believe that all danger from philosophy
falsely so called had passed away?
    DARWIN. No; but they thought, I imagine, that philosophy falsely so
called could be exposed as false.[1]

If ever there was a contradiction in terms it must surely have been
Christian Darwinism. What concourse could a mere theory of biological development have with a religion which proclaimed God to be
Maker of heaven and earth? What constructive relationship could
possibly subsist between a theory which taught the survival of the
fittest 'fortuitous' variations in a brutal struggle for existence and a
theology which taught God's designing providence in a creation that
he saw was 'good'? What conceivable logic could there be in uniting a
theory of mankind's physical and psychical evolution from lower
animals with a belief that human beings were uniquely created in the
image of God? The name Christian might have been annexed to anti-
Darwinism or perhaps to some version of evolution which did honour
to the purposes and character of the Creator, but never, surely, to that
theory set forth by the agnostic naturalist Charles Darwin.
    Yet Christian Darwinism did exist – the appellation was used as
early as 1867 – and its representatives on both sides of the Atlantic
were among the ablest and most orthodox of the post-Darwinian
controversialists.[2] In Great Britain two theologians, James Iverach and
Aubrey Lackington Moore, found room for Darwin's science in a fresh
understanding of the historic doctrine of a triune God.

Iverach, Moore, and the doctrine of God

The Scottish contribution to Christian Darwinism was made by JAMES IVERACH (1839–1922). A native of the parish of Halkirk, Caithness, Iverach entered the University of Edinburgh in the same session which saw the publication of the *Origin of Species*. Four years later, having taken honours in the mathematical and physical sciences, he followed in the Free Church tradition of his youth and began preparation for the ministry at New College. In 1869 Iverach was ordained to the charge of West Calder near Edinburgh, where he identified closely with the mining community, and in 1874 he was translated to Ferryhill in Aberdeen. At Ferryhill he worked long and hard, building up a flourishing congregation and in his spare time studying science and philosophy in an effort to relate his faith to modern thought. His labours were rewarded in 1887 by an appointment to the chair of apologetics in the Free Church College, Aberdeen. Over the next three decades Iverach occupied in turn the chairs of dogmatic theology and New Testament exegesis, and from 1905 to 1907 he served as principal of the College. In retrospect, however, there can be little doubt that his best work was done in defending and restating Christian truth in the face of scientific avowals of naturalism, agnosticism, and speculative philosophy. 'The brightness of his Christian faith, inherited and then made his own', wrote one appreciative student, 'could not be disintegrated by the ever-growing complexities of the Spencerian evolution, nor obscured by the dazzling cloudbanks of Hegelian speculation'. Among Iverach's several works dealing with philosophical theology the most important are two 'Present Day Tracts' published by the Religious Tract Society, *The Philosophy of Mr Herbert Spencer Examined* (1884) and *The Ethics of Evolution Examined* (1887), and three longer monographs, *Is God Knowable?* (1884), *Christianity and Evolution* (1894), and *Theism in the Light of Present Science and Philosophy* (1899).[3]

Iverach was convinced that Darwin's theory 'may be held in such a form as to have no dangerous consequences for philosophy or theology'. He was no less certain that Spencer's law of evolution, with its 'far reaching consequences', is 'altogether different from the scientific theory of Mr Darwin, with its limited range and carefully guarded statements'. Darwin's theory is a 'good working hypothesis', albeit an hypothesis 'attended with many difficulties'; Spencer's law has such difficulties that it 'may be disproved, and shown to be an untenable hypothesis'.[4] The criticisms Iverach made were neither the carping of a threatened cleric nor the niggling of a scholastic theologian. He boldly tore apart the philosopher's vacuous definitions, disputed fairly the facts they were supposed to explain, denied Spencer's assumption that the effect is more complex than the cause, insisted, above all, on the recognition of irreducible rational self-consciousness as prior to all scientific understanding, and relented somewhat only when it became apparent that Spencer was willing to concede some positive statements about the 'Infinite and Eternal Energy' which pervades the cosmos.[5]

Thus the form in which Iverach thought Darwin's theory might safely be held was not one of agnostic evolution. Or, to be more precise, it was not one of agnostic evolution according to an all-embracing law. For while cautioning that 'any tenable view of the nebular hypothesis, or any view consistent with the facts, has presented that hypothesis in a form which is not available for the purposes of evolution', Iverach was quite willing to encompass all things, from nebular beginnings to a biological theory characterised, as was Darwin's, by a 'limited range and carefully guarded statements', in an 'evolution' which begins and proceeds in accordance with free and rational purpose. Evolution can be neither universal nor true if it is supposed to get the determinate from the indeterminate, intelligence from the unintelligible, something from nothing. But evolution, conceived simply as the orderly succession of causes which work out the sovereign purposes of the Rational Power from whom all things proceed, is most certainly true and necessarily universal.[6] It is the inescapable postulate of minds which know from their own experience that intelligence can be the only source of order. It is not a religious alternative to a scientific conception of the world but an interpretation of scientific findings from a religious point of view. 'We do not interfere, in any way, with the work or the method of mechanical science', said Iverach, 'when we take their results and show that they may be read in another fashion'. On the contrary, the work of science is strengthened by the religious conviction that nothing occurs by chance. 'The idea of law and uniformity is...quite consistent with the idea of purpose', for 'purpose excludes arbitrariness and irregularity'.[7]

Darwin's theory of biological evolution by natural selection could therefore be seen from two points of view, one strictly scientific and the other teleological. From the first, natural selection offers only a fragmentary conception of nature, abstracting organisms from the concatenation of causes which have produced them, ascribing their variations to unknown laws (or sometimes, wrongly, to chance) and their survival to such variations as have putative utility in the struggle for existence. From the second, the religious viewpoint, natural selection is seen as it truly is: as an expression of the sum total of causes, internal and external, which have transpired to the end that just those forms of life which are currently observed should exist. Interpreted thus, natural selection does not damage the argument from design but actually strengthens it. Objecting to Huxley's use of a rifle bullet fired straight at a mark to describe the creation of organisms by design, and grape shot their creation by natural selection, Iverach pointed out that

it is not necessary to teleology to suppose that 'each organism is fired straight at a mark'. What is necessary is that the organism hits the mark. If the hitting of the mark is accomplished by a persistent process prolonged throughout the centuries, implying completeness of arrangement and adjustment of means to ends in a complicated series, then the result is not against teleology; on the contrary, it simply heightens our view of the skill of the teleologist.[8]

'The argument from order to intelligence is much more cogent than it was in Paley's time', Iverach declared. 'No one ever strengthened the argument as Darwin has done.' Darwin's evidence for organic evolution is 'irresistible'; the reality of natural selection is undeniable; and even the creation of the eye – Paley's favourite illustration – as described by Darwin need not be termed improbable. 'What is essential', said Iverach, 'is that we maintain and vindicate the continued dependence of all creation on its Maker, and that if things are made so as to make themselves, God is their Maker after all'.[9]

But of course the Darwinian theory was not without its difficulties. Iverach had a large and largely accurate understanding of the controversies that raged about natural selection in the last decade of the century – the three-sided debate between Weismann, Spencer, and Romanes and the opposition of American Neo-Lamarckians – so that under the circumstances his adherence to Darwinism was only as great as could be expected from any non-specialist. The 'factors of evolution', he realised, were much in dispute. He had 'tried to read with an open mind what has been written on natural selection' and was unable to see that its supporters 'succeeded in using the phrase in a consistent manner'. Sometimes it is assigned a constructive role; on other occasions its function is said to be simply destruction. This is misleading, for natural selection, Iverach agreed, can produce nothing by itself; scientifically its function can only be described as elimination of the unfit. At the same time, however, those who speak of natural selection inconsistently – the 'pure Darwinians of every shade' – are obliged to find utility everywhere in nature, and in so doing they perform the good service of strengthening the conviction that natural selection is but the expression of a rational teleology. In the literature of Darwinism, Iverach observed, 'it looks sometimes as if...we had a teleology run mad. No Bridgewater treatise is so teleological as almost any Darwinian book we may happen to open.' Natural selection may be described as a metaphor, 'but as soon as we begin to work with it its metaphorical character disappears, and it becomes intensely real, and is quite capable of doing anything. It has the character constantly ascribed to it both of a directing agency and of a presiding intelligence; and it does seem as if both were needed if evolution is to be an intelligible process.'[10]

Since Iverach held that, scientifically speaking, natural selection cannot play a constructive part in evolution, the real question was the origin and nature of the variations selected. Here the suspended judgement of Romanes was his example. Were variations in some measure the result of the inheritance of acquired characters? 'We may wait until the controversy is settled', he replied, adding prophetically, 'the issue may be decided in the next century'.[11] For the present it was enough to acknowledge that 'the idea of indefinite variation is becoming antiquated, and that of definite variation coming more and more to the front'. This will not impair the action of natural selection, said Iverach, citing Huxley in support, but will encourage the search for the unknown laws of variation which has been frustrated by repeated appeals to fortuitous and indefinite variation. And these laws, in turn, when discovered, will 'help us to a new conception of order and stability, and give a new meaning to design'. They will lead us to see 'the working out of the wonderful unity of plan in the millions of diverse living constructions, and the modifications of similar apparatus to serve diverse ends'. They will give evidence that 'the Power to which the plan is due is never absent from the working out of it' and will show how this Power 'delivers us from the tyranny of chance'.[12]

Therefore, according to Iverach's rational teleology, the Darwinian mechanism is a proximate explanation of the purposeful guidance by which God's will is realised throughout the organic world. 'From one point of view', he realised, 'natural selection gives that guidance'; but then 'natural selection is itself nothing but a set of conditions which may be dealt with quantitatively and mechanically'. Doubtless there is a selective power in life, a power of 'almost infinite precision' which is unremitting in its conservation of evolutionary gains and pays 'unremitting attention to the movement of life in relation to wider unities and large meanings'. The question is, 'Can you rationally predicate such qualities of a series or congeries of varying agencies hypostatized under the name of natural selection?' Iverach was not inclined to think so because, with Darwin, he refused to believe that life's manifold and marvellous adjustments 'have emerged as the outcome of an infinite series of trials and errors'. 'If these adjustments are there now in almost infinite precision', he said, 'we naturally think that the steps which led up to them were not lacking in precision. If the outcome of the process is full of such interrelations as are described in every book that treats of evolution, surely we may infer that the processes are also intelligible.'[13]

Iverach's anthropology, like his cosmology and evolutionary teleology, was founded on rational self-consciousness. Mankind has fully evolved from other forms of life; neither physically nor psychologically is it possible to say where the differences which separate *Homo sapiens* from the lower animals begin to emerge. Yet the result of human evolution is a creature whose rationality and self-consciousness are utterly unique. It is for this reason, Iverach believed, that 'comparative psychology can make little progress'. Human beings are too apt to read their own rationality into what they observe. In his works on mental evolution Romanes 'has done more than any other man in the attempt to prove that the intelligence of animals is the same in kind as the intelligence in man, though he admits a difference in degree', and his attempt is without doubt 'the most valuable even to those who disagree with him'.[14] But in fact he has failed to make his case, for there is no means in principle by which a difference in kind might ever arise. Romanes attributes a difference in kind to a difference of origin. Theologians, however,

know of only one origin for the universe. ... They do not distinguish between man and the lower animals by a difference of origin; for all derived existence must, they believe, trace its origin to God. ... As far as the question of origin is concerned, there is for the theologian no question of difference in kind. ... Nor for the evolutionist can there be. ..any difference of kind; for all things are from the primal source of being whatever that may be, and all things are what they are by the same kind of process.[15]

What, then, is a difference in kind and why has Romanes failed to perceive it between mankind and other animals? According to Iverach, a difference in kind is simply a very large difference in degree. Romanes acknowledges the rational self-consciousness which distinguishes the human species but he calls it a difference in degree because he has not reflected sufficiently on its uniqueness: because, in other words, he is too ready to find its adumbrations in the rest of the animal kingdom. But 'we shall not quarrel about the phrase, if we get the thing', Iverach conceded. 'We say it is a great distinction, call it as we please.' There is 'nothing like it in the world beneath' and, so far as evolution is at present concerned, 'it is just bound to accept it, and to accept it without explanation'.[16]

This was not of course to imply that just any explanation of mankind's uniqueness might, in the long run, suffice. It would have to be an explanation 'such as will not break up the unity of human nature, and assign the origin of [man's] body to one set of causes and his mind to another; and it must not bring in a cause here which operates only at this point or at a few other points in the whole history of the earth'. On neither count did Iverach find Wallace's approach to human evolution acceptable. By bringing into action 'some new cause or power' at each of three stages in the development of the world – the transition from inorganic matter to organic, the genesis of sensation and consciousness, and the introduction of the higher human faculties – Wallace had given expression to 'a certain kind of deism'. 'But deism', said Iverach, 'is a superannuated form of thought which cannot be resuscitated at the present hour'. 'Are we to hold that only at these three stages can we find anything that points to a world of spirit?' he demanded. 'Are we to bring in the world of spirit only where our favourite theory fails? If there are breaks like these in the theory of evolution, is it not time to revise our theory?' Moreover, if God is the author of evolution, 'is there no way of conceiving of the Divine presence and power in the world save that of continual interference?' 'Is it not more reasonable, as it is certainly more Scriptural, to trace the origin of man, body, soul, spirit, as a unity, to the creative power of God?'[17]

Iverach believed that 'it is not possible to think that God is ever absent from his creation, or we must think that He is always absent'. 'To me', he confessed,

creation is continuous. To me everything is as it is through the continuous power of God; every law, every being, every relation of being are determined by Him, and He is the Power by which all things exist. I believe in the immanence of God in the world, and I do not believe that he comes forth merely at a crisis.... Apart from the Divine action man would not have been, or have an existence; but apart from the Divine action nothing else would have an existence.[18]

The immanence to which Iverach referred was not the world-soul of the Stoics and Spinoza, nor the 'monstrous self-consciousness of the Hegelian school', nor the impersonal will of Schopenhauer and the unconscious intelligence of Hartmann. Much less was it Spencer's Unknowable. It was that rational and conscious immanence to which the Apostles first bore witness, the 'creative and sustaining activity of the Logos', the second person of the Trinity, the Lord Jesus Christ. 'Modern science and philosophy will have done us a great service', said Iverach, 'if it will force students of theology to go back not merely to the history of theology, but to the New Testament itself, to search out its meaning, to gather together and to set forth in order and method its profound teaching on the relation of God to the world'.[19]

In the last two decades of the nineteenth century AUBREY LACKINGTON MOORE (1843–1890) was 'the clergyman who more than any other man was responsible for breaking down the antagonisms toward evolution then widely felt in the English Church'.[20] From an ecclesiastical viewpoint Moore's credentials were impeccable. His father was the Reverend Daniel Moore, vicar of Holy Trinity, Paddington, and prebendary of St Paul's. His education was obtained first at St Paul's School, then at Oxford, the haven of the High Church party, where in 1871 he gained a First Class in Moderations and in the Final Schools. Moore thereafter maintained the closest of relationships with the University and its students. In 1873 he was elected to a Fellowship at St John's College. There he served as a tutor until 1876, and at Magdalen and Keble Colleges from 1881 until his death. He lectured on ecclesiastical history, almost monopolising the Reformation period, and in his work as a parish priest, an examining chaplain, and an honorary canon of Christ Church he set his students an example of churchmanship at its best. 'No one', wrote E. S. Talbot, sometime warden of Keble College, 'was more respected among undergraduates'.[21] What impressed them most was Moore's intelligent and fearless faith. For while the Christian religion was his by inheritance, Moore had also 'accepted it with his full-grown mind, at a time when all such acceptance had to be won through the stress of conflict, because it satisfied his passion for righteousness and his intellectual craving for unity'.[22]

From a scientific viewpoint Moore's credentials were also impressive. Though only an amateur, he acquired an extensive knowledge of botany, partly through a friendship cultivated in the Keble Common Room with the young Darwinian, E. B. Poulton. Such was his expertise that he was offered the curatorship of the Botanical Gardens, a post he occupied in 1887. But Moore's qualifications to discuss scientific matters were acquired primarily through 'the stress of conflict', through wide and careful reading in science, history, and philosophy, and through continual interaction on related subjects with students of science. 'I am bound to say that I never met any man who combined such large stores of knowledge in all these several departments', wrote a friend, George Romanes, after Moore's death. He was 'equally at home' among Oxford scientists and theologians, serving as a 'link of union' between these members of the University which could 'ill be spared by either'. 'If it were true that no one was more fully imbued with the faith spiritual', Darwin's *protégé* reflected, 'it was quite as true that no one was more fully imbued with the faith scientific, or the assured conviction that, lead where she may, it is the first intellectual duty of a man to follow the guidance of science, leaving all "consequences" to take care of themselves'.[23]

Unlike many theologians of his generation, Moore learnt to understand the scientific enterprise as scientists themselves understood it. For this he deserves the recognition which thus far he has received for his essay, 'The Christian doctrine of God', contributed in 1889 to that manifesto of progressive Anglo-Catholicism, *Lux Mundi.* Though 'remarkable in its originality of thought and brilliancy of style', to use the words of Romanes, the essay but laid the groundwork for a larger reconciliation of Christianity and evolutionary science.[24] We turn therefore to Moore's articles on evolution – tokens of the *magnum opus* he might have lived to complete – which were gathered up in two volumes, *Science and the Faith* (1889) and *Essays Scientific and Philosophical* (1890).

Moore began writing on evolution in a paper, 'Recent advances in natural science in their relation to the Christian faith', delivered before the Church Congress at Reading in 1883. There, in bold and epigrammatic language, he expressed a theology that was to determine his approach to the subject for the remaining seven years of his life. It was a theology which refused to connect the Christian faith necessarily with evolution or the denial of evolution, but which held that evolution should be 'specially attractive to those whose first thought is to hold and to guard every jot and tittle of the Catholic faith'. Faith, said Moore, is not dependent on any particular understanding of organic origins, for whatever science may reveal as the method of origination, it is, after all, only a revelation of God's method of creation. Evolution *or* creation is thus a false antithesis. There is a *tertium quid* which a believer may refer to indifferently as 'supernatural evolution' or 'natural creation'. If Christians baulk at this alternative it is because they 'have come to acquiesce in a sort of unconscious Deism', to separate the natural and the supernatural, to equate creation with a series of supernatural interferences in the course of nature. It is because they have failed, in other words, to understand God's relation to his world. 'There are not, and cannot be, any Divine interpositions in nature', Moore declared, 'for God cannot interfere with Himself. His creative activity is present everywhere. There is no division of labour between God and nature, or God and law. . . . *For the Christian theologian the facts of nature are the acts of God.*' To hold otherwise would be decidedly unorthodox, and if evolution, by its 'natural' explanation of the divine method of creation, has made heterodoxy less viable, then it may be welcomed as an ally of the Catholic faith. Positively, on the other hand, by demonstrating the marvellous unity of nature, evolution helps to 'restore the belief that real knowledge, the knowledge of God and his working, is possible'.[25]

Two years later, in 1885, Moore reviewed Frederick Temple's Bampton Lectures, *The Relations between Religion and Science.* While finding much in them worthy of praise, he could not help but regret 'the sharp severance between the physical and the moral, which shows itself all through the Bishop's arguments, and at times becomes almost deistic'.[26] To rehabilitate the argument from design is a commendable undertaking and to do so on the basis of evolution is the only way at present that it can justifiably be done; but to argue that God makes things 'make themselves' through 'one original impress' of his will on nature 'seems to imply, however little it was intended, that God withdraws Himself from his creation, and leaves it to evolve itself, though according to a foreseen and fore-ordered plan'. And this, clearly, is to evoke the spirit of deism, the consequences of which for Christian theology are very dire indeed. 'It is of the first importance', said Moore, 'that a Christian apologist should not use language which seems to invest the world with a power of self-unfolding, for it is this, more than any theory of evolution, which contradicts belief in God'. Instead of saying that the stress of the argument from design is transferred from visible adaptations to 'the original properties impressed on matter from the beginning' which make adaptation possible, one should state that 'every adaptation, however minute, is in itself a new proof of purpose, design, and plan'. For only then will the knowledge of God, his wisdom, power, and goodness, be finally and ultimately secured. 'If Christianity is to hold its own as a true philosophy of the universe', Moore concluded, 'it must abandon explicitly and implicitly the Kantian dualism'.[27]

Spencer's proposed 'reconciliation' of science and religion, needless to say, was even less acceptable than Temple's. The merciless critique to which Moore in 1885 subjected part six of the *Principles of Sociology*, the *Ecclesiastical Institutions*, merely echoed his basic dissatisfaction with what he took to be Spencer's larger enterprise.[28] ' "Give us the Knowable, and you shall have the rest, which is far the larger half" ', he objected,

sounds like a liberal offer made by science to religion, till we remember that every advance in knowledge transfers something from the side of the unknown to the side of the known, in violation of the original agreement. . . . Every triumph of science on this theory. . .becomes a loss, not a gain, to religion. The very existence of God is bound up with that part of His work in nature which we cannot understand, and, as a consequence, we reach the paradox that the more we know of His working, the less proof we have that He exists.[29]

Moore had no patience with such a philosophy. Save for critical comments, he virtually ignored Spencer and at no time showed any enthusiasm for cosmic evolutionary speculation. All his interest was reserved for Darwin and the Darwinians. In reviewing their works Moore exhibited a deep and sympathetic understanding, not because Darwin's theory had no effect on the Christian faith, but because he believed that what it removed it replaced with something far better.

Of all Moore's writings on the subject of evolution, his essay 'Darwinism and the Christian faith', published in 1888, was the longest and most comprehensive. Its occasion was the appearance of Darwin's *Life and Letters*; its audience were readers of the *Guardian*, the periodical which most clearly reflected High Church opinion; its purpose was to explain for an 'ordinary Churchman', who both 'accepts the dogmatic position of the English Church' and 'believes the doctrine of evolution to be the truest solution yet discovered by

science of the facts open to its observation', how the new truth relates
to the old and 'what reconstruction of traditionally accepted views and
arguments is necessary and possible'. 'Such a man, accepting Darwinism', Moore forewarned, 'will expect not only that a reconstruction, or
at least a resetting, of his beliefs will be necessary, but also that real
effort, moral and intellectual, will be required for the work'. New
truth cannot, without such effort, be related to truth already appro-
priated, and the more far-reaching the new truth, the greater will be
the effort required. 'This is why the inrush of new truth means un-
settlement, and perhaps, in the reconstruction, a renouncing of something which has been associated with spiritual truth, though not of the
essence of the truth itself.'[30]

Moore certainly did not refer to the first article of the Creed. On
considering the views of Huxley, Clifford, and Spencer on ultimate
origins, and the development of Darwin's own belief in God, he could
assert without fear of contradiction that 'evolution neither is, nor
pretends to be, an alternative theory to original creation'. His reference
was rather to the docrine of special creation, something with which
spiritual truth had long been associated and to which evolution was
indeed an alternative. 'It is a question between two views as to secondary creation', he declared, 'or, more strictly, between a theory, and
the denial of the possibility of a theory, as to the method of this
creation'. There was no need to apologise for renouncing the latter
interpretation after Moore had finished with the subject. Nowhere
else in the literature of the post-Darwinian controversies does a theologian treat the species question with such penetrating insight. Moore's
discussion stands alone, comparable only to passages in the essays of
Huxley and Asa Gray. 'The dead hand of an exploded scientific
theory rests upon theology', he lamented. Although Darwin's work
destroyed once and for all the belief in organic fixity, 'Christians in
all good faith set to work to defend a view which has neither Biblical,
nor patristic, nor mediaeval authority.' 'If the theory of "special
creation" existed in the Bible or in Christian antiquity', he allowed,
'we might bravely try and do battle for it. But it came to us some two
centuries ago from the side of science, with the *imprimatur* of a Puritan
poet [Milton].' Thus 'it is difficult *à priori* to see how the question,
except by a confusion, becomes a religious question at all'.[31]

What Darwinism destroyed, however, it replaced, and replaced
with something far better. To Moore the Darwinian theory was
'infinitely more Christian than the theory of special creation' because

'it implies the immanence of God in nature, and the omnipresence of his creative power'. 'Those', he said, 'who opposed the doctrine of evolution in defence of "a continued intervention" of God, seem to have failed to notice that *a theory of occasional intervention implies as its correlative a theory of ordinary absence'* – a doctrine which 'fitted in well with the Deism of the last century'. 'For Deism, even when it struggled to be orthodox, constantly spoke of God as we might speak of an absentee landlord, who cares nothing for his property so long as he gets his rent.' 'Yet anything more opposed to the language of the Bible and the Fathers', Moore declared, 'can hardly be imagined'. 'Cataclysmal geology and special creation are the scientific analogue of Deism. Order, development, law, are the analogue of the Christian view of God.'[32]

Just as the concept of secondary creation has been recast in terms of evolution, so, in Moore's judgement, the argument from design has been transformed by the theory of natural selection. Here, too, a doctrine based on the Christian view of God supersedes a mere adjunct of spiritual truth. 'The old and rapid argument from nature to an omnipotent and beneficent author', said Moore, 'was never logically valid. To a thinking man its death-knell was sounded by Kant long before the death-blow was given by Darwin.' An 'Architect of the world' could never become the Creator–God of Abraham, Isaac, and Jacob, the God and Father of the Lord Jesus Christ, and 'the old *couleur de rose* view of nature', which informed the argument for his existence as developed by Paley, has taken on a different aspect in the light of the struggle for existence. Of course Christians are still bound to believe that the world is the work of an almighty, all-wise, and beneficent Creator, but their faith cannot gain support from an old and discredited teleology. A new teleology is needed, and this is just what Darwinism helps to provide.[33]

Darwin could not help but acknowledge 'the endless beautiful adaptations' everywhere in nature and 'the extreme difficulty or rather impossibility of conceiving this immense and wonderful universe, including man with his capacity of looking far backwards and far into futurity, as the result of blind chance or necessity'. 'The mind refuses', he said, 'to look at this universe, being what it is, without having been designed'. Professor Kölliker promptly accused Darwin of being 'in the fullest sense of the word a teleologist' and Huxley, in defending Darwin, was 'driven to distinguish between the teleology of Paley and the teleology of evolution'. Gray, on the other hand, acclaimed

'Darwin's great service to natural science in bringing back to it Teleology' and Darwin wrote in response that the statement 'pleases me especially', adding, 'I do not think any one else had ever noticed the point.'[34] Thus the Darwinians themselves testify to the 'elimination of chance' by natural selection, to the fact that utility or purpose governs the phenomena of life. Every species is both a means and an end in order that the whole of nature may increasingly manifest a more comprehensive and consistent pattern of design. Every organ, which on the old theory of special creation was considered useless, is seen to have an explanation in the past or in the future, according as it is rudimentary or nascent. 'There is nothing useless', Moore declared, 'nothing meaningless in nature, nothing due to caprice or chance, nothing irrational or without a cause, nothing outside the reign of law. This belief in the universality of law and order is the scientific analogue of the Christian's belief in Providence.'[35]

Not unpredictably, with so much staked on the principle of utility, Moore was disposed to side with those biologists who, in the growing controversy over the factors of organic evolution, assigned a prominent role to natural selection. He knew that in evolutionary theory there was 'much. . .unsettled' and that the doctrines of Weismann, Spencer, and Romanes were mutually incompatible. He also understood that Darwinism consisted of two elements 'which are by no means necessarily connected, the one the Lamarckian theory of descent, the other the more strictly Darwinian theory of natural selection'.[36] Being thus well apprised of the issues at stake in the controversy, he assumed the task in 1889 of reviewing Weismann's *Essays on Heredity* and Wallace's *Darwinism*.

Moore praised the former book as 'the most important contribution to speculative biology which has been made since the "Origin of Species" was published'. It contains, he said, 'a complete and coherent biological theory' which is of special importance for two reasons: first, because Weismann 'attempts to explain the existence of variation, and to show that the variations, though innumerable and *practically* infinite, are neither really infinite nor accidental, but from first to last are subject to law'; and second, because the theory, if true, involves 'the triumph of the Darwinian principle of natural selection' over modern champions of Lamarckism. That Moore not only foresaw, but perhaps even anticipated, this eventuality is indicated by the note he appended to his review, exploring the theological implications of a theory of heredity which does not involve the inheritance of acquired characters. As for Wallace's book, apart from certain critical remarks (to which we shall return), Moore gladly signalised its contention that all specific characters are either useful or correlated with useful characters. To him the 'bearing of this upon the question of teleology' was as 'obvious' as its bearing upon the views of Spencer and the American Neo-Lamarckians, and he expressed that Wallace's criticisms on these views had been made successfully.[37] But notwithstanding such open sympathy with the Neo-Darwinians, in the last result Moore took a position with regard to natural selection more nearly like that of his friend Romanes. The resemblance was probably not coincidental for, in reviewing Weismann's Essays, Moore expressed Romanes' sentiments exactly: 'We cannot help feeling, what was indeed a priori probable in a theory which was directly opposed to Lamarckianism, that too hard a line is drawn between the reproductive and the somatic cells.' The inheritance of acquired characters 'may be a slower process than we had imagined, but it is hard to believe that it does not take place'.[38]

On the subject of human evolution Moore maintained, as always, that a Christian's controversy with a Darwinian agnostic is 'a controversy with his agnosticism, not with his Darwinism'. Darwinism does not degrade mankind by relating the species to the rest of the creation, at least not more than it is already degraded. For what has been called a 'gospel of dirt' in fact appears in scripture as a 'Gospel of grace'. Darwinism teaches human beings afresh that they owe their constitution neither to some innate superiority to other forms of life, nor to their own efforts, but to the special gift of Almighty God. Moreover, by 'the application of its own methods and its own tests', Darwinism has recognised the biblical truth that mankind is 'the roof and crown of all things visible' and in so doing 'has rendered any form of nature-worship henceforth impossible'. If Darwinism be true, human beings can no longer worship organic nature, for they are the latest and highest of its creatures.[39]

But this is on the physical side. What bearing does Darwinism have on the spiritual nature of mankind? The human soul, like primary creation, Moore pointed out, lies beyond the range of science, while even theology cannot speak on the subject without uncertainty. Historically, Christians have been either traducianists, believing in the physical derivation of the soul, or creationists, regarding the soul as uniquely created in each individual. If the latter doctrine is to be maintained, it will need 'resetting' in the light of evolution; its separation of body and soul, which 'neither the theology nor the science of to-day will find...easy to accept', will have to be reinterpreted.[40] If, on the other hand, traducianism – a doctrine with more scientific appeal – is to prevail, it will have to be held despite the fact that it cannot account for the origin of the first human soul. Clearly, neither option was quite satisfactory. Thus Moore recommended patience and caution.

If we believe that man, as man, is an immortal soul, though we cannot say when he became so, or that strictly speaking, he ever did *become* so, we need not be surprised to meet the difficulty again in the evolution of man from lower forms.... The two questions are so closely bound together, that we feel that a theory which is to be true of either must be applicable to both. We have, probably, as much to learn about the soul from comparative psychology, a science which as yet hardly exists, as we have learned about the body from comparative biology, and any theory of the origin of the soul in the individual, and still more in the first man, whether suggested from the side of theology or of science, must be tentative and provisional, and will be in danger of losing one truth in its anxiety to preserve another.[41]

At two points, however, it was not necessary to be tentative. In the first place Moore stated that 'the soul cannot be a "special" creation, whether in Adam or in his children'. 'There is no "species" of soul', he said. 'We may call it, if we will, an "individual" creation, but is not all creation individual creation, from the religious point of view?'[42] In this connexion Wallace went astray by supposing that the spiritual nature of mankind was 'superadded' to the physical nature by 'spiritual influx' rather than by natural selection. Even on his own terms, assuming this spiritual influx to be present in all degrees below mankind and to work uniformly there by natural selection, why should its law not be the same in its highest operation? Or if it works by some other law, why should we not be able to trace the rudiments of its operation in the rest of the living world?[43] But if Wallace erred by creating a mental and spiritual dichotomy between mankind and other animals, Romanes erred in the opposite direction, by endeavouring to eliminate any difference in kind which separates them. In the second place, therefore, Moore had to emphasise that a difference in kind, whether to the scientist or to the Christian theologian, is merely an enormous difference in degree. It stands for the fact that in ontogeny one thing is from the first potentially that which another thing cannot become – the difference, for example, between the reason of a child and the reason of an animal as opposed to the difference between the reason of a child and the reason of an adult. Of course 'if difference in kind means, what Mr Romanes wants to make it mean, difference of origin', said Moore, 'there is no such thing as difference of kind either for idealist or realist, for Pantheist, Materialist, or Christian, and Mr Romanes has only given the *coup de grâce* to a moribund Deism'.[44]

A moribund deism indeed. Moore's thesis in his best-known and most profound essay, his contribution to *Lux Mundi*, was that the Christian revelation of God is both final and progressive: 'final, for Christians know but one Christ, and do not "look for another"; progressive, because Christianity claims each new truth as enriching our knowledge of God, and bringing out into greater clearness and distinctness some half-understood fragment of its own teaching'. From the outset, Moore argued, the Christian doctrine of a personal, triune God represented the highest attainments of conscience and speculative reason. It embraced the ethical ideals of the Greek, fulfilled the hope of the Jew, and thus, by 'taking up and assimilating all that was best and truest in non-Christian ethics', revealed a personal, moral relation between God and mankind in the divine man, Jesus Christ. It learnt from the Greek the impossibility of abstract unity and the idea of immanent reason, from the Jew the necessity of monotheism, from apostolic experience the reality and power of the Resurrection, and thus, by a tri-union of the divine transcendence with the divine immanence, the religious with the philosophic idea of God, became 'the only safe-guard in reason for a permanent theistic belief'.[45]

The Christian doctrine of God, according to Moore, has continued to be enriched and renewed by the most profound moral and philosophical attainments of mankind. In the sixteenth century an intense ethical revolt called the Church back to an apprehension of the divine morality that was lost in the Middle Ages. And in the nineteenth century another revolt – a scientific revolution – has helped the Church to recover an understanding of God's triune nature that was obfuscated by the deism of the Enlightenment. 'Science', said Moore,

had pushed the deist's God farther and farther away, and at the moment when it seemed as if He would be thrust out altogether, Darwinism appeared, and, under the guise of a foe, did the work of a friend. It has conferred upon philosophy and religion an inestimable benefit, by showing us that we must choose between two alternatives. Either God is everywhere present in nature, or He is nowhere. He cannot be here, and not there. He cannot delegate his power to demigods called 'second causes'. In nature everything must be His work or nothing. We must frankly return to the Christian view of direct Divine agency, the immanence of

Divine power in nature from end to end, the belief in a God in Whom not only we, but all things have their being, or we must banish him altogether.

Thus Moore called the Church to a 'fearless reassertion of its doctrine of God'. Darwin was urging him on. 'It seems', he said, 'as if, in the providence of God, the mission of modern science was to bring home to our unmetaphysical ways of thinking the great truth of the Divine immanence in creation'.[46]

Gray, Wright, and the Calvinist tradition

In the United States Christian Darwinism was advanced by two theologically minded naturalists, the botanist Asa Gray and the geologist George Frederick Wright. Their endeavours to win acceptance for Darwinism among American Protestants were unified by a working partnership that lasted from 1874 until 1881 and a friendship that ended only with Gray's life. Their enterprise was sustained by a common commitment to the evangelical Calvinism which had dominated New England theology for more than two hundred years.

ASA GRAY (1810–1888), professor of natural history at Harvard from 1842 to 1873 and thereafter director of the University Herbarium, was the foremost defender of Darwinism in America. A first-rate scientist in his own right, Gray placed the descriptive botany of North America on a firm basis by the publication of numerous popular botanical guides and textbooks, an exceptionally accurate *Manual of the Botany of the Northern United States* (1847), and a monumental *Synoptical Flora of North America* (1878). With J. D. Hooker in England and Alphonse de Candolle in France, he pioneered the study of plant geography, contributing an analysis of the affinities of the flora of Japan and eastern North America which told heavily on the side of evolution. But despite the abiding importance of his own scientific work, Gray is remembered chiefly as the one American naturalist who identified himself from the outset with the promotion of Darwin's. He entered into correspondence with Darwin in 1855, supplied him at his request with botanical information bearing on the question of transmutation, and received from Darwin in 1857 an abstract of the theory of natural selection which would later help to establish his priority over Wallace. By Christmas 1859, when he obtained a copy of the *Origin of Species*, Gray had in principle adopted the Darwinian theory.[47]

Being thus accurately and sympathetically acquainted with Darwin's argument, Gray determined to secure it a wide and serious hearing. He did his best to make equitable arrangements for an American edition of the *Origin* and succeeded, initially at least, in narrowing the field of would-be pirates to D. Appleton and Company, who issued an authorised edition early in 1860. In March of the same year he reviewed the book in the *American Journal of Science and Arts*, comparing Darwin's views with those of Louis Agassiz in a most telling manner. In April and May he faced Agassiz and his henchmen in person, before audiences at the American Academy of Arts and Sciences, and made strong the case for transmutation which the Neo-Platonists could not accept. In September Gray again entered the pages of Silliman's *Journal*, this time in a dialogue concerning 'design vesus necessity', and in October he published in the *Atlantic Monthly* the last of a series of three articles arguing that natural selection is 'not inconsistent with natural theology' – a series Darwin found so useful that he proposed and underwrote its publication as a pamphlet in 1861.[48] Darwin, clearly, was delighted with Gray's support. The two did not see eye to eye on the theological implications of natural selection, nor would they ever, but Darwin declared that Gray knew the *Origin* as well as he himself did, that his friend seemed a wonderful hybrid, 'a complex cross of lawyer, poet, naturalist and theologian', who never said a word or used an epithet which did not 'express fully my meaning'.[49]

As if it were not sufficient that a prominent naturalist of Darwin's own generation, a generation which condemned evolution by acclamation, should advance his cause in America, and that almost single-handedly, this naturalist was also a devout and orthodox Christian believer. Raised in the rural Presbyterianism of upstate New York, Gray drifted steadily into rationalism and materialism after his graduation from medical school in 1831. However, in 1835, under the influence of his elder friend and fellow-botanist, John Torrey, he regained his first faith and joined a Presbyterian church of the New School, that sector of the denomination which sought to temper the understanding of divine sovereignty and human freedom characteristic of the older Calvinism. The reality of this conversion was proved when Gray went to Harvard in 1842. Parting from most of his colleagues, who attended Unitarian services in the College chapel, he transferred his membership to the Congregational church in Cambridge. And there he continued to attend, despite his marriage in 1848 to Jane Loring, whose faith was a creedless, conservative Unitarianism.[50] Between husband and wife there never was apparently any friction over religious matters. Indeed, Jane Gray seemed pleased to report that her husband's view of biblical inspiration was not a narrow one and that, although his temper in theological discussions, which he greatly enjoyed, was 'naturally conservative', his religious sentiments grew 'broader and sweeter' as the years went by.[51] Yet toleration did not preclude orthodoxy, and certainly not in the person of Asa Gray. In the midst of an entrenched Unitarianism, a growing liberalism in Congregational theology, and a vigorous movement of scientific naturalism, he remained a moderate Calvinist and an adherent of the fundamental doctrines of evangelical Christianity.[52]

Committed to an orthodox theological tradition and also to the promotion and defence of Darwinism, Gray endeavoured to show that his commitments were compatible. Such a demonstration, he believed, would itself advance Darwin's cause, for the main obstacles confronting his theory were theological in character. 'You see', he told Darwin, 'I am determined to baptize [the *Origin of Species*], nolens volens, which will be its salvation'.[38] Gray performed the baptism with an effusion of periodical articles and two books. First came the review, the dialogue, and the series on natural selection and natural theology which Darwin had reprinted as a pamphlet. All but the review appeared anonymously. Then between 1860 and 1875 there followed an assortment of essays, most of which were written from a scientific standpoint, though each had a theological aspect, and two of which were anonymous reviews of works on evolution and theology. One of these reviews caught the eye of George Frederick Wright, a Congregational minister who was also an aspiring geologist. By enquiring of the publisher Wright was put in contact with the author, whom he soon persuaded to disclose his identity by reprinting his 'essays and reviews pertaining to Darwinism' in a separate volume. Gray, with Wright's help, prepared a concluding essay on 'evolutionary teleology' and in the autumn of 1876 the book was published by D. Appleton and Company under the title *Darwiniana*. Finally, in 1880 Gray stated once again, and with great candour, his synthesis of Darwinism and Christian orthodoxy in two lectures before the theological students of Yale College. The lectures appeared later in the same year in a modest volume entitled *Natural Science and Religion*.

Evolution to Gray was exactly what it was to Darwin: a 'natural theoretical deduction from accepted physical laws' which 'serves to connect and harmonize' the facts of nature in 'one probable and consistent whole'.[54] As an hypothesis, or 'congeries of Hypotheses', it was 'strictly' and 'purely' scientific. 'Deductive evolution', wrote Gray in the preface to *Darwiniana*, 'lay beyond the writer's immediate scope' and is a subject 'which neither the bent of his mind nor the line of his studies has fitted him to do justice'. By this he referred not to legitimate deductions from limited hypotheses such as those by which Darwin constructed his theory of natural selection, but speculative deductions from first principles which lie beyond the immediate data of biological science. Naudin's theory that species 'wear out' as their residual 'evolutive force' is dissipated, a deduction from the doctrine of the equivalence of force, Gray pointed out, 'is singularly unlike Darwin's in most respects, and particularly in the kind of causes invoked and the speculations indulged in'.[55] Spencer's philosophy, which deduced organic evolution 'ex necessitate rei', he cared for even less, for it belonged to the tradition of Agassiz. Both Agassiz and Spencer offered philosophic world-views, the one Neo-Platonic idealism, the other agnostic realism. Both reached conclusions about nature which followed necessarily from their metaphysical premises rather than from empirical investigation, the one removing the origin and distribution of species from the range of science altogether, the other deducing organic evolution from a universal mechanistic law. Neither Agassiz nor Spencer therefore was strictly scientific. By interpreting the facts of nature within the constraints of a speculative philosophical system, each had denied the metaphysical neutrality of scientific knowledge, which was the distinctive premise of Darwinism.[56]

Concerning religious as well as philosophical beliefs, Gray and Darwin were quite at one. Evolution *per se* offered no sure proof of theism, else Darwin would not have been an agnostic, but neither did it destroy theism. 'Certainly I agree with you', wrote Darwin to Gray, 'that my views are not at all necessarily atheistical'.[57] Like physical theories generally – the theory of gravitation and the nebular hypothesis, for example – evolution dealt with efficient causes. First and final causes lay beyond its scope. Order and not origins, method and not meaning, were its concerns. Thus Gray never tired of stressing that 'the adoption of a derivative hypothesis, and of Darwin's particular hypothesis, . . .would leave the doctrines of final causes, utility, and special design, just where they were before'. It would not bring in 'any new kind of scientific difficulty' – any, that is, 'with which philosophical naturalists were not already familiar'. For since the factual basis of the argument from design remains the same, regardless of the process by which the facts were brought into existence, the teleological inferences from these facts are unchanged.[58]

But Gray was not interested simply in maintaining the teleological *status quo*. The argument from design had never been conclusive, or so it seemed to many, because nature contains numerous apparently purposeless and pernicious phenomena. Nor for the same reason, he believed, would the argument ever seem conclusive so long as it remained tied to a belief in the special creation of immutable species. Once admit the evolutionary origin of species, however, and the difficulties diminish; the argument becomes more 'relevant', if not more 'cogent', than ever. In an evolutionary teleology useless organs find their *raisons d'être* in the past or in the future, according as they are vestigial or incipient; the seeming waste of life and matter becomes part of the general 'economy of nature'; and the imperfections and failures of life are explained, in the Darwinian scheme, as the necessary result of competition, struggle, and natural selection, without which there would be none of the diversification that leads up to 'higher and nobler forms'. 'The most puzzling things of all to the old-school teleologists are the *principia* of the Darwinian', Gray declared. 'In this system the forms and species, in all their variety, are not mere ends in themselves, but the whole a series of means and ends, in the contemplation of which we may obtain higher and more comprehensive, and perhaps worthier, as well are more consistent, views of design in Nature than heretofore.' Although Darwinism gave no proof of theism, and left the question of design entirely untouched, it did coincide with a theistic view of nature and, by eliminating 'common objections', permitted the construction of a stronger and broader teleology than was ever before devised.[59]

For Darwin, however, there were objections which Gray could not eliminate. Darwin was not merely troubled by the amount of misery that he perceived in nature; he also found himself unable to follow Gray in believing that design extended from the whole of nature to all its parts. With some difficulty he could accept that the laws of nature were designed; but to believe that God purposes a man to be killed by lightning, that particular swallows are appointed to snap up particular gnats, that man's rudimentary mammae or the shape of his own nose was designed – this seemed utterly impossible. And if these phenomena were not designed, Darwin saw no need to believe that the evolution of life, which involves nothing but similar, apparently random occurrences, was itself specially designed.[60]

Gray, on the other hand, holding that Darwin had left the doctrines of final causes, utility, and special design 'just where they were before', had somehow to account for the evidences of design in nature despite Darwin's misgivings. At first, in his review of the *Origin*, he stated that 'the origination of. . .improvements, and the successive adaptations to meet new conditions or subserve other ends, are what answer to the supernatural, and therefore remain inexplicable'. 'Wisdom foresees the result' of each development, which 'circumstances and. . .natural competition' bring to pass. In his dialogue on design and necessity Gray was more explicit. '*Variation* itself', he said, 'is of the nature of an origination'. While it may hereafter be 'shown to result from physical causes', these causes will always resolve themselves into two factors: 'one, the immediate secondary cause of the changes; . . .the other an unresolved or unexplained phenomenon' which will itself contain an irreducible factor. God himself is the very last, irreducible causal factor and, hence, the source of all evolutionary change.[61]

Again, in the third article of his series on natural selection and natural theology, Gray made the same point, this time, however, by means of a helpful metaphor.

So long as gradatory, orderly, and adapted forms in Nature argue design, and at least while the physical cause of variation is utterly unknown and mysterious, we should advise Mr Darwin to assume, in the philosophy of his hypothesis, that variation has been led along certain beneficial lines. Streams flowing over a sloping plain by gravitation (here the counterpart of natural selection) may have worn their actual channels as they flowed; yet their particular courses may have been assigned; and where we see them forming definite and useful lines of irrigation, after a manner unaccountable on the laws of gravitation and dynamics, we should believe that the distribution was designed.[62]

'I cannot believe this', Darwin at once replied. One 'would have to believe that the tail of the Fantail was led to vary in the number and direction of its feathers in order to gratify the caprice of a few men'. On further reflexion he wrote again, 'I. . .grieve to say that I come to differ more from you. It is not that designed variation makes, as it seems to me, my deity "Natural Selection" superfluous, but rather from studying, lately, domestic variation, and seeing what an enormous field of undesigned variability there is ready for natural selection to appropriate for any purpose useful to each creature.'[63]

To illustrate the point Darwin composed a metaphor of his own in the concluding pages of *The Variation of Animals and Plants under Domestication* (1868). So crucial is this metaphor for understanding the differences between Darwin and the Christian Darwinians that we reproduce it at length:

If an architect were to rear a noble and commodious edifice, without the use of cut stone, by selecting from the fragments at the base of a precipice wedge-shaped stones for his arches, elongated stones for his lintels, and flat stones for his roof, we should admire his skill and regard him as the paramount power. Now, the fragments of stone, though indispensable to the architect, bear to the edifice built by him the same relation which the fluctuating variations of organic beings bear to the varied and admirable structures ultimately acquired by their modified descendants. . . .

The shape of the fragments of stone at the base of our precipice may be called accidental, but this is not strictly correct; for the shape of each depends on a long sequence of events, all obeying natural laws. . . . But in regard to the use to which the fragments may be put, their shape may be strictly said to be accidental. And here we are led to face a great difficulty, in alluding to which I am aware that I am travelling beyond my proper province. An omniscient Creator must have foreseen every consequence which results from the laws imposed by Him. But can it be reasonably maintained that the Creator intentionally ordered, if we use the words in any ordinary sense, that certain fragments of rock should assume certain shapes so that the builder might erect his edifice? If the various laws which have determined the shape of each fragment were not predetermined for the builder's sake, can it be maintained with any greater probability that He specially ordained for the sake of the breeder each of the innumerable variations in our domestic animals and plants; – many of these variations being of no service to man, and not beneficial, far more often injurious, to the creatures themselves? . . . But if we give up the principle in one case, . . .no shadow of reason can be assigned for the belief that variations, alike in nature and the result of the same general laws, which have been the groundwork through natural selection of the formation of the most perfectly adapted animals in the world, man included, were intentionally and specially guided. However much we may wish it, we can hardly follow Professor Asa Gray in his belief that 'variation has been led along certain beneficial lines', like a stream 'along definite and useful lines of irrigation'. If we assume that each particular variation was from the beginning of all time preordained, then that plasticity of organisation, which leads to many injurious deviations of structure, as well as the redundant power of reproduction which inevitably leads to a struggle for existence, and, as a consequence, to the natural selection or survival of the fittest, must appear to us superfluous laws of nature. On the other hand, an omnipotent and omniscient Creator ordains everything and foresees everything. Thus we are brought face

to face with a difficulty as insoluble as is that of free will and predestination.[64]

'I found your stone-house argument unanswerable in substance (for
the notion of design must after all rest mostly on faith, and on accumulation of adaptations, etc.)', Gray replied. 'I understand your argument
perfectly, and feel the might of it.'[65]

Consequently, in the essay on evolutionary teleology with which he
concluded *Darwiniana*, Gray adopted a new metaphor, one that
allowed a biblical quotation expressive of divine sovereignty to be
substituted for a statement on the nature of designed variation.

Natural selection is not the wind that propels the vessel, but the rudder
which, by friction, now on this side and now on that, shapes the course.
The rudder acts while the vessel is in motion, effects nothing when the
vessel is at rest. Variation answers to the wind. 'Thou hearest the sound
thereof, but canst not tell when[ce] it cometh and whither it goeth.' Its
course is controlled by natural selection, the action of which at any given
moment, is seemingly small or insensible; but the ultimate results are
great.[66]

Beneath the new figure, however, were the old convictions. Gray
reverted to them in his lectures at Yale. The 'proximate causes' of
variation will certainly come to be known. The first and final causes
will not. 'In each variation lies hidden *the mystery of a beginning*' and
in this inscrutable inception (Gray implied but did not state) lies the
creative power of God. Variation under divine guidance (again
implied, not stated) is directed towards particular ends which natural
selection by itself, operating on omnifarious and indiscriminate
variations, could probably never achieve.[67]

The time had long passed when Darwin and Gray would argue over
evolutionary teleology. Each had made his case and each understood
the other's. Darwin grew into a deeper agnosticism as a result of the
debate and Gray cautiously but steadfastly retained his old convictions.[68] There matters stood at Darwin's death in April 1882. Then, in
October of the same year, the debate renewed with the publication by
Romanes of a reply to criticism of his *Scientific Evidences of Organic
Evolution* (1882). Romanes entitled his essay, which appeared in the
Contemporary Review, 'Natural selection and natural theology'.
When Gray had read it his pen flew into action. Why, if he and Darwin
and Romanes agree that science and theology are separate departments
of thought, Gray wrote to the editor of *Nature* – why does Romanes

betray the agreement by claiming that natural selection destroys all evidence of special design? He admits that the evidence was good under the theory of special and instantaneous creation; will he now say that it fails because creation has been discovered to be evolutionary and gradual? Surely this is a wrongful incursion by science into the territory of theology.[69]

Romanes replied that one who believes there is no logical connexion between natural science and natural theology should not be deemed inconsistent 'merely because he endeavors to show the fictitious character of the logical connection which has been erroneously supposed to exist'. The evidence of special design, he argued, could only constitute evidence under a theory in which no physical cause was supposed to explain the origin of organic structures and their progressive adaptations. Now that a physical cause, namely natural selection, has been given, no teleological inferences can be made – none, that is, unless it could be shown that the selected variations always take place 'in the directions required for the operation of the physical cause in question'. But here the burden of proof lies with the natural theologian.[70]

Gray took up the burden at once, pointing out, as he had to Darwin, that natural selection, which results only in the destruction of less favourable variations, is no substitute for intelligence, since it cannot account for the favourable variations that survive. The variations, he maintained, must have been guided intelligently in adaptive directions. Romanes, with all deference to a distinguished naturalist nearly forty years his senior, simply disagreed. Natural selection, he said, presupposes the existence of promiscuous variations, which only seem to occur in special and advantageous ways on account of all other variations being eliminated. This Gray could not accept. 'Omnifarious variation is no fact of observation, nor a demonstrable or, in my opinion, even a warrantable inference from observed facts', he returned. 'It is merely an hypothesis, to be tried by observation and experiment.' One thing, however, he would admit: that 'if variation in animals and plants is lawless, of all kinds and in all directions, then no doubt the theory of natural selection may be "the substitute of the theory of special design"'.[71]

At last the question of an evolutionary teleology had been resolved to a simple matter of observation. Darwin's observations, Romanes replied, were the most authoritative of all. He hammered home Darwin's mature judgement that variation is orderly but aimless and clinched the point by quoting in full the 'stone-house argument' which Gray had long ago found 'unanswerable'. And so, evidently, he found it again, for with Romanes' letter the exchange ended.[72]

Both Darwin and Romanes pointed out that divinely directed variation makes natural selection superfluous. Had Gray agreed there would be little reason to call him a Darwinian. But instead, while maintaining that natural selection is unintelligent and thus incapable by itself of producing nature's contrivances, Gray held that it is the primary mode in which intelligence comes to expression in nature. He defended natural selection, in other words, much as did Darwin and Romanes. The Neo-Lamarckians, it will be recalled, charged that Darwin's reliance on use and disuse and the direct action of the environment as causes of variation – causes which were supposed to lead, in some cases, to large-scale adaptive change – had made natural selection superfluous. Darwin and Romanes repudiated this interpretation and insisted that natural selection was 'the most important, but not the exclusive, means of modification'. Similarly, Gray would not allow the intelligent and inscrutable First Cause of variation, or any other transcendental cause, to replace natural selection as the mechanism of evolution. He could not see, for example, that the evolutionists of the 'Hyatt–Cope school' contributed 'any *vera causa* at all'. 'As the conception of the derivation of one form from another is the only distinctly-pointed alternative to specific supernatural creation', he declared, 'so the principle of natural selection, taken in its fullest sense, is the only one known to me which can be termed a real cause in the scientific sense of the term. Other modern hypotheses assign metaphysical, vague, or verbal causes, such as development, anticipation, laws of molecular constitution, without indicating what the special constitution is, – none of which have much advantage over the "*nisus formativus*" of earlier science.'[73]

Furthermore, in his interpretation of variation Gray made assumptions which have led some to rank him as an advanced Darwinian and a precursor of contemporary evolutionists. Although the search for precursors is a dubious enterprise, in this instance it may have some value in showing that one's theology need not predestine one's science for the scrap-heap of discarded theories. Gray left the proximate causes of variation an open question, eschewing the Lamarckian causes adopted by Darwin and his followers and insisting that variations originate 'not from without but from within' the organism. At the same time he rejected both Darwin's canon *Natura non facit saltum* and the idea of transmutation *per saltus* as advocated by Owen and Mivart. Indeed, seven years before Fleeming Jenkin delimited variation within a 'sphere', a decade before Galton conceived variation by analogy with the facets of a 'stone', and more than half a century before micromutation was recognised as the main source of evolutionary change, Gray wrote: 'Even if we regard varieties as oscillations around a primitive center or type, still it appears from the readiness with which such varieties originate that a certain amount of disturbance would carry them beyond the influence of the primordial attraction where they may become new centers of variation.'[74]

Thus it is not enough to say with Hooker that Gray understood Darwin's theory clearly, and in the sixties, while seeking to 'harmonize it with his prepossession', did so 'without disturbing its physical principles in any way'. Nor does it suffice to add, in the words of a reviewer in the *Popular Science Monthly*, that in 1876 *Darwiniana* was 'perhaps the fullest and most trustworthy exposition and illustration of what is to be properly understood by "Darwinism" that is to be found in our language'. All this was true (*pace* some of Gray's interpreters) and more.[75] For the Darwinism of Gray was a Darwinism with a future. 'Those parts of Darwinism which survive', his biographer states, '. . .were precisely those parts which Gray accepted' when he called himself a Darwinian.[76]

It remains, however, to determine Gray's approach to human evolution. At first his Darwinism did not extend this far. In 1860, lacking evidence of intermediate links between the Bimana and the Quadrumana, Gray could only 'believe in the separate and special creation of man, however it may have been with the lower animals and with plants'. Then in 1871 Darwin sent him a copy of the *Descent of Man*, though fearing, he said, that 'parts, as on the moral sense, will. . . aggravate you'. Gray, on reading the book, replied, 'Almost thou persuadest me to have been "a hairy quadruped, of arboreal habits, furnished with a tail and pointed ears".'[77] Five years later, it seems, the persuasion was nearly complete. In the concluding essay of *Darwiniana* Gray mentioned in passing the difficulty of 'drawing a line between the simpler judgments and affections of man and those of the highest-endowed brutes'. But it was not until 1880, in his Yale lectures, that Gray fully overcame his long-standing reticence to speak of mankind's place in nature. There he declared that man 'is as certainly and completely an animal as he is certainly something more'. His origin is no less divine because it involved the 'transformation of a

brute mammal' than if it had been directly and immediately supernatural. Nor is the origin of mankind less natural because the psychological gradations which bridge the gap between *Homo sapiens* and the
higher animals cannot account for 'the transcendent character of the
superadded', the immortality that neither science nor philosophy can
explain.[78] Lest these statements should seem offensively paradoxical,
Gray reminded his audience that 'one of the three old orthodox
opinions, – the one held to be tenable, if not directly favored by
Augustine, and most accordant to his theology, as it is to observation –
is that souls as well as lives are propagated in the order of Nature'.
The doctrine, he admitted, is controversial. The origin of each individual soul has puzzled theologians as much as the origin of species
has baffled naturalists. But perhaps for this very reason, and also
because 'the Darwinian and the theologian (at least the Traducian)
take similar courses to find a way out of their difficulties, they might
have a little more sympathy for each other'. Indeed, Gray concluded,
'the high Calvinist and the Darwinian have a goodly number of points
in common'.[79]

The reference was doubtless to himself. Gray was a Darwinian and
a Calvinist, though not in fact a thoroughgoing high Calvinist. He
shunned 'deductive evolution' and the speculative theology it implied,
preferring rather the empiricism of Darwin and the doctrinal formulations of traditional faith. He extended evolution to mankind, body
and soul, and found warrant for this in a doctrine held for centuries
by believers in original sin. Above all, he defended an evolutionary
teleology, trusting the sovereignty of God in the face of a difficulty
regarding the causes of variation which seemed 'as insoluble as is that
of free will and predestination'. To Gray an evolutionary teleology was
but the human conception, a conception thus fraught with enigma and
mystery, of the continued and orderly outworking of God's sovereign
purposes in nature.

Christian Darwinism in America was as much the special creation of
GEORGE FREDERICK WRIGHT (1838–1921) as of Asa Gray. Brought up
in the midst of conservative Congregationalism, militant abolitionism,
and the revivalism of Charles Grandison Finney, Wright attended
Oberlin College, the institution in which these movements converged.
He graduated in 1859 but remained at the College to complete the
theological course, absorbing the keen gaze and relentless logic of
President Finney and immersing himself in the New School Calvinism

and 'perfectionism' of the 'Oberlin Theology'. In 1862 he assumed the pastorate of the Congregational church in Bakersfield, Vermont and there continued his education. Wright was probably the only minister on either side of the Atlantic who, while fulfilling his clerical duties, read the Bible through in the original languages, translated Kant's *Critique of Pure Reason*, studied the philosophical works of Mill, Hamilton, and Noah Porter, and read appreciatively the *Origin of Species* and Lyell's *Antiquity of Man*. And doubtless he was the only minister anywhere who found the time, while engaged in such pursuits, to become an authority on the glacial geology of his region.[80]

In 1872 Wright took the pulpit of the Free Christian Church in Andover, Massachusetts. Andover was then the intellectual centre of Congregationalism – its Seminary the home of Edwards Amasa Park, Joseph Henry Thayer, and Austin Phelps; its Association of Congregational Ministers the meeting-place of Francis Howe Johnson, Theodore Thornton Munger, and Henry A. Coit – and Wright became a participant in theological discussions which gave direction to all his subsequent work.[81] His orthodoxy won the confidence of Park, the last great representative of the 'New England Theology'. On retiring in 1881 Park urged him to write against Andover's 'new departure', a belief in 'future probation' for unbelievers coupled with an increasingly critical attitude towards the Bible, which would later result in the trial of five professors for heresy. Further, in 1883, after forty years as editor of the learned and urbane theological monthly, *Bibliotheca Sacra*, Park consigned his responsibilities to Wright, who occupied the post thereafter until his death.[82]

The theology of Andover was not all that made Wright's years there so eventful. Its topography was equally important. There were gravel ridges running beside the Free Church parsonage, and, with an eye practised to observe evidence of glaciation in the Green Mountains of Vermont, Wright identified them as a moraine. He argued his case in 1875 before the Essex Institute in Salem, challenging the accepted interpretation that the ridges had a marine origin and predicting similar glacial effects elsewhere in New England. A year later, having verified his prediction, Wright described his findings before the Boston Society of Natural History. In 1879 he again reported before the Society, this time with the endorsement of James Dwight Dana and Clarence King, and in the next year he began to follow the gravel ridges southward through New Jersey and into Pennsylvania.[83]

As Wright's geological interests moved west so did he. Oberlin offered him the chair of New Testament language and literature, and in 1881, over protests from Dana, he accepted the post, knowing it would release him from a minister's busy schedule for the pursuit of his scientific and theological work. Wright had discovered the great terminal moraine of the last glacial retreat in North America. Now it was his privilege to trace this dramatic line of demarcation, almost singlehandedly, from Pennsylvania to the Mississippi River. By 1884 the task was completed, its findings were reported in scholarly journals, and Wright became recognised as one of America's leading authorities in glacial geology. His Lowell Lectures in 1887, published by D. Appleton and Company as *The Ice Age in North America* (1889), received praise from Dana, King, Joseph Le Conte, and Alexander Winchell, and for years remained the standard text on the subject. His Lowell Lectures in 1892, which appeared in the International Scientific Series under the title *Man and the Glacial Period* (1892), though rather more controversial, were also highly spoken of. Indeed, it is a tribute to Wright's reputation that when his book came under sharp criticism from members of the United States Geological Survey, who contrived to make his theology reflect upon his science, his geologist colleagues rallied round him, defending the legitimacy of his views on glacial man even when they themselves could not agree. Wright, for his part, remained candid and courteous throughout the affair and thus emerged more respected than ever before.[84]

The critics did make at least one valid point however: Wright's science never lost touch with his theology. From the outset his geological investigations were motivated by a desire to shed light on the biblical account of the origin of mankind. And the reading in philosophy and science that he undertook in his Bakersfield pastorate was intended to help him find a Christian approach to scientific method. Darwin had focussed attention on these two crucial questions – mankind and method – and Wright, who had responded favourably to the *Origin of Species*, was convinced that neither should be answered lightly. Late in 1871, on the eve of his departure from Bakersfield, Wright published in *The New Englander* a trenchant essay entitled 'The ground of confidence in inductive reasoning'. Then at Andover, having obtained access to current periodicals, his thoughts gained strength and direction from a certain anonymous writer of essays on evolution and theology. That writer, Asa Gray, had meanwhile been impressed with Wright's essay on inductive reasoning and had made enquiries as to what sort of man the author might be. Gray got his answer and Wright discovered his anonymous mentor when, in 1874, Wright wrote to *The Nation* concerning a review of Hodge's *What Is Darwinism?* He received a polite letter from Gray, the reviewer, and before long the two were sharing a common concern that Darwin should obtain a full and sympathetic hearing from American Protestants.[85]

Like father and son – twenty-eight years separated them – Gray and Wright formed a partnership which owed its success to their kindred spirit. No two Christian men on either side of the Atlantic were more determined to advance the cause of Darwinism.[86] For the duration of Wright's Andover pastorate they worked together, remaining fast friends thereafter until Gray's death in 1888. Wright, who had been much influenced by Gray's anonymous essays and had perceived their utility, not only as an answer to infidel evolutionists, but as a response to anti-Darwinians such as Charles Hodge and J. W. Dawson, urged Gray in 1875 to issue the essays in a separate volume. Gray was reluctant because he could not be sure there would be a demand for the book and because, in any event, he would dislike the publicity. More important, he would have to interrupt his other work to prepare the volume, chiefly to add some comments on 'pending questions' related to Darwinian teleology (doubtless involving Darwin's 'stone-house argument'), which he did not, off-hand, know how to state conclusively. Thereupon Wright canvassed theological opinion at Andover, ascertained that the book would meet a widespread need, and placed his services as editor and consulting theologian at Gray's disposal. A year later *Darwiniana* was published, by all rights the most important book on evolution and theology written by a professional scientist in Great Britain and America during the last third of the nineteenth century.[87]

Gray and Wright collaborated in 1877 and 1878 to rebut the scepticism of the astronomer Simon Newcomb and to expose Boston's reverend philosophical charlatan, Flavius Josephus Cook. In 1879 and 1880 they consulted over the preparation of Gray's Yale lectures, *Natural Science and Religion.*[88] Gray, however, was not the sole beneficiary of the partnership. While his lectures and essays profited from Wright's theological expertise, his own mature understanding of evolutionary theory aided Wright in preparing a worthy companion to *Darwiniana.* Just before their partnership was formed, Wright had been asked to contribute to *Bibliotheca Sacra* a series of articles stating the pros and cons of Darwinism. Under the heading 'Recent books bearing upon the relation of science to religion' five articles appeared, the first in July 1875, the last in January 1880. Each met with Gray's approval and some, Wright recalled, 'were indebted to him for much of their form of statement'.[89] As a whole, in fact, the series was quite without parallel. The bibliographies prefixed to four of the articles showed that Wright had virtually mastered the literature of the subject, both scientific and theological. Moreover, the breadth and candour of his approach and his thorough understanding of the points at issue made the series a paradigm of fair-minded and intelligent discussion in the midst of controversy. Now Wright, having seen the value of Gray's collection of essays, several of them rather old, found justification for an up-to-date collection of his own. He added to his articles an essay on prehistoric man and another on the relation of the Bible to science, dedicated the manuscript to Gray, and published it in 1882 as *Studies in Science and Religion*. The book was probably the most important of its kind written by an American clergyman in the last third of the nineteenth century.

Wright's *Studies* was not, strictly speaking, a companion to *Darwiniana* but to his *Logic of Christian Evidences*, published two years before. Together, said Wright in the preface to *Studies*, they might be entitled 'The Unity of Method in Science and Religion'. Both, being the outcome of his study of inductive logic begun fourteen years earlier, were illustrations of a long-standing conviction: that in science and religion there is but a single method by which to discover truth. The *Logic* was an illustration of the inductive method applied to the proof of Christianity; his *Studies* would apply the same method to an evaluation of Darwinism. 'The best defenses of Christianity, so far as it is a system of positive revelation', said Wright, 'are not to be found in speculative philosophy'. Nor are the best defences of Darwinism. 'Christianity in its appeal to historical evidence allies itself with modern science' in resisting 'the glittering generalities of transcendentalism'.[90]

Just what kind of defence science and religion could expect from the inductive method was the theme of Wright's first article in *Bibliotheca Sacra*, 'The nature and degree of scientific proof'. Consisting largely of quotations from *The Principles of Science*, a 'treatise on logic and scientific method' published in 1874 by the Owens College professor, W. Stanley Jevons, the article was omitted from *Studies* in deference to Wright's earlier essay on inductive reasoning. None the less, it reveals an advanced concept of scientific explanation. Jevons, like philosophers of science before him, stressed the role of hypothesis as an organ of induction and compared the Baconian method unfavourably to the Newtonian. Wright applauded. Already, in his interpretation of the gravel ridges at Andover, he had learnt the importance of 'having a clew or hypothesis to direct our observations'. Where Jevons made an advance over his predecessors was in his clear statement of the relation between induction and probability, and of its bearing on verification. Again Wright expressed his approval. Induction, he agreed, can never be perfect. And the results of imperfect induction are never more than probable. Neither the defender of the Bible nor the physical scientist can be 'absolutely certain' as to the validity of his convictions. Neither can establish his beliefs beyond a 'high degree of probability'. 'Doctors of science as well as doctors of theology', said Wright, 'walk by faith, and not by sight'.[91]

But why, if the results of induction are never more than probable, should anyone have faith in them? Wright answered in the first chapter of *Studies*, 'The ground of confidence in inductive reasoning'. He began by dispensing with inadequate reasons for faith in induction. Distinctions between essential and inessential characters, between substance and attribute, as grounds of confidence in induction, he said, simply beg the question, for only induction itself can make such distinctions. In any event they tend 'to entangle the real problem with useless questions concerning ontology'. The assumption that the uniformity of nature, which science approximates by induction, is the real ground of confidence overlooks the fact that 'absolute uniformity is not proved to be a law of nature anywhere in the physical world.' More important, it overlooks the dependence of the universe on its Creator: that is, on 'the existence and discoverability of final causes in the constitution and laws of the physical creation'.[92]

With the mention of final causes Wright took up his own argument. First he separated induction from deduction. The former deals with particular facts that 'depend on the action of a personal Creator', the latter with intuitions which 'have no dependence upon a personal Creator' – time, space, pure mathematics, and moral philosophy – and from which, therefore, one cannot 'evolve the whole conception of nature'. Second, having undercut the basis of rationalism in science, Wright stated the Christian premise of inductive reasoning: 'that the Creator works in accordance with the highest wisdom for the highest good of being'. This, he declared, furnishes the 'absolute, final cause of the creation, both when considered as a whole and in its several

parts. . . . No single and limited good can be assigned as the final cause
of any contrivance in nature. The real final cause of any contrivance
. . .is the sum of all the uses to which it is ever to be put.' To say less, as
finite beings in a world characterised by moral disorder, would be to
exceed 'the evident limit of human acquirements'. Third, having
stipulated that 'God's choice of the good of being and his wisdom in
promoting it' invest nature with purpose in part and whole, Wright
affirmed that 'the universe is a "solidarity" – that nothing is made in
vain – that every part is a complement to every other part'. To determine the significance of this thing or that, to declare this beneficial or
that detrimental, without reference to the end – the good of being –
for which the whole is conceived, would again be presumptuous.⁹³

Finally, given a *universe*, one which is real and in which benevolence
is ultimately and consistently pursued, Wright argued as a corollary of
God's love that '*the exhibition of his veracity will be of such kind as
to give us opportunity to supply our wants – the wants which he has
created*'. One of these is 'a tolerable degree of uniformity in the
material and spiritual facts' that serve to build the human character –
a 'tolerable degree', of course, from the human point of view, for the
inductions on which life depends can have 'only a greater or less
degree of probability'. Yet what are called probabilities suffice for
those who believe that the phenomena, one and all, partake in a
benevolent end. The conviction that '*one part of what God does truly
represents the whole of what he does*', Wright concluded, 'lies at the
bottom of all our confidence in inductive reasoning'. Far from hindering such reasoning, it establishes and secures it. The conviction 'leaves
range enough to satisfy the roving propensities of any reasonable
explorer in the realms of science. It allows Darwin and his opponents
to fight out on purely scientific grounds the battle between their
theories.'⁹⁴

In chapter two of *Studies*, 'Darwinism as an illustration of scientific
method', Wright set forth the case for Darwinism on 'purely scientific
grounds'. His strategy was first to state the species problem and then
to show the relevance of Darwin's solution. 'Species', he began, is an
'ambiguous and ill-defined word'. It refers not so much to facts of
nature as to the fallible judgements of naturalists. What naturalists
discern to be species may indeed be arranged in a hierarchy of orders,
but differing judgements make the classification uncertain. There is
also the question of the dispersion of species. Some naturalists, believing in multiple creations, deny an organic connexion between identical

forms that are distantly removed; others, accepting migration as an explanation, affirm it. And what of the distribution of species in space and time? Some naturalists account for the unique and striking adaptations of species to their environments by positing separate creations; others explain the distribution of species through 'descent with modification'. Species formerly thought to be distinct have been joined by intermediate links; but these, too, may be interpreted either as distinct creations or as ancestors of the contemporary forms. Finally, naturalists are confronted with the phenomena of homologous and rudimentary structures, embryology, and analogous variation. Again they have to choose between special, supernatural creation and an explanation based on organic evolution.[95]

'To cut the Gordian knot with the simple assertion, "so God has made it"', Wright declared, '...would be suicidal to all scientific thought, and would endanger the rational foundation upon which our proof of revelation rests'. Yet this is, in other words, precisely what the special creationist does. Refusing to find an explanation of the origin of species in 'secondary causes', he abandons the method by which alone science can survive and Christianity can establish its basis in historical fact. Darwin, on the other hand, without claiming to offer a 'final solution of the problems of nature' (in this respect he is 'much more cautious than some of his followers'), adheres to the inductive method of reasoning. From the 'elasticity of species' coupled with the effects of a struggle for existence over an extended period of time he concludes that all species may have originated, one from another, by natural selection. His theory cannot, of course, be proved by direct observation. But then 'we are permitted to determine very few things by direct observation. . . . We go beyond observation whenever we try to prove anything.' Darwinism, like Christianity, must be judged by its general agreement with the facts; its proof must be left to 'our confidence in the fundamental principle of induction'. And here, for the special creationist, there was an additional ominous implication. Since 'the conformity of complicated facts to theories concerning the operation of natural forces may as effectually involve the testimony of God to the truth of those theories, as the agreement of a signature with a business man's known hand-writing may connect the two together, and prove the genuineness of a document', to reject Darwinism was not merely to jeopardise the argument from historical 'evidences' for the truth of the biblical revelation. It was implicitly to impugn nothing less than the goodness and veracity of God.[96]

Darwin was pleased with Wright's statement of his theory. On reading it in *Bibliotheca Sacra* he told Wright that it was 'most clear' and asked him to send a copy of the next article in the series, the one in which the theory's difficulties would be discussed. If, as seems likely, Darwin perused 'Objections to Darwinism and the rejoinders of its advocates', he could not have helped but be impressed.[97] After making short shrift of the notion that Darwinism is a 'mere theory', Wright tackled the toughest problems first – the very problems he was best qualified to treat. To objections that certain species appear abruptly in geological time and that intermediate forms are frequently absent in single formations, he responded by explaining clearly how elevation, subsidence, and denudation have rendered the geological record, 'even in the best preserved sections, . . .poor and beggarly beyond description'. To the objection raised by William Thomson that the lapse of time has been insufficient for natural selection to do its work, he answered boldly that neither the rate of change among wild species nor the validity of the physicist's mathematical calculations was conclusively established. Assuming that time has been short, however, Wright saw that 'natural selection must either have been incompetent for the results, or have worked the faster'. In such a case he was inclined to accept the latter solution. One who 'believes in a providential Ruler', he said, 'can easily grant that the Creator, through the combination of the forces which produce natural selection, may hasten the development of a variation even more rapidly and surely than man can do by his combination of these forces'.[98]

In the remainder of the chapter Wright took up the objections of Darwin's other troublesome critics, Fleeming Jenkin and St George Mivart. He quoted and condensed Darwin's own replies, stressing wherever possible that the emendations, concessions, or clarifications involved only served to solidify the agreement of natural selection with a theistic view of nature. When, for example, Darwin sought to overcome the swamping effects of blending inheritance – Jenkin's objection – by multiplying the number of variant individuals, Wright noted that 'large numbers of individuals do not vary at the same time and in the same direction, by chance' and that the tendency to variation itself 'still remains among the mysteries of the creation'. Or when, in response to Mivart's objection that natural selection, operating on fortuitous variations, could never evolve independent similarities of structure such as are present in the wings of birds and bats, Huxley replied that 'variation is neither indefinite nor fortuitous, nor does it take place in all directions, in the strict sense of these words', Wright featured it as a statement 'with which no theist need quarrel'.[99]

Here, indeed, was the heart of Wright's rejoinders and an idea to which he, like Gray, inevitably recurred: the ultimate causes of variation are inscrutable and thus, by implication, divine. 'To realize how indeterminate the problem...is, even after Mr Darwin leaves it', said Wright, after quoting the introductory sentences of the conclusion to the sixth edition of the *Origin of Species*,[100] 'we need to combine the definite quantities which are assumed'.

First, variation is produced by action of the 'conditions of life' (a term as complex as all nature) upon the 'individual organism' (another term of equal complexity). This raises our quantity to the second power. Secondly, we must introduce 'natural selection' (a term as broad as that of both the others combined). In considering any specific result in nature, we find ourselves in the presence of an indefinitely large indetermination, raised to the fourth power. In other words, we cannot tell deductively what variations will arise, unless we know all about the constitution of the individual, and all about the outward circumstances that act upon it to produce variation; and we cannot know what variations will be perpetuated till we know how each is related to the whole system of nature. It would seem that such an hypothesis left God's hands as free as could be desired for contrivances of whatever sort he pleased.[101]

Wright was a Calvinist as well. His primary object in expounding Darwinism and replying to its critics was to clarify 'the logical relation of [Darwin's] principles to the system which without peradventure sets God on a throne of supreme authority'. Little wonder, then, that he concluded his third chapter by citing Gray's metaphor: the sailing vessel whose rudder is natural selection and whose 'propelling agency' is the sovereign wind of God.[102]

The fourth chapter of *Studies*, 'Concerning the true doctrine of final causes or design in nature', was the counterpart to the essay on evolutionary teleology in which Gray's nautical metaphor appeared. Wright had assisted Gray, in writing his essay, to treat of some 'pending questions' related to Darwinism and design.[103] Now he would treat the questions himself, but at a theological depth where Gray was unable to operate. 'The inference of design in nature is drawn from complexity and nicety of adaptation', said Wright.

This inference need not be affected by any new view of the mode of origination, and cannot be rebutted, except by assigning a sufficient physical cause, irrespective of intelligence. If any one asserts that these adaptations arise from necessity, he is bound to show by what necessity.

Until that is shown, the inference of an intelligent cause is as good as it ever was, however much our conception of nature's intricate machinery may be enlarged.[104]

But while Wright did not consider Darwinism inimical to the Paleyan argument rightly understood, he held that Darwinism does modify 'in some degree' the Paleyan interpretation of design, and at just the point where it is 'open to criticism'. The watch found lying on a heath, from which Paley inferred the existence of a watchmaker, 'reveals two separate things which we are likely to confound, namely, design and man's method of executing design'. Accordingly, Paley's argument from the watch to a watchmaker and, by analogy, from contrivance in nature to a Contriver, 'incurs the danger of encouraging conceptions of God which are too anthropomorphic, both as to the narrowness of the design contemplated and as to the means to attaining the end'. It is nothing but 'arrogance' which follows Paley in pronouncing on the ultimate reasons which actuated the divine mind in bringing certain things into existence or in choosing the manner in which their existence has been realised. The world is not so simple as to allow finite creatures that privilege of judgement. Where, for example, is the benevolent design in the waste and apparent failures and imperfections of nature, Wright demanded – where is it, that is, 'apart from the general system in which they are introduced'? 'The truth is', he declared, 'that the rose-colored views of many evolutionists, and of still more of the pietistic interpreters of natural theology, are built upon a very narrow basis of facts, to the exclusion of another class of facts which abound in startling number'.[105]

Darwinism affects the older teleology by accentuating this 'other class of facts'. By making waste and apparent failures and imperfections integral, not incidental, to the development of life, it absolutely requires an enlarged conception of design. Wright, returning to a theme in his first chapter, found that conception in the comprehensive theory of virtue elaborated by the elder Jonathan Edwards and his successors. Virtue consists in choosing the 'good of being', according to the New England theologians. God, in whom choice and action coalesce, manifests this virtue in working all things together for good. The good of being can thus be considered the absolute final cause of the creation, both as a whole and in its several parts. Since human beings, however, must necessarily have a limited understanding of the good of being, they should not presume to give an exhaustive interpretation of any part of the creation. A full and correct interpretation belongs only to the Creator, who sees each part in its relation to all the other parts and in relation to the purpose of the whole. In this connexion Wright had formerly stated that 'the real final cause of any contrivance. . .is the sum of all the uses to which it is ever to be put'. Now he argued the corollary that 'the use to which we may put a thing is never more than a fragment of the final cause of its existence'.[106]

With this axiom the stage was set for a reply to Darwin's 'stonehouse argument'. Gray had found the argument 'unanswerable' and had sought refuge in a nautical metaphor which expressed rather simply a belief in the sovereignty of God. Wright found the argument 'worthy of the most careful study of the theologian' and responded with a metaphor which expressed in detail his enlarged, Calvinistic conception of teleology. He took as his illustration a saw-mill, 'the main object of whose construction is the production of timber'.

A combination of reasons, no single one of which may have been sufficient alone, accounts for the existence of each particular saw-mill. The price of labor, the facility to a market for the principal production, the obstacles to be overcome in getting the raw materials to the mill, and, finally, the use that can be made of the refuse, or incidental production of the establishment, may, any one of them, come in as the determining reason. All the profits of the mill may be in the sale of the slabs, or in economizing them and the sawdust as fuel. The uses the miller's children may make of the refuse for play-houses, and the miller's wife for kindling, are none of them so insignificant as not to be taken into account. The children very naturally might, at a certain age, fix upon their incidental advantage as the main object, or final cause, for which the mill existed. And their error may not be half so ludicrous as that we make in assigning the temporary advantages we derive from them as the exhaustive reason for the existence of the several parts of the universe that come within the range of our limited observation. Indeed we may well suppose that the highest conception of the perfection and design of the divine workmanship which our imagination can compass, is but a partial appreciation of the utility of the chips that have fallen off incidentally in the process of rearing the walls of the city of God. We are living in the quarry, and are concerned with the fragmentary pieces of emerald and sardonyx and topaz that are scattered thickly about the region where God's hand is at work.[107]

The final cause of wood slabs or (in Darwin's metaphor) stone fragments is not fully revealed in the fact that they are found useful for the constructive purposes of mankind. Both the slabs and the fragments are part of a larger system in which they serve, perhaps in many ways, to accomplish a larger end. 'The use to which we put a thing is never more than a fragment of the final cause of its existence.' Likewise the capacity for variation in animals and plants, while offering 'a wide range of uses subservient to the purposes which men may cherish, whether benevolent or otherwise', is the source from which, through natural selection, God's benevolent purposes are realised in the whole creation.[108]

Darwin had argued that either all variations are guided or none is; that total guidance would involve a predetermination of the variations employed by breeders; and that, since such total guidance is incredible, variation must be wholly aimless. Wright, without referring to variation as guided but as indeterminate, argued that the final cause of variation consists of all the ends which it ever serves, the ends selected by breeders being thus but a fragment of the final cause; that this final cause, the 'good of being', is the divine purpose for all creation; and therefore that variation, whether selected by breeders or by the powers of nature, participates in God's universal benevolent design. It was an argument for total guidance, to be sure, but it was constructed to make such guidance – 'a difficulty as insoluble as is that of free will and predestination' – appear more credible. Darwin had raised 'no new questions regarding final causes'. Wright needed no new theology to answer them.[109]

The relevance of a Calvinistic theology to Christian Darwinism was one of Wright's old convictions. Intimations of it occurred in his article on inductive reasoning that appeared in 1871. These, however, were only echoes of an unpublished lecture, entitled 'Darwinism and Calvinism', which he delivered before a group of New England ministers in the previous decade. Deploring the clerical opposition to Darwin, and especially that of the Oberlin theologian Asa Mahan, who had merely repeated all the arguments formerly employed against Vestiges without discerning that a new theory was afoot, Wright maintained that Darwin's work actually allies itself with the Reformed faith in discouraging romantic, sentimental, and optimistic interpretations of nature. Far from threatening orthodox beliefs, Darwinism and those who represent it, he predicted, will likely lead Christians to an outlook 'eminently favorable to sound doctrine, and will lay the ground for a most telling additional chapter to Butler's Analogy'.[110] Wright wrote better than he knew. A 'most telling additional chapter' was indeed the result, if not for Joseph Butler's Analogy of Religion (1736) then certainly for his own Studies in Science and Religion.

Confronted with rationalistic deism, Bishop Butler drew a detailed analogy between revealed religion and the course and constitution of nature. By showing that revealed religion is open only to such objections as can also be brought against natural religion, he captured the defences of the deists for the cause of orthodox faith. Wright, confronted with the anti-Darwinian Calvinism of Hodge and Dawson, drew 'Some analogies between Calvinism and Darwinism'. By showing that Darwinism is open only to such objections as can also be brought against Calvinism, he captured the theological defences of the anti-Darwinian Calvinists for the faith of those who accept evolution by natural selection. The Darwinian, in other words, he said, 'may shelter himself behind Calvinism from charges of infidelity'. And

the student of natural history who falls into the modern habits of speculation upon his favorite subject may safely leave Calvinistic theologians to defend his religious faith. All the philosophical difficulties which he will ever encounter, and a great many more, have already been bravely met in the region of speculative theology. The man of science need not live in fear of opprobrious epithets; for there are none left in the repertory of theological disputants which can be specially aimed at the Darwinian advocate of continuity in nature. The Arminian, the Universalist, and the Transcendentalist long ago exhausted their magazines in their warfare against the lone camp of the Calvinist; while the Calvinist has stood manfully in the breach, and defended the doctrine that method is an essential attribute of the divine mind, and that whatsoever proceeds from that mind conforms to principles of order; God 'hath foreordained whatsoever comes to pass'. The doctrine of the continuity of nature is not new to the theologian. The modern man of science, in extending his conception of the reign of law, is but illustrating the fundamental principle of Calvinism.[111]

Clearly, Wright was employing 'Calvinism' in a broad sense. 'Augustinianism', for his purposes, would do as well. What required emphasis was the 'fundamental principle' of divine sovereignty, a sovereignty which not only sees the sparrow fall and numbers the hairs of the head, but which 'comprehends' free human actions and ordains the system of the universe in part and whole. No body of doctrine is so generally characterised by this principle as the theology associated with the name of John Calvin.[112]

Proceeding with the analogies, Wright noticed first that 'Darwinism conforms to the facts both of nature and of the Bible in not being a theory of invariable and progressive development'.[113] As natural selection comprehends both the degradation and extinction of species as well as their advancement, so, according to the catechism, human beings fell from their first estate, involving all nature to some extent with them, but they retain a moral nature which may both hinder and help their spiritual condition.

In the second place, Darwinism and Calvinism agree that mankind is genetically one. As natural selection depends on the inheritance of favourable variations, so the Calvinist holds that corruption has been transmitted from Adam to all his descendants. Notwithstanding the 'hereditary transmission' of sinful conditions, however, the Calvinist thinks he can absolve God from '*direct* responsibility' for human sin either by supposing (with the New School party) that depraved tendencies rather than sinful qualities are transmitted or by 'introducing a counter mystery, analogous to that entertained by the Darwinians' in the principle of heredity: namely, the belief – compatible with the teaching of both Augustine and the elder Edwards – that the soul is propagated by natural generation.[114]

Thirdly, Wright again pointed out that 'the adjustment of the doctrines of foreordination and free-will occasions perplexity to the Calvinist in a manner strikingly like that experienced by the Darwinian in stating the consistency of his system of evolution with the existence of manifest design in nature'. And as 'the Calvinist assumes that the highest good of the whole [creation] is consistent with that constituted order of things in which sin is allowed to exist, and in which the freedom that makes sin possible and actual may be put to good use, and even the wrath of man may be made to work God's praise', so the Christian Darwinian 'is compelled to assume that the revelation of *method* and *order* in nature is a higher end, and so a more important factor in the final cause of the creation than are the passing advantages which organic beings derive from it as the scheme of nature is unfolding'.[115]

Fourthly, Darwinism and Calvinism are alike in the limits they assign to speculative reason. Each is an hypothesis; each is founded on 'probable or moral evidence'; each is proved in so far as it 'explains or co-ordinates complicated phenomena which otherwise are confused and unintelligible', the one the phenomena of organic nature, the other the phenomena of scripture and human nature. 'Darwinism is a powerful protest against unrestricted *a priori* methods.' It 'does not propose to explain ultimate facts, but only to interpret their significance regarding the mode or laws of the Creator's action'. Likewise Calvinism, in emphasising the frailty of human reason, 'rejects "absolute" religion' and 'insists upon anchoring its speculations to a solid body of facts'. It does not try to penetrate the mysteries surrounding the mode of the divine existence and the manner in which moral character has been transmitted from Adam to his posterity, but simply acknowledges that these mysteries involve insoluble 'questions of ontology' which arise inevitably in any ultimate inquiry. Thus the Darwinian and the Calvinist alike 'accept the humble role of the interpreter of God's revealed systems'.[116]

Finally, Wright returned to that 'fundamental principle' held in common by Darwinian scientists and Calvinistic theologians: the sovereign rule of law throughout nature. Both in the history of creation and in the history of divine revelation God has 'transmitted his action' mainly by reliance on 'what are called natural means'. The evolution of life has had no more need of miracles than has the preservation of the text of scripture and the evidence by which its genuineness is established. For miracles are meant to serve moral ends; and if the Creator has 'relied in so large a degree upon natural means for the dissemination of the moral forces of his spiritual kingdom', there can be 'no a priori presumption against his having relied wholly upon such means in the development of the lower kingdoms of organic life'. 'We may conclude', said Wright, bringing his chapters to a close, 'that Darwinism has not improperly been styled "the Calvinistic interpretation of nature"'.[117]

While Wright's analogies were intended primarily to convince orthodox believers that Darwinism was compatible with their faith, they also served to show liberal Christians that the best science of the age was on the side of orthodoxy. This was at first an incidental advantage, but after the publication of *Studies in Science and Religion* it became a major theme in Wright's efforts to ally Darwinism and the Christian faith. Between his assumption of the editorship of *Bibliotheca Sacra* in 1883 and his appointment in 1892 as professor of the harmony of science and revelation in Oberlin College – a decade bisected, not insignificantly, by the death of Gray – Wright shifted his theological stance. He became less concerned to commend modern science to believers in revealed theology than to defend revelation from the advocates of liberal theology by showing its agreement with modern science. Gray was gone, the higher criticism was rife, and Wright's reputation, both in scientific and religious circles, was such that he could command the hearing which Gray had obtained twenty years before. Boldly and resolutely he exploited his advantage, publishing two books, *Scientific Aspects of Christian Evidences* (1898), the Lowell Lectures for 1896, and *Scientific Confirmations of Old Testament*

History (1906), and more than forty articles on apologetic subjects concerned with the relations of science and faith. Throughout these writings he upheld the inspiration of scripture and the historicity of its saving events, insisting that the 'theory' of orthodoxy is the one which best fits the facts of science, history, and human experience. For his forthright conservatism, however, Wright himself was somewhat exploited. He placed his pen at the disposal of lesser men and at length, as we have seen in an earlier chapter, he became a contributor to *The Fundamentals* and a prominent figure on the right wing of a polarisation in American religious life, which came to be known as Fundamentalism.[118]

It was an ignominious end. Wright, needless to say, deserves rather to be remembered for his partnership with Gray and for his remarkably consistent support of Darwinism. In 1888 he renewed his commitment to the science and theology of *Darwiniana* in an article entitled 'The debt of the church to Asa Gray'. In 1889, drawing on Darwin's recently published *Life and Letters*, he demonstrated with perfect clarity just 'how far apart' were Darwin and Spencer 'in their whole plane of movement'. In his Lowell Lectures seven years later he showed himself fully apprised of the debate over the factors of evolution, distinguishing carefully between the views of Spencer, Weismann, and Darwin, and maintaining as always that Darwinism and design are reconciled in the divine superintendence of variation. In 1900 Wright lashed out at the current 'evolutionary fad', the popular belief in a 'deductive, *a priori* process' of the 'Spencerian variety' as the rationale for human progress. 'The cast off clothing of the evolution philosophy of fifty years ago', he lamented, 'is now extensively being picked up and put on by many of the religious philosophers and biblical critics of the day'. Christian theists, on the other hand, could have 'no well-grounded objections to that enlargement of the sphere of the action of secondary causes which was involved in the simple statement of the Darwinian theory' – despite its excessive draught on time and its excessive emphasis on the minuteness of variations.[119]

In 1909 Wright took it upon himself to temper the 'indiscriminate laudation' of the Darwin centenary by emphasising 'the mistakes of Darwin and his would-be followers'. Again he pointed out that Darwin erred in his assumption of excessive geological time and in his adherence to the canon *Natura non facit saltum*. But he also had strong words for evolutionists who 'have none of the modesty of their leader', who believe Darwinism is 'applicable to human history', and who 'ride their hobby rough-shod through every department of human thought'. He concluded by reaffirming that 'no doctrine of theology is affected to any appreciable extent by the indefinite theory of the origin of species through natural selection' and that, on the contrary, the Darwinian theory supports orthodox theology by giving rise to 'numerous new analogies. . .between religion, natural and revealed, and the constitution and course of nature'. This, in turn, was the theme – the theme now more than forty years old – of Wright's second article of the centennial year, 'Calvinism and Darwinism'. 'Calvinism', he wrote, 'is comprehensive enough to shelter any reasonable system of evolution under its ample folds. If only evolutionists would incorporate into their system the sweetness of the Calvinistic doctrine of Divine Sovereignty, the church would make no objection to their speculations.'[120]

In 1912 Wright published his last scientific work, *The Origin and Antiquity of Man*. There he summed up and defended the positions he had arrived at after a lifetime of geological investigation: a unitary theory of the glacial period, with its close at a relatively recent date; the emergence of *Homo sapiens* during this period, probably not more than fifteen thousand years ago; and the possession by human beings, early in their history, of their distinctive intellectual characteristics. From these beliefs one might conclude that Wright was no Darwinian at all. One might, in fact, adduce as collateral evidence his 'provisional theory' (Wright ever admired the spirit of pangenesis) that the climatic conditions of Central Asia at the culmination of the glacial epoch may have provided 'the original paradise of the human race'.[121] But this would be to judge rashly. Regardless of the disparities among his beliefs, scientific and religious, Wright laboured to bring them together under the conviction that the reign of God is fundamentally a reign of law, in the evolution of mankind no less than in that of the lower creation.

So far as the physical organism is concerned, Wright had long been convinced that 'man is genetically connected with the highest order of the Mammalia, but. . .not descended from any existing species of that order'.[122] As for the mental endowments of mankind, he was rather less certain. He had long been impressed with Wallace's views on the subject and thus, without stating his own position firmly, had always left room for 'the direct interference of the Creator'. At the same time, however, he had clung to the analogy of mental evolution and traducianism – evasively, perhaps, in his third series of Lowell Lectures, but now with full candour in the *Origin and Antiquity of Man*. Wright maintained that the impartation of mankind's higher mental qualities, which 'have but at least a rudimental development in the highest of the lower animals',[123] was concurrent with the 'finishing off of man's physical organization'.

How this impartation took place it may not be possible for us to comprehend, but that it did take place, through creative interference or creative prearrangement, at a definite epoch in history, is as easily comprehensible as that the germ in which we each as individuals originate is quickened into true spiritual life, and becomes endowed with reason, at a definite point in its existence. When the embryo really becomes human and is endowed with the prerogatives of immortal existence is as much a mystery to the Christian philosopher as the question, When in the line of development did the natural ancestry of man become endowed with its higher human prerogatives?[124]

But if in fact human beings received their higher capacities through 'natural ancestry' and by 'foreordination', these were none the less 'a divine gift inwrought into an orderly system'. Mental evolution by 'divine appointment' presents 'no greater philosophical difficulty... than there is in the well-known fact of the evolution of the individual soul from its parents'.[125] In these statements Wright entertained views which were at least compatible with Darwin's anthropology, if not altogether openly favouring it. Had he held otherwise he would surely have found better cause to take Romanes to task than merely for minimising the intellectual differences between mankind and lower animals. Tinged with equivocation though it may at times be, Wright's conception of human origins was certainly evolutionary and, in contrast to the Fundamentalist movement growing up about him, unmistakably Darwinian.[126]

And so, we may thus conclude, was Wright's conception of evolution as a whole. 'Throughout the conflicts of his later years and despite his hostile response to liberalism', writes one able historian, 'he maintained almost intact most of the major positions of Christian Darwinism'.[127] An understatement, perhaps, but indubitably true – and truer still if 'despite' be made to read 'because of'. For as Wright laboured forty years to explain, there were resources in orthodox Calvinist theology which facilitated Christian Darwinism as fully and inevitably as they contradicted the 'glittering generalities' of liberal evolutionary speculation.

12

DARWINISM AND DARWINISTICISM
IN THEOLOGY

It is no vulgar 'act of faith' that is at issue here, no ignoble acquiescence
in orthodoxy or submission to an establishment. What is at issue is the
faith in science itself, or in what passes as the necessary logic of science.
The theory of natural selection is in many respects almost the ideal
scientific theory: it is eminently naturalistic, mechanical, objective,
impersonal and economical. . . . [Darwin] never doubted that he was a
passive, disinterested observer accurately recording the laws revealed in
nature. . . .
  It was for this reason that Darwinism did not turn out to be the
implacable enemy of religion that was first suspected. For Darwinism
shared with religion the belief in an objective knowledge of nature. If
religion's belief was based on revelation and Darwinism on science, with
good will the two could be – as indeed they were – shown to coincide. The
true challenge to orthodox religion came with the denial of the possibility
of all objective knowledge. . . . Pre-Kant and pre-Kierkegaard, Darwinism
appears as the citadel of tradition.

Gertrude Himmelfarb[1]

For more than forty years after the publication of the *Origin of Species*
evolutionary thought in Britain and America was in a state of great
confusion. Theory vied with theory, philosophy with philosophy, to
explain the manner in which life and matter had assumed its present
forms. The theory of natural selection, which in Darwin's view
described 'the most important, but not the exclusive', means of organic
modification, was subjected not only to base and baseless theological
attacks, but to some well-founded scientific criticisms. Continuous
variation has limits; blending inheritance cancels favourable variants;
utility cannot be found in some morphological features, certainly not
in the incipient stages of many useful structures; time has been inadequate for natural selection to do its work; and appeals to the
causes of variation – the direct action of the environment and the use
and disuse of parts – to resolve these fundamental difficulties only
prove that the real causes of evolution are, after all, just what the

300 THEOLOGY AND EVOLUTION

pre-Darwinian evolutionists said they were. So argued Jenkin, Thomson, and Mivart. So echoed critics by the score. And though the theories to which they turned were less well founded than their objections to Darwinism, the theories managed to prevail. Deriving in part from Lamarck's account of transmutation, they all nevertheless explained the diversity and directionality of life by means of an innate progressive tendency and the adaptive effects of the inheritance of acquired characters. But depriving natural selection of its influence, these theories had a pronounced affinity for religion. Cope's panpsychic theism was a case study in what Neo-Lamarckism could become. Spencer's *Synthetic Philosophy* had even greater appeal, offering cosmic scope, deductive certainty, assurances of progress, and a crypto-theistic agnosticism which could become all things to all men. Opposing both versions of evolution was the Neo-Darwinism of Wallace and Weismann, but in its *a priori* rejection of the inheritance of acquired characters and in Wallace's spiritualistic account of human development it opposed Darwinism as well. A new factor was needed before natural selection could be vindicated and its rivals decisively reproved. This factor, 'the law of variation for some one species of plant', which Huxley sought in 1861, was only furnished in 1900 with the rediscovery of the work of an obscure Augustinian monk named Gregor Mendel.[2]

To the historian studying the post-Darwinian controversies amid the confusion of post-Darwinian evolutionary thought, one question presses on him more than any other. It has not to do with the many who opposed evolution in the interests of Christian faith, for though their efforts are intrinsically worth attention, they bore little relation to the ebb and flow of theoretical debate. Nor does the question concern those who embraced evolution and gave up Christianity. Their number was small and their careers, subsequent to the loss of faith, were in most cases little influenced by changes in religious thought. The question which intrigues the historian has to do with the majority of leading controversialists who were both Christians and evolutionists, those who, in one degree or another, were attentive to developments both in theology and in evolutionary theory. Why, one would like to know, did some become Darwinians and others Darwinists? Why was it that a few remained loyal to Darwin, despite the travails of his theory, while the many, aping and abetting the critics of natural selection, took up other versions of evolution?

There would appear *prima facie* to be no easy answer to this two-sided question but only answers as many and as complex as the

Darwinians and Darwinists themselves. Each had his own religious disposition towards evolutionary thought and each evidently expressed it in his own peculiar way. Moreover, since each disposition and the struggle involved in expressing it were facts of a personal nature, the answers must be doubly difficult to obtain. On second thought, however, one might attempt an answer, not merely for every individual but for Darwinians and Darwinists as a whole, if there were some theoretical access to their dispositions and struggles, some construct that could describe the manner in which historical individuals come to terms with challenges to their prevailing intellectual commitments. For then, from what is known of the individual responses, some general explanation of Christian Darwinism and Christian Darwinisticism might reasonably be inferred.

In analysing Christian responses to Darwin we have been aided by such a construct. We have learnt from Leon Festinger's theory of 'cognitive dissonance' to look for evidence of 'dissonance reduction' – ideas about evolution and conceptions of Darwinism adopted for the purpose of relieving the tension involved in arbitrating between Darwinian and Christian beliefs. Discovering this evidence and, in the case of Christian Darwinisticism, describing its schemes of providence and progress has revealed how a significant number of controversialists came to terms with Darwin. But it has not exhausted the utility of Festinger's theory. *Why* Christians came to terms with Darwin in the manner they did, some as Darwinians, others as Darwinists, is a question as interesting as *how* they did, and a question which arises with special urgency when the post-Darwinian controversies are studied as a constitutive part of the history of post-Darwinian evolutionary thought. In answering this question as well, the theory of cognitive dissonance will prove instructive.

The theological paradox

Given that dissonance results from decisions between Darwinian and Christian beliefs, and given that dissonance reduction must involve a change in the dissonant cognitions themselves or a change in their relationship by the addition of cognitions which dilute their dissonance or reconcile them, it would seem possible to offer at least one explanation of the responses of Christian Darwinians and Christian Darwinists: namely, that those whose theological cognitions as a whole were most consonant with their cognitions of Darwinism would have been able to preserve their beliefs regarding the creation of life and mankind without modifying Darwinism; and, conversely, that those whose theological cognitions, taken as a whole, were least consonant with their cognitions of Darwinism would have tended to preserve their religious beliefs by accepting some other version of evolution. Or to be more precise: those whose theology possessed the resources by which dissonance could be reduced, consonance achieved, would not have needed to reduce their dissonance at the expense of Darwinism; whereas those whose theology lacked these resources would have had to reduce their dissonance by modifying Darwinism to one degree or another.

Now this surely is a reasonable generalisation. It explains Christian responses to Darwin not at an empirical but at a metaphysical level – just where historians have usually, though often crudely, sought an explanation. And it tallies with what we know already of the religious difficulties of Christian Darwinists. A theological analysis, we may thus expect, will reveal why some Christians became Darwinists and others Darwinians.

In proceeding with this analysis, a further generalisation would seem quite reasonable as well. One might predict with some assurance that it was those liberal-minded Christians who had abandoned orthodox theology who could most readily accept Darwinism; and, on the other hand, that it was Christians who clung doggedly to orthodoxy who were unable to reconcile Darwinism with their theology, but, if they could countenance evolution at all, had to accept it in other forms. Indeed, if it has been said once it has been scores of times in the literature of the post-Darwinian controversies: the 'orthodox' would have nothing to do with Darwin. Conservatives, traditionalists, and 'fundamentalists' fought evolution tooth and claw while the enlightened representatives of liberal Christianity welcomed the new science with open arms. 'Protestants of Calvinist, Lutheran, or more evangelical persuasion', it is claimed, '...rebuffed and long resisted the theory of evolution, meeting it with a withering fire of denunciation and vilification. Liberal Protestants, led by a Henry Ward Beecher or a Lyman Abbott, were to do much in silencing the guns of theologians and scientists who so rashly put on their armor and answered the call of battle.'[3]

Disregarding the metaphor, it would be idle to deny this interpretation out of hand. In one sense it is just. Conservatives and traditionalists in theology whose devotion to pre-Darwinian natural philosophies was as great as their reverence for the literal letter of the Bible certainly did resist all forms of phylogenetic evolution. And liberal theologians, typified by Beecher and Abbott, were notorious for glorying in evolution and an evolutionary faith. In another sense, however, the received interpretation is seriously misleading. It obscures one truth with another, a subtle and thus perhaps more important truth with one which is only too obvious: the obscured truth being that what liberals took as 'the theory of evolution' was no more Darwinism than what most conservative anti-evolutionists understood by 'the theory of evolution'. Neither party, in fact, could reconcile Darwinism with their beliefs. More thoroughly still is this truth obscured by the assertion that 'through the seventies most Protestants (other than advanced religious liberals) held to the eventually damaging position that Darwin's theories could *never* be reconciled with religion. Only a few far-sighted divines were saying that natural selection could, if proved true, be interpreted as a part of the Divine method.'[4]

Indeed, an interpretation which would thus correlate Darwinism inversely with orthodox theology, implying that liberals were Darwinians and the orthodox either anti-Darwinians or Darwinists, not only obscures the evident truth that Darwinism was incompatible with liberal theology; it neglects entirely what may well be the central and regulative paradox of the post-Darwinian controversies: namely, that it was only those who maintained a distinctly orthodox theology who could embrace Darwinism; liberals were unable to accept it. Christian Darwinism was a phenomenon of orthodoxy, Christian Darwinisticism, on the whole, an expression of liberalism. The correlation between Darwinism and orthodoxy was not inverse but direct.

Such, at any rate, is the paradox which emerges with remarkable clarity from our study of the leading controversialists in Great Britain and America. The Darwinians to a man remained closely in touch with established theological traditions. They were not so much innovators as those who undertook to preserve and defend received truth and to communicate it authoritatively in a form cogent to their contemporaries. James Iverach was a sturdy evangelical Scottish Free Churchman, an opponent of the 'New Theology', and the author of an admiring biographical portrait of the elder Jonathan Edwards. Aubrey L. Moore was raised an Anglo-Catholic and educated at Oxford, the haven of High Churchmen, where he lectured until his untimely death. His essays bristle with references to the Fathers and in the introductory pages of his *Science and the Faith* he frankly confessed that he thought it 'impossible to defend Christianity on anything less than the whole of the Christian creed'. Asa Gray, a conservative Congregationalist and moderate Calvinist, described himself in the preface of his *Darwiniana* as 'one who is scientifically, and in his own fashion, a Darwinian, philosophically a convinced theist, and religiously an acceptor of the "creed commonly called the Nicene", as the exponent of the Christian faith'. George Frederick Wright, also a Congregationalist and Calvinist of the New School, was for thirty-seven years editor of *Bibliotheca Sacra*, the foremost journal of conservative theological scholarship in America. Zealous for a high doctrine of scripture, he allied himself with those who became Fundamentalists in the twentieth century. In addition to his numerous contributions to the literature of glacial geology and his many theological writings, he penned a standard biography of the American revivalist and Oberlin theologian, Charles Grandison Finney.[5]

The Darwinists, in contrast, were overwhelmingly committed to theological reconstruction. They were neologians, striving to assimilate faith to scientific speculation and religious experience, discarding traditional views whenever expedient and rendering the residue in a form acceptable to their contemporaries. St George Mivart's autonomous rationalism earned him the epithet 'modernist' and brought about his expulsion from the Roman Catholic Church. Frederick Temple fared somewhat differently in the Church of England, becoming archbishop of Canterbury, but this despite his consistent latitudinarianism, from his contribution to *Essays and Reviews* onwards. John Bascom rejected the 'perverse theory' of Calvinism on which he had been reared and elaborated a philosophy which 'belonged to the most liberal form of Christian apologetics'. Joseph Le Conte was – to use his own words – at 'first orthodox of the orthodox; later, as thought germinated and grew apace, I adopted a liberal interpretation of orthodoxy; then, gradually I became unorthodox; then, in deep sympathy with the most liberal movement of Christian thought; and finally, to some extent, a leader in that movement'. Thomas Howard Mac-Queary's evolutionary and iconoclastic approach to Christian doctrines cost him his clerical career in the first outcome of a heresy trial in the history of the Protestant Episcopal Church. Lyman Abbott, believing the New Theology to be an advance over the New School Calvinism he had formerly accepted, was America's leading representative of evangelical liberalism. Francis Howe Johnson, perhaps the weightiest of the 'Andover liberals', also moved far from his Calvinist heritage in assimilating Christian doctrine to a personalist philosophy. George Matheson, a 'conspicuous representative of liberal theology', retained the doctrinal forms of conventional Calvinist orthodoxy but invested them with a new Hegelio–Spencerian meaning. 'I am as broad as can be', he said, 'but it is a broad positive'. Henry Ward Beecher was similarly expansive, though with neither the philosophical sophistication nor the consistently positive approach. 'I am an evolutionist', he declared, 'and that strikes at the root of all medieval and orthodox modern theology'. Minot Judson Savage and John Fiske, like Beecher, jettisoned Calvinism for the theological liberties of the *Synthetic Philosophy* but both became Unitarians in the process. Henry Drummond was twice subjected to charges of heresy, in 1892 and 1895, and on the latter occasion no less than twelve overtures were made to the General Assembly of the Free Church of Scotland regarding his *Ascent of Man*. Drummond's writings, observed one liberal critic, are 'the anaesthetic which a clement Providence has administered to orthodoxy during the excision of its diseased doctrines'.[6]

Exceptions to this paradoxical pattern – orthodox Darwinians and liberal Darwinists – there may be, but those who first come to mind prove fairly unconvincing. The Reverend Baden Powell (1796–1860), Savilian Professor of Geometry at Oxford and a Fellow of the Royal Society, was, like Temple, a contributor to *Essays and Reviews* and an exponent of liberal divinity, especially in opposition to the Tractarians. Yet he showed every sign of becoming a Darwinian. In a remarkable essay on the 'philosophy of Creation', published in 1855, he argued for organic evolution as a 'probable conjecture', subject to laws as yet unknown. In his *Order of Nature*, which appeared in 1859, he wrote again of the transmutation question as 'one simply of rational philosophical conjecture as to the most probable mode in which, conformably to natural analogies, it might be imagined to have taken place'. And in his contribution to *Essays and Reviews* a year later he was quick to acknowledge that natural selection provided the law and analogy by which the problem had been solved. Regrettably, Powell died on 11 June 1860, before he could read Darwin's commendation of his work in the 'Historical sketch' prefixed to the third edition of the *Origin of Species*. Although his theology of nature may have been harmonious with Darwinism as he understood it, Powell did not express his views on evolution in the decades when natural selection was assailed – in the crucible where Darwinians and Darwinists were proved.[7]

Charles Kingsley (1819–1875) would appear to be another obvious exception: a disciple of F. D. Maurice, a Christian Socialist, and a canon of Dean Stanley's Westminster on the one hand; and, on the other, a steadfast follower of Darwin and Gray. It was his glowing response to a pre-publication copy of the *Origin* which Darwin quoted, as from 'a celebrated author and divine', in the second and subsequent editions of the book, in order to allay religious fears. No doubt Darwin was also pleased to have Kingsley's support against the attacks of Jenkin and the Duke of Argyll, and had he read of his clerical friend urging Wallace to 'extend to all nature the truth you have so gallantly asserted for man' he might have approved as well. As for Gray, it was his pamphlet *Natural Selection Not Inconsistent with Natural Theology* – the one Darwin proposed and underwrote – which Kingsley described to Maurice as 'by far the best forward step in Natural Theology' and which, he added, says 'better than I can all that I want to say'.[8] In this juxtaposition of Maurice and Gray, however, there is a hint that Kingsley may be only an apparent exception to the rule that Christian Darwinians were orthodox Christians. For whatever Kingsley's relationship with the former (and it was probably not a simple one), the latter had a theology that was congenial to his own. Like Gray, Kingsley was no 'verbal-inspiration-monger', but neither was he a radical.[9] He deplored *Essays and Reviews* for parading 'doubts, denials, destructions' and reaffirmed his intention to 'keep to the orthodox formulae'. He believed in 'reverent and rational liberty in criticism (within the limits of orthodoxy)' and thought that Colenso's *Pentateuch* violated it. Though often associated with the Broad Church, he referred to himself as an 'old-fashioned High Churchman'. Indeed, in lecturing to theological students in 1871 on 'the natural theology of the future', Kingsley described himself 'as (I trust) an orthodox priest of the Church of England', one who believes its theology to be 'eminently rational as well as scriptural' and who holds that the present 'divorce between Science and Christianity' would not exist had the Church's 'orthodox thinkers' for the last hundred years 'followed steadily' in the steps of Bishop Berkeley, Bishop Butler, and Archdeacon Paley, the 'three greatest natural theologians'.[10]

Admittedly, the exceptions among Christian Darwinists seem rather more substantial. George Henslow was a member of that conservative evangelical alternative to existing scientific societies, the Victoria Institute, although his faith did not preclude a deep interest in spiritualism during his later years. The Duke of Argyll, a member of the Church of Scotland, seems to have remained fairly loyal to the established form of belief. But against these apparent exceptions must be weighed the salient point that the Darwinists nearest Darwin in their conception of evolution were the most orthodox Darwinists of all. From what we know already of Christian Darwinians this would hardly seem a surprise. Joseph S. Van Dyke and James McCosh were Princetonians, the former a graduate and tutor of the College, the latter its president. Both were Calvinists of the Old School: Van Dyke studied under Charles Hodge at the Seminary; McCosh served under Hodge at the College. *Theism and Evolution*, Van Dyke's contribution to the post-Darwinian controversies, bore the *imprimatur* of Hodge's son and successor in Princeton's chair of theology. *The Religious Aspect of Evolution* and all McCosh's other publications on the subject bore the authority of one who had been a leader in the Scottish Free Church and had earned a favourable reputation on both sides of the Atlantic for evangelical 'common sense' in philosophy.

Therefore the paradox fundamentally remains: orthodox theology was the correlative of Darwinism, liberal theology the concomitant of Darwinisticism. While for some it may have been possible to be orthodox and a Darwinist (just as it was also possible to be orthodox, in one sense, and opposed to evolution), there is little evidence that it was possible to be both liberal and a Darwinian. In this theological paradox lies an explanation of Christian responses to Darwin.

The orthodoxy of Darwinism

According to the generalisation with which we began, the theology of Christian Darwinians must have been uniquely congruent with Darwinism and capable itself of reducing the dissonance involved in arbitrating between Christian beliefs about the creation of life and mankind, and the Darwinian causal explanation of that event. Now we have identified this theology as orthodox theology. That is to say, we have pointed out that those Christians who accepted Darwinism, or who nearly did so, were adherents of established theological traditions and guardians of historic deposits of truth. But what, it will be asked, was there about orthodoxy – what notions were there in that fund of theological truth held in common by Van Dyke, McCosh, Iverach, Moore, Gray, and Wright – which made Darwinism acceptable? Or what was there about Darwinism that made it acceptable to orthodox theology? These questions, as we shall see, are both related and distinct.

As such we shall answer them separately, the latter first, and show their points of contact in an interim analysis of the decline of Darwin's theology.

It has been argued persuasively that the triumph of Darwinism was 'the triumph of a Christian way of picturing the world over the other ways available to scientists' in the nineteenth century. The ideas that Darwin took for granted in formulating the theory of evolution by natural selection defined the universe as it had been 'elaborated by orthodox naturalists and natural theologians' since the time of John Ray. It was 'fundamentally different from the universes proposed by some other nineteenth-century scientists'.[11] The Platonic universe of Louis Agassiz and the British paleontologist Edward Forbes was static and irrelative to the search for efficient causes. Darwin described a world in which change is wrought constantly by causes accessible to investigation. The universe of Charles Lyell involved constant change as well, but only by uniformitarian oscillations about an eternal mean. Darwin conceived a world of historic process, unique, creative, and unrepeatable. The universe of Jean Baptiste Lamarck also embodied a creative process, but one which moved inherently and progressively towards mankind, the measure of all things. Darwin portrayed a world in which mankind is the latest and noblest species to be evolved, but by no means the ultimate species for which all things have inherently and progressively come into existence. For Darwin as for Christians, the world is a real historical place; its events are a meaningful and unrepeatable sequence; its purpose includes human beings but is not fully realised in them.[12]

The agreement of Darwinian and Christian cosmologies was hardly coincidental. Darwin derived his universe in large measure from those 'orthodox naturalists and natural theologians' who had dominated British science until his time. Not least among them were his professors at Cambridge, the Reverend John Stevens Henslow and the Reverend Adam Sedgwick, the Christian philosophers John Herschel and William Whewell, whose writings Darwin read early in his career, and the authors of *The Bridgewater Treatises on the Power, Wisdom and Goodness of God as Manifested in the Creation*, which appeared between 1833 and 1836, and came to Darwin's attention soon after he returned from the voyage of the *Beagle*.[13] More influential on Darwin than any of these, however, were two Anglican clergymen whose books not only defined the universe in which his evolution took place, but contributed substantially to the mechanism of that evolution. We refer to the Reverend William Paley (1743–1805), archdeacon of Carlisle, and the Reverend Thomas Robert Malthus (1766–1834), the incumbent of Albury in Surrey.[14]

For the Cambridge B.A. examination, which he passed in 1831, Darwin was required to master Paley's *Principles of Moral and Political Philosophy* (1785) and his *View of the Evidences of Christianity* (1794). So impressed was he with these works, evidently, that he determined to master Paley's *Natural Theology* (1802) as well. The *Evidences* and *Natural Theology* gave him as much pleasure as Euclid had ever done. He was 'charmed and convinced' by their long lines of argument. 'The careful study of these works', Darwin wrote in 1876, '. . .was the only part of the Academical Course which, as I then felt and as I still believe, was of the least use to me in the education of my mind'.[15] To verify this statement in detail would require a separate study, but it would not be a very difficult one. Even the casual reader of Paley and Darwin must be struck by their common dialectic: the rhetorical questions, the familiar examples, the heaping up of facts, and the pleading in the face of contrary evidence which their arguments alike contain. A closer reader, especially of the *Natural Theology*, the *Origin of Species*, and the *Descent of Man*, would notice further similarities. For Darwin, who 'hardly ever admired a book more than Paley's "Natural Theology"' and 'could almost formerly have said it by heart', was influenced by the substance of its arguments as well as by their form.[16]

Paley took the eye as the first and foremost 'application of the argument' from design. Darwin took it as the *pièce de résistance* for an omnivorous natural selection. Paley devoted much of one chapter to the instincts of birds and insects. Darwin did the same. Paley made much of the purposeful relations between the wax, the honey, and the sting of bees. Darwin showed how natural selection bears on each. Paley, recognising that cases of apparent inutility in nature are but 'extremely rare', declared that 'true fortitude of understanding consists in not suffering what we know to be disturbed by what we do not know'. Darwin at first adopted an identical point of view and only later, in conceding the existence of useless structures, admitted that it was his 'former belief, then almost universal, that each species had been purposefully created', which had led him to assume 'that every detail of structure, excepting rudiments, was of some special, though unrecognised, service'.[17] Paley employed the term 'relation' to designate 'the fitness of. . .parts or instruments to one another, for the purpose of producing, by their united action, [an] effect', and the term 'compensation' to designate the fitness of parts or structures to supply what others lack. Darwin referred to the coordinated appearance and accumulation of variations in an organism as 'correlation of growth' (later as 'correlated variation') and to the balance that obtains among these variations as 'compensation'.[18] Paley, in his constant imputation of human qualities to lower forms of life, showed that he was as certain he could understand the instincts and feelings of animals as he was convinced he could rightly interpret the intentions of God in creation and providence. Darwin's discussions of sexual selection and the expression of the emotions in mankind and animals are reminiscent of Paley, drawing out the obvious implication that the difference between mankind and other animals is one of degree and not of kind.[19]

The dialectic of natural theology and the perception of adaptive purpose extending from the eye, through the unity of somatic structures, and up to the highest expressions of instinct and emotion, were two important aspects of Darwin's debt to Paley. Without both, the argument for natural selection, grounded in thorough inductions and founded on the principle of utility, could hardly have been constructed. But there was, arguably, a third aspect of Darwin's indebtedness which was no less crucial to his theory. Natural selection gave a strictly empirical account of organic change. It had nothing to do with occult qualities, metaphysical entities, or vital forces. And if there was one thing that Paley opposed, it was the endowment of nature with such self-organising and self-operating properties. A watch, he declared, is not made by a 'principle of order'. To say so is merely to substitute 'words for reasons, names for causes'. A 'principle of order' is nothing if it is not the actual intelligence of the watchmaker, his 'mind and intention' working through ordinary causes and adapting means to an end. Likewise a watch is not made by 'law'. To say so is again merely to beg the question. 'It is a perversion of language', Paley insisted, 'to assign any law, as the efficient, operative cause of any thing. A law presupposes an agent; for it is only the mode, according to which an agent proceeds: it implies a power; for it is the order, according to which that power acts. Without this agent, without this power, which are both distinct from itself, the *law* does nothing; is nothing.'[20]

The object of these arguments was to undermine at a stroke the Lucretian account of creation, the nomistic theology of the deists, and the latest speculations in natural history. Paley had little patience with the French naturalist Buffon. His belief in spontaneous generation was plainly uncritical and his 'internal molds' which govern the production of living forms from raw matter were gratuitous fictions, too Platonic to be intelligible. Nor did Paley respond favourably to the system of 'appetencies' which had recently been put forward by his poetic compatriot Erasmus Darwin. Granted, the system is not necessarily atheistic. The appetencies which produce habit, change, and evolution by the inheritance of acquired changes are appointed by an intelligent and designing Creator. But the system coincides with atheism in that it does away with final causes. Form proceeds from use, not use from form. The appetencies, acting independently, as if they were something in themselves, are supposed to have created every vegetable and animal type, quite apart from the Creator's particular design.[21] Is it any wonder that the grandson of Erasmus Darwin, imbued with Paley's philosophy, found Agassiz 'wild and paradoxical in all his views', Owen the author of 'miserable inconsistencies and rubbish', and the Neo-Lamarckians incomprehensible?[22]

Darwin's encounter with Malthus came seven years and a circumnavigation after he was examined on Paley's works. Having spent fifteen months after his voyage on the *Beagle* making notes on the transmutation of species, he sat down on 28 September 1838 to read 'for amusement' the sixth edition (1826) of Malthus' *Essay on the Principle of Population*.[23] Immediately, on page six of the first volume, he discovered the main idea of natural selection, a concept he had come across many times before but which now, after years of thoughtful observation, he could recognise as the driving force behind adaptive change in nature: the struggle for existence, consequent on the geometrical multiplication of population and the arithmetical increase of the means of sustenance. 'One may say', Darwin noted, 'there is a force like a hundred thousand wedges trying [to] force every kind of adapted structure into the gaps in the oeconomy of nature. or rather forming gaps by thrusting out weaker ones.'[24] Now it will not do to overestimate the influence of Malthus on Darwin. The rudiments of natural selection were present in Darwin's thought well before he came upon the principle of population. Malthus' *Essay* was as much the occasion of the theory as its cause.[25] But neither would it be right to think that Darwin, flushed with the thrill of discovery, got no farther than page six of volume one. Always keen to pick up useful information, he continued reading the *Essay* on into the next month, noting relevant passages and gaining for Malthus a deep and abiding respect.[26]

What impressed Darwin about Malthus, beyond the principle of population itself, one can only surmise, but it would seem to have been connected with his power of reasoning. After the *Origin of Species* was published Darwin took comfort from the fact that his own critics were unable to understand Malthus either. 'As he sneers at Malthus', Darwin wrote of Samuel Haughton, 'I am content, for it is clear he cannot reason'.[27] The misunderstandings were not, in fact, unrelated. Malthus had a theory as well. Like Darwin, he had to defend it from obscurantism on the right and distinguish it from fanciful hypotheses on the left. 'Practical' persons who issue 'declamations against theory and theorists', he declared, 'may be classed among the most mischievous theorists of their time'. While inveighing against speculation, they base their own arguments on a few feeble facts, thereby prejudicing and misleading the many who 'do not stop to make the distinction between that partial experience which...is no foundation whatever for a just theory, and that general experience, on which alone a just theory can be founded'.[28] Those, on the other hand, who seek to enlarge the scope of human science by indulging in 'improbable and unfounded hypotheses', Malthus stated, only succeed in contracting it. His entire *Essay* was conceived as an argument against such theorists, and against the utopian political philosophers of the time, William Godwin and the Marquis de Condorcet, in particular. Godwin and Condorcet believed in mankind's indefinite and inevitable progress, partly through the inheritance of acquired perfections. No constraints of space, time, society, or human nature, they maintained, could ultimately withstand the triumphant march of reason. Malthus countered this 'wide and unrestrained speculation' with its 'wild flights and unsupported assertions', not by disallowing a theoretical approach to human problems, but by demonstrating 'a perfect readiness to adopt any theory warranted by sound philosophy'. Such a theory, derived through 'patient investigation and well-supported proofs', was the principle of population.[29]

The principle of population gave the lie to notions of limitless and automatic progress. By showing that human populations tend to outstrip their food supply as a geometrical series does an arithmetic, it placed mankind squarely within the context of nature, subjecting the species to forces which utopian speculation could hardly hope to transcend. 'Vice', 'misery', and a struggle for the means of existence, according to Malthus, must result from the disparity between population and food. These serve to check the growth of population, but only at an expense too great for the believer in human perfectibility. Obviously, however, there would be some question as to whether the expense was not also beyond the means of the believer in a beneficient and omnipotent God. Did not the principle of population belie the Christian faith as well? In the first edition of his *Essay*, published in 1798, Malthus devoted the two concluding chapters to answering this question. In later editions his arguments reappeared in conjunction with an appeal for 'moral restraint' from marriage as 'the proper check to population'.[30]

Unlike Paley, who stressed that evil is necessarily involved in mankind's temporal probation, Malthus conceived a theodicy in which natural and moral evils contribute to a process of preparation for the future life. They are 'the instruments employed by the Deity in admonishing us to avoid any mode of conduct which is not suited to our being, and will consequently injure our happiness'.[31] If, for example, both population and food were to increase in the same proportion, there would be no 'motive...sufficiently strong to overcome the acknowledged indolence of man, and make him proceed in the cultivation of the soil'.[32] But the evils attendant on the principle of population provide such a motive and thus keep man 'more alive to the consequences of his actions' and give his faculties 'greater play and opportunity of improvement' than could the conditions under some more harmonious arrangement of nature. Similarly, without the threat of overpopulation and its attendant miseries, early marriages and indigent families would create havoc in society. The principle of population encourages moral restraint from marriage until one has a fair prospect of being able to support one's children. Its purpose is thus benevolent, for it promotes virtue as it conduces to 'the formation and improvement of the human mind'.[33]

The 'final cause' of natural selection, as Darwin described it on 28 September 1838, was 'to do that for form, which Malthus shows is the final effect (by means however of volition) of...populousness on the energy of man'.[34] In one case as in the other, 'formation and improvement' would result. In neither case, however, would the development be inevitably and indefinitely progressive. The Malthusian theodicy did not involve a system of universal progress. It consisted in showing that the evils of overpopulation are inseparable from the virtues required to mitigate it. Without the former the latter would not exist, but only other evils in their stead. Indeed, as human volition intervenes, even if understanding of the principle of population were widely diffused, the virtues could at best be proximately secured. 'On the whole, therefore', Malthus wrote in the concluding paragraph of his *Essay*, '...our future prospects respecting the mitigation of the evils arising from the principle of population may not be so bright as we could wish'. 'Yet', he added,

> they are far from being entirely disheartening, and by no means preclude that gradual and progressive improvement in human society, which, before the late wild speculations on this subject, was the object of rational expectation.... A strict inquiry into the principle of population obliges us to conclude that we shall never be able to throw down the ladder, by which we have risen to this eminence; but it by no means proves, that we may not rise higher by the same means.[35]

This 'rational expectation', the legacy of a Christian view of history, was Darwin's as well. While convinced that progress in civilised nations is 'no invariable rule', in the last paragraph of the *Descent of Man* he excused man 'for feeling some pride at having risen, though not through his own exertions, to the very summit of the organic scale' and suggested reservedly that 'the fact of his having thus risen...may give him hope for a still higher destiny in the distant future'.[36]

In no mere external sense, therefore, was Darwinism related to the orthodox theology of the day. If the universe of Darwinism owed much to the natural theologians, so, in large measure, did its mechanism. From Paley, Malthus, and others Darwin derived a real historic world, one in which empirical methods reveal efficient causes through well-supported proofs. But from Paley and Malthus in particular he learnt how these causes produce organic change. Paley stressed adaptation to environment through divine contrivance; Malthus stressed 'formation and improvement', through struggle, without inevitable progress. Darwin synthesised the two in the theory of natural selection: struggle yields adaptive improvement without inevitable progress. Only one element was missing – divine contrivance – and in its loss Darwin lost his faith.

The decline of Darwin's theology

The change in Darwin's religious views is well known, having been enlisted as often by infidels to accredit their unbelief as it has been adduced by believers to discredit evolutionary science: from an early orthodoxy Darwin passed in middle life into a liberal form of theism, and thence, in his last decades, into an agnosticism tending at times towards atheism. Until recent years the main sources available for studying this development were Darwin's expurgated autobiography and his numerous published letters. The former contained his censored reflexions late in life and the latter revealed his religious views almost entirely in connexion with the controversies subsequent to 1859. Thus, though much could be said about Darwin's growing agnosticism, little could be known of his earlier break with orthodoxy.[37] But with the appearance of a complete edition of the autobiography and the publication of several important manuscripts, including the contents of six notebooks written by Darwin between 1837 and 1840, the defect in evidence has been repaired. We can now trace Darwin's departure from the theology of those orthodox naturalists and natural theologians who contributed so much to his thought.

Born into a family noted for its liberality in politics and religion, Darwin was christened in the Church of England and sent to a school conducted by the Unitarian minister whose chapel his mother attended. In 1828, after an abortive attempt at Edinburgh to follow his father in the medical profession, he entered the University of Cambridge, where his father had decided that he should prepare for the ministry. The thought of being a country clergyman was not unattractive, and soon, with the aid of Pearson's *Exposition of the Creed* and Bishop Sumner's *Evidence of Christianity Derived from Its Nature and Reception* (1824), Darwin abandoned whatever were his scruples about professing belief in all the doctrines of the Church.[38] As in his earlier studies, however, the lure of natural history and an adventuresome life was more than his filial devotion could withstand. On graduation in 1831 he took, not holy orders, but a post aboard H.M.S. *Beagle*. On 27 December of the same year, in the full flower of Christian faith and with 'a burning zeal to add even the most humble contribution to the noble structure of Natural Science', Darwin set sail on a voyage which was to pluck faith's flower and restructure the sciences of natural history.[39]

Within five years Darwin saw what few naturalists see in a lifetime. The world was his museum, the islands and shores of the southern hemisphere its chief exhibits. Much that he encountered was new and wonderful, and his letters teem with enthusiastic descriptions.[40] Much also was problematic. Face to face with a mountain or a coral reef, the biblical chronology seemed nonsense. The sheer variety of flora and fauna, exquisite in form and colour, could only make one 'wonder that so much beauty should be apparently created for such little purpose'. And in reflecting on the geographical distribution of species, the differences between animals on continent and continent, island and island, Darwin wrote in his diary that 'an unbeliever in everything beyond his own reason might exclaim, "Surely two distinct Creators must have been at work" '.[41]

Most striking of all creatures, and without a doubt the one Darwin found most disturbing, was mankind. The reception of himself and his mates by the native inhabitants of Tierra del Fuego was the 'most curious & interesting spectacle' he had ever beheld. Between them the differences seemed greater than the differences between wild and domesticated animals.[42] Yet when he observed the shocking behaviour of civilised people towards indigenous populations he was forced to think again. The fruits of imperialism, Spanish and British, included cruel slavery and cold-blooded massacre. 'Who would believe in this age in a Christian civilized country that such atrocities were committed?' an outraged Darwin demanded. Argentina 'will be in the hands of white Gaucho savages instead of copper coloured Indians. The former being a little superior in civilisation, as they are inferior in every moral virtue.'[43] Perhaps, then, primitive people were not so different after all. Had not the three Anglicised Fuegians aboard the *Beagle* completed in a single generation the circuit from savage to European and back, resettling happily among their miserable and backward tribes? 'One's mind hurries back over past centuries, & then asks, could our progenitors be such as these?' Darwin mused on the homeward voyage. 'Men, – whose very signs and expressions are less intelligible to us than those of the domesticated animals; who do not possess the instinct of those animals, nor yet appear to boast of human reason, or at least of arts consequent on that reason.'[44]

The lowliness of human reason, its development 'over past centuries' from something less than the instinct of domesticated animals, shook Darwin's confidence in the eminently rational conclusions of orthodox naturalists and natural theologians.[45] These conclusions depended for their validity on a lofty estimate of mankind's intellectual powers, an estimate which in turn was based on an elevated notion of mankind's origin. But, wrote Darwin in mid-1838,

let man visit Ourang-outang in domestication, hear expressive whine, see its intelligence when spoken [to], as if it understood every word said – see its affection to those it knows, – see its passion and rage, sulkiness & very extreme of despair; let him look at savage, roasting his parent, naked, artless, not improving, yet improvable and then let him dare to boast of his proud preeminence.[46]

Such pride was, in Darwin's opinion, that of the slave-trader, even in those who abhor the practice.

Has not the white man, who has debased his nature by making slave of his fellow Black, often wished to consider him as other animal. – it is the way of mankind & I believe those who soar above such prejudices yet have justly exalted nature of man. like to think his origin godlike.[47]

The belief in mankind's special creation, therefore, is as presumptuous as the practice of slavery. Both are species of megalomania. To think that Whewell, the great historian of science, could claim that 'length of days adapted to duration of sleep of man!!! whole universe so adapted!!! & not man to Planets'. This surely is an 'instance of arrogance!!' Darwin scribbled in disgust. Elsewhere in his notebooks he confided to himself: 'Man in his arrogance thinks himself a great work worthy the interposition of a deity. more humble & I believe truer to consider him created from animals.'[48] And in another place Darwin wrote: 'Why is thought being a secretion of the brain, more wonderful than gravity a property of matter? It is our arrogance, . . .our admiration of ourselves.'[49]

Darwin's revulsion at human pride extended from natural theologians' reasoning on the creation of their own species to their reasoning on creation in general. Late in 1838, having got from Malthus the inspiration of natural selection, Darwin read some passages in one of the most formidable recent works of natural theology, John Macculloch's three-volume *Proofs and Illustrations of the Attributes of God*. As usual he made notes, and in them for the first time his theory distinctly emerged, tested against Macculloch's belief in divine contrivance and special creation. When we venture to say why this or that was created, Darwin declared, 'we sink into contemptable quiries' and 'lower the creator to the standard of one of his weak creations'. When Cuvier, cited by Macculloch, speaks of the vertebrate plan and of abortive bones as evidence of divine intention to pursue the plan to its utmost exhaustion and abandonment, it is human design determining a Godhead. As if one could speak of 'the designs of an omnipotent creator, exhausted & abandoned'. 'Such', Darwin fumed, 'is man's philosophy, when he argues about his Creator!!' Finally, when Macculloch submitted the marsupial bones of the opossum as a special adaptation to protect the young, Darwin could bear it no longer. 'Good God', he remarked, '. . .what trash'.[50]

There had to be some other explanation of creation and for Darwin it had to be in terms of law. The 'theory' of special creation, he argued, is not science at all.

> The explanation of types of structure in classes – as resulting from the *will* of the Deity, to create animals on certain plans, – is no explanation – *it has not the character of a physical law* & is therefore utterly useless. – it foretells nothing because we know nothing of the will of the Deity, how it acts & whether constant or inconstant like that of man.[51]

The appeal of special creation lies not in its scientific explanations but in its flattery of human pride. Unable to predict by law, it assumes that God's will is predictable, 'like that of man'. 'As long as we consider each object an act of separate creation', Darwin stated, 'we admire it more, because we can compare it to the standard of our own minds'. Humility would adopt a lower estimate of mankind's intellectual powers and 'consider [the] Creator as governing by laws'.[52] Humility would thus render a truer and nobler account of the origin of species.

If Darwin was concerned to protect God from the godlike pretensions of his highest creature, he was no less intent on shielding him from the aspersions of his lower creation. While on the *Beagle* Darwin found it increasingly difficult to attribute everything he observed to a superintending Providence. The wastage of superfecundity and the peculiarities of biogeography seemed beneath the dignity of the divine creative power. At the same time, in reading Lyell's *Principles of Geology* (1830–3) and, no doubt, in the recollection of his grandfather's deistic *Zoonomia* (1794–6), Darwin saw the potential of 'secondary causes', the large scope of creation which might be left to 'law'. Gradually, therefore, his universe – hitherto a thoroughly Christian universe – was left to function on its own. The humility which, in acknowledging the lowliness of human reason, forsook miraculous explanations of creation became the *hubris* which, in pursuing scientific explanations and a 'grander' theology, sought to relieve God of his world.

Already in Darwin's first notebook on the transmutation of species, begun within the year after his voyage, the process was well under way. God and law are two means of creation. Where one leaves off the other takes up the job. In this connexion Darwin conceived the analogy which he would pursue for years to come: as planets move in their orbits by law so created forms are adapted to their environments.

Astronomers might formerly have said that God ordered each planet to move in its particular destiny. In same manner God orders each animal created with certain form in certain country, but how much more simple and sublime power let attraction act according to certain laws, such are inevitable consequences – let animal be created, then by the fixed laws of generation, such will be their successors. Let the powers of transport be such, and so will be the forms of one country to another. – Let geological changes go at such a rate, so will be the number and distribution of the species!!⁵³

Whether Darwin understood Newton's theology of nature, which he had studied in the works of Herschel and Whewell, is less important than the fact that he thought its simplicity and sublimity characterised his idea of creation by law.⁵⁴ A similar idea – a 'grand idea' Darwin called it – had been expressed by the naturalist Etienne Geoffroy Saint Hilaire: 'God giving laws and then leaving all to follow consequences'.⁵⁵ Darwin did not forget it. In his second notebook he wrote of 'the wonderful power of adaptation given to organization'; and, in toying with a materialistic interpretation of the relation between brain structure and hereditary mental traits, he exclaimed, half-reproachfully, 'Love of the deity effect of organization, oh you materialist!'⁵⁶

During the summer of 1838, when he opened the first of two notebooks on man, mind, and materialism, Darwin found the 'grand idea' again in the Positive Philosophy of Auguste Comte. Comte taught that each branch of knowledge passes through three theoretical stages: first, one in which theological or fictitious explanations are given for natural phenomena; then, one in which metaphysical or abstract explanations are given; and finally, one in which scientific explanations prevail. On learning of Comte's doctrine from a review by David Brewster, Darwin immediately saw its implications. That 'the fixed laws of nature should be universally thought to be the *will* of a superior being' whose nature is akin to mankind's not only shows that science is yet in its theological state; it makes one suspect, from a scientific point of view, that 'our will may arise from. . .fixed laws of organization'. Both the concept of God and the concept of a contriving God may be simply the product of mankind's biological structure. Comte therefore, Darwin must have realised, could make one a predestinarian atheist. But as his idea of a theological state of science was also a 'grand *idea*', it could lead to a nobler conception of God.[57] The philosopher errs, Darwin declared, 'who says the innate knowledge of creator...has been implanted in us...by a separate act of God, & not as a necessary integrant part of his most magnificent laws. which we profane in thinking not capable to produce every effect of every kind which surrounds us'.[58]

So far had God receded from the world by the time Darwin had done with Comte that final causes began to disappear. Previously Darwin wrote of teleology almost out of habit, with only the occasional oblique reference to purposelessness in nature.[59] Now his chief end was to do away with ends. In the search for some '*law of variation*' to explain the plumage of birds he noted how the Peahen takes the feathers and spurs of the Peacock, remarking, '*No final cause here.*' He wrote of 'chance & unfavourable conditions', of '*accidental* hardy seedlings', and of ' "fortuitous" ' infinitesimal advantages.[60] 'No one can be shocked at absence of final cause', he said, and offered as proof – irony of ironies – the mammae of the male, the one structure for which Paley confessed himself 'totally at a loss to guess...the reason, final or efficient'.[61] To be sure, Darwin did at one point conceive of 'the formation of laws invoking laws & giving rise at last even to the perception of a final cause'. But it could not have been more than a few days later, in reading indignantly Macculloch's natural theology, that he at last became conscious of how far he had ventured from his old point of view. 'Is it anomaly in me to talk of Final Causes: consider this?' he asked.[62]

Final causes or no final causes, however, Darwin still believed that his idea of creation was far grander than the conventional one. 'What a magnificent view one can take of the world', he exulted.

Astronomical causes modified by unknown ones, cause changes in geography & changes of climate suspended to change of climate from physical causes, – then suspended changes of form in the organic world, as adaptations, & these changing affect each other, & their bodies by certain laws of harmony keep perfect in these themselves. – instincts alter, reason is formed & the world peopled with myriads of distinct forms from a period short of eternity to the present time, to the future. – How far grander than idea from cramped imagination that God created (warring against those very laws he established in all organic nature) the Rhinoceros of Java & Sumatra, that since the time of the Silurian he has made a long succession of vile molluscous animals. How beneath the dignity of him...[of] whom it has been declared 'he said let there be light & there was light' – bad taste.[63]

This was written on 16 August 1838. It remained Darwin's private vision thereafter until 1859, when it became a matter of public concern. In his 'Sketch of 1842', in his 'Essay of 1844'; and in all the editions of the *Origin of Species* 'the grandeur of this view of life' is the concluding thought: the earth goes cycling on according to the 'fixed law' of gravity while, through resulting changes, the whole variety of life is evolved in accordance with other appointed laws. Only that which Darwin found 'derogatory' to the Creator is removed from subsequent renderings of the 1838 passage, its place being given to the thought that life seems 'ennobled' when viewed, not as the outcome of special creations, but as 'the lineal descendants of some few beings which lived long before the first bed of the Silurian system was deposited'.[64]

Darwin not only retained his deistic vision of creation. He carried through its rationale to the end of his life. The lowliness of human reason is the principal handicap of theological speculation on origins. In speaking of special creation one explains nothing, but merely restates facts in brazenly dignified language. One has no right to assume that the Creator works by intellectual powers like one's own. Indeed, since these powers have been evolved from unreasoning lower animals by fixed biological laws, one cannot help but experience a 'horrid doubt' as to the trustworthiness of every metaphysical conviction.[65] If, however, the limitations of human reason could somehow be transcended, there would still be the dysteleologies, complexities, and fortuities of nature for theology to reckon with. The best one might do under these circumstances is 'to look at everything as resulting from designed laws, with the details, whether good or bad, left to the working out of what we may call chance'.[66] But in practice this viewpoint is difficult to adopt. No astronomer, for example, 'in showing how the movements of the planets are due to gravity, thinks it necessary to say that the law of gravity was designed that the planets should pursue the courses which they pursue'. Thus law must reign supreme, producing effects in nature which by chance strike mankind as having been contrived. 'There seems to be no more design in the variability of organic beings and in the action of natural selection, than in the course which the wind blows. Everything in nature is the result of fixed laws', even to the formation of man's rudimentary mammae.[67] To say otherwise (with Herschel and Gray) is to show that the subject in one's mind is 'in Comte's theological stage of science'.[68] To believe in miraculous creations or in the 'continued intervention of creative

power', said Darwin, is to make 'my deity "Natural Selection" superfluous' and to hold *the* Deity – if such there be – accountable for
phenomena which are rightly attributed only to his magnificent laws.[69]

What remained of Darwin's orthodoxy was its universe. Within a
decade of his graduation from Cambridge and within three years of
the *Beagle* voyage the providential God of orthodox natural theology
had vanished behind an impenetrable barrier of laws.[70] Ordinarily this
theological decline would be sufficiently clear from the evidence
presented thus far. But as we have been concerned to show that Darwin
was specially influenced by Paley and Malthus, it is necessary to demonstrate how their theological influence did in fact recede. This can be
done in a most telling manner.

Although at times his terminology is ambiguous and his statements
seem inconsistent, Paley believed firmly in a 'ruling Providence', in
God's 'creative, . . .continuing care' for the world.[71] The laws of nature
are only the modes of divine action. Being universal, they show that
God is omnipresent. He 'upholds all things by his power', a power
'penetrating the inmost recesses of all substance'. Of course God and
the world must not be confused. 'Neither the *universe*, nor any part of
it which we see, can be he.' Yet, while natural theology may speak
only of a 'virtual presence', the language of scripture, Paley stated,
'seems to favour' a metaphysical rendering of God's relation to the
world as one of 'essential ubiquity'.[72]

In view of this doctrine Paley was particularly intent on depriving
nature of all self-organising and self-operating properties. A 'principle
of order' or a 'law' is nothing, he declared, if it is not the mode of
God's immediate providence. To show how one might conceive of an
omnipotent God so limiting himself as to work according to general
laws, Paley wrote in his *Natural Theology* as follows:

It is as though one Being should have fixed certain rules; and, if we may
so speak, provided certain materials; and, afterwards, having committed
to another Being, out of these materials, and in subordination to these
rules, the task of drawing forth a creation: a supposition which evidently
leaves room, and induces a necessity, for contrivance. . . . We do not
advance this as a doctrine either of philosophy or of religion; but we say
that the subject may safely be represented under this view, because the
Deity, acting himself by general laws, will have the same consequences
upon our reasoning, as if he had prescribed these laws to another.[73]

Now Darwin, who claimed to have learnt the *Natural Theology* almost by heart, expressed himself similarly in the fragmentary sentences of his 'Sketch of 1842'.

If every part of a plant or animal was to vary,....and if a being infinitely more sagacious than man (not an omniscient creator) during thousands and thousands of years were to select all the variations which tended towards certain ends.... Who, seeing how plants vary in garden, what blind foolish man has done in a few years, will deny an all-seeing being in thousands of years could effect (if the Creator chose to do so), either by his own direct foresight or by intermediate means – which will represent the creator of this universe. Seems usual means.[74]

Here Paley's 'materials' are the 'variations' from which a 'being' – the personification of natural selection – effects adaptations. The 'certain ends' for which the variations are selected answer to the 'contrivance' in Paley's illustration. Paley insists, however, that his 'Being' is only a representation of God's immediate orderly action. Darwin divorces the two. His being is pointedly 'not an omniscient creator'. The 'Creator' may choose to have the being act either by 'direct foresight' (a difficult term) or by 'intermediate means'. But as intermediate means are (the being's?) 'usual means' of operation, they represent the 'creator' of the universe.[75]

The ambiguities in this hastily written passage are reduced to a single large ambiguity in Darwin's more detailed 'Essay of 1844'.

Let us now suppose a Being with penetration sufficient to perceive differences in the outer and innermost organization quite imperceptible to man, and with forethought extending over future centuries to watch with unerring care and select for any object the offspring of an organism. ... I can see no conceivable reason why he could not form a new race... adapted to new ends....

Seeing what blind capricious man has actually effected by selection during the few last years, and what in a ruder state he has probably effected without any systematic plan during the last few thousand years, he will be a bold person who will positively put limits to what the supposed Being could effect during whole geological periods. In accordance with the plan by which this universe seems governed by the Creator, let us consider whether there exists any *secondary* means in the economy of nature by which the process of selection could go on adapting, nicely and wonderfully, organisms...to diverse ends.[76]

The 'being' of 1842 has now become a 'Being'. And as 'blind capricious man' is spoken of as not having had a 'systematic plan' of selection, one naturally assumes that the Being would have such a plan for the accomplishment of his 'ends'. But no, it is the 'Creator' who has the 'plan' and '*secondary* means' which are brought in to execute it. Thus the Being is not the Creator. He is secondary means. But if the Being is secondary means then he is not, properly speaking, a Being – not even a 'being'. The whole point of Paley's illustration, on the other hand, is that a Being may represent the divine government because that government, after all, consists of God's immediate personal providence.

Darwin seems to have spotted this difficulty, for in his unpublished 'big book' on species, written between 1856 and 1858, he replaced the 'Being' with 'nature' in the parallel passage, explaining that 'by nature, I mean the laws ordained by God to govern the Universe'.[77] Although this statement does not appear in the 'abstract' of the big book, published in 1859 as On the Origin of Species, Darwin continued to depersonalise his demiurge in the first edition.

As man can produce and certainly has produced a great result by his methodical and unconscious means of selection, what may not nature effect? Man can act only on external and visible characters: nature cares nothing for appearances, except in so far as they may be useful to any being. She can act on every internal organ, on every shade of constitutional difference, on the whole machinery of life. .... How fleeting are the wishes and efforts of man! how short his time! and consequently how poor will his products be, compared with those accumulated by nature during whole geological periods. Can we wonder, then, that nature's productions should be far 'truer' in character than man's productions; that they should be infinitely better adapted to the most complex conditions of life, and should plainly bear the stamp of far higher workmanship?[78]

The 'Being' of 1844 remains identified with 'nature'; 'ends' have been replaced with utility; and the 'Creator' is merely reflected in the 'far higher workmanship' of nature's productions. A consistent deism is finally within reach.[79]

But instead of achieving this consistency by further depersonalising natural selection, in the second edition of the Origin Darwin reinstated the ambiguity of 1844. He dignified his depersonalised demiurge with a proper name, making 'Nature' the selector. Together with a very un-deistic quotation from Butler's Analogy, inserted on the verso of the half-title page at the instigation of Gray, and a quotation from Kingsley's commendatory letter added to the last chapter, this altera- tion left the second edition of the Origin, beginning, middle, and end, rather more orthodox than its author.[80] Darwin must soon have felt the discrepancy. In preparing the third edition a year later he retained Butler, Kingsley, and 'Nature' as a selector, but inserted an explicit disclaimer.

It has been said that I speak of natural selection as an active power or Deity; but who objects to an author speaking of the attraction of gravity as ruling the movements of the planets? Every one knows what is meant and is implied by such metaphorical expressions; and they are almost necessary for brevity. So again it is difficult to avoid personifying Nature; but I mean by Nature, only the aggregate action and product of many natural laws, and by laws the sequence of events as ascertained by us. With a little familiarity such superficial objections will be forgotten.[81]

In point of fact everyone did not understand the significance of scientific metaphors; their objections to Darwin's metaphor were not altogether superficial; and they certainly did not forget them. But in this statement of thoroughgoing nominalism, coupled with his earlier remarks, Darwin had at last eliminated the possibility that his God might play an active role in the world. His break with Paley's theology was unequivocally complete.[82]

Darwin's break with the theology of Malthus is easier to trace. It first appears in the week after his crucial reading of the *Essay on the Principle of Population.* In the last chapter of the first volume of the *Essay* Malthus revealed that his doctrine of providence was not unlike that of Paley. He wrote:

It accords with the most liberal spirit of philosophy to believe that no stone can fall, or plant rise, without the immediate agency of divine power. But we know from experience, that these operations of what we call nature have been conducted almost invariably according to fixed laws.[83]

On reading the passage Darwin reproduced it in his notebook with embellishments:

It accords with the most *liberal*! spirit of philosophy to believe that no stone can fall, or plant rise, without the immediate agency of the deity. But we know from *experience*! that these operations of what we call nature, have been conducted *almost*! invariably according to fixed laws.[84]

Now Malthus had been leading up to the assertion that the causes of population and depopulation have been about as constant as any law of nature with which human beings are acquainted. Darwin seized on this assertion and applied it not just to the population and depopulation of one species but to the 'extermination & production of new forms'.[85] At the same time, however, he let it be known by his embellishments that his application of the Malthusian principle would not be encumbered by the Malthusian theology. The spirit of this theology was not in fact liberal, its method was not sufficiently empirical, and its conception of nature left open the possibility that a superintending deity might violate the 'fixed laws' by which the 'extermination & production of new forms' proceeds.

In writing his 'Sketch of 1842' Darwin worked closely from his notes, paraphrasing them in many instances. Thus it seems likely that he had his transcription of Malthus in mind when, in conclusion, he set forth the twin rationale of his deistic doctrine of creation.

It accords with what we know of the law impressed on matter by the Creator, that the creation and extinction of forms, like the birth and death of individuals should be the effect of secondary...means. It is derogatory that the Creator of countless systems of worlds should have created each of the myriads of creeping parasites and...worms which have swarmed each day of life on land and water.... It accords better with...the lowness of our faculties to suppose each [organism] must require the fiat of a creator, but in the same proportion the existence of such laws should exalt our notion of the power of the omniscient Creator.[86]

The same passage, with slight modifications, occurs at the end of the 'Essay of 1844'. But as if to leave no doubt that the theology of Malthus was the foil for his 'grander' view of the divine government, Darwin precedes his concluding paragraphs with the statement:

For my part I could no more admit the...proposition [that 'each of... three species of rhinoceros, were separately created with deceptive appearances of true relationship'] than I could admit that the planets move in their courses, and that a stone falls to the ground, not through the intervention of the secondary and appointed law of gravity, but from the direct volition of the Creator.[87]

According to Malthus, a stone falls by 'the immediate agency of divine power'. For Darwin a stone must fall through 'the intervention of... law'. Although this passage was omitted from the *Origin* and only a moderate vestige of the former passage was retained, Darwin's theology henceforth remained the same.[88] It was not the theology of Malthus, Paley, and those orthodox naturalists and natural theologians whose universe Darwin had fitted out with a mechanism of adaptive change. It was a theology which would prove exceedingly difficult to sustain.

The consonance of orthodoxy and Darwinism

Having uncovered the orthodox theological roots of Darwinism and traced the decline of Darwin's theology, we must return to our earlier question: What was there about orthodoxy itself – what concepts were there in that fund of doctrines held in common by Van Dyke, McCosh, Iverach, Moore, Gray, and Wright – which made Darwinism acceptable? A partial answer has been implied in our discussions of the teleology of Paley, the theodicy of Malthus, and the doctrine of providence to which both these clergymen subscribed. Now a fuller answer must be obtained from the Christian Darwinians themselves in order that we may see the relevance of their orthodoxy to the theological difficulties encountered by Darwin.[89]

The theological roots of those who adopted Darwin's understanding of evolution lay in the 'biblical' or classical Christian conception of God as Creator. This conception, as has been well shown, was an important factor in the rise of modern science.[90] From Greek antiquity to the end of the Middle Ages nature was looked upon as a semi-independent power; whatever happened according to nature followed a pattern that seemed rational to the human mind. Nature, in this Aristotelian world, accomplished 'the full being of her immanent Forms' by means of efficient and final causes. Forms (or essences) and causes alike could be fully discerned by mankind. Gradually, however, the biblical view of nature was revived, and during the Reformation, in large measure, the fictions of Aristotelian philosophy gave way to the Creator–God of scripture. The doctrine of a divine Creator was seen to imply both the absolute dependence of all created things on him and their total differentiation from him. God is immediately active in all events; natural objects are nothing but his instruments. 'The order of nature is founded not on an immanent logic, but on God's care for his creatures. God does not intervene in an order of nature which is semi-independent; He acts either according to a regular pattern, or else in a more exceptional way, or even in a unique way.'[91] Nature therefore is to be regarded as fully contingent, the product of God's voluntary activity, not to be confused with him but neither to be divorced from his sustaining and controlling presence.

In consequence of this doctrine natural philosophers came to accept that all reliable knowledge of nature must be acquired *a posteriori*. If the creation is fully contingent, if there is nothing in the nature of things to restrict or regulate God's activity, then human beings, who were themselves created, waste their time in trying to deduce the manner in which the divine purposes proceed. Indeed, the Bible enjoins this attitude again and again in declaring that God's ways are not those of mankind, that his purposes may in fact appear as foolishness.

Scripture also says that human beings are made in God's image, but this does not entitle them to attempt by reason to rise above the world which they have been given and of which they are a part. Through a rational empiricism mankind may find order in nature and so discern the will of God. When nature does not appear rational to them, however, this too they must accept as God's design. They must become as little children, said Bacon in his *Advancement of Learning*; they must be well studied in 'the book of God's word' and in 'the book of God's works', allowing their attitude towards the former to become their outlook on the latter. These of course were words which Darwin quoted at the outset of the *Origin of Species*.[92]

Bacon was not the only seventeenth-century philosopher who represented the classical Christian point of view. Basso, Boyle, Malebranche, and Berkeley, though differing in other respects, were at one in protesting the deification of nature. Pascal, Hooke, and Newton meanwhile insisted that the will of God in a contingent universe is known only by empirical methods.[93] In the eighteenth century Bishop Butler was the most notable defender of this by now well-established outlook. 'The only distinct meaning of the word "natural"', he declared in a quotation Darwin also included at the outset of the *Origin*, 'is *stated, fixed, or settled*; since what is natural as much requires and presupposes an intelligent agent to render it so, *i.e.*, to effect it continually, or at stated times, as what is supernatural or miraculous does to effect it for once'. Elsewhere in his *Analogy of Religion* (1736) Butler decried the *a priori* method: 'Forming our notions of the constitution and government of the world upon reasoning, without foundation for the principles which we assume', he said, '...is building a world upon hypothesis, like Descartes'.[94]

At the dawn of the nineteenth century Paley and Malthus, as we have seen, were among the advocates of a free and perpetual Providence, the contingency of nature, and empirical methods in science. The generation of naturalists and natural theologians which came after them paid lip service to their theology, but in arguing for or against uniformitarianism in geology many accepted tacitly a deistic view of the world. These 'semi-deists' – especially Buckland, Sedgwick, Conybeare, Murchison, Whewell, and Lyell – believed, or wrote as if they believed, that God may 'intervene' in the course of nature. To deists or practical deists such as Robert Chambers and Charles Darwin this doctrine was of course no more acceptable than orthodox theology. Others, however, found the doctrine unacceptable because orthodox theology taught that God is never absent from nature so as to be able to intervene in it. Taking the best from deism – its perception of uniformity – and attributing the uniformity of nature to God's orderly providence; taking the best from semi-deism – its belief in divine sovereignty – and attributing all events, regular, exceptional, and unique, to the sovereign will of God, they maintained the classical outlook of Paley, Malthus, and the Christian founders of modern science. In the first half of the century the Scottish naturalist and Free Churchman, Hugh Miller, was fundamentally of this persuasion. In the post-Darwinian period orthodoxy had prominent representatives in Van Dyke, McCosh, and the Christian Darwinians.[95]

Like Darwin himself, the orthodox Christians who followed him were great admirers of Paley. Aware of the limitations of his *Natural Theology* – its antiquated devotion to specific forms, its romantic view of nature, and its presumption to specify final causes – they took the evidence it presented and placed it in the service of a better teleology. If Van Dyke looked for a 'second Bishop Butler' in whose intellectual grasp 'much of the *a priori* reasoning of modern science would become...mere hay, wood, and stubble', he also foresaw 'some future Paley' showing that through natural selection 'the Creator has left upon His handiwork innumerable traces of intelligent design'. The new teleology would not be weaker than the old, he said, for 'it cannot be proved that this complicated series of events results independent [*sic*] of a continued divine agency'.[96] McCosh took a similar view, deploring Spencer's speculations, upholding the design argument with Darwinian facts, and stating his conviction that 'the supernatural power is to be recognized in the natural law', that 'supernatural design produces natural selection'.

The Darwinians proper were less concerned with proofs than with discovering purposeful order in nature. Each demurred at philosophical speculation, whether of an idealistic or an agnostic variety. Each affirmed or implied the metaphysical neutrality of scientific knowledge. Yet each was convinced that Darwin had strengthened and enlarged the teleology which Paley had formerly made so cogent. Iverach stripped natural selection of its metaphorical character, finding in its organic utilities evidence 'of a directing agency and of a presiding intelligence'. 'What is essential', he said, 'is that we maintain and vindicate the continued dependence of all creation on its Maker'. Moore credited Darwin with the 'elimination of chance' and the elevation of purpose, through natural selection, to the dominant feature of life. The unity, orderliness, and purposefulness of natural processes he attributed to the constant action of an immanent Providence. Gray, to whom Van Dyke, McCosh, Iverach, and Moore, as well as Wright, were variously indebted,[97] insisted that Darwin had left 'the doctrines of final causes, utility, and special design, just where they were before'. On the basis of natural selection, however, he saw fit to construct an evolutionary teleology in which Paley's dysteleologies could be explained as part of a larger and grander design. Here, as in Paley's view, 'natural law. . .is the human conception of continued and orderly Divine action'.[98] In Wright's work, finally, the science of Darwin and the theology of Paley merged into one. Dismissing 'useless questions concerning ontology', Wright found justification for hypo-thetico-deductive reasoning and an evolutionary teleology in a sovereign God who constantly chooses and constantly acts in the interests of 'the highest good of being'. The uniformity of God's choice is the ground of confidence in scientific method; the benevolence of his choice, which dictates uniformity, sets in perspective the dysteleologies that Paley could not explain; and the sovereignty of his choice stands as a permanent rebuke to the 'arrogance' of those who presume to specify the one final cause of any part of creation.

Not surprisingly, those who saw in natural selection the out-working of an enlarged Paleyan teleology were also those who combined their acceptance of evolution with a Malthusian theodicy. Natural selection, operating in Paley's world, produced perfect adaptations without the aid of immanent metaphysical causes; operating in the world of Malthus, it produced these adaptations without a metaphysic of inevitable progress. Orthodox Christians who accepted natural selection as a function of the Paleyan and Malthusian world, and this world as a consequence of their theology, were thus as insistent on an empirical theodicy as on an empirical account of teleology.

Van Dyke's object throughout his work was to argue against every form of metaphysical evolution save that system of universal providence revealed in Holy Scripture. The last chapter of *Theism and Evolution*, 'Science and the Bible; no conflict', stood in sharp contrast to the optimistic progressivism of the time, upholding on the basis of scriptural and thermodynamic arguments the ominous doctrine that 'the present economy will have an end'. McCosh expanded on this view in a manner which Van Dyke also doubtless found congenial. In his *Method of the Divine Government* he pointed out the 'traces of original grandeur and subsequent ruin' which in the world mingle together almost harmoniously and serve as the conditions for mankind's 'training to virtue' and 'preparation to a final judgment and consummation'. Neither nature nor the human mind is without evidence that the earth 'is to be visited by a brighter and more glorious era', McCosh added; but this evidence is 'not very clear' and thus would seem to require the light of revelation 'to give the true explanation'. The same interpretation, as we have seen, appears in his *Religious Aspect of Evolution*.[99]

Iverach, like McCosh, held out faith and hope as the remedies for a mind perplexed by dysteleology. Concerning apparent inutilities and abortive developments he wrote that 'enough is known to me of the wisdom of that power made manifest in the relations of living beings to enable me to trust that wisdom is manifested here also'. And in another place he expressed confidence that, though one cannot 'know the final outcome of things', the Power 'which has brought the nebula to the stage where life with its thought, its morality, and religion exists in the earth, will continue to work in such a way as to bring it further issues yet, and to an end worth all the cost'.[100]

Moore's theodicy was constructed on the premise that a Christian is not bound to show, or to say that he can show, that marks of beneficence are everywhere apparent in nature, and that 'still less is he bound to assert, as the old teleology did, that he can demonstrate the wisdom and goodness of God from nature alone'. To resolve the moral difficulties posed by nature Moore recommended the spirit which in science approaches the unknown with the conviction that order and law will ultimately prevail. For 'what our rational nature resents', he said, 'is not the existence of facts which *we* cannot explain, but of facts which *have no* explanation; and what the moral nature rebels at is not suffering and pain, but needless – *i.e.* meaningless – pain, suffering which might have been avoided'. 'And here', he added, 'Darwinism gives us a hint, if it is but a hint. "Natural selection works solely by and for the good of each being." The arrangement of the work is "generally beneficent", and tends to progress towards, or to maintain perfection.'[101]

The 'good of each being', the perfect adaptation of corporeal and mental endowments, is the implicit theodicy of the *Origin of Species*.[102] By a slight change in perspective and wording it became the theodicy of Gray and Wright. God's purposes, they maintained, are comprehensive, far-reaching, and intricately interrelated. Not one of them can be specified unambiguously by a finite creature in a world characterised by moral disorder. Thus the beneficence of apparently purposeless and pernicious natural phenomena can only be seen in relation to all the ends which such phenomena are ever made to serve and as a consequence of God's wisdom in thereby promoting 'the highest good of being'.[103] Indeed, Wright believed that 'God's choice of the good of being and his wisdom in promoting it are the only actual and absolute uniformities' in the world. One can no more encompass divine purposes in an evolutionary scheme, even if to vindicate their benevolence, than one can assign the final cause of any single phenomenon. In his article, 'The evolutionary fad', published in 1900, Wright expressed this conviction in words which might have been repeated by any of those orthodox Christians who looked at evolution from Darwin's point of view. Having explained on the basis of the 'Malthusian law' the development of population, language, and race, and having described the 'counteracting agencies' which have caused linguistic and racial evolution to cease, he wrote:

The course of human history [has not] run so smoothly that any natural law of evolution can be used either in forecasting the future of the human race or in interpreting the records of the past. The only ground of hope that the world is to continue to improve arises from the evidence, dimly written in the past and clearly revealed in the Christian revelation, that there is 'a power above us working for righteousness', and that there is a divine plan of progress with which we are permitted to coöperate, and a divine spirit who is ready to coöperate with us. If there is anything which history teaches, it is that man, left to himself, degenerates; that the light which is shining brighter and brighter in our advancing civilization is borrowed light.[104]

For Darwin's orthodox followers progress was no necessity. It was, as Malthus and Darwin saw it, the object of 'rational expectation'. God's vindication lay in purposes beyond human comprehension and such progress as might occur but served to fulfil the biblical promise that the Creator works all things together for good and brings history to the conclusion of his free and benevolent choice.

What relevance, then, did orthodoxy have to a Christian acceptance of Darwinism? Its relevance consisted first of all in its continuity with the tradition of natural theology from which Darwinism itself emerged. The teleology of Van Dyke, McCosh, and the Christian Darwinians was the teleology of Paley chastened and enlarged. Their theodicy, like that of Malthus, recognised the subordination of dysteleologies to larger ends. And their doctrine of creation involved that biblical conception of God and nature which had been held by orthodox naturalists and natural theologians since the dawn of modern science. Now, in the second place, it remains to show that orthodox theology was relevant to a Christian acceptance of Darwinism, not only by reason of its continuity or consonance with the universe which Darwinism presupposed, but because of its capacity to reduce the dissonance stemming from conflicts between Darwinian and Christian beliefs.

Dissonance reduction and orthodox theology

To regard the phenomena of nature 'with a constant reference to a supreme intelligent Author', wrote Paley in the conclusion of his *Natural Theology*, is to lay 'the foundation of every thing which is religious. . . . Whereas formerly God was seldom in our thoughts, we can now scarcely look upon any thing without perceiving its relation to him.'[105] Darwin proved the converse true. Unable to regard nature with a constant reference to a supreme intelligent Author, he laid the foundation of everything which was irreligious. Whereas in his early years God was often in his thoughts, in later life he could scarcely look upon any thing and perceive its relation to him. Why was Darwin unable to regard nature with a constant reference to a supreme intelligent Author? Because, for one reason, as we have argued, there were phenomena in nature which seemed beneath the dignity of the divine providence. Chief among these were the variations which fed the mechanism of natural selection and thus lay at the source of adaptive change.

To Lyell, who, like Gray and Wright, was much concerned about the causes of variation, Darwin wrote:

Do you consider that the successive variations in the size of the crop of the Pouter Pigeon, which man has accumulated to please his caprice, have been due to 'the creative and sustaining powers of Brahma?' In the sense that an omnipotent and omniscient Deity must order and know everything, this must be admitted; yet, in honest truth, I can hardly admit it. It seems preposterous that a maker of a universe should care about the crop of a pigeon solely to please man's silly fancies. But if you agree with me in thinking such an interposition of the Deity uncalled for, I can see no reason whatever for believing in such interpositions in the case of natural beings, in which strange and admirable peculiarities have been naturally selected for the creature's own benefit.

In a postscript Darwin added that 'such a question, as is touched on in this note, is beyond the human intellect, like "predestination and free will"'. Later he told Lyell that if one should attempt to resolve the question by saying that all phenomena were 'ordained before the foundations of the world', then 'the subject has no interest for me'. And on another occasion he wrote that 'the saying seems to me mere verbiage. It comes to merely saying that everything that is, is ordained.' Questions such as these are 'beyond the human intellect; and the less one thinks on them the better'. To Gray, however, he confessed a few weeks later, 'I know I am in the same sort of muddle. . .as all the world seems to be in with respect to free will, yet with everything supposed to have been foreseen or pre-ordained.' Later still he admitted, 'I cannot keep out of the question.' Finally, in 1868, Darwin undertook to rule out belief in the providential origin of species with his 'stone-house argument'. If variations are divinely superintended, he reasoned, then the phenomena of natural selection depend on 'superfluous laws of nature' and God becomes the author of 'many injurious deviations of structure' and the 'redundant power of reproduction'. Yet the fact that 'an omnipotent and omniscient Creator ordains everything' could not be overlooked and once again Darwin was 'brought face to face with a difficulty as insoluble as is that of free will and predestination'.[106]

But, though Darwin found the difficulty 'insoluble' and a hindrance to theistic faith, it did not appear as such to the adherents of that orthodox theological tradition which was inaugurated by Augustine, systematised by John Calvin, preserved in Scotland, transplanted in New England, and represented on both sides of the Atlantic in the nineteenth century by theologians of the Presbyterian and Congregational churches. The outstanding feature of this tradition was its ability to hold in tension the doctrines of free will and predestination, to reconcile 'chance' and providence, 'second causes' and a *prima causa omnium*. With Calvin it could ascribe all things to the 'directly upholding and governing hand of God', even those events which seemed independent of or irreconcilable with divine purposes.[107] Thus for Van Dyke and McCosh, both Old School Presbyterians, Iverach, a leader in the Free Church of Scotland, and Gray and Wright, conservative Congregationalists, Darwinism raised no problem which had not been encountered many times before, nor any problem which could not be dealt with by appeal to the sovereign purposes of God. Darwin might regard their belief in providence as 'mere verbiage' but to them it was the central doctrine in an orthodox Calvinistic theology.[108]

The relevance of a Calvinistic theology to the reconciliation of providence and natural selection is nowhere more apparent than in the writings of McCosh and Wright. In his *Method of the Divine Government*, first published in 1850, McCosh maintained that God governs the world not only by laws, but by a 'complication' of laws which produces 'fortuities'. The prevalence of 'accident', he said, 'cannot be accidental'.

It is in the very constitution of things. It is one of the most marked characteristics of the state of the world in which our lot is cast. It is, in fact, the grand means which the Governor of the world employs for the accomplishment of his specific purposes, and by which his providence is rendered a particular providence, reaching to the most minute incidents, and embracing all events and every event. It is the especial instrument employed by him to keep man dependent, and make him feel his dependence.[109]

Among the other 'purposes served by the complication and the fortuities of nature' was the giving of 'variety' to the works of God. Little wonder, then, that McCosh so readily came to terms with Darwin; that he could write in his *Religious Aspect of Evolution*, 'The design is to be seen in the mechanism. Chance is obliged to vanish because we see contrivance.... Supernatural design produces natural selection.'

Wright made the relevance of a Calvinistic theology even more explicit. The Darwinian, he said, 'may shelter himself behind Calvinism from charges of infidelity' and the would-be Darwinian 'may safely leave Calvinistic theologians to defend his religious faith'. Drawing on the doctrine central to the whole New England Theology, from Edwards through Hopkins and Taylor to Park and Finney, that the divine moral attributes were 'comprised in love, which was a choice, and a choice of something which we can understand, viz., the highest good of being', Wright became the only theologian on either side of the Atlantic to frame a full and cogent reply to Darwin's 'stone-house argument' while retaining a Darwinian conception of evolution.[110] His 'saw-mill argument' was constructed to show that, since God's choice of the highest good of being invests natural phenomena with purposes which human beings cannot exhaustively comprehend, the variations in plants and animals may be useful for both artificial and natural selection and yet partake in a universal benevolent design. In the face of 'injurious deviations of structure' and the 'redundant power of reproduction' the Christian Darwinian, Wright explained, 'is compelled to assume that the revelation of *method* and *order* in nature is a higher end, and so a more important factor in the final cause of the creation than are the passing advantages which organic beings derive from it as the scheme of nature is unfolding'.[111]

Thus, by grounding providence in a high view of the sovereignty of God, a Calvinistic theology enabled Christians to dilute the dissonance arising from the conflict of teleology and natural selection, and so to reconcile the central doctrine of Darwinism with their faith. There was another Darwinian doctrine, however, which this orthodox theology also helped to make acceptable. Indeed, the reconciliation was a gift from none other than Darwin himself in the concluding pages of the *Descent of Man*. The 'arrogance' of believing in one's own special creation, it will be recalled, encouraged Darwin to adopt the 'more humble' view that human beings had been 'created from animals'. This, in turn, led him to doubt whether any metaphysical convictions, including the belief in Paley's 'supreme intelligent Author', could ever be trusted. Yet one conviction he did provide for, the very one by which human beings most commonly distinguished themselves from other forms of life, and thereby he did implicitly secure a basis in human existence for trusting other metaphysical beliefs. 'Few persons', said Darwin, 'feel any anxiety from the impossibility of determining at what precise period in the development of the individual, from the first trace of a minute germinal vesicle, man becomes an immortal being; and there is no greater cause for anxiety because the period cannot possibly be determined in the gradually ascending organic scale'. 'I am aware', he added,

that the conclusions arrived at in this work will be denounced by some as highly irreligious; but he who denounces them is bound to shew why it is more irreligious to explain the origin of man as a distinct species by descent from some lower form, through the laws of variation and natural selection, than to explain the birth of the individual through the laws of ordinary reproduction. The birth both of the species and of the individual are equally parts of that grand sequence of events, which our minds refuse to accept as the result of blind chance. The understanding revolts at such a conclusion, whether or not we are able to believe that every slight variation of structure...and other such events, have all been ordained for some special purpose.[112]

Precisely. In describing the mechanism of evolution Darwin raised no new problem for believers in divine providence; in extending evolution to mankind he raised no new problem either. On the contrary, he showed theologians that the dissonance involved in believing in mankind's unique immortality while accepting that the difference between mankind and other animals is one of degree and not of kind could be weakened by referring to the historic debate between creationists and traducianists over the origin of the human soul. Van Dyke and McCosh refused to follow Darwin's lead, in part perhaps out of loyalty to Calvin's strict creationism.[113] But others, who identified with the broader views of Augustine or who traced their theological ancestry to Jonathan Edwards, were able to use the analogy between traducianism and evolution to justify a Darwinian doctrine of mankind. Moore held that the questions of psychogenesis in the race and in the individual are 'so closely bound together' that a theory which is to be true of either – and he looked to comparative psychology to provide such a theory – 'must be applicable to both'. Wright, followed by Gray, argued that the evolution of mankind's higher nature by 'divine appointment' presents 'no greater philosophical difficulty...than there is in the well-known fact of the evolution of the individual soul from its parents'.[114]

With the analogy between traducianism and evolution, however, we emerge from Calvinistic theology into that larger province of Christian orthodoxy where another theological tradition made its contribution to the reduction of dissonance between Darwinian and Christian beliefs. We refer to Anglo-Catholic theology and its doctrine of the divine immanence as set forth by Moore. Calvinism to Moore was 'an immoral travesty of the gospel of Christ', predestination an 'awful and anti-Christian doctrine'. The Christian doctrine of God, he believed, does indeed reconcile science and faith, but not by sacrificing morals to a one-sided emphasis on the divine sovereignty.[115] Rather the reconciliation comes in a fresh appreciation of God's triune nature and a 'fearless reassertion' of 'the old almost forgotten truth of the immanence of the Word, the belief in God as "creation's secret force"'. No less a doctrine will accommodate both Darwinism and theistic belief. 'Slowly and under the shock of controversy', said Moore, 'Christianity is recovering its buried truths, and realizing the greatness of its rational heritage'.[116]

The dilemma presented by Darwinism, according to Moore, was much like the dilemma confronted by the ancient world: religion demanding the transcendence and unity of God, philosophy demanding the immanence and unifying power of reason; religion unable to satisfy philosophy without lapsing into pantheism, philosophy unable to satisfy religion without giving up the order of nature; and neither religion nor philosophy capable of reconciling the undifferentiated unity, which it sought, with mankind's instinctive belief that the Maker and Sustainer of the world is a personal being. To this dilemma the Church responded with the doctrine of the Trinity. It fused the religious with the philosophic idea of God, the divine transcendence with the immanence of the divine *Logos*, and thereby witnessed to that diversity in unity which invests the Godhead with eternal personality. Now the same doctrine, Moore argued, is able to 'claim and absorb the new truths of our scientific age'. Religion, influenced by deism as well as the Judeo-Christian tradition, demands that a transcendent personal God furnish nature with an external particular design. Science, on the other hand, 'demands a unity in nature which shall be not external but immanent, giving rationality and coherence to all that is, and justifying the belief in the universal reign of law'.[117] The Christian doctrine of God meets both demands. It brings together the teleology of Paley and Huxley's 'wider teleology' while guarding against Darwin's deism and the near-pantheism of Spencer and Fiske. 'All and more than all that philosophy and science can demand, as to the immanence of reason in the universe, and the rational coherence of all its parts, is included in the Christian teaching: nothing which religion requires as to God's separateness from the world, which He has made, is left unsatisfied.' Natural selection thus becomes the method and design the meaning of God's immanent activity in creation.[118]

Moore's reassertion of orthodox theology was truly 'fearless'. 'Either God is everywhere present in nature', he declared, 'or He is nowhere. ... Everything must be His work or nothing. We must frankly return to the Christian view of direct divine agency, the immanence of Divine power in nature from end to end, ...or we must banish him altogether.' The same boldness, the same fearless statement of theological alternatives in the face of Darwinism, was characteristic of another prominent Anglican clergyman, Charles Kingsley. Like Moore, Kingsley would give no quarter to Calvinism: to him it was not a theology but a 'demonology'.[119] Like Moore, or rather, like his master Maurice, he was profoundly Trinitarian. Amid the heated controversies of the sixties, with Darwin 'conquering everywhere...by the mere force of truth and fact', Kingsley pointed out that scientists 'find... they have got rid of an interfering God...[and] have to choose between the absolute empire of accident, and a living, immanent, ever-working God'. Soon he was urging clergymen to tell these scientists, as he himself had done, that 'the unknown *x* which lies below all phenomena, which is for ever at work on all phenomena, on the whole and on every

part of the whole, . . .is none other than that which the old Hebrews
called. . .The Breath of God; The Spirit who is The Lord and Giver
of Life'.[120] The alternatives were clearly defined: 'God is great, or else
there is no God at all.' For Kingsley as for Moore, Darwin had settled
the question.[121]

The immanence of God in the world, a corollary of the Trinitarian
doctrine accepted throughout Christendom, was also available to those
who approached the problem of providence and natural selection
from the Calvinistic tradition. Although none employed it with the
bold consistency of Moore and Kingsley, none overlooked it either.
Van Dyke, while holding to mankind's creation by supernatural intervention, believed that 'in strictness of speech, there are no secondary
causes', that God is 'in nature' as well as above it. 'To combine two
conceptions, the immanency and the transcendency, regarding God as
immanent in nature and at the same time infinitely superior to nature',
he said, '. . .is what we understand as christian theism'. McCosh likewise spoke of the incursion of miraculous powers into the world but
believed none the less that 'God is present among all his works', that
'as he fills universal space, . . .it is reasonable to think that he pervades
the universe as an active agent'. Iverach was more circumspect in his
views. He believed in 'the immanence of God in the world', not that
God 'comes forth merely at a crisis' but that everything, including
mankind's higher nature, 'is as it is through the continuous power of
God'. True, he could refer to the divine immanence as 'Energising
Reason' and as an 'immanent rational principle', but not without the
proviso, safeguarding nature from deification, that 'it is too early to ask
if the living power we see at work in the world of life is also a transcendent power, which means something for itself. As far as we have
yet looked at the world and the phenomena presented to us by it, we
have no data even for the consideration of such a question.'[122]

Gray reserved his severest criticism for two fellow-Calvinists, Charles
Hodge and J. W. Dawson, neither of whom could conceive of creation
taking place except by interference with the course of nature. Although
he sometimes spoke of 'secondary causes', Gray was careful to distinguish the semi-deism of Hodge and Dawson from his own belief in a
concursive providence. The intervention of the Creator, he said, may
be regarded 'either as. . .*done from all time*, or else as *doing through
all time*. . . . We much prefer the second of the two conceptions of
causation, as the more philosophical as well as Christian view.'[123] In
the last article he ever wrote, a review of Darwin's *Life and Letters*,

Gray upheld the 'Christian view' with greater theological insight, suggesting 'that the prevalent Latin-Church conception of transcendent Divinity – of a God apart from the world and operating extraneously – may have operated as unhappily in natural as in Christian theology; and that the fuller recognition of Divine immanence, of the Divine presence in nature and in the course of nature, might help to lessen some of these difficulties'.[124]

Wright, on the other hand, was never enthusiastic about the doctrine of divine immanence. Scientifically he saw it as a throw-back to the idealism of Agassiz and a threat to a causo-mechanical account of evolution. Religiously he thought it a harbinger of pantheism and the negation of free will.[125] Indeed, so wary was he of the doctrine that he (perhaps unconsciously) inverted Gray's views on the subject, stating after his colleague's death that Gray had conceived the Creator as acting neither from all time *nor* through all time, but as 'only now and then' putting his hand 'directly to the work' – an idea which, in these very words, Gray specifically refused to identify as Christian.[126] But in the last result, and notwithstanding all his fears, Wright had to admit that it was the doctrine of divine immanence 'as set forth by its extreme advocates' which he opposed, that in reality 'the doctrine. . .as held by the majority of its advocates does not differ essentially from the theory which supposes the Deity to be both the creator and the supervisor of the universe'. Neither conception of providence interferes with the 'continuity of nature'; neither posits 'a break in the course of nature at any point'. The 'web of nature' remains intact, but 'additional threads' – the variations on which natural selection depends – may be spun from the substratum of immanent creativity and 'inserted' in the fabric in order to 'increase the complexity and add to the beauty' of the weave.[127]

Liberalism and Darwinisticism

Our discussion of Darwinism and Darwinisticism in theology began with the question which, more than any other, must press upon the historian who studies the post-Darwinian controversies as a constitutive part of the history of post-Darwinian evolutionary thought: Why did some Christians become Darwinians despite the travails of Darwinism, while others, aligning themselves with Darwin's detractors and critics, became Darwinists? On the basis of the theory of cognitive dissonance we sought an answer in the realm of theology and discovered there the paradox that Christian Darwinians were notably orthodox in their beliefs, and Christian Darwinists were notably liberal. Turning then to the first half of the question with which we began, the more difficult and more interesting part, we endeavoured to explain how it was their orthodox theology, in fact, which determined that some Christians could become Darwinians. We showed, first, that Darwinism as a theory was steeped in the orthodoxy of Paley and Malthus; second, that Darwin himself, though deeply influenced by their teleology and theodicy, could not reconcile their doctrine of providence with the phenomena his theory presupposed; and finally, that the orthodox theology of the Christian Darwinians was relevant to an acceptance of Darwinism not only by virtue of its consonance with the orthodoxy of Paley and Malthus, but because of the ability of its Calvinistic and Trinitarian doctrines to reduce the dissonance which led Darwin to abandon the Christian faith.

Having thus replied to the first part of our initial question, however, what can be said in response to the second? That Christian Darwinists were as numerous as Darwin's critics comes as no surprise, but is it possible to explain their existence theologically, and not merely as a reflexion of prevailing views in science? Indeed, considering that several of the Christian Darwinists – Mivart, Henslow, Drummond, and Le Conte – were practising naturalists themselves, may it not be possible to indicate in general the theological affinities of non-Darwinian evolutionary thought?

Here the answer is not so straightforward as for Christian Darwinism. From start to finish Christian Darwinians retained an orthodox theology which Darwinism seems to have strengthened and confirmed. Christian Darwinists, without exception, began with orthodoxy, but the great majority ended in liberalism, making the transition under conditions which are necessarily obscure. It is difficult, if not impossible, for example, to say exactly what theology an individual held at the time he first encountered Darwinism, or whether, in fact, he first encountered Darwinism or some other version of evolution. Only the outcome of his theological transition can be known with any certainty and this knowledge only becomes available about the time his works on evolution and religion were published. Thus, if these works are to be regarded as the product of dissonance reduction and as evidence, in particular, of the kind of evolution that liberal theology could embrace, then it is necessary to assume: first, that an individual had actually learnt of Darwinism before his works were written; and second, that his theology had by then acquired sufficient definition to influence what he wrote. Neither assumption, however, seems unwarranted. On the contrary, each describes what appears to have been the experience of the more astute individuals who put forth the effort to make a book-length contribution to the post-Darwinian controversies. For them, as for all the controversialists, the reduction of dissonance between Darwinian and Christian beliefs was not an isolated event but a dialectical process which culminated often once, sometimes twice, and in a few cases three or more times in the publication of a book. So it must be primarily from these volumes, rather than from detailed knowledge of the struggles which underlay them, that we infer the manner in which liberalism and Darwinisticism were connected.[128]

Perhaps the most obvious theological conclusion to be drawn from the works we have reviewed is that the efforts of Christian Darwinists to preserve the purposes and character of God in the face of Darwinism were informed by a doctrine of creation which was held neither by Paley and Malthus nor by the Christian founders of modern natural science. Two presuppositions about nature, according to M. B. Foster, had to be displaced before modern science could begin: the Aristotelian doctrine that natural objects are endowed with an 'active potency' to realise their forms; and the belief of Greek religion that natural objects are but 'the appearance of [a] god', their growth or motion 'the manifestation of a divine activity'. In each instance, Foster argues, it was the classical Christian doctrine of creation which effected the displacement and made empirical science possible: first, by the 'attribution to God of an activity of will' which eliminated 'the Pagan conception of nature as self dependent'; and second, by a 'far deeper distinction between God and nature' which rendered natural objects both material and contingent, and not merely the 'apparent' objects of a priori knowledge.[129] Yet, ironically, the Greek ideas which orthodox doctrine displaced were virtually what the Christian Darwinists presupposed in their efforts at dissonance reduction. Some, by imposing an evolutionary agency between the Creator and his creation – an innate tendency, internal directing force, protoplasmic power, or natural law – invested nature with self-operating properties and thereby did implicitly, if not intentionally, release it from dependence on the active will of God. Others, by attributing creative processes to immanent divine power or to the divine immanence itself, blurred the distinction between God and nature in such a way as to obstruct the search for the mechanical verae causae of organic evolution. In each instance, moreover, by preserving teleology through a non-empirical orthogenesis, Christian Darwinists provided the metaphysics for a theodicy of universal and automatic progress. However, since neither their teleology nor their theodicy was derived from an orthodox doctrine of creation, neither was compatible with the universe described by Paley and Malthus. And since this universe was the one which Darwin presupposed, we may conclude that liberal Christians were led into Darwinisticism by the basic lack of consonance between Darwinism and their theology. In the last result Darwin and the Christian Darwinists lived in different worlds.

But consonance was not all that the theology of Christian Darwinists lacked. In many cases the resources by which the dissonance between Darwinian and Christian beliefs could be reduced were also in short supply. One obvious feature of Darwinistic theology was its disregard for the Reformed or Calvinistic tradition. Mivart was a Roman Catholic and a liberal one at that. Temple and Henslow were Anglicans, MacQueary an American Episcopalian, and each no doubt looked upon Calvinism as did Moore and Kingsley. Bascom, Abbott, Johnson, Beecher, Savage, and Fiske were the products of New England Congregationalism and leaders among the multitude who turned against its Calvinism in the later nineteenth century.[180] Drummond, Matheson, and the Duke of Argyll were Scottish Presbyterians whose Calvinism had also been so eroded by new intellectual currents – by Hegel, Spencer, and the *Naturphilosophie* – as to have no bearing on their conception of evolution. To be sure, most of these Darwinists did at least nominally retain a Trinitarian creed. And this might have served them well in the encounter with Darwinism had it not been coupled with an heterodox doctrine of creation. But there were some, acting perhaps more consistently on that doctrine's deistic or pantheistic implications, who rejected even this vestige of orthodoxy. Savage and Fiske were conspicuous Unitarians, MacQueary became one in time, and Le Conte, a lapsed Presbyterian, was deeply sympathetic with this 'most liberal movement of Christian thought'.

'Creative activity in God, material substance in nature, empirical methods in natural science – how closely each of these three involves the other, ...' writes Foster. 'A defect in the philosophical conception of God is reflected in corresponding defects both in the doctrine of nature and in the theory of natural science.'[181] If Christian Darwinists illustrate this evaluation it is not merely as representatives of theological liberalism in the later nineteenth century. Representing as they also did the theories of Darwin's detractors and critics, they indicate the essential heterodoxy of the whole tradition of evolutionary speculation to which Darwinisticism belonged. Lamarck was a thoroughgoing deist, his theory of evolution 'a striving toward perfection; an organic principle of order over against brute nature; a life process as the organism digesting its environment; a primacy of fire seeking to return to its own; a world as flux and as becoming', much of which bears about the same relation to the Darwinian theory as does Aristotelian to Galilean mechanics.[182] Chambers also revealed himself as a deist, clearly in the early editions of his *Vestiges of the Natural History of Creation*, implicitly in the rest. He explained evolution by the mediate operation of an invariable natural law and sought to reconcile the evil it involved with the 'recognised character of the Deity' by supposing 'that the present system is but a part of a whole, a stage in a Great Progress, and that the Redress is in reserve'.[183] Owen was a Broad Churchman whose theology and biology alike were conformable to Oken's *Naturphilosophie*. Evolution he delegated to a 'constantly operating secondary creational law'. Spencer shared the deism of Lamarck until the later 1850s, when, as J. D. Y. Peel has pointed out, he reified Calvinistic doctrines in the necessitarian and progressive evolution of the sovereign Unknowable. The Neo-Lamarckians in America and Samuel Butler in Britain, on the other hand, chose spirit over matter and free will over necessity. Cope found the Unitarians more receptive to his panpsychic evolution than the Quakers; Butler left the Church of England, disdaining original sin, only to have his theory of evolution by animal 'cunning' – his 'biological Pelagianism' – rejected by orthodox Darwinians.[184]

Of course it would be idle to pretend that Darwin was not theologically heterodox as well. Determined to shield God from the pretensions of human science and the aspersions of the lower creation, he embraced a 'grander' theology which amounted to little more than deism. At length, having regarded nature for decades without 'a constant reference to a supreme intelligent Author', he could scarcely accept even this conception of God. All he could believe in was 'my deity, "Natural Selection"'. However, in committing himself thus to a causo-mechanical account of evolution, to 'material substance in nature' and 'empirical methods in natural science', Darwin revealed where his ultimate loyalties lay. In a sense natural selection was *the* Deity in the universe of orthodox theology. Order, purpose, historic progression without inevitable progress – these were the doing of the

Christian God in the world of Paley and Malthus. In this world, as Kingsley observed, either 'God is great, or else there is no God at all'. That Darwin was in a 'muddle' over the question may be a sure sign of liberalism, but that he refused to resolve it at the expense of natural selection shows that his 'liberalism' was incidental to the development of his theory. The orthodoxy of Darwinism was that, not of its author, but of the theology of nature which his theory presupposed.

CONCLUSION:
ON COMING TO TERMS WITH DARWIN

It has been a matter of controversy. . .whether evolutionary theory
demonstrates the need for a new religion to include the new idea of an
evolving Universe or whether nothing more is needed than a transformed
– or for the first time clearly understood – Christianity.

John Passmore[1]

Darwin's theological perplexity and his eventual uneasy agnosticism
might be better construed as evidence of the incompleteness of his
work than as proof of its essential heterodoxy. Such at any rate is an
interpretation which Darwinian commentators, old and new, would
seem to allow. Huxley thought that evolution by natural selection is
'neither Anti-theistic nor Theistic', that 'it simply has no more to do
with Theism than the first book of Euclid has'. Chauncey Wright
argued forcefully for the metaphysical 'neutrality' of the theory. David
Hull has remarked that 'the development of evolutionary theory since
Darwin might well be described as a scientific theory in search of a
metaphysics'. And in his *Providence Lost* Richard Spilsbury has
written of contemporary Neo-Darwinians in much the same vein in
which Geoffrey Wells wrote of Darwin forty years ago. 'The fragmentary man', said Wells, 'can only manifest a fragmentary truth'.
Darwin was 'a specialist getting on with his job, in a sociological,
political, and religious vacuum'. He was 'incomplete', and 'Darwinism accordingly inadequate as a philosophy by which men may live'.[2]

Of course Darwinism has never been held in a 'vacuum', apart
from ultimate beliefs, nor has it ever been purged of its metaphysical
susceptibilities. As a 'philosophy by which men may live', however,
Darwinism has always claimed more for itself than its founder did.
Darwin made untenable that view of the world – by no means a distinctly Christian view – which sought ultimate certainty through inductive inferences and perceived an essential fixity in biological
species. His theory also created unavoidable conflicts for those who
proposed to uphold a Christian conception of the purposes and character of God. But while Darwin was quite aware of the bearings of his

work on Christian theology, he refused to exploit them, as others did, in a clergy-baiting campaign. On the contrary, he seems to have done all within his power to provide for a theological setting of his 'fragmentary' theory. The *Origin of Species* began with quotations from the natural theologians Bacon, Butler, and Whewell and closed with the remarks of 'a celebrated author and divine'. It personified natural selection, admired its 'far higher workmanship', and commended Asa Gray's pamphlet on natural selection and natural theology. It even undertook to glorify the Creator, not simply in a concluding description of the 'grandeur in this view of life, with its several powers. . . originally breathed into a few forms', but in the pious language of natural theology employed through and through – employed, as it were, 'preliminary. . .to a genuine transformation or rebirth of the terms', according to one rhetorical analysis of the book. Darwin's 'natural adaptations were "beautiful and exquisite" precisely because the processes which produced them were more than merely mechanical and more than merely cruel'.[3] In the *Variation of Animals and Plants under Domestication* Darwin cushioned the force of his 'stone-house argument' by referring to the conundrum of free will and predestination. And in the *Descent of Man* he allowed for belief in human immortality by recalling the debate between creationists and traducianists over the origin of the soul.[4]

To say that Darwin's theological statements were disingenuous is to show a callous disregard for the depth and affliction of his metaphysical perplexity. To say they were merely a sop to his critics is to overlook the fact that they recurred long after it had become evident that his theory would receive a sympathetic hearing.[5] Far better would it be to interpret Darwin's theologising, with all its ambiguities, in the same manner which does justice to his tortuous defence of natural selection: to recognise that, having once taken a stand on an issue, Darwin was not about to foreclose the possibility of further explanations. Honest to a fault, he kept the options open for others and for himself, whether concerning the causal factors of evolution or the relations of evolution and theology. Perhaps, indeed, the reason why the *Origin of Species* practically 'invited. . .theistic accommodation'[6] was that Darwin, sliding slowly from deism into agnosticism, knew full well that Darwinism is 'inadequate as a philosophy by which men may live'. Helpless in his increasing desiccation, he tried as only he could do, in a fragmentary way, to help others supply the deficiency.[7]

What form of 'theistic accommodation', if any, Darwin had in mind is impossible to say. He set forth only the terms it had to meet: acceptance of organic evolution, of natural selection as 'the most important, but not the exclusive, means of modification', and of human descent from lower animals through a difference 'of degree and not of kind'. Or to be more precise: acceptance of the limitations of hypothetico-deductive reasoning, the utility and perfection of natural adaptations, the historic progression of organic forms without inevitable progress, and a 'more humble' approach to the origin of mankind. But such terms, as we have seen, defined by implication the theology that could accommodate them. To make this implication clear was the work of a better theologian than Darwin.

Thomas Henry Huxley, that inveterate old warrior who did so much for the military metaphor, also did more than all commentators, save the Christian Darwinians, to show the theological affinities of Darwinism. As priest of the 'Church scientific', preacher of 'lay sermons', and protagonist of a 'New Reformation', Huxley had learnt his apologetics from the 'unassailable' Bishop Butler.[8] His favourite ploy in controversy was to turn his opponents' 'guns' upon themselves: to play off Suarez against Mivart on evolution, Genesis against Gladstone on geology, and Jesus against Principal Gore on the Deluge. This was a clever and effective device, and by no means entailed any commitment to the opposing views. When, however, his own religious convictions, or lack thereof, came under fire from the Positivist Frederic Harrison, Huxley did a most remarkable thing. Instead of appealing to Auguste Comte or some other approved interpreter of liberal irreligion, he couched his defence in the phrases of orthodox Christian faith – and this, not as a concession to a religious audience, but as a virtual confession to the readers of that repository of British freethought, the *Fortnightly Review*. Said Huxley:

If the doctrine of a Providence is to be taken as the expression. . .of the total exclusion of chance from a place even in the most insignificant corner of Nature; if it means the strong conviction that the cosmic process is rational; and the faith that, throughout all duration, unbroken order has reigned in the universe – I not only accept it, but I am disposed to think it the most important of all truths. . . . If, further, the doctrine is held to imply that, in some indefinitely remote past aeon, the cosmic process was set going by some entity possessed of intelligence and foresight, similar to our own in kind, however superior in degree; if, consequently, it is held that every event, not merely in our planetary speck, but in untold millions of other worlds, was foreknown before these worlds were, scientific thought, so far as I know anything about it, has nothing to say against that hypothesis. It is in fact an anthropomorphic rendering of the doctrine of evolution.[9]

Huxley went on to explain how such a providence is 'wholly incompatible with the notion of "special Providences"', how it is 'undoubtedly as responsible for the phenomena of human existence as for any others', and how, in a 'new departure', it 'generated morality' in man, distinguishing him with a 'changed destiny' and determining that 'that which he would not he does...and that which he would he does not'. 'If a genuine, not merely a subjective immortality, awaits us', Huxley added, 'I conceive that, without some such change as that depicted in the fifteenth chapter of the second [sic] Epistle to the Corinthians, immortality must be eternal misery'.[10]

But the real climax of Huxley's apologia, and possibly the most revealing religious statement made by a Darwinian in the nineteenth century, was his frank admission of the theological affinities of his scientific faith. Referring to his views on providence and morality, he wrote:

It is the secret of the superiority of the best theological teachers to the majority of their opponents, that they substantially recognize these realities of things, however strange the forms in which they clothe their conceptions. The doctrines of predestination; of original sin; of the innate depravity of man and the evil fate of the greater part of the race; of the primacy of Satan in this world; of the essential vileness of matter; of a malevolent Demiurgus subordinate to a benevolent Almighty, who has only lately revealed himself, faulty as they are, appear to me to be vastly nearer the truth than the 'liberal' popular illusions that babies are all born good and that the example of a corrupt society is responsible for their failure to remain so; that it is given to everybody to reach the ethical ideal if he will only try; that all partial evil is universal good; and other optimistic figments, such as that which represents 'Providence' under the guise of a paternal philanthropist, and bids us believe that everything will come right (according to our notions) at last.[11]

This may not have been pure orthodoxy but it certainly was not liberalism. If Huxley, an adherent of 'scientific Calvinism' for forty years, had anything to say, there could be little doubt as to what kind of theology was best equipped to come to terms with Darwin.[12] As Daniel Day Williams put it half a century later, 'Darwinism applied to human life might have been made the basis of a radical Calvinism or other theology of human sin.... The liberals tended to forget Darwinism, and to use the idea of evolution as an undergirding of the idea of progress in which emphasis is not upon human failure, but upon an increasing human achievement of moral worth.'[13]

Liberal Christians 'tended to forget Darwinism' because, as we have argued, their theology was unable to receive it. Instead they were attracted to theories which could transform Darwinian evolution in accordance with their conceptions of the purposes and character of God. That the world is the outcome of God's omnipotent and beneficent design all Christians, orthodox and liberal, were agreed. Confronted with Darwinism, however, they differed over how this belief was to be maintained. Those who counted themselves members of established theological traditions, who concerned themselves with preserving historic deposits of truth, had at their disposal the unique resources for accepting evolution on Darwin's terms. Thus equipped, with Huxley they could say, 'Not a solitary problem presents itself to the philosophical Theist, at the present day, which has not existed from the time that philosophers began to think out the logical grounds and logical consequences of Theism.'[14] Those, on the other hand, who turned against established theological traditions, who took scant notice of historic doctrines of creation and providence, cut themselves off from Darwin's world and from the resources by which, if Darwinism were true, it could be kept a Christian world. Left to the devices of modernity, they solved their theological problems with concepts of divine immanence, human goodness, and social and religious progress, only to have their evolutionary speculations embarrassed and undermined by future turns of events. 'Many of the present difficulties of liberalism', wrote Williams on the eve of the Second World War, 'are explainable by the fact that in the nineteenth century special theories of development were accepted as final which have later had doubt thrown upon them or have been rejected altogether'.[15]

But it would be naïve to think that orthodoxy has not been embarrassed and undermined as well. Within forty years of the publication of the *Origin of Species* the doubt thrown upon the scriptures, not by evolution but by historical criticism alone, had called into question the validity of any theology which would constrain them to testify in unison to doctrines of creation and providence whatever. No less important, the political economy, or theodicy, inherent in those doctrines in their traditional form seemed to many, not merely irrelevant, but utterly pernicious as a premise for dealing with the social and economic iniquities of modern industrial society. Meanwhile, in meeting the scientific challenge, the Christian Darwinians were never more than a tiny minority, their writings never more than marginally effective against the rising tide of evolutionary naturalism and evolutionary liberalism and the strong undercurrent of popular anti-evolutionary beliefs. They failed to persuade the greatest minds of their generation; they failed to impress the least. 'The traditional concepts of Christian metaphysics were strong enough to adjust and adapt to the new facts', as one critic has pointed out; but to many this seemed a 'superficial accommodation'. 'Perhaps a goodly number of alternative metaphysical schemes could adjust to the facts of evolution with equal ease.'[16] On this premise twentieth-century philosophers and theologians have taken up the burden of the post-Darwinian controversies. Whether their labours have supplied what Darwinism lacked remains for each person to decide. The struggle to come to terms with Darwin has not yet ceased.

Perhaps we need to be much more radical in the explanatory hypotheses considered than we have allowed ourselves to be heretofore. Possibly the world of external facts is much more fertile and plastic than we have ventured to suppose; it may be that all these cosmologies and many more analyses and classifications are genuine ways of arranging what nature offers to our understanding, and that the main condition determining our selection between them is something in us rather than something in the external world. . . . Perhaps it is the rise to a dominating ambition of the human need to control nature's processes, and to do so as exactly as possible, that accounts for our modern preferences in this matter. In this case a historical analysis of the growth of this need, an investigation of the corresponding motives which underlay the assumptions of earlier thought, and a systematic enquiry into the factors which have conditioned the rise and fall of these interests, would certainly be needed if we are to hope for any mature insight into their respective promise of permanence, the possibility of their reconciliation, and the relative plausibility of cosmological constructions which emphasise one or the other of these approaches.

Edwin Arthur Burtt, *The Metaphysical Foundations of Modern Physical Science*

NOTES TO THE TEXT

References to the variorum text of the *Origin of Species*, edited by Morse Peckham (Darwin, 1959), are followed by a notation in parentheses giving the chapter and sentence or variant numbers assigned by Peckham. References to the published transcriptions of Darwin's notes and notebooks (De Beer *et al.*, 1960–1, 1967; Gruber and Barrett, 1974) are followed by a notation in parentheses giving the page numbers in the original manuscripts.

Introduction

1. T. H. Huxley, 1887, p. 204. Cf. L. Huxley, 1903, I, 344–5.
2. Darwin to H. Fawcett, 18 September 1861, in F. Darwin and Seward, 1903, I, 195.
3. See R. M. Young, 1973, p. 365.
4. M. McGiffert, 1958, p. 1; Simonsson, 1958, p. 23; Boromé, 1960, p. 191; Metzger, 1961, p. 71, n. 96; W. Coleman, 1971, p. 175; Bowler, 1976a, p. 144, n. 40.
5. Adeney, 1901; Macran, 1905, chaps. 5–6; Waggett, 1905 and 1909; Clodd, 1907; Bryce, 1909; Davis, 1909; Tennant, 1909; Elliott, 1912; J. A. Thomson, 1923; J. M. Wilson, 1925; Delaney, 1927; Macpherson, 1927; Lunn, 1931; Coffin, 1940; Raven, 1943, chaps. 3–4; Pantin, 1950–1; Raven, 1951; Raven, 1953, I, chap. 9; Bowle, 1954; Ramm, 1955 and 1963; G. S. Carter, 1957; Niebuhr, 1958; Fitch, 1959; Stackhouse, 1959; Zimmerman, 1959, chaps. 1, 6; Boromé, 1960; Brett, 1960; Dillenberger, 1960, chap. 8; Shideler, 1960; Wernham, 1960; Greene, 1961; Lack, 1961; R. D. Baker, 1962; Kent, 1966; Vanderpool, 1973a; Cupitt, 1975.
6. Schlesinger, 1967.
7. See also J. M. Turner, 1944 and the narrower ethical studies in Gantz, 1937 and 1939; and Parkinson, 1942. The dates given in the text are those of first publication. For theses and dissertations the dates of submission are specified.
8. See also the relevant chapters in S. Warren, 1943; Elliott-Binns, 1946 and 1956; May, 1949; and Wood, 1955.
9. See also Kennedy, 1957; Persons, 1959; Weisenburger, 1959; Dillenberger, 1960; Willey, 1961; and Greene, 1967.
10. Loewenberg, 1959a and 1965.
11. See also Boller, 1970; P. A. Carter, 1971; Reardon, 1971; Wagar, 1972; Saffin, 1973; and Vanderpool, 1973b.
12. See also the related but more specialised studies in MacLeod, 1970; C. H. Peterson, 1970; Jensen, 1970 and 1971–2; Barton, 1976; and Brock and Macleod, 1976.

13. E.g., Conrad Wright, 1941; Gillispie, 1959a; Millhauser, 1959; De Jong, 1962; Cannon, 1960, 1964a, and 1964b; Sherwood, 1969; Guralnick, 1972, 1974, and 1976; McElligott, 1973; Greene, 1974; Hovenkamp, 1976; Yule, 1976; Bozeman, 1977; Brooke, 1977a; and Numbers, 1977.

14. The following references are only to titles which have some immediate bearing on the post-Darwinian controversies. On Lyman Abbott, see I. V. Brown, 1953 and McCrossin, 1970. On Louis Agassiz, see Lurie, 1959 and 1960; and 'Agassiz, Darwin and evolution' (1959), in Mayr, 1976. On the Duke of Argyll, see Bowler, 1977a and Gillespie, 1977. On Henry Ward Beecher, see McLoughlin, 1970. On Samuel Butler, see Willey, 1960; Mudford, 1966; F. M. Turner, 1974a, chap. 7; and Breuer, 1975. On Edward Drinker Cope, see Osborn, 1931 and Bowler, 1977b. On James Dwight Dana, see Sanford, 1965. On John William Dawson, see C. F. O'Brien, 1971. On Henry Drummond, see McIver, 1958 and Schott, 1972. On John Fiske, see Pannill, 1957; Higgins, 1960; Berman, 1961; and McCrossin, 1970. On Philip Henry Gosse, see Ross, 1977. On Asa Gray, see M. McGiffert, 1958 and Dupree, 1959. On John Thomas Gulick, see A. Gulick, 1932 and Lesch, 1975. On Thomas Henry Huxley, see H. Peterson, 1932; Bibby, 1959a and 1959b; Jensen, 1959; Mudford, 1966; Randel, 1970; Murphy, 1973; Bartholomew, 1975; Rehbock, 1975; Rupke, 1976; and Helfand, 1977. On Charles Kingsley, see A. Johnston, 1959. On Charles Lyell, see Bartholomew, 1973 and 1974. On James McCosh, see Gerstner, 1945. On St George Mivart, see Gruber, 1960. On Richard Owen, see MacLeod, 1965 and Brooke, 1977b. On George John Romanes, see F. M. Turner, 1974a, chap. 6 and Lesch, 1975. On Minot Judson Savage, see McCrossin, 1970. On Newman Smyth, see Gentle, 1976. On Herbert Spencer, see R. M. Young, 1967; J. D. Y. Peel, 1971; Sharlin, 1976; and Wiltshire, 1978. On Alfred Russel Wallace, see George, 1964; McKinney, 1972; R. Smith, 1972; Kottler, 1974; and F. M. Turner, 1974a, chap. 4. On Alexander Winchell, see Davenport, 1948 and 1951. On James Woodrow, see Eaton, 1964 and Gustafson, 1964. On Chauncey Wright, see Wiener, 1945 and 1972, chap. 3. On George Frederick Wright, see M. McGiffert, 1958 and Morison, 1971.

15. On British denominations, see Hay Watson Smith, 1922 (Free Church of Scotland); Ellegård, 1958a; and Holifield, 1972 (English Methodist). On American denominations, see Roberts, 1936; E. L. Clark, 1952 (Southern Baptist); Dietz, 1958 (Lutheran); Street, 1959 (Southern Presbyterian); Illick, 1960 and D. F. Johnson, 1968 (Old School Presbyterian); Reist, 1970 and 1975 (Northern Baptist); Numbers, 1975 (Seventh Day Adventist); and Numbers, 1977, app. 1. See also Simonsson, 1971, which touches on Jewish responses, and Bezirgan, 1974, which deals with responses in the Islamic world.

16. Pearson, 1907; Cratchley, 1933; Ellegård, 1957b; Pelikan, 1960; Lack, 1961; Benz, 1967; Overman, 1967; Haber, 1971.

17. C. M. Williams, 1893; Mackintosh, 1899; Tufts, 1909; Dewey, 1910; Gantz, 1937 and 1939; Fisch, 1947; Quillian, 1950; Ellegård, 1957a; Blau, 1952, chap. 5; Blau, 1959; Fulton, 1959; Girvetz, 1959; Kultgen, 1959; Passmore, 1959; Collins, 1960; Goudge, 1961; Randall, 1961 and 1977; J. B. Wilson, 1965; Deely and Nogar, 1973; Hull, 1973a and 1973b.

18. On Germany, see Mullen, 1964; Altner, 1965; Hübner, 1966; Dörpinghaus, 1969; Simonsson, 1971; and Montgomery, 1974a and 1974b. On France, see Begouën, 1945; Stebbins, 1965 and 1974a; L. L. Clark, 1968; Conry,

1974; Farley, 1974; and Paul, 1974 and 1979. On Russia, see Kline, 1955; S. S. White, 1968; and J. A. Rogers, 1973, 1974a, and 1974b. On other countries, see Simonsson, 1958 and 1971 (Sweden); Glick, 1974b (Spain); Bulhof, 1974 (The Netherlands); Moreno, 1974 (Mexico); and Roome, 1976 (Canada). The proposal made by Eggleston, 1935, p. vi and Mendelsohn, 1964, p. 50 for a comparative study of the reception of Darwinism has been fulfilled in Glick, 1974a, which includes a synoptic essay on comparative historiography in Hull, 1974.

19. See Greene, 1957, p. 68.
20. Of the twenty-eight controversialists studied in part III, twenty-one saw their monographs on evolution published on both sides of the Atlantic.
21. See the observations in Ratner, 1936, pp. 108, 111; Blau, 1959, p. 142; Betts, 1959; p. 163; and Ahlstrom, 1961, p. 286.
22. E.g., Chadwick, 1966–70, II, 23ff.
23. See the bibliography in the author's doctoral thesis, completed at the University of Manchester in 1975.
24. Schlesinger, 1967, pp. 2–3; Loewenberg, 1941, pp. 339–50; Loewenberg, 1933, p. 689; Loewenberg, 1935, p. 233. Cf. Atkins, 1932, pp. 112–25.
25. Pfeifer, 1957, pp. 76–83. Cf. F. H. Foster, 1939, p. 43. For the separate chronology of responses in the southern United States, see Eaton, 1964, pp. 441–8.
26. Ellegård, 1958a, pp. 24ff; Chadwick, 1966–70, II, 23–8.
27. Among general apologetic works on the Catholic Church and science are Ronayne, 1879; Walsh, 1908; Kneller, 1911; Windle, 1924 and 1927; and Agar, 1940. For philosophical and theological studies of evolution, see Zahm, 1896; Dorlodot, 1922; Muckermann, 1928; Messenger, 1931 and 1949; J. A. O'Brien, 1932; Hauber, 1942; Begouën, 1945; Ong, 1960; Fothergill, 1961; and Deely and Nogar, 1973. The principal historical studies are found in Rádl, 1930, chap. 8; Eggleston, 1935, chap. 14; Ebenstein, 1939; Morrison, 1951; Ellegård, 1958a; Betts, 1959; Dörpinghaus, 1969; Simonsson, 1971; Lyon, 1972; Root, 1974; and Paul, 1974 and 1979.
28. Ellegård, 1958a, pp. 37–8; Schneider, 1945, p. 10; Persons, 1950a, p. 425.
29. M. McGiffert, 1958, p. 3. See also D. B. Wilson, 1977.
30. F. M. Turner, 1974a and 1974b; R. M. Young, 1969a, 1969b, 1970a, 1971a, and 1973.
31. 'The nature of the Darwinian revolution' (1972), in Mayr, 1976, p. 293; R. M. Young, 1973, p. 384. On the latter point, see also Brooke, 1977a, pp. 223–5.
32. Kuhn, 1970, pp. x, 86. Cf. Ghiselin, 1972, pp. 114, 123, 133 and Gruber and Barrett, 1974, pp. 256–7.
33. See Greene, 1971.
34. McDonagh, 1976, pp. 61–5 suggests other ways in which the theory of cognitive dissonance may interpret the Kuhnian 'revolution'.
35. Peckham, 1970. Cf. Persons, 1950a, pp. 439ff.
36. T. H. Huxley, 1887, p. 203.
37. Passmore, 1972, p. 259.

Chapter 1

1. G. K. Clark, 1962, p. 10.
2. P. Spencer, 1955.

3. Sarno, 1969; Saffin, 1973; Brush, 1974–5; Rosen, 1975; Barton, 1976.
4. Fiske, 1894; Haar, 1948; Leverette, 1963 and 1965.
5. Fleming, 1950, p. 76.
6. *Ibid.*
7. Fiske, 1894, pp. 169–70.
8. Draper, quoted in Fleming, 1950, p. 5.
9. *Ibid.*, p. 31.
10. *Ibid.*, p. 42.
11. *Ibid.*, p. 44.
12. Or perhaps from the nativism of his friend and colleague Samuel F. B. Morse? See Billington, 1964, pp. 122–5 and Fleming, 1950, p. 30.
13. Draper, 1875*a*, p. vi.
14. *Ibid.*, pp. x–xi.
15. *Year of Preparation*, 1869, pp. xxiii, xxv, xxxvii; McNabb, 1907, pp. 31–2, 47.
16. Draper, 1875*a*, p. 363.
17. Vidler, 1961, chap. 3.
18. Draper, 1875*a*, pp. 328ff.
19. Draper, quoted in Fleming, 1950, p. 124; Döllinger, 1891, p. 103. At the conference which he proposed should hear his case at the forthcoming meeting of German bishops, Döllinger requested 'that a man of scientific training of my own choice may be allowed to be present' (*ibid.*, p. 86). Could he have had Draper in mind?
20. Draper, 1875*a*, p. 332.
21. Vidler, 1961, p. 152.
22. Hales, 1958, p. 140.
23. Draper, 1875*a*, pp. 352, 361–2.
24. Mowat, 1939, p. 122.
25. Kneller, 1911, p. 390.
26. Chadwick, 1966–70, II, 14. See also Chadwick, 1975, pp. 161–4.
27. Overton, 1925, p. 73; Roberts, 1936, p. 44.
28. Fleming, 1950, p. 134; Glick, 1974*b*, pp. 339–40.
29. Fleming, 1950, p. 131. For the review literature, see *ibid.*, pp. 193–4 and Roberts, 1936, p. 44, n. 21. Fleming credits G. F. Wright, 1876*c* as 'incomparably the keenest review on either side'. Fisher, 1883 was probably the most competent historical critique. See also Youmans, 1875*a*, 1875*b*, and 1876.
30. Draper, 1875*a*, p. ix.
31. Nevins, 1927, pp. 286–9; Schlesinger, 1967; Hopkins, 1967, pt 1; May, 1949, pts 2–3; H. K. Beale, 1941, pp. 202–7; Schmidt, 1953; Schmidt, 1957, pp. 161–7; Metzger, 1961, pp. 46–70.
32. Goblet d'Alviella, 1886, chaps. 9–10; S. Warren, 1943; Gabriel, 1956, chap. 15; Sopka, 1972; Jensen, 1959 and 1967; Randel, 1970; W. J. Baker, 1974; Roberts, 1936, pp. 55ff.
33. Rudy, 1951; Veysey, 1965.
34. Schmidt, 1930, p. 184; McGrath, 1936.
35. Schmidt, 1930, p. 228; Veysey, 1965, pp. 1–18, 25–32, 203–12. Cf. Guralnick, 1976.
36. See G. E. Peterson, 1964; Metzger, 1961, p. 64, n. 72; and Berman, 1961, p. 78.
37. A. D. White, 1905, I, 11, 251, 397.
38. *Ibid.*, I, 17.

39. *Ibid.*, I, 19–25.
40. *Ibid.*, I, 25–42.
41. *Ibid.*, I, 276.
42. *Ibid.*, I, 278–81.
43. *Ibid.*, I, 287–300.
44. *Ibid.*, I, 300.
45. W. P. Rogers, 1942, pp. 50–1, 211.
46. A. D. White to G. Smith, 1 September 1862, quoted in *ibid.*, p. 52. Cf. Guralnick, 1974, p. 358.
47. A. D. White, 1905, I, 422ff; W. P. Rogers, 1942, pp. 75–8.
48. A. D. White, 1905, I, 402–5.
49. On White's ' "exceeding broad" ' churchmanship, cf. *ibid.*, I, 403 with W. P. Rogers, 1942, pp. 81–2.
50. A. D. White, 1905, I, 425. Cf. White to W. T. Hewett, 30 April 1894, quoted in W. P. Rogers, 1942, p. 83.
51. White to E. Cornell, 3 August 1869, quoted in Bishop, 1962, p. 191.
52. A. D. White, 1869, 1876*a*, and 1876*b*. *The Warfare of Science*, a volume of 150 pages, was issued by Daniel Appleton in New York and Henry S. King in London, the publishers of the International Scientific Series. A sizable sale of the British edition was ensured by adding to the text a prefatory note by John Tyndall, whose address at the Belfast meeting of the British Association in August 1874 earned him the notoriety which White enjoyed in America. On the cover 'Tyndall and White' are specified as the authors.
53. A. D. White, 1905, I, 425–6; A. D. White, 1876*b*, pp. 5–6; A. D. White, 1896, I, viii–ix.
54. A. D. White, 1876*b*, pp. 144–5.
55. *Ibid.*, pp. 7–8.
56. Quoted in Schmidt, 1957, p. 176.
57. A. D. White, 1905, II, 494. See Bordin, 1958. Later as a Cornell professor Burr 'was easily the leading authority in Europe or America on the history of toleration' (Barnes, 1962, p. 263).
    As university librarian Burr came to know John R. Mott, a promising young student and president of the Cornell Young Men's Christian Association. So impressed was Burr with Mott's scholarship that he invited Mott to accompany him, with all expenses paid, for a year of 'original research in German and Latin books and manuscripts in the libraries of Berlin, Bonn, Dresden, Heidelberg, Zürich, Paris, and London...in preparation for a *magnum opus* on which ex-president White was then closely engaged' (Mathews, 1934, p. 61). After painstaking consideration of the offer, in which the nature of the *magnum opus* does not seem to have been a factor, Mott decided not to interrupt his studies. He had set his face towards a career in the Christian ministry at D. L. Moody's Mount Hermon conference the year before, where he had been chosen to serve on a missionary deputation to the colleges and universities of America. During the year he might have spent abroad, Mott was appointed national secretary of the Intercollegiate Y.M.C.A., an office which led him to the general secretaryship of the World's Student Christian Federation, the chairmanship of the International Missionary Council, and, in 1948, the honorary presidency of the World Council of Churches (Latourette, 1969, IV, 505).
58. S. Warren, 1943, p. 63. The book emerged from the *Warfare of Science* by a series of additions and expansions published in the *Popular Science*

Monthly between 1885 and 1892 under the rubric 'New Chapters in the Warfare of Science'. Each instalment was reprinted separately under that title. Translations of the completed work were published at Paris in 1899, at Turin in 1902, and at Leipzig in 1911. Reprints have been issued at New York by George Braziller and at London by Arco Publishers, both in one volume in 1955, at New York by Dover Publications in two volumes in 1960 (see Bube, 1960) and by the Free Press in one volume in 1965. Until his death in 1938 Burr laboured periodically on a thorough revision of the book, engaging an assistant for a year when, in the wake of the Scopes Trial of 1925, Appleton, the publisher, suggested a definitive edition. But so great were the difficulties that Burr had in identifying himself with the work, as he differed with its basic assumptions, that the task was never completed (Bainton, 1942, pp. 52–3).

59. A. D. White, 1896, I, ix. Cf. Guralnick, 1974, pp. 354–7.
60. A. D. White, 1896, I, 217, 224.
61. A. D. White, 1896, I, vii. Cf. G. L. Burr, 1936, p. 92 and Schmidt, 1957, p. 169.
62. Draper, 1875a, p. 367; A. D. White, 1896, I, xii.
63. A. D. White, 1905, II, 495.
64. A. D. White, 1910, pp. 158, 160–1 (White's emphasis).
65. E.g., Flint, 1899, pp. 316–19; J. Y. Simpson, 1925, pp. 96–100, 104–11; Bainton, 1942, chap. 2; and C. A. Russell *et al.*, 1974, pp. 5–50.
66. W. P. Rogers, 1942, p. 53.
67. Metzger, 1961, pp. 66–7.
68. A. D. White, 1905, I, 436–7 does mention that Russell was 'lacking in his handling of delicate questions something of the *suaviter in modo*'. Bishop relates both incidents in some detail (Bishop, 1962, pp. 192ff, 213ff) but neither he nor White discloses that the free-thinker Francis Ellingwood Abbot was refused an appointment at Cornell because, said President White, a 'broad churchman' was sought. See Abbot's paper *The Index*, 2 (14 January 1871), 12. On White's conservatism, see Veysey, 1965, pp. 82–3.
69. Dillenberger, 1960, p. 14.
70. Quoted in G. L. Burr, 1936, p. 89; A. D. White, 1896, I, xi.
71. *Ibid.*, I, 410.
72. E. A. White, 1952, p. 33. Cf. Fosdick, 1957, p. 52.
73. S. Warren, 1943, p. 64. Cf. Barnes, 1962, p. 311.
74. Benn, 1906, II, 162, 164, 395, 438–9, 441.
75. Robertson, 1929, II, 318; B. Russell, 1935, pp. 7, 19, 75; Homer W. Smith, 1953, pp. 371, 481.
76. Greene, 1967, p. 12; Coleman, 1971, p. 87.
77. Taylor, 1939, p. 232; Dampier, 1929, pp. 297–302; Drachman, 1930, pp. 215, 236, 237; Hardin, 1960, pp. 96, 102.
78. Loewenberg, 1941, p. 350; S. Warren, 1943, p. 66; Schneider, 1945, p. 4; Barzun, 1958, p. 37; Hayes, 1941, p. 126. See also Jensen, 1959 and J. A. Campbell, 1968, chap. 4.
79. Harper, 1935, p. 105. The parallel between Shields and Herbert Spencer, the English philosopher of evolution, is too close to overlook. Each conceived, when in middle life, a unification of all knowledge based on a reconciliation of science and religion (see H. Spencer, 1862–96, I [1900], pt I). Shields' proposal for a *Philosophia Ultima* appeared in 1861, when he was thirty-six years of age; Spencer's original programme for *A System of*

Synthetic Philosophy was drawn up in 1858, in his thirty-eighth year, and printed in 1860 (see Duncan, 1911, chaps. 8–9 and app. B; and H. Spencer, 1904, II, app. A). Each man laboured on his unification for forty years, Shields' work appearing in three large volumes between 1877 and 1905, the year after his death, Spencer's appearing in ten volumes between 1862 and 1896, seven years before his death. Each enterprise was subsidised and supported by a wealthy clientele and each, when complete, was a monumental anachronism. Shields, 1877, pp. 220, 297 gives an unfavourable opinion of Spencer's philosophy.

80. Shields, 1877, p. 23.
81. *Ibid.*, p. 430.
82. *Ibid.*, pp. 586–7.
83. *Ibid.*, pp. 16–17.
84. Writing 'mit besonderer Rücksicht auf Schöpfungsgeschichte', Zöckler provides the nearest we possess to a history of the Christian doctrine of creation, a work which is 'much more balanced than the more familiar and vastly more partisan book by Andrew Dickson White' (Pelikan, 1960, p. 30). For Zöckler's views on evolution, see Hübner, 1966, pp. 34–8 and Altner, 1965, pp. 6–12. On Zöckler's apologetic theology, see Elert, 1921, pp. 244–8.
85. Zöckler, 1877–9, I, 1–2, 13. See also Ronayne, 1879 and Walsh, 1908.
86. A. C. McGiffert, 1916, p. 322; J. Y. Simpson, 1925, p. vii.
87. Carlyle, 1878, p. 226; Loraine, 1903, pp. xiv, 52–3.
88. Macpherson, 1927, pp. 186, 188; Tennant, 1909, pp. 419–20; Major, 1927, p. 60.
89. Elliott-Binns, 1946, pp. 163, 166; Elliott-Binns, 1956, p. 51; Reardon, 1971, pp. 13, 14. Cf. Elliott, 1912, pp. 138, 147.
90. Roberts, 1936, pp. 1–8, 21, 38–9, 64, 73; Ramm, 1955, pp. 15, 18, 19.
91. Coulson, 1959, p. 281; Overman, 1967, pp. 71, 74.

Chapter 2

1. 'Psychosis', 1884, p. 49.
2. Bainton, 1960, pp. 64–5.
3. Trevelyan, 1922, p. 302.
4. Anderson, 1971, p. 66. See also Summerton, 1977.
5. Sidgwick and Sidgwick, 1906, p. 187. See R. V. Sampson, 'The limits of religious thought: the theological controversy', in Appleman *et al.*, 1959, pp. 63–80; Hinchliff, 1964; McCabe, 1908; and Arnstein, 1965.
6. Crowther, 1970, chaps. 4–5; Willey, 1956, chap. 4; Chadwick, 1966–70, II, 75–90.
7. Robertson, 1933, p. 9.
8. Anderson, 1971, pp. 69–70.
9. Beahm, 1941, p. 185. Cf. G. K. Clark, 1962, p. 189.
10. Carlyle, 1878, pp. 224–5, 229–30. See Andrews, 1898, pp. 185ff and the historical background in Sandeen, 1970. See also Houghton, 1957, pp. 204–14.
11. Anderson, 1971, p. 70.
12. Bellah, 1968.
13. Handy, 1971, p. 65; T. L. Smith, 1965, pp. 216, 219; Bushnell, quoted in W. S. Hudson, 1970, pp. 82, 83. Cf. Bellah, 1968, p. 12.

14. Anderson, 1971, p. 71.
15. Pusey, 1878; Carlyle, 1878, p. vi.
16. T. H. Huxley, 1893–4, IX, 188–319; L. Huxley, 1903, III, 176–80.
17. Benn, 1906, II, 193–4.
18. Cf. Savage, 1892, p. 104.
19. Raven, 1943, p. 51.
20. *Ibid.*, p. ix.
21. Baillie, 1951, pp. 10–11; Wood, 1956, pp. 283–4.
22. Quoted in Houghton, 1957, p. 34. Cf. Livingston, 1974.
23. T. H. Huxley to the Bishop of Ripon (W. B. Carpenter), 16 June 1887, in L. Huxley, 1903, III, 20.
24. T. H. Huxley, 1887, p. 183; T. H. Huxley, 1893–4, V, 131.
25. L. Huxley, 1903, I, 217–18.
26. Huxley to E. R. Lankester, 6 December 1888, in L. Huxley, 1903, III, 93.
27. T. H. Huxley, 1893–4, IV, 216.
28. See Millhauser, 1959. Huxley's own review of the tenth edition (1853) of *Vestiges* was, he later confessed, the 'only review I ever had qualms about, on the ground of needless savagery' (quoted in L. Huxley, 1920, p. 37). See Bartholomew, 1975, pp. 526–7.
29. Huxley to Darwin, 23 November 1859, in F. Darwin, 1887, II, 232. See Bibby, 1959a.
30. T. H. Huxley, 1898–1902, II, 393–4.
31. T. H. Huxley, 1893–4, II, 52–3. See T. H. Huxley, 1870, p. viii and T. H. Huxley, 1893–4, V, x.
32. Irvine, 1972, p. 7.
33. L. Huxley, 1903, II, chap. 14; Himmelfarb, 1968a, pp. 287–94.
34. A. D. White, 1896, I, 70–1; Ellegård, 1958b, p. 380.
35. J. R. Green to W. B. Dawkins, 3 July 1860, in F. Darwin, 1887, II, 323 (cf. the abridgement in Stephen, 1901, pp. 42–5); F. Darwin, 1887, II, 320; Darwin to Huxley, 11 April 1880, in *ibid.*, III, 241.
36. See *inter alia* Rádl, 1930, p. 96 and H. Ward, 1927, p. 306 (cf. pp. 299–301).
37. Cornish, 1910, II, 224; Bibby, 1959b, p. 69 (cf. Himmelfarb, 1968a, p. 482, n. 8); Grønbech, 1964, p. 160 (cf. Boromé, 1960, p. 171); Fothergill, 1952, p. 115.
38. Cf. Abelès *et al.*, 1961, p. 547 and the translation by A. J. Pomerans, *Science in the Nineteenth Century* (London: Thames & Hudson, 1965), pp. 477–8.
39. Stackhouse, 1959, p. 945.
40. See also Cross, 1930, pp. 43–4; Lack, 1961, p. 11; I. T. Ramsey, 1964, p. 1; and Overman, 1967, pp. 72–4.
41. Huxley to C. Kingsley, 23 September 1860, in L. Huxley, 1903, I, 320.
42. T. H. Huxley, 1893–4, III, 120–1.
43. *Ibid.*, III, 210; *ibid.*, V, 290–1; T. H. Huxley, 1870, p. 173. See Eisen, 1964 and Cashdollar, 1978, p. 75.
44. T. H. Huxley, 1893–4, II, 147, 149; L. Huxley, 1903, II, 63.
45. T. H. Huxley, 1893–4, IV, 230; II, 150.
46. Huxley to Lord Farrer, 6 December 1885, in L. Huxley, 1903, II, 427.
47. L. Huxley, 1903, III, 400. On the other occasion Richard Owen was probably involved. But cf. L. Huxley, 1920, p. 37.
48. Trevelyan, 1922, p. 397; Harvie, 1976, pp. 299ff; Helfand, 1977.
49. Huxley to J. D. Hooker, 8 October 1868, quoted in Bibby, 1959b, p. 26; Huxley to Hooker, 2 December 1890, in L. Huxley, 1903, III, 182.

50. Huxley to Lord Farrer, 6 December 1885, and Huxley to J. Skelton, 8 January 1886, in L. Huxley, 1903, II, 427, 436.
51. Huxley to J. Knowles, 15 January 1886, in L. Huxley, 1903, II, 428.
52. Gladstone, 1897, p. 98. Largely in consequence of this review, which circulated widely as a pamphlet, *Robert Elsmere* 'became the talk of the civilized world' (*Encyclopaedia Britannica*, 11th edn, s.v. 'Ward, Mary Augusta'). The author, Mrs Humphrey Ward, was the sister-in-law of Huxley's eldest surviving son, Leonard.
53. T. H. Huxley, 1893–4, V, 366–7.
54. *Ibid.*, V, 413.
55. L. Huxley, 1903, III, 397.
56. T. H. Huxley, 1893–4, V, 14, 19.
57. *Ibid.*, V, 32. Cf. L. Huxley, 1903, III, 213.
58. Huxley to his daughter, 1 March 1895, in L. Huxley, 1903, III, 356.
59. In an appendix to H. Peterson, 1932.
60. Bibby, 1959b, p. 66.
61. Grønbech, 1964, p. 154.
62. Huxley to J. Morley, 7 February 1878, and Huxley to H. Spencer, 7 November 1886, in L. Huxley, 1903, II, 241, 470; Huxley to E. Haeckel, 20 May 1867, in L. Huxley, 1903, I, 416. Cf. T. H. Huxley, 1893–4, V, 1–2.
63. See Jensen, 1959 and 1967. Irvine, 1972, pp. 117–21, 138–9 is the *locus classicus*. On other abettors of the military metaphor, see Dupree, 1959, pp. 280–3; C. F. O'Brien, 1971, p. 183; and Barton, 1976, p. 71.
64. Loewenberg, 1935, pp. 232, 233; Commager, 1950, pp. 178, 181; Hofstadter, 1955, p. 25. See also Loewenberg, 1934, *passim*; Eggleston, 1935, pp. 9, 49, 59, 65; Kramer, 1948, *passim*; G. S. Carter, 1957, pp. 66–7; Ellegård, 1958a, pp. 156, 172, 218; and Willey, 1959, pp. 61–2.
65. Sandeen, 1968, pp. 21–2; Sandeen, 1970, chap. 8; Marsden, 1975, p. 126. Historians differ as they seek to account for the movement's origins, growth, and decline or persistence by focusing variously on social and political circumstances or on the theological background of the doctrines which it championed. Sandeen, who has faulted Cole, 1931 and Furniss, 1954 for accepting an oversimplified sociological definition of Fundamentalism, explains the movement as a twentieth-century alliance between two novel nineteenth-century theologies: the Princeton Theology of Alexander, Hodge, and Warfield, and the dispensationalism of J. N. Darby. But L. Moore, 1968, p. 202 objects that the alliance does not take account of individuals such as A. H. Strong, who allied themselves with Fundamentalists but were neither Princeton Calvinists nor dispensationalists. More important, Moore asks why Sandeen is so concerned with Fundamentalism (in Sandeen's words) 'until about 1918'. As Marsden, 1971, p. 145 points out, it was the Fundamentalism of the twenties, 'when the term was first used, that should provide the definition of whatever we choose to call "the Fundamentalist movement"'. For collections of primary documents, see Vanderlaan, 1925 and Gatewood, 1969. On the history of Fundamentalism after 1930, see Casper, 1963 and Marsden, 1975.
66. L. Moore, 1968; Reist, 1970; Henry, 1951.
67. Warfield, 1932b, p. 238; Warfield, 1915, p. 209. See D. F. Johnson, 1968, chap. 5.
68. There were three editors, A. C. Dixon (vols. 1–5), Louis Meyer (vols. 6–10), and R. A. Torrey (vols. 11–12), who presided over an editorial committee of well-known conservative clergymen and divines (Sandeen, 1968, pp. 18–

19; *The Fundamentals*, XII, 3–5). If the foreword to volume 11 is any indication, the contents of each volume were first screened by the editorial committee. Thus there is reason to think that these Fundamentalist fathers were at least tolerant of theistic evolution. Cf. Stackhouse, 1959, p. 945.

69. Orr, 1897, p. 99; Orr, 1905, pp. 87–8; *The Fundamentals*, IV, 97, 100–3; *ibid.*, VI, 94–6. See Orr, 1910, chaps. 5–6 and Neely, 1960.

70. G. F. Wright, 1898, pp. 106–7; *The Fundamentals*, VII, 10. Dixon, the editor, expected Wright to 'make the matter so clear that the vagaries of Evolution shall be driven from the minds of thousands'. Whether the 'vagaries' the essay dispelled were the ones Dixon had in mind is a moot point, but the fact remains that it appeared in volume 7 and that Wright contributed another essay to volume 9, subsequent to the republication of his evolutionary views in *The Origin and Antiquity of Man* (1912). For these remarks in context, see Morison, 1971, pp. 414ff.

71. Sandeen, 1968, p. 21.

72. See P. A. Carter, 1968. Present-day fundamentalists are so proud of their militant heritage that they insist on projecting it back into the nineteenth century. See Dollar, 1966 and Dollar, 1973, where the military metaphor achieves its consummate sanction in the white-washing of the Reverend J. Frank Norris, who shot an unarmed man to death in his study in 1926.

73. Bailey, 1953; L. Johnson, 1954; Gatewood, 1966; V. Gray, 1970; Meadows, 1972; Farmer, 1974; Amick, 1975. See the bibliographic essay in Gatewood, 1969.

74. Furniss, 1954, p. 17; Leuba, 1921, pp. 184–218, 219–80. Cf. William Jennings Bryan's use of the statistics in Kennedy, 1957, pp. 23–9.

75. J. P. Campbell, 1891 had been available for all to read. C. H. Peterson, 1970 points out that the decline of opposition to evolution in American higher education was due, not to the triumph of science over religion as envisaged by Draper and White, but to opposing forces at work within science itself. He adds that there was no simple relation between the rise of evolutionary biology and a decline in the religious life of the colleges (pp. 195–6). On evolution in the secondary schools, see Skoog, 1969.

76. Hofstadter, 1963, p. 123.

77. *Ibid.*, pp. 122–5.

78. Furniss, 1954, pp. 23–6; Hofstadter, 1963, pp. 132–6.

79. Quoted in Furniss, 1954, pp. 36, 41, 53, 57.

80. See *ibid.*, chap. 4.

81. Gatewood, 1969, p. 36.

82. Cf. Riddle, 1954. On Shipley, see Amick, 1975.

83. De Camp, 1968 and Ginger, 1969 are the most important studies of the trial climaxed by this scene. Contemporary accounts abound; the best book-length treatments are L. H. Allen, 1925; Osborn, 1926; and Lippman, 1928. On Bryan, see Levine, 1965 and Coletta, 1964–9. For the defendant's recollections, see Scopes and Presley, 1967.

84. Arthur Garfield Hays, a defence attorney at the trial, quoted in Kennedy, 1957, pp. 48, 49.

85. Dillenberger, 1960, p. 236.

Chapter 3

1. Croce, 1963, pp. 40–1.
2. A. R. Wallace to Darwin, 2 July 1866, in F. Darwin and Seward, 1903, I, 268. See R. M. Young, 1971a.
3. Passmore, 1966, p. 93.
4. Fischer, 1970, pp. 246–7. See Leuchtenberg, 1973.
5. Metzger, 1961, chap. 2. Kuhn, 1970, pp. 92ff offers justification for this metaphor.
6. Black, 1962, pp. 44–5.
7. Ibid., p. 46.
8. A point clearly made in C. A. Russell et al., 1974, pp. 5–50.
9. Boutroux, 1909, pp. 347, 349.
10. F. M. Turner, in the Yale Ph.D. dissertation, 1971, p. 37, published as F. M. Turner, 1974a.
11. Cf. Somervell, 1929, pp. 131–9 and Barton, 1976, pp. 5–6, 53–65, 71–2.
12. Ramm, 1955, p. 43.
13. Robertson, 1929, II, 323–4. Cf. Benn, 1906, II, 156–7.
14. A. Sedgwick to Darwin, 24 December 1859, in F. Darwin, 1887, II, 248, 249. Sedgwick's review is reprinted in Hull, 1973b, pp. 155–70.
15. Darwin, 1958, p. 65. See Barlow, 1967 and '[Recollections of Professor Henslow]' (1862), in Darwin, 1977, II, 72–4.
16. Cf. J. S. Henslow to J. D. Hooker, 10 May 1860, in Barlow, 1967, pp. 205–6 with Himmelfarb, 1968a, p. 271. See Blomefield, 1862 for the touching account of Sedgwick's visit to Henslow on his deathbed a year later.
17. Baillie, 1951, pp. 6, 9. See D. B. Wilson, 1977.
18. See D. E. Allen, 1976.
19. Galton, 1874, pp. 126–7, 135–6; Hilts, 1975, p. 30.
20. Brock and MacLeod, 1976, pp. 40–4, 51–7; Bill, 1954–6. The Declaration of Students of the Natural and Physical Sciences was printed for its signatories and the original document was deposited at Oxford in the Bodleian Library. Brock and MacLeod count sixty-five Fellows of the Royal Society, whereas Kinns, 1884, pp. 479–98, which reprints the entire Declaration, specifies seventy-seven, of whom six were elected between 1865 and 1884, five had not been elected at all, and one, Charles Woodward, had been elected as early as 29 April 1841. Thus the figure of sixty-six.
21. Daubeny, quoted in Brock and MacLeod, 1976, p. 45; Herschel, in 'Science and Scripture', The Athenaeum, no. 1925 (17 September 1864), 375.
22. Ellegård, 1958a, p. 337.
23. L. Huxley, 1903, I, 368–77; H. Spencer, 1904, II, 116; Hirst, quoted in Brock and MacLeod, 1976, p. 50; MacLeod, 1970, pp. 311–12; Jensen, 1970, p. 70; Jensen, 1971–2. Barton, 1976, chaps. 1–2 describes the liberalism which united the members' disparate metaphysics.
24. Journal of the Transactions of the Victoria Institute or Philosophical Society of Great Britain, 1 (1867), 1–36.
25. Himmelfarb, 1968a, p. 268. Cf. Keith, 1927, pp. 15–16.
26. Scott, 1970, pp. xi–xiv.
27. T. H. Huxley, 1887, p. 186. See De Beer, 1963, pp. 157–79 and Barton, 1976, pp. 163–75.
28. Darwin to T. H. Huxley, 25 November 1859, in F. Darwin, 1887, II, 232–3.
29. T. H. Huxley, 1887, p. 188.
30. Bartholomew, 1973, pp. 290–303.

31. Darwin to W. B. Carpenter, 3 December 1859, in F. Darwin, 1887, II, 239. Cf. Darwin to T. Davidson, 30 April 1861, in F. Darwin, 1887, II, 369. Darwin's casual use of the military metaphor should be understood in the light of his feelings of insecurity and embattlement.

32. *Pace* F. Darwin, 1887, II, 308n., Clark did oppose Darwin's theory. See Himmelfarb, 1968a, p. 480, n. 10.

33. For Henslow's moderate view, see F. Darwin, 1887, II, 327n.

34. Bibby, 1959b, pp. 72–7.

35. Owen's review (reprinted in Hull, 1973b, pp. 175–213) was the 'most influential' of all the reviews of the *Origin* (Ellegård, 1958a, p. 188).

36. Darwin to C. Lyell, 10 April 1860, in F. Darwin, 1887, II, 301. Cf. Darwin to A. Gray, 23 July 1862, in F. Darwin and Seward, 1903, I, 203.

37. MacLeod, 1965, p. 278.

38. Geikie, Shaler, and Weismann, quoted in Himmelfarb, 1968a, pp. 295–6; Haeckel, quoted in Bölsche, 1906, p. 133; Poulton, 1909, p. 21 on Westwood; F. Darwin, 1887, II, 261n. on Whewell.

39. Kennedy, 1957, p. vii.

40. Hannah, 1867, pp. 8–9 and Hannah, 1868, pp. 398–9 testify to this effect.

41. Temple, 1860, pp. 8–10, 14–15. *Contra* Somervell, 1929, p. 133, the sermon did not mention the Darwinian theory.

42. Powell, 1861, p. 139; Powell, quoted in Darwin to Lyell, 15 February 1860, in F. Darwin, 1887, II, 285. The passage on the eye is in Darwin, 1859, chap. 6. Darwin returned the compliment in the 'Historical sketch' prefixed to the *Origin*'s third edition (1861) by referring to Powell's evolutionary interpretation of creation as 'masterly' (Powell, 1856, essay III; Darwin, 1959, p. 69 [66]). See Darwin's letters to Powell in De Beer, 1959, pp. 51–4.

43. Maurice, 1884, II, 452, 608.

44. Liddon, 1872, p. 56n.; Liddon, 1880, p. 49; Liddon, 1893–7, IV, 332–6. See Pusey, 1878, pp. 13–14, 52 and the responses in Darwin to C. Ridley, 28 November 1878, in F. Darwin, 1887, III, 235 and Darwin to J. B. Innes, 27 November 1878, in Stecher, 1961, p. 244. Liddon and Pusey differed over the matter of membership on the Darwin Memorial Committee established at the time of Darwin's death in 1882. As for himself, Pusey believed that 'he should not join' but that his personal judgement would not apply to his younger colleague, whose 'responsibilities might be different'. Liddon nevertheless declined membership 'on the true ground – a wish not to vex Dr Pusey' – for his own conviction was that 'we owe Darwin much for his courageous adherence to Theistic truths under a great deal of pressure' (J. O. Johnston, 1904, pp. 275–6). Cf. Bowen, 1968, pp. 178–84.

45. Tristram, 1859, pp. 429, 431–2. Tristram still believed in the special creation of many species. See Tristram, 1858–60, pp. 219–28 for a review of the *Origin*. The Darwin–Wallace communication to the Linnean Society is reprinted in Darwin and Wallace, 1958. It is discussed as an historical 'nonevent' in Moody, 1971.

46. Hort, 1896, I, 414. Carpenter, 1933, p. 514 reports that Hort examined in the Cambridge Natural Science Tripos in 1872–3, the year in which he also delivered the Hulsean Lectures, *The Way, the Truth, and the Life* (1893).

47. C. Kingsley to Darwin, 18 November 1859, in F. Darwin, 1887, II, 287; Kingsley to H. W. Bates, [1863], in F. Kingsley, 1877, II, 175. In the second edition (1860) of the *Origin* Darwin inserted a sentence from Kingsley's

letter to him (attributing it to 'a celebrated author and divine') in order to show that there was no good reason for his doctrines to 'shock the religious feelings of any one' (Darwin, 1959, p. 748 [XIV:*183.2–3:b*]).

48. R. W. Church to Mrs [J. L.] Gray, in J. L. Gray, 1893, II, 751; B. A. Smith, 1958, p. 136. Cf. Church to A. Gray, 12 March 1860, in M. C. Church, 1894, p. 184, where Church agrees with Gray's understanding of the theological implications of the *Origin*.

49. Reardon, 1971, pp. 430–46.

50. P. C. Simpson, 1909, I, 285. See Hay Watson Smith, 1922.

51. McCosh, Wright, Savage, Johnson, Abbott, and Beecher are discussed in part III. On Abbot, see Persons, 1950a, pp. 428–33 and Blau, 1952, chap. 5. On the Congregationalists, see Buckham, 1919; A. Gulick, 1932; Bacon, 1934; F. H. Foster, 1939, chap. 9; Behney, 1941; D. D. Williams, 1941; H. S. Smith, 1955, chap. 8; Persons, 1961; and Gentle, 1976.

52. Mrs C. B. Booth (niece of Youmans) to W. H. Roberts, 28 August 1933, in Roberts, 1936, p. 122, n. 7; E. L. Youmans to H. Spencer, n.d., in Fiske, 1894, p. 266; Le Conte, 1903, pp. 288–90. Mrs Booth, who was in New York with her uncle at the time, adds that 'he was very much amused by it and also very much interested. When he would be leaving he would say, "Now I am going to *preach* to the *parsons.*"'

53. Ellegård, 1958a, p. 332.

54. Elliott-Binns, 1956, pp. 35–6. Cf. Vidler, 1961, p. 118 and Taylor, 1939, pp. 234–5, which neglects this point. The response of Mandell Creighton in a letter of 1 May 1871 illustrates it precisely (Creighton, 1913, I, 93).

55. Cross, 1930, p. 43; Hardwick, 1920, pp. 93–4.

56. Ellegård, 1958a, p. 29; Himmelfarb, 1968a, p. 281; Bryce, 1909, p. x; Huxley to C. H. Middleton, 2 June 1863, quoted in Raven, 1943, p. 47. This letter is no longer preserved in the library of Christ's College, Cambridge. See also Holifield, 1972, pp. 14–16.

57. Darwin to A. Gray, 18 February 1860, and Darwin to L. Jenyns, 7 January 1860, in F. Darwin, 1887, II, 286, 288; Blomefield, 1889, pp. 35–6, 78; Innes, quoted in Stecher, 1961, p. 256. On Jenyns (later Blomefield), see Teidman, 1963. The presence of Aveling and Büchner can be inferred from Aveling, 1883 and Stecher, 1961.

58. Farrar, 1904, pp. 104, 107–10.

59. Irvine, 1972, p. 133; Huxley to Kingsley, 23 September 1860, in L. Huxley, 1903, I, 313–20. Apparently Kingsley's letter has not survived.

60. Waugh, quoted in L. Huxley, 1903, II, 44; Farrar, 1904, pp. 110–11.

61. Quoted in A. W. Brown, 1947, p. 29; W. Ward, 1893, pp. 315, 316–17.

62. Eve and Creasey, 1945, pp. 40, 62, 124. See Tyndall, 1894, pp. 42–8, 175–8.

63. Tyndall, 1874. On the impact of the 'Belfast Address', see Eve and Creasey, 1945, chap. 15 and Barton, 1976, pp. 248–56 for Britain; and Roberts, 1936, pp. 44ff for America.

64. Eve and Creasey, 1945, pp. 126–7. See F. M. Turner, 1974b, pp. 57–8 and Brush, 1974–5.

65. Eve and Creasey, 1945, pp. 206–7, 225.

66. A. W. Brown, 1947, pp. 20ff.

67. Magee to his wife, 11 February 1873, in *ibid.*, pp. 31–2 (Brown's additions). Magee announced his acceptance of evolution in a sermon preached in December 1885 (Chadwick, 1966–70, II, 24).

68. A. W. Brown, 1947, pp. 23–4, 31. Herbert Spencer, John Stuart Mill, and John Henry Newman were three of the few notable non-members.

69. *Ibid.*, pp. 30, 91. See R. M. Young, 1969*b*, pp. 27–32 for a promising interpretation of the Society. Cf. Livingston, 1974, pp. 17–35.

Chapter 4

1. Altick, 1974, p. 231.
2. Tillich, 1951–63 and 1959; Turbayne, 1970; Douglas, 1970 and 1975.
3. Thus attempts at demythologising the military metaphor through the sociology of science (see, for example, F. M. Turner, 1974*b* and 1974*a*, chap. 2) or the sociology of knowledge (see, for example, R. M. Young, 1973) are definitely not precluded.
4. See Baumer, 1960, chap. 3 for the background.
5. Rice, 1904, p. 252; Loewenberg, 1941, p. 358; Houghton, 1957, pp. 58ff, 71. See also Whitehead, 1926, p. 224; Wood, 1955, p. 18; Elliott-Binns, 1956, pp. 12–13; Vidler, 1961, pp. 112–13; R. D. Baker, 1962, pp. 31ff; Davies, 1962, pp. 173ff; and Chadwick, 1966–70, II, 22.
6. See Eggleston, 1935, pp. 75–6; Himmelfarb, 1968*a*, pp. 450–2; and Budd, 1967.
7. Loewenberg, 1934, pp. 250ff; Jacks, 1917, I, 137ff, 309ff. Cf. J. B. Wilson, 1965.
8. Maitland, 1906, pp. 133, 139, 151; cf. 489. See Stephen, 1873; Annan, 1951, pp. 162–6; and Livingston, 1974, pp. 8–10, 24–7.
9. Maitland, 1906, p. 146.
10. Perry, 1935, I, 209, 217.
11. H. James, 1920, I, 145; Perry, 1935, I, 233, 301, 322–4.
12. M. White, 1975, p. 99. On Darwinism in James' philosophy, see Perry, 1935, I, 263, 265, 469–70, 490 and Wiener, 1972, chap. 5.
13. Willey, 1960, p. 80. See F. M. Turner, 1974*a*, chap. 7; B. Coleman, 1974; Breuer, 1975; and Mudford, 1966.
14. Bartholomew, 1974, chap. 2, pp. 47–8, 61–3; C. Lyell to J. D. Hooker, 9 March 1863, in K. M. Lyell, 1881, II, 326.
15. Entry of 1 November 1858, in L. G. Wilson, 1970, p. 196; Bartholomew, 1974, pp. 65–8, 284–5.
16. Bartholomew, 1974, pp. 286, 324, 380, 396; C. Lyell, 1863, pp. 405ff, 469; Darwin to Lyell, 6 March 1863, in F. Darwin, 1887, III, 11–12; Lyell to Darwin, 11 March 1863, in K. M. Lyell, 1881, II, 364.
17. Bartholomew, 1973, pp. 276–7, 303; C. Lyell, 1867–8, II, 492.
18. F. M. Turner, 1974*b*; Brush 1974–5.
19. G. J. Romanes, 1874.
20. F. Darwin and Seward, 1903, I, 352–4; Carroll, 1976, letters 455, 465, 474, 477; E. Romanes, 1896, pp. 38, 48; Duncan, 1911, p. 181; French, 1970, pp. 254–61.
21. E. Romanes, 1896, p. 79.
22. G. J. Romanes, 1878, pp. 113, 114.
23. E. Romanes, 1896, pp. 82–3. For the probable influence of the American missionary and naturalist John Thomas Gulick, see 'Correspondence between Mr Romanes and Mr Gulick', 1896 and J. T. Gulick, 1896.
24. E. Romanes, 1896, pp. 334, 339. Turner, 1974*a*, chap. 6 is the best recent discussion of Romanes' spiritual pilgrimage. Doubtless there is much truth in Turner's contention that neither Gore nor Mrs Romanes was theologically

disinterested in preserving the memory of a scientist returned to faith. Indeed, it may well be that Romanes was 'somewhat chameleon-like in his tendancy [sic] to gauge his opinions to those of the intellectual circle with whom he was involved at the moment' (p. 141). But one cannot help but wonder whether Turner has forced his subject to fit his thesis by portraying Romanes' final reconciliation with the Church as a mere 'act of pure agnosticism' (p. 162) and not at the same time sufficiently stressing the judgement of Gore and Waggett, who ministered to Romanes for years before the end came: namely, that through perceiving 'the positive strength of the historical and spiritual evidences of Christianity' Romanes 'lived to find the freer faith for which process and purpose are not irreconcilable, but necessary to one another', and, at last, returned to 'that full, deliberate communion with the Church of Jesus Christ which he had for so many years been conscientiously compelled to forego' (Gore's 'Concluding note by the editor', in G. J. Romanes, 1895b, p. 184; Waggett, 1909, p. 486). See also Waggett, 1905, pp. 101–2 and Nias, 1961, pp. 53–5.

25. Cf. Aveling, 1883 and Foote, 1889 with Warfield, 1889 and 1932a; Myers, 1893; Symonds, 1893; Luzzatti, 1901; Nash, 1928; A. J. Russell, 1934; and Mandelbaum, 1958. None of these studies contains a fully satisfactory interpretation of Darwin's religious views.

26. On Huxley, see T. H. Huxley, 1893–4, IX, 46–116; Mudford, 1966, p. 197 and passim; and Helfand, 1977, pp. 172–3. On Wyman, see Dupree, 1953, p. 244. On Dana, see Sanford, 1965, p. 531; 'James Dwight Dana', 1895; and Gilman, 1899, p. 255. On Sidgwick, see D. G. James, 1970, pp. 11–12; Sidgwick and Sidgwick, 1906; F. M. Turner, 1974a, chap. 3; and Schneewind, 1977. On Marshall, see Keynes, 1972, pp. 167–71. On Clifford, see Frederick Pollock's introduction in Clifford, 1879, I, 31ff. On Moule, see MacDonald, 1922, p. 33. On Lankester, see E. R. Lankester to A. L. Moore, 16 February [1888], in Moore Autograph Collection, B/M/781, American Philosophical Society Library, Philadelphia. On the New England writers, see J. M. Turner, 1944, chap. 14. Chadwick, 1966–70, II, 112ff. and Livingston, 1974 contain a sympathetic extension of this discussion.

27. E. A. White, 1952, p. 37.
28. Cf. G. K. Clark, 1962, pp. 13–14.
29. Himmelfarb and Eagly, 1974, p. 17.
30. Festinger, 1959, pp. 3–13. Cf. Festinger et al., 1964a, chap. 2.
31. Festinger et al., 1964a, p. 157; Festinger, 1959, pp. 18–24, 42–7.
32. The theory of cognitive dissonance is fully applicable to the ideas and behaviour of historical individuals. Indeed, some of its earliest support was drawn from the history of religious enthusiasm. See Festinger et al., 1964b and the summary in Festinger, 1959, pp. 252ff. For a detailed historical application, see Källstad, 1974.
33. For general discussions of Darwin's impact on Christian doctrines, see Waggett, 1905 and 1909; A. C. McGiffert, 1915, pp. 175–86; Roberts, 1936, pp. 129–50; Lack, 1961, chaps. 6–9; and Barbour, 1966, pp. 88–9.
34. Ellegård, 1958a, p. 98.
35. Festinger, 1959, p. 47.
36. Ellegård, 1958a, p. 21. On the distinction between evolution and the Darwinian mechanism of natural selection – a distinction which historians have not always observed – see Lovejoy, 1909; Waggett, 1909, p. 484; Persons, 1959, p. 5 (cf. Persons, 1950a, p. 237); and Wichler, 1961, pp. 219, 222.
37. The theory of cognitive dissonance requires that the conflicting cognitions

be fairly well defined. But since the conflicts varied somewhat from individual to individual, we can do no more at present than specify in general the several most important ones. A fuller delineation of Darwinian and Christian cognitions must await later chapters. The theory, as amended by subsequent research, also requires that a 'definite commitment' result from a decision if there is to be evidence of dissonance reduction (Festinger, 1964, p. 155). In the present application, however, both the decision and the commitment are necessarily obscure. The decision is a private event that cannot possibly be abstracted from the dissonance reduction in which it is imbedded. The commitment is probably a public event – a conversation, sermon, letter, or book – but evidence of it has not always survived and the evidence would undoubtedly show signs of dissonance reduction if it had. Nevertheless, the decision and the commitment are quite reasonable postulates, both of which serve to account for the verbal behaviour that we are considering as a manifestation of dissonance reduction.

38. Peckham, 1970, p. 191. Peckham limits consideration to the *Origin of Species* as a document defining Darwinism and oversimplifies both Darwinism and Darwinisticism, disregarding the various unresolved issues in post-Darwinian evolutionary theory which led Darwin, out of caution and a due regard for evidence, to retain a streak of Lamarckism in the *Origin* and led others, not to a 'perverting and self-deluding acceptance of the work' (p. 191), but to an acceptance conditioned in many cases by their solutions to the unresolved issues. Moreover, Peckham errs in thinking that it was the 'Radical Romantic' who 'could most readily. . .accept the *Origin*' (p. 196). Yet we maintain that the distinction between Darwinism and Darwinisticism is fundamentally sound and that it can be salvaged conveniently by means of Festinger's theory.

39. R. M. Young, 1970a, p. 13; cf. p. 23. In discussions of the post-Darwinian controversies Simonsson, 1958, chap. 2 and Simonsson, 1971, chap. 1 analyse responses to 'contradictions' between Christianity and natural science but fail to show any interest in what features of Darwinism *per se* were offensive to Christians.

40. J. W. Gruber, 1960, chap. 1.

41. Quoted in *ibid.*, p. 31.

42. Quoted in *ibid.*, pp. 36–7. This encounter occurred less than a fortnight after Mivart's election, supported by Darwin himself *inter alia*, to a Fellowship of the Royal Society (*ibid.*, p. 231, n. 10).

43. *Ibid.*, pp. 49, 50; cf. p. 116. See Root, 1974.

44. See chap. 2 above, the text at n. 44, and F. Darwin, 1887, III, 143ff.

45. J. W. Gruber, 1960, p. 111.

46. Quoted in *ibid.*, p. 209.

47. We have by no means explored all the implications of the theory of cognitive dissonance. For example, dissonance reduction interpreted as a social phenomenon would help to account for the 'Declaration' on science and scripture, the X Club, the Victoria Institute, and possibly the Metaphysical Society (see chap. 3 above, the text at nn. 20–4, 66–9). Says Festinger: 'If a cognitive element that is responsive to reality is to be changed without changing the corresponding reality, some means of ignoring the real situation must be used.' The change is often made easier if one is 'able to find others who would agree with and support his new opinion. In general, establishing a social reality by gaining the agreement and support of other people is one of the major ways in which a cognition can be changed when the pressures

to change it are present' (Festinger, 1959, p. 21). Cf. Barber, 1961 for a similar view.

Chapter 5

1. Roger, 1976, p. 484.
2. Darwin, 1859, pp. 126–8. Cf. Darwin, 1959, pp. 270–2 (IV:384–94).
3. Ruse, 1975c and 1975b. See also Cramer, 1896, chap. 15; Crombie, 1959, p. 360; Ghiselin, 1969, p. 63; and Gruber and Barrett, 1974, p. 173.
4. This of course is a reconstruction of Darwin's argument. Ruse, 1971 shows that the *Origin* contains at least three arguments in behalf of a struggle for existence and two related arguments for natural selection, none of which is formulated with great rigour.
5. Darwin, 1859, p. 63. See R. M. Young, 1969a; Gale, 1972; and Ruse, 1973.
6. See Hull, 1967; Ghiselin, 1969, chap. 3; Ellegård, 1957a; 'Agassiz, Darwin, and evolution' (1959), in Mayr, 1976, pp. 251–76; and MacLeod, 1965.
7. Darwin, 1959, p. 317 (V:285–6).
8. Darwin, 1859, pp. 194, 206. Cf. Darwin, 1959, pp. 361 (VI:173) and 378 (VI:264–5).
9. Darwin, 1959, p. 185 (IV:125).
10. Vorzimmer, 1972, pp. 43–69, 122–4; Bowler, 1974; Darwin to A. R. Wallace, 22 January and 2 February 1869, in F. Darwin, 1887, III, 107; Darwin, 1959, pp. 178–9 (IV:95.4–11:e). Jenkin's review is reprinted in Hull, 1973b, pp. 303–44.
11. Darwin, 1959, p. 169 (IV:41.2–3:c).
12. *Ibid.*, p. 117 (1:310.3:e). See Darwin to J. D. Hooker, 7 August 1869, in F. Darwin and Seward, 1903, I, 314 and Vorzimmer, 1972, pp. 127–57.
13. Darwin, 1859, p. 43; cf. pp. 52–3, 167–8.
14. See Vorzimmer, 1972, pp. 71–95.
15. Darwin, 1959, p. 86 (1:57).
16. Vorzimmer, 1972, pp. 27–30, 44–5.
17. *Ibid.*, pp. 145–7.
18. See Olby, 1966, chap. 3.
19. Vorzimmer, 1972, p. 126; see also pp. 21–42, 97–126.
20. Darwin, 1959, pp. 367 (VI:207–8), 372 (VI:221, 223), 373 (VI:228:f).
21. *Ibid.*, pp. 363–4 (VI:190–1), 367–8 (VI:206–14).
22. Mivart, 1871, chap. 2.
23. Darwin, 1959, pp. 367–8 (VI:210—210 + 11 + 12:f).
24. *Ibid.*, p. 369 (VI:219—219 + 20:f); cf. p. 234 (IV:VII.382. 65.0.12.5:f).
25. *Ibid.*, p. 243 (IV:VII.382.65.0.50.23:f). Cf. Darwin to Hooker, 16 September 1871, in F. Darwin and Seward, 1903, I, 332–3 and Vorzimmer, 1972, pp. 213–24.
26. Darwin, 1859, p. 287; Darwin, 1959, pp. 483–5 (IX:57–71). See Burchfield, 1974, pp. 302–7.
27. Darwin, 1959, pp. 478 (IX:31 and 31.e), 485 (IX:74 and 74:[e]).
28. See Haber, 1959, chaps. 4–5; Toulmin and Goodfield, 1965, chap. 7; and Gillispie, 1959a, chaps. 3–5.
29. Burchfield, 1975, p. 37.
30. *Ibid.*, chap. 2. The articles and address are reprinted in W. Thomson, 1889–94, vol. 2.
31. Burchfield, 1975, pp. 70, 81.

32. Darwin to T. H. Huxley, 19 March 1869, in F. Darwin, 1887, III, 113; T. H. Huxley, 1893–4, VIII, 329.

33. W. Thomson, 1889–94, II, 89–90. Although Thomson's theories were deliberately grounded in an orthodox theology of nature, it is questionable whether the theology demanded the theories. If so, it was through the medium of Paley's 'solid and irrefragable' argument in his *Natural Theology* (1802), which Thomson believed to be opposed to natural selection (though not necessarily to evolution) because natural selection is 'too like the Laputan method of making books' and too little cognisant of a 'continually guiding and controlling intelligence' (Thomson's presidential address to the British Association for the Advancement of Science, 1871, in Basalla *et al.*, 1970, p. 128; for Darwin's response, see Darwin to Hooker, 6 August 1871, in F. Darwin and Seward, 1903, I, 329–30). On Thomson's religious and theological views, see A. G. King, 1925, pp. 28–31; D. B. Wilson, 1974; Burchfield, 1975, pp. 47–50; and C. Smith, 1976.

34. Darwin to Hooker, 1867, in F. Darwin and Seward, 1903, II, 5–7. See Darwin, 1959, p. 596 (XI:*229.1–3:d*). Jenkin was a colleague and close friend of Thomson's (Stevenson, 1904, p. 99).

35. Darwin to J. Croll, 31 January 1869, in F. Darwin and Seward, 1903, II, 163–4; Darwin, 1959, pp. 480 (IX:*35.6:e*), 482 (IX:49–51), 485 (IX:72–4), 485–7 (IX:*75.6–19:e*), 513 (IX:219), 749 (XIV:190), 757 (XIV:254).

36. Darwin to Hooker, 24 July 1869, in F. Darwin and Seward, 1903, I, 314; Darwin–Wallace letters in Marchant, 1916, I, 242, 248–51; Mivart, 1871, chap. 6; Darwin to Huxley, 19 March 1869, in F. Darwin, 1887, III, 113; Hooker to Darwin, 1870, in F. Darwin and Seward, 1903, II, 6–7.

37. Darwin, 1959, p. 513 (IX:*219.4.1:f*). See Darwin to Wallace, 5 January 1880, in Marchant, 1916, I, 304 and Burchfield, 1974, pp. 307–21.

38. Darwin, 1959, p. 728 (XIV:*55.1:f*).

39. Darwin, 1874, pp. 30ff, 44, 62, 241, 422–3, 425, 607.

40. See Vorzimmer, 1972, pp. 233ff.

41. Darwin, 1874, p. 62; see also pp. 30, 608. Cf. Darwin to M. Wagner, 13 October 1876, with Darwin to K. Semper, 19 July 1881, both in F. Darwin, 1887, III, 159, 344–5.

42. Darwin, 1874, p. 607.

43. *Ibid.*, pp. 32–5. Cf. Darwin to F. Galton, 7 November 1875, in F. Darwin and Seward, 1903, I, 360 and Darwin, 1958, p. 89n.

44. Darwin, 1874, p. 61.

45. Darwin to Huxley, 5 November 1880, in F. Darwin and Seward, 1903, I, 389.

46. Eiseley, 1961, p. 242. See Darwin to Hooker, 16 and 22 January 1869, in F. Darwin and Seward, 1903, II, 379.

47. Vorzimmer, 1972, p. 239.

48. Darwin, 1874, pp. 61, 469; Darwin, 1959, pp. 379 (VI:*270:f*), 253 (IV:VII.*382.65.0.50.165:f*).

49. Darwin, 1959, p. 75 (50:e). The first edition reads 'main' for 'most important'.

Chapter 6

1. Conn, 1900, p. 41.

2. Mivart, 1871, p. 23.

3. Darwin to A. R. Wallace, 2 September 1872, in Marchant, 1916, I, 278.

4. Cf. Vorzimmer, 1972, pp. 250–4 with Ghiselin, 1969, *passim*. When Vorzimmer quotes (p. 251) as evidence of Darwin's retirement in a 'state of frustrating confusion' his comment to Chauncey Wright (3 June 1872), 'I have resolved to waste no more time in reading reviews of my works or on evolution', he makes Darwin support his own views by dropping the last clause of the sentence: viz., 'excepting when I hear that they are good and contain new matter' (F. Darwin, 1887, III, 164).

5. Darwin, 1958, pp. 136–7.

6. Darwin to E. Haeckel, 27 December 1871, in F. Darwin and Seward, 1903, I, 335–6. Cf. Lyell's recollection of Darwin's advice, reprinted in Bartholomew, 1976a, p. 216.

7. J. A. Thomson, 1899, p. 225. Cf. Ellegård, 1958a, p. 334 and chap. 12.

8. E. S. Russell, 1916, chap. 13; Gillispie, 1959b, pp. 270ff; Wilkie, 1964, pp. 289ff; M. J. S. Hodge, 1971, pp. 327–8, 343–4; Omodeo, 1971, p. 20; 'Lamarck revisited' (1972), in Mayr, 1976, pp. 233ff; R. Burkhardt, 1977, chaps. 5–6. For the two components in post-Darwinian theories of heredity, see Ostoya, 1951, p. 171.

9. Bourdier, 1969, pp. 44–52; Rudwick, 1972, pp. 150–3, 175.

10. Lovejoy, 1959; Millhauser, 1959.

11. Chambers, 1969, pp. 222–31, Cf. *ibid.*, 11th edn, 1860, pp. 160–1, where Chambers is more generous to Lamarck.

12. *Ibid.*, 10th edn, 1853, p. 155, quoted by Darwin in the 'Historical sketch' prefixed to the third (1861) and later editions of the *Origin of Species* (Darwin, 1959, p. 64 [33]). An abbreviated version of this passage appeared as early as the fifth edition of *Vestiges*, 1846, p. 231. See Ogilvie, 1973.

13. In the appendix to the eleventh edition of *Vestiges*, 1860, pp. lv–lvi, Chambers regarded his views as an improvement on Lamarck's 'imperfect' theory.

14. Richard Owen, 1848, pp. 171, 172; Owen, 1849, pp. 85–6; Owen, 'Darwin on the origin of species' (1860), reprinted in Hull, 1973b, pp. 180, 181, 184, 188, 212.

15. Richard Owen, 1866–8, III, 797. On Chambers and Owen, cf. Rev. Richard Owen, 1894, I, 252ff, 310–11 with Brooke, 1977b.

16. E. S. Russell, 1916, p. 215. Cf. Gillispie, 1959b, p. 275 with M. J. S. Hodge, 1971, pp. 347–52. See also MacLeod, 1965, pp. 264–74 and E. S. Russell, 1916, chap. 8. 'The looking to "natural laws" and "secondary causes" for the "progression" of "organic phenomena" is the substantial acceptance of evolution, as set forth by Goethe, Oken, Lamarck, and Geoffroy' (T. H. Huxley, 1894, p. 318).

17. Samuel Butler was undoubtedly Britain's most determined and belligerent advocate of a revived and enlightened Lamarckism. There is little evidence, however, that his books, *Life and Habit* (1877), *Evolution, Old and New* (1879), *Unconscious Memory* (1880), and *Luck or Cunning?* (1885), had any constructive effect on religious thinkers, much less on the scientific establishment. On Butler's Lamarckism, see E. S. Russell, 1916, chap. 19; on his relations with Darwin and the Darwinians, see Willey, 1960; on his religious views, see chap. 4 above, the text at n. 13.

18. Pfeifer, 1965 and 1974, pp. 198–201; Packard, 1901, pp. 382ff; Cope, 1896, pp. 8–10; Packard, 1877.

19. 'On the origin of genera' (1868), in Cope, 1887a, p. 43. As late as 1871 Cope had not read Lamarck ('On catagenesis' [1884], in Cope, 1887a,

pp. 422–3), though within five years he was fairly acquainted with his work ('On the theory of evolution' [1876], in Cope, 1887a, p. 124).

20. Vorzimmer, 1972, p. 237.
21. 'The method of creation of organic forms' (1871), in Cope, 1887a, pp. 174–5.
22. Cope, 1896, p. 201.
23. Chambers, 1969, pp. 215–16, 222–3 (emphasis added). See M. J. S. Hodge, 1972.
24. Osborn, 1931, pp. 527–44 and Bowler, 1977b review Cope's contributions to evolutionary theory.
25. Cope, 1896, p. vi, chaps. 5–6.
26. *Ibid.*, pp. 473–92.
27. *Ibid.*, p. 453.
28. Cope, 1887a, p. 7; 'On catagenesis' (1884), in *ibid.*, pp. 424, 425.
29. 'On archaesthetism' (1882), in Cope, 1887a, p. 405; Cope, 1896, chap. 10.
30. Cope, 1896, chap. 9, p. 516.
31. Osborn, 1931, pp. 544–5.
32. Cope, 1887b, pp. 23, 28, 29, 31; Cope's letter of 1 October 1888, in Osborn, 1931, p. 544. This point supplements Bowler, 1977b, pp. 261, 264–5. On Cope's controversy over his *Theology of Evolution* with Edmund Montgomery, 'one of the most erudite and enthusiastic members of a group of biologists...who laid the foundations of emergent, organic evolutionism and of empirical naturalism' (Schneider, 1963, p. 321), see Keeton, 1947.
33. *American Naturalist*, 1 (March 1867), 2. At this time E. S. Morse and F. W. Putnam served as editors with Hyatt and Packard.
34. C. King, 1877, p. 470. See Burchfield, 1975, p. 117 and Wilkins, 1958, pp. 208–11.
35. Packard, 1880.
36. 'On the hypothesis of evolution: physical and metaphysical' (1870), in Cope, 1887a, pp. 154, 168.
37. See Cope, 1887b, pp. 36–9; Osborn, 1931, pp. 564ff; and Haller, 1975, pp. 187–202.
38. On Lamarck, see Darwin, 1959, pp. 60–1 (9, *13*.3:d*); comments in Darwin's letters, 1844–63, in F. Darwin, 1887, II, 23, 29, 39, 215; III, 14 and in F. Darwin and Seward, 1903, I, 41, 153; and Egerton, 1976. On *Vestiges*, see Darwin, 1959, p. 64 (36, 38) and comments in Darwin's letters, 1849–60, in F. Darwin, 1887, I, 333, 344; II, 39 and in F. Darwin and Seward, 1903, I, 85; II, 75, 136. On Owen, see Darwin to T. H. Huxley, 3 January 1861, in F. Darwin, 1887, I, 178 and Darwin, 1959, pp. 64–6 (40—*45.13:e*).
39. Darwin, 1959, p. 349 (VI:*160.0.1–8:f*); Darwin–Hyatt letters, 10 October 1872 to 13 February 1873, in F. Darwin and Seward, 1903, I, 338–48.
40. Darwin to W. E. Darwin, 1876, in Carroll, 1976, letter 502; Darwin to E. S. Morse, 23 April 1877, in F. Darwin, 1887, III, 233; Darwin to A. Hyatt, 8 May 1881, in F. Darwin and Seward, 1903, I, 393.

Chapter 7

1. A. C. Armstrong, 1904, pp. 160–1.
2. Darwin, 1958, p. 140.
3. *Ibid.*, p. 141. On Darwin's confusion about his method, see Feibleman, 1959, p. 14.

NOTES TO PAGES 155-61 373

4. Barrett, 1974, p. 149.
5. Darwin, 1859, p. 488. Cf. Darwin to L. Jenyns (Blomefield), 7 January 1860, in F. Darwin, 1887, II, 263-4.
6. De Beer *et al.*, 1960-1, II, 109 (*C*, 223). See Darwin, 1958, pp. 130-1; Gruber and Barrett, 1974, chap. 2; and Herbert, 1977, pp. 190ff. On the state of the question of human origins during Darwin's early years, see Mandelbaum, 1957, pp. 351-7; Eiseley, 1972; Bynum, 1974; and Gruber and Barrett, 1974, chaps. 9-10.
7. Darwin, 1874, pp. 65, 126.
8. *Ibid.*, pp. 97, 124-5, 126.
9. *Ibid.*, pp. 113n., 126, 131, 144.
10. *Ibid.*, pp. 126, 613; see also p. 180. Cf. Gruber and Barrett, 1974, p. 216.
11. Darwin, 1874, pp. 128, 132, 133.
12. *Ibid.*, pp. 134, 140, 141, 142, 143.
13. McConnaughey, 1950, p. 412; Conry, 1974, pp. 414-15. The latter point is made forcibly in Jones, 1974, pp. 98-134 and Greene, 1977.
14. Darwin, 1874, pp. 145, 619. On Darwin's hopeful belief in long-term progress, see also Darwin to C. Lyell, 27 August 1860, in F. Darwin and Seward, 1903, II, 30 and Darwin, 1958, p. 92. At least twice in each edition of the *Origin of Species* Darwin asserts the independence of natural selection from any 'necessary', 'universal', or 'fixed' law of progressive development (Darwin, 1959, pp. 223 [IV:382.18:c and e], 523 [X:21], 567 [XI:41—41:f]). On the concluding passage in the *Origin*, where Darwin states that, 'as natural selection works solely by and for the good of each being, all corporeal and mental endowments will tend to progress towards perfection' (Darwin, 1959, p. 758 [XIV:266]), cf. Darwin to Lyell, 25 October 1859, in F. Darwin, 1887, II, 176-7; the use of 'perfect' in all its forms throughout the *Origin* (Darwin, 1859, pp. 186-94, 459-60, 472, 475 etc.), especially the rejection of Lamarck's 'innate and inevitable tendency towards perfection' (Darwin, 1959, p. 223 [IV:382.16:c]); and 'the differentiation and specialisation of organs as the test of perfection' in the *Descent of Man* (Darwin, 1874, p. 91). See also Thoday, 1958 and G. G. Simpson, 1974.
15. Darwin, 1874, pp. 110, 113n., 124-5, 134-40. See C. M. Williams, 1893; Mackintosh, 1899; Tufts, 1909; Gantz, 1937 and 1939; Parkinson, 1942; and Quillian, 1950.
16. Darwin, 1874, pp. 135, 182, 618.
17. Mandelbaum, 1971, pp. 85, 87, 109.
18. Darwin, 1874, p. 142. Cf. Darwin to G. A. Gaskell, 15 November 1878, in F. Darwin and Seward, 1903, II, 50. See Greene, 1959b, pp. 333-6.
19. Darwin, 1874, pp. 141-2, chap. 7; Darwin to W. Graham, 3 July 1881, and Darwin to Lyell, 11 October 1859, in F. Darwin, 1887, I, 316; II, 211. Darwin refers the 'extinction', 'extermination', or 'elimination' of races to disease and a decline in fertility – causes consequent on the presence, rather than the policies, of civilisation. On Darwin's aversion to racial slavery, see Gruber and Barrett, 1974, pp. 65-8, 181-5.
20. Darwin, 1874, p. 618.
21. *Ibid.*, pp. 143, 618. Cf. Mandelbaum, 1971, p. 230. On post-Darwinian theories of progress based on struggle and conflict, see Wagar, 1972, chaps. 4, 7.
22. E.g., Shaw, 1921, pp. lvi-lvii; Parkinson, 1942, chap. 4; Barzun, 1958, pp. 87-126; and R. E. D. Clark, 1967, chap. 6. On the less prominent

coöperative and reform movements which may also go under the name of Social Darwinism, see Montagu, 1952; Hofstadter, 1955; and Loewenberg, 1957. On the opprobrious misuse of the term, see Bannister, 1970.

23. It is easy to fault Darwin for naïveté of foresight. Such was his confidence in 'the permanence of virtue on this earth' that he could not imagine an advanced society, modelled on that of the hive, in which the principles of social duty, though 'acquired for the good of the community', would be reversed (Darwin, 1874, p. 99n.). Truthfully did he admit to having 'never systematically thought much...on morals in relation to society' (Darwin to F. E. Abbot, 16 November 1871, in F. Darwin, 1887, 1, 306). On natural selection in human society, see Bock, 1953, p. 123; Stark, 1961, p. 56; and Leeds, 1974.

24. J. A. Rogers, 1972, p. 268. See also Barker, 1915, p. 133 and Herbert, 1977, pp. 195-6.

25. Darwin to H. Spencer, 25 November 1858, in F. Darwin, 1887, II, 141; 2 February 1860, in Duncan, 1911, p. 98; 9 December 1867, in F. Darwin and Seward, 1903, II, 442. See also Darwin to Spencer, 10 June 1872, in F. Darwin, 1887, III, 165-6; and 31 October 1873, in F. Darwin and Seward, 1903, I, 351-2.

26. H. Spencer, 1904, II, 27-8. See Wiltshire, 1978, p. 68.

27. Darwin to Lyell, 25 March 1865, in Carroll, 1976, letter 307; Darwin to F. M. Balfour, 4 September 1880, in F. Darwin and Seward, 1903, II, 424-5; Darwin to J. D. Hooker, 30 June 1866, in F. Darwin and Seward, 1903, II, 235; Darwin to Hooker, 10 December 1866, in F. Darwin, 1887, III, 55-6; Darwin to L. H. Morgan, in Stern, 1928, p. 181. See also the correspondence in F. Darwin, 1887, III, 120, 193; F. Darwin and Seward, 1903, I, 368; II, 48; Marchant, 1916, I, 175-6, 191, 283; and Carroll, 1976, letters 201, 446.

28. Darwin, 1958, p. 109. Darwin's memory was perhaps a bit 'hazy' on the question of Spencer's influence on him. See Darwin to A. R. Wallace, 5 July 1866, in F. Darwin and Seward, 1903, I, 270-1 and Vorzimmer, 1972, p. 86.

29. R. M. Young, 1967, p. 274. See Plochmann, 1959 and Freeman, 1974.

30. Spencer, 'The filiation of ideas' (1899), in Duncan, 1911, p. 536. See J. D. Y. Peel, 1971, chap. 1 and Wiltshire, 1978, chaps. 1-2.

31. J. A. Thomson, 1932, p. 103.

32. 'Filiation of ideas' (1899), in Duncan, 1911, pp. 539, 545. Of Chauncey Wright, Spencer's American disciple John Fiske remarked, 'I never knew an educated man who set so little store by mere reading, except Mr Herbert Spencer' (Fiske, 1885, p. 107).

33. 'Filiation of ideas' (1899), in Duncan, 1911, pp. 541ff, 546. See Bowler, 1975, pp. 106-8.

34. The essays on development and progress are reprinted in H. Spencer, 1890, vol. 1. On the *Principles of Psychology*, see R. M. Young, 1970b, chap. 5. Of all Spencer's writings, the first and second editions of this book, together with the *Principles of Biology*, were the most useful to Darwin. See Darwin, 1959, p. 757 (XIV:256:f); Darwin, 1874, p. 67; Darwin, 1872, pp. 227n., 263; and Greene, 1977, p. 6.

35. 'Filiation of ideas' (1899), in Duncan, 1911, pp. 550-1. Spencer's prospectus, issued in 1858, is reprinted in Rumney, 1934, pp. 297-303.

36. Darwin to Lyell, 20 October 1859, in F. Darwin, 1887, II, 175. For the unfinished manuscript of the 'larger book', see Darwin, 1975.

37. With reference to his work on the circulation of sap in plants, probably the

only original research he ever carried out, Spencer boasted: 'My argument was wholly inductive and unguided by hypothesis: for, until observations and experiments had suggested one, no view at all was entertained by me' (H. Spencer, 1904, II, 127). Cf. Meldola, 1910, app.

38. See Duncan, 1911, pp. 236, 258, 313, 555 and H. Spencer, 1862–96, I (5th edn, 1884), v–vi. Cf. Darwin, 1872, p. 10, n. 11.

39. H. Spencer, 1862–96, I (5th edn, 1884), 396. See *ibid.*, I (6th edn, 1900), 321 for the final minor revisions. The much-quoted parody of Spencer's law was composed not, as is sometimes claimed, by William James (who only quoted it approvingly; see Perry, 1935, I, 482 and Plochmann, 1959, p. 1455, n. 2) but by the mathematician and rector of Croft near Warrington, Thomas Penyngton Kirkman: 'Evolution is a change from a nohowish, untalkaboutable all-alikeness, to a somehowish and in-general-talkaboutable not-all-alikeness, by continuous somethingelsifications and sticktogetherations' (Kirkman, 1876, p. 292). That Spencer found little humour in it is evident from the three pages he devoted to a refutation in the appendix to the fourth edition (1880) of *First Principles*.

40. H. Spencer, 1904, II, 11; 'Filiation of ideas' (1899), in Duncan, 1911, p. 555. Cf. Meldola, 1910, pp. 34–5 with W. H. Hudson, 1904, pp. 36–9; Plochmann, 1959, pp. 1454–5; Freeman, 1974, pp. 213–20; J. D. Y. Peel, 1971, chap. 6; and Wiltshire, 1978, chap. 3.

41. H. Spencer, 1904, II, pts 8–10. See Wiltshire, 1978, chap. 4.

42. R. A. Armstrong, 1905, p. 48; Clodd, 1907, p. 184. On the Darwin–Wallace paper, see chap. 3 above, n. 45.

43. See T. H. Huxley, to F. C. Gould, 1889, in Clodd, 1902, pp. 220–1; Benn, 1906, II, 204–35; and Eisen, 1968. James Martineau and Robert Flint were outstanding among theological critics.

44. Quoted in Clodd, 1907, p. 186. See Barker, 1915, chap. 8.

45. Hofstadter, 1955, pp. 32–4. On Spencer's place in American philosophy, see Fisch, 1947, pp. 360–2; Goudge, 1947, p. 140; Wiener, 1972; and Russett, 1976, chap. 3.

46. H. Spencer, 1862–96, I (5th edn, 1884), 113; Goblet d'Alviella, 1886, p. 219. Spencer's philosophical ideas were religious in origin and theological in content; his sociology expressed the values of English middle-class Nonconformity. See T. H. Huxley to F. C. Gould, 1889, in Clodd, 1902, pp. 220–1; Chauncey Wright, quoted in Wiener, 1972, p. 62; G. J. Romanes, 1895a, p. 117n.; W. H. Hudson, 1904, pp. 114–16; Benn, 1906, II, 225; Barker, 1915, chap. 4; Sharlin, 1976, p. 463; and Wiltshire, 1978, pp. 207–9. On theological criticism in the United States, see Roberts, 1936, pp. 53–4 and Cashdollar, 1978, pp. 75–8.

47. H. Spencer, 1862–96, VII (1882), 603–67; VIII (1896), chap. 13.

48. *Ibid.*, VII (1882), 639.

49. H. W. Beecher to Spencer, June 1866, in Duncan, 1911, p. 128; *Atlantic Monthly*, quoted in Hofstadter, 1955, p. 33. See Barker, 1915, pp. 130–1; Hofstadter, 1955, chap. 2; Persons, 1958, pp. 223ff; Molloy, 1959; Fine, 1964, chap. 2; Boller, 1969, chap. 2; Russett, 1976, chap. 4; and Wiltshire, 1978, pp. 96–7. *Illustrations of Universal Progress* (1864), a collection of Spencer's essays, appeared only in the United States. It went through numerous editions.

50. 'Mr Spencer's address', in Youmans, 1887, p. 32.

51. Darwin, 1874, pp. 143, 618.

52. H. Spencer, 1887, p. 74.

53. H. Spencer, 1904, II, 50. See also H. Spencer, 1862–96, III (1867), 500n.; H. Spencer, 1887, p. 9; and Spencer to E. Fry, 3 November 1894, in Duncan, 1911, p. 351.
54. H. Spencer, 1904, II, 100. Cf. H. Spencer, 1868, pp. 74–80.
55. H. Spencer, 1862–96, II (1864), 468–9. Cf. H. Spencer, 1887, p. 33.
56. Darwin, 1875, II, 328n.
57. Darwin, 1959, pp. 225–6 (IV:*382.35–37:c*). See n. 14 above.
58. Over the years Spencer became less sanguine about the immediacy and rapidity of progress. His fears culminated in the conclusion of the last volume of the *Principles of Sociology* (H. Spencer, 1862–96, VIII [1896], 599–601). Thus both Darwin and Spencer came to think of progress as a long-term eventuality. However, for Darwin progress was a belief about history which his theory could not validate; for Spencer progress – whether immediate or eventual, rapid or prolonged – was a fact which his law revealed to be an immanent physical necessity. Cf. Rumney, 1934, chap. 10 and Bowler, 1975, p. 114 with Greene, 1959b, p. 328; Greene, 1961, pp. 96–8; and Greene, 1975, pp. 256–7. See also Passmore, 1959, pp. 46–7; Greene, 1959a, pp. 437–41; Burrow, 1970, pp. 198ff; and Wiltshire, 1978, chap. 8.
59. E.g., see H. Spencer, 1862–96, II (1864), 244–52, 455–7; III (1867), 166–8, 195–201; IV (3rd edn, 1880), 421ff; VI (3rd edn, 1885), 37, 90–1; IX (1892), 99, 471.
60. H. Spencer, 1887, p. iii.
61. Delage and Goldsmith, 1920, p. 201. See Kellogg, 1907, p. 391, n. 4; Bliakher, 1973; and Churchill, 1978.
62. H. Spencer, 1862–96, II (2nd edn, 1898), 621, 650. Spencer's articles are reprinted in *ibid.*, pp. 602–91.
63. Wallace to R. Meldola, 10 June 1893, in Marchant, 1916, II, 56; Bourne, 1910, p. 34. On the decline of Lamarckism in American sociology, see Lane, 1950 and Stocking, 1968.
64. Hofstadter, 1955, p. 32.
65. Barzun, 1958, p. 80. See also May, 1949, pp. 47, 142 and Commager, 1950, p. 85.

Chapter 8

1. Darwin, 1959, pp. 747–8 (XIV:183—*183.0.0.4:f*).
2. In the passage Darwin no doubt referred primarily to the misrepresentation of Mivart. See Vorzimmer, 1972, p. 243 and cf. G. J. Romanes, 1889, p. 248.
3. See Darwin to T. H. Huxley, 11 May 1880, and Darwin to the editor of *Nature*, 5 November 1880, in F. Darwin and Seward, 1903, I, 387, 389 for typical late expressions of the tension in Darwin's views.
4. Kellogg, 1907, p. 157, n. 4.
5. See H. Spencer, 1862–96, II (1864), 449; H. Spencer, 1887, p. 75; Cope, 1896, pp. 5–7, 476; and Packard, 1901, pp. 384–5.
6. 'The origin of species' (1860), in T. H. Huxley, 1893–4, II, 77. Cf. 'Obituary' (1888), in T. H. Huxley, 1893–4, II, 291.
7. Provine, 1971, pp. 14–24. See Olby, 1966, pp. 70–83; Cowan, 1972a, pp. 403ff; and Forrest, 1974, chaps. 8, 14.
8. Vorzimmer, 1972, pp. 112–14; Provine, 1971, p. 23.
9. 'Obituary' (1888), in T. H. Huxley, 1893–4, II, 292; Huxley to H. Spencer, 31 January 1886, in Duncan, 1911, p. 270.

10. 'Evolution in biology' (1878), in T. H. Huxley, 1893–4, II, 223; T. H. Huxley, 1887, p. 196; 'The Darwinian hypothesis' (1859), in T. H. Huxley, 1893–4, II, 19–20. Cf. Bartholomew, 1975, pp. 534–5. On Huxley's quest for proof, see T. H. Huxley, 1893–4, II, vi, 73–5, 463–4 and T. H. Huxley, 1887, p. 198. See also Poulton, 1896, chap. 18 and Poulton, 1908.

11. Darwin to G. Bentham, 22 April 1868, in F. Darwin and Seward, 1903, II, 371; Darwin, 1875, II, 370. Pangenesis was neither conceived to meet the criticisms of Jenkin, as has sometimes been assumed (see, for example, Eiseley, 1961, pp. 216–17; cf. Vorzimmer, 1972, pp. 120–1), nor was it otherwise connected with the elaboration or defence of the theory of natural selection. In fact it was 'never more than a speculation designed to explain a number of empirical observations' (Ghiselin, 1969, p. 183). On Darwin's growing insecurities about pangenesis, see the correspondence in F. Darwin, 1887, III, 72–5, 78–84, and p. 195, where Darwin admits that 'its life is always in jeopardy'. On Darwin's place in the history of the idea, see Zirkle, 1935, 1936, and 1946.

12. Galton, 1908, pp. 296–8; Galton, 1876, p. 345. See Vorzimmer, 1972, pp. 254–61; Cowan, 1972a; and Forrest, 1974, chap. 8.

13. G. J. Romanes, 1892–7, II, 40, 42, 137, 156; G. J. Romanes to E. B. Poulton, 11 November 1889, in E. Romanes, 1896, pp. 223–5.

14. G. J. Romanes, 1892–7, II, 9.

15. *Ibid.*, I, 374–6.

16. Darwin, 1959, pp. 747 (XIV:*183:f*); letters on isolation, 1868–78, in F. Darwin, 1887, III, 157–62; G. J. Romanes, 1892–7, III, 101–11. See Vorzimmer, 1972, chap. 7.

17. G. J. Romanes, 1886; Lesch, 1975, pp. 486–9.

18. Lesch, 1975, pp. 489–97.

19. G. J. Romanes, 1892–7, I, 374. See E. Romanes, 1896, pp. 209–17.

20. Wallace, 1889, pp. viii, 444.

21. Wallace, 1891, pp. 34–90, 338–94; Wallace, 1889, chap. 10; Vorzimmer, 1972, chap. 8. Further, Wallace agreed with Huxley and Galton in questioning the smallness and rarity of favourable variations and with the latter in rejecting Darwin's hypothesis of pangenesis. See A. R. Wallace to Darwin, 12 July 1871, in Marchant, 1916, I, 267 and Wallace, 1905, II, 21–2. On the background to Wallace's thought, see McKinney, 1972.

22. Wallace, 1889, pp. 142–50.

23. *Ibid.*, pp. 137, 141, 183, 437. See also Lesch, 1975, pp. 490–1.

24. 'Mimicry, and other protective resemblances among animals' (1867), in Wallace, 1891, p. 35.

25. See Wallace, 1889, p. viii.

26. G. J. Romanes, 1892–7, II, 181–2. See the response in Wallace, 1900, II, 379, 392.

27. 'On heredity' (1883), in Weismann, 1891–2, I, 78; see also pp. 81, 101.

28. *Ibid.*, I, 105.

29. 'The significance of sexual reproduction in the theory of natural selection' (1886), in Weismann, 1891–2, I, 277ff.

30. 'The continuity of the germ plasm as the foundation of a theory of heredity' (1885), in Weismann, 1891–2, I, 176; 'On heredity' (1883), in Weismann, 1891–2, I, 81.

31. G. J. Romanes, 1892–7, II, 51.

32. G. J. Romanes, 1893, p. 107; cf. p. 170.

33. On Galton, see Darwin, 1874, pp. 617–18; Darwin to W. R. Greg, 31

December 1878, in Carroll, 1976, letter 557; Darwin to F. Galton, 3 December [1869], in Galton, 1908, p. 290; letters from Darwin to Galton, 1873–5, in F. Darwin and Seward, 1903, I, 360–2; II, 43–4; and Cowan, 1972b. On Romanes, see Lesch, 1975, pp. 497–500; Huxley to J. D. Hooker, 9 March 1888, in L. Huxley, 1903, III, 62; and T. H. Huxley, 1893–4, II, 288ff. On Weismann, see Darwin's 'Prefatory notice', in Weismann, 1882, I, vi; letters from Darwin to R. Meldola, 1877–82, in Poulton, 1896, pp. 205–10, 217; Darwin to A. Weismann, 22 October 1868, in F. Darwin and Seward, 1903, I, 311; E. Romanes, 1896, p. 347; and Duncan, 1911, p. 356.

34. Wallace, 1905, II, 22; Darwin, 1874, p. 126.
35. 'The development of the human races under the law of natural selection' (1864), in Wallace, 1891, pp. 183–5.
36. Ibid., pp. 179, 185.
37. Kottler, 1974, pp. 162ff; R. Smith, 1972, pp. 190–1.
38. Darwin to Wallace, 14 April 1869, in Marchant, 1916, I, 243.
39. 'The limits of natural selection as applied to man' (1870), in Wallace, 1891, pp. 202, 203.
40. R. Smith, 1972, pp. 184–6.
41. 'Limits of natural selection' (1870), in Wallace, 1891, pp. 187, 204–5.
42. Wallace, 1889, pp. 463–4, 474–7.
43. Wallace, 1905, II, 17.
44. G. J. Romanes, 1889, p. 245; G. J. Romanes, 1890, p. 831; G. J. Romanes, 1888, pp. 2, 3n.
45. G. J. Romanes, 1888, pp. 16–20, 395.
46. Ibid., p. 12. Romanes here alludes to 2 Timothy 1:10.
47. G. J. Romanes, 1888, pp. 389, 430.
48. Ibid., p. 432.
49. G. J. Romanes, 1890, p. 831.
50. Darwin to Wallace, 12 July 1881, in Marchant, 1916, I, 319.
51. See Carroll, 1976, letters 488, 495, 513, 514, 548, 624 and Darwin to Romanes, 4 June 1876, in E. Romanes, 1896, pp. 60–1.
52. Romanes to Wallace, 1880, in Wallace, 1905, II, 313; Romanes to Darwin, 22 April 1880, in E. Romanes, 1896, p. 97.
53. E. Romanes, 1896, p. 48. See G. J. Romanes, 1895b, p. 109.
54. Wallace to Romanes, 18 July 1890, in Wallace, 1905, II, 318. See Kottler, 1974, pp. 180–2.
55. Wallace, 1905, II, 326. Cf. E. Romanes, 1896, p. 90.

Chapter 9

1. Dewey, 1910, pp. 2, 3.
2. Boller, 1969, chap. 2. Ross, 1977 shows how another anti-Darwinian, Philip Gosse, has been badly misrepresented.
3. A. D. White, 1896, I, 70.
4. Losee, 1972, pp. 64–7.
5. Herschel, 1830, pp. 200, 204; Mill, 1872, II, 12; Whewell, 1967, pt I, bk I, chaps. 2–4. See Losee, 1972, pp. 127, 154; Hull, 1973a; Hull, 1973b, pp. 16–28; Ellegård, 1957a, pp. 365–71; Letwin, 1965, pp. 272–80; Letwin, 1973, pp. 312–18.
6. Darwin, 1859, p. 482.
7. Darwin, 1977, II, 89–137; Darwin to T. Jamieson, 6 September 1861, in

De Beer, 1959, p. 38; Darwin, 1958, p. 84; 'The parallel roads of Glen Roy' (1876), in Tyndall, 1899, I, 205–28; Barrett, 1973; Hull, 1973b, pp. 25–6; Rudwick, 1974, a superlative study.

8. De Beer et al., 1960–1, III, 142 (D, 117); Darwin to J. D. Hooker, 23 April 1861, in F. Darwin, 1887, II, 362. Cf. Darwin to unidentified recipient, 14 March 1861, in De Beer, 1958, p. 113 and Darwin to G. Bentham, 22 May 1863, in F. Darwin, 1887, III, 25.

9. Ellegård, 1958a, pp. 189, 191. Cf. Ellegård, 1957a, pp. 364–5, 382ff.

10. M. C. F. Morris, 1897, p. 213. Cf. Mullens and Swann, 1917, pp. 415–18 and D. E. Allen, 1976, p. 178.

11. F. O. Morris, 1877, p. 4. It was this very passage or one much like it (there were many) which prompted Darwin to write to his friend, the former vicar of Down, the Reverend J. Brodie Innes (27 November 1878): 'If you were to read a little pamphlet which I [received] a couple of days ago by a clergyman, you would laugh & admit that I had some excuse for bitterness; after abusing me for 2 or 3 pages in language sufficiently plain and emphatic to have satisfied any reasonable man, he sums up by saying that he has vainly searched the English language to find terms to express his contempt of me & all Darwinians' (Stecher, 1961, p. 244; cf. F. Darwin, 1887, II, 288–9).

12. M. C. F. Morris, 1897, p. 216. Darwin deplored cruelty to animals but he defended vivisection undertaken in the interests of physiological research. In 1875 he gave evidence to this effect before the Royal Commission on Vivisection and thus no doubt incurred a double portion of Morris' wrath. See F. Darwin, 1887, III, 199–210; F. Darwin and Seward, 1903, II, 435–41; and French, 1975, 70ff, 364–5.

13. F. O. Morris, 1869, pp. viii, 46–7, 56. Cf. M. C. F. Morris, 1897, pp. 217–18.

14. Loewenberg, 1935, p. 244. See Bacon, 1929.

15. W. S. Tyler, 1895, p. 171; Le Duc, 1946, pp. 83–4. On Hitchcock, see Guralnick, 1972. Cf. Conrad Wright, 1941.

16. P. Smith, 1935, p. 398; W. S. Tyler, 1895, p. 171.

17. E. F. Burr, 1873, pp. 9, 10–18, 21, 25.

18. E. F. Burr, 1870, pp. 257, 262, 263–4.

19. Ibid., p. 266.

20. Knudson, 1936; Townsend, 1883, p. 3.

21. Townsend, 1883, pp. 24–5, 30, 269–80.

22. Townsend, 1896, pp. 48–9, 73–5.

23. Darwin to Hooker, 15 January 1861, in F. Darwin, 1887, II, 358; Darwin to J. S. Henslow, 26 October 1860, in Barlow, 1967, p. 213.

24. Darwin–Wallace letters, 27 July to 4 August 1872, in Marchant, 1916, I, 271–3; Darwin to the editor of Nature, 3 August 1872, in Darwin, 1977, II, 168. The review and letters appear in the issues of 25 July and 1 and 8 August 1872. Cf. Darwin to H. T. Stainton, 28 September 1881, in De Beer, 1958, p. 109.

25. Bree, 1872, pp. 11–13, 95, 98–9, 125, 162, 185.

26. Chadwick, 1966–70, I, 441.

27. G. B. Smith, 1908. The liberals might have reflected that barely five years had passed since Birks was ostracised by fellow-evangelicals for accepting a doctrine of eternal punishment not unlike that which in 1853 cost Maurice his professorship in King's College, London. See Rowell, 1974, pp. 123–9.

28. Birks, 1872, pp. 209–10. See H. Spencer, 1862–96, I (5th edn, 1884), 580–6 and Spencer to E. L. Youmans, 7 October 1876, in Duncan, 1911, p. 188.

29. Birks, 1872, pp. 227, 230–1, 240–1.
30. Birks, 1882, pp. 272, 275–85, 289.
31. *Ibid.*, pp. 295, 296, 309–10. See Darwin, 1959, p. 165 (IV:*14.6–9:c*).
32. From its earliest days Darwinism has engaged the dialectical skills of the legal profession. Charles Lyell and St George Mivart, both trained at Lincoln's Inn, were among the first to contest Darwin's case for natural selection. Robert Anderson, a Plymouth Brother and well-known criminal lawyer, dismissed evolution altogether in *A Doubter's Doubts about Science and Religion*, a slim volume which appeared anonymously in 1889, the year after Anderson's appointment as head of Scotland Yard. In the United States the Fundamentalist movement retained two well-known advocates of anti-Darwinism, Philip Mauro, whose *Evolution at the Bar* appeared in 1922, and William Jennings Bryan, whose last litigation was the prosecution of John T. Scopes in 1925 for teaching evolution in a public school. As late as 1971 another American legal critic of Darwinism, Norman Macbeth, produced a bill of indictment under the title *Darwin Retried*.
33. Fish, 1930; Curtis, 1887, p. 43.
34. Curtis, 1887, p. ix.
35. *Ibid.*, pp. 16–17.
36. *Ibid.*, pp. 93, 94, 101.
37. *Ibid.*, p. 20.
38. Sandeen, 1962, p. 310. Cf. Vernon, 1921, p. 203. The title posed a question which Hodge thought had been evaded in an extempore debate during the General Conference of the Evangelical Alliance, held at New York City in the autumn of 1873. For the remarks of Hodge *inter alia* on that occasion, see 'Discussion on Darwinism and the doctrine of development', 1874.
39. C. Hodge, 1874, pp. 11–12, 26, 27.
40. *Ibid.*, pp. 144–5, 160–1; C. Hodge, 1872–3, I, 10, 11, 573. See D. F. Johnson, 1968, p. 46 and Illick, 1960. For critiques of Hodge, see Hicks, 1883, pp. 309–25; F. H. Foster, 1939, chap. 3; and Pfeifer, 1957, pp. 131, 144ff.
41. Collard, 1942; C. F. O'Brien, 1971, p. 30; Levere and Jarrell, pp. 1–24, 109–10; Roome, 1976, pp. 189–91.
42. C. F. O'Brien, 1971, p. 61. See also *ibid.*, chap. 6 and C. F. O'Brien, 1970.
43. 'Discussion on Darwinism and the doctrine of development', 1874, p. 320 (emphasis in original); Dawson, 1880*b*, pp. 317, 329; Dawson, 1890, pp. 13, 65, 67, 119. Dawson agreed with Hodge that Darwinism was 'practically atheistic' (Dawson, 1880*b*, p. 348; cf. C. Hodge, 1874, pp. 119–25, 155ff).
44. Hull, 1973*b*, pp. 22–3, 68ff. However, according to Glass, 1959*a*, p. 37, Bacon himself believed in the mutability of species.
45. Lovejoy, 1960, pp. 227–8; Eiseley, 1961, pp. 23–6; Hull, 1967, pp. 309–12; Glass, 1959*a*, pp. 30–6; Glass, 1959*b*, pp. 144–51; Goudge, 1973, pp. 174–5; Deely and Nogar, 1973, pp. 29–62.
46. See Farber, 1972.
47. Hull, 1973*b*, p. 326.
48. Cuvier carefully guarded his statements about creation. He did not seek their confirmation in the Bible. See Von Hofsten, 1936, p. 92 and W. Coleman, 1964, pp. 170ff.
49. W. Coleman, 1964, chap. 2, pp. 170–86; Greene, 1959*b*, pp. 123–5, 169–73; E. S. Russell, 1916, chaps. 3, 5–7. On Cuvier's and Lamarck's philosophies of science, see Hull, 1967, pp. 326–7.
50. Lurie, 1960, pp. 62, 63. See 'Agassiz, Darwin, and evolution' (1959), in Mayr, 1976, pp. 251–60.

NOTES TO PAGES 207–13

51. For Agassiz's opposition to Darwin's philosophy of science, see his 'Evolution and permanence of type' (1874), in Hull, 1973*b*, pp. 434, 445.
52. Agassiz, 1866, p. 3. See Bowler, 1976*a*, pp. 47–53.
53. Agassiz, 1866, p. 117. See 'Agassiz, Darwin, and evolution' (1959), in Mayr, 1976, *passim.*
54. Agassiz, 1962, p. 9. Cf. Gode-von Aesch, 1941, chaps. 6, 8.
55. Lurie, 1960, pp. 61–2, 264–5. See Haller, 1975, pp. 76–8, 84–6.
56. Lurie, 1960, p. 100. As Lurie masterfully portrays him, Agassiz seems far too impressed with his own judgement to take very seriously the claims of religious authority. 'Never faithful in church attendance, he had made a comfortable adjustment to the Boston Unitarianism of his wife's family which allowed him the latitude he demanded of religion. His New England friends admired his courage in debating against dogmatic interpretations of the Scriptures, even though some of them were made uncomfortable by the implications of his argument for the equality of man. But this was Agassiz speaking, and Agassiz was a symbol of scientific authority' (p. 262).
57. Darwin to unidentified recipient, 14 March 1861, in De Beer, 1958, p. 113; Agassiz, quoted in Lurie, 1960, p. 297; Darwin to A. Gray, 11 August 1860, in F. Darwin, 1887, II, 333.
58. Darwin, 1859, pp. 485, 486.
59. Birks, 1882, pp. 242, 278–85, 308. For Beale's explicitly religious theory of vitality, see L. S. Beale, 1871 and Geison, 1969, pp. 285–90.
60. E. F. Burr, 1883, pp. 166, 167, 169. On Agassiz, see also E. F. Burr, 1873, pp. 178ff.
61. Bree, 1872, pp. 2, 6, 76, 333, 363, chap. 27.
62. Curtis, 1887, pp. xviii, xix, 114–18.
63. Townsend, 1869, chap. 4; Townsend, 1881, pp. 26–7.
64. C. Hodge, 1874, pp. 132, 141, 145, 152ff, 161; C. Hodge, 1872–3, I, 222. Cf. *ibid.*, II, 78ff. The semi-centenary of Hodge's professorship was held at Princeton on 24 April 1872. Following the customary tributes and an address by the Reverend Joseph T. Duryea, 'The title of theology to rank as a science', Hodge rose and responded to the honours and affection showered on him that day.

> The law of the fixedness and transmissibility of types pervades all the works of God. . . . The same law controls the life of institutions. What they are during their forming period, they continue to be. This is the reason why this Institution owes its character to Dr Alexander and Dr Miller. . . . Drs Alexander and Miller were not speculative men. They were not given to new methods or new theories. They were content with the faith once delivered to the saints. I am not afraid to say that a new idea never originated in this Seminary [A. A. Hodge, 1880, pp. 519, 521].

After Hodge's death in 1878, however, fixity no more prevailed at Princeton than in natural history. All three of Hodge's successors in the chair of theology – his son Archibald Alexander Hodge, Benjamin B. Warfield, and his grandson Caspar Wistar Hodge, Jr – were 'willing to accept as compatible with Christian theology a theory of evolution which was properly limited' (D. F. Johnson, 1968, p. 284). Little did Charles Hodge realise the transmutation that was in store.
65. See Bowler, 1976*a*, pp. 53, 79–83.
66. Dawson, 1880*a*, pp. 345, 349, 376. For the influence of Agassiz, see *ibid.*, pp. 82ff, 246–7, 350–1.
67. Dawson, 1890, pp. 16, 40, 93, 103–4, 218–19, 226.

68. Ellegård, 1958a, p. 203. Cf. Eiseley, 1961, pp. 95–7.
69. See Ellegård, 1958a, pp. 296–7, 332ff; Roberts, 1936, p. 117; and Furniss, 1954, p. 19.
70. Dewey, 1910, pp. 2, 3. Dewey's own rationale for this statement is obviously unacceptable.
71. A critical reading of Dewey, 1929, chaps. 2–3 further illumines these points.
72. Gode-von Aesch, 1941, p. 205; cf. pp. 36, 197. See also Ritterbush, 1972 and Dewey, 1929, chap. 11.
73. See Gilkey, 1965, chaps. 1–2, 5. On the effects of realistic and idealistic philosophies in nineteenth-century theology, see Webb, 1933; Henry, 1951; and Ahlstrom, 1955.
74. See Numbers, 1977, chaps. 7–8.
75. Dawson, 1890, p. 182.

Chapter 10

1. 'Science and pseudo-science' (1887), in T. H. Huxley, 1893–4, V, 115–16.
2. Dillenberger, 1960, p. 224.
3. Atkins, 1932, p. 124.
4. Gillispie, 1959a, p. 223.
5. Ellegård, 1958a, pp. 332–3.
6. Per contra, see Kennedy, 1957, p. vii; Dillenberger, 1960, p. 223; Lack, 1961, p. 80; Kent, 1966, p. 11; Holifield, 1972, pp. 17ff; and Brooke and Richardson, 1974, p. 109.
7. These judgements are supported by Gillispie, 1959a, p. 220 and Ellegård, 1958a, chap. 6, pp. 332–3 as well as by Raven, 1951, pp. 142ff; Pfeifer, 1957, chap. 3, p. 182; M. McGiffert, 1958, pp. 176, 199ff; Persons, 1959, p. 9; Cannon, 1961, pp. 118–27; R. M. Young, 1970a, pp. 21–7; Cupitt, 1975, pp. 125–8; and Numbers, 1977, p. 118.
8. Before 1872 this approach was 'probably the majority opinion among both scientists and the general public' (Ellegård, 1958a, p. 136; cf. pp. 270–4).
9. Temple, 1885, pp. 113–14, 167. See Sandford, 1906, I, 583 and McPheeters, 1948, pp. 353ff, 395ff.
10. McLoughlin, 1970, pp. 4, 39. See Duncan, 1911, p. 100; Beecher, quoted in E. L. Youmans to H. Spencer, November 1864, in Fiske, 1894, p. 201n.; Beecher, quoted in Hibben, 1942, pp. 301–2; and 'Mr Beecher's remarks', in Youmans, 1887.
11. Beecher, 1885, pp. 113, 115.
12. Henslow, 1895, p. 1; Henslow, 1888, pp. vii, xi, 335. Henslow was the son of Darwin's old professor, John Stevens Henslow, and the brother-in-law of J. D. Hooker. See 'Rev. George Henslow', 1926; L. Huxley, 1918, I, 374, 452; Barlow, 1967, p. 193; and 'Fertilisation of plants' (1877), in Darwin, 1977, II, 191.
13. Henslow, 1888, p. vii; Henslow, 1908, p. vii; Henslow, 1904, pp. 53, 56–7, 147, 161ff, 174, chap. 8. See the Romanes–Henslow correspondence in E. Romanes, 1896, pp. 327–41; G. J. Romanes, 1892–7, II, 20; and Wallace, 1900, I, 305–14.
14. G. Peel, 1909 and Geikie, 1909. For Owen's influence, see G. D. Campbell, 1868, pp. 30–1, 197–8; I. Campbell, 1906, I, 411, 573, 581; and Bowler, 1977a, pp. 37–9.
15. G. D. Campbell, 1898, pp. 44ff, 64, 149, 155, 162. See also G. D. Campbell,

1884, pp. 262ff. Cf. 'Creation by law' (1867), in Wallace, 1891, pp. 141–66 with the Duke's reply in G. D. Campbell, 1868, pp. 393-7.

16. Mivart, 1871, pp. 277, 314, n. 2.

17. Mivart, 1876, pp. 274–5. Cf. Bowler, 1977a, pp. 39–40.

18. Mivart, 1876, pp. 276, 277. Cf. Mivart, 1871, pp. 207–8, 248–9, 257ff.

19. Bascom, 1871, pp. 228–9. On Bascom's career and influence, see Bascom, 1913; Robinson, 1922; Bates, 1929; May, 1949, p. 146; Veysey, 1965, pp. 217–20; and Dorn, 1966, p. 26.

20. Bascom, 1880, pp. 128, 145; Bascom, 1897, pp. 10–11.

21. Bascom, 1897, pp. 13, 16; Bascom, 1880, pp. 133, 144–5, 168.

22. G. A. Smith, 1899, pp. 30–2, chap. 4; Napier, 1901, pp. 17–18, chaps. 8, 18; J. Y. Simpson, 1901, pp. 30–1; McIver, 1958; Schott, 1972. See Spencer to Youmans, 17 May 1883, in Duncan, 1911, p. 232.

23. Drummond, 1894, pp. 15, 16, 44–5, 414, 418, 429, 435, 436. See Spencer to E. L. Linton, 6 June and 3 September 1894, in Layard, 1901, pp. 310, 312 and Duncan, 1911, p. 363.

24. Le Conte, 1891, pp. 8, 9–29. See Le Conte, 1903; Pfeifer, 1965, pp. 161ff; Haller, 1975, pp. 154–66; and Stephens, 1976.

25. Le Conte, 1891, pp. 258ff, 301.

26. Le Conte, 1903, pp. 288–90. Romanes seems to have been one of those touched by the book. See G. J. Romanes to J. Le Conte, 7 May 1888, in E. Romanes, 1896, pp. 234–5 and G. J. Romanes, 1892–7, I, 412n.

27. *The Churchman*, 64 (3 October 1891), 423. See MacQueary, n.d.; 'The Sentence of Howard MacQueary', 1891; 'End of the MacQueary Case', 1891; and 'MacQueary, Thomas Howard', 1942. On other heresy trials of the period, see Shriver, 1966; Eaton, 1962; and Gustafson, 1964.

28. MacQueary, 1891a, pp. x, 16, 229–30. MacQueary dedicated his second book to another liberal Episcopalian, Andrew Dickson White, who had supported him in his ecclesiastical litigation. See MacQueary, 1891b and White's letter of 12 January 1891, reprinted in the preface to MacQueary, n.d.

29. I. V. Brown, 1953, p. 123.

30. Abbott, 1892, pp. 1, 3; Abbott, 1897, pp. iii, 6, 7, 19, 20, 176. See also Abbott, 1915, pp. 449–50, 458.

31. Abbott, 1892, pp. 8–9; Abbott, 1897, p. 96. Cf. McCrossin, 1970, pp. 133–4 and H. S. Smith, 1955, p. 178.

32. Abbott, 1897, pp. 10, 15; Abbott, 1892, pp. 246–7.

33. Schneider, 1945, p. 11; F. H. Johnson, 1891, pp. 188–9 (cf. pp. 51ff). See D. D. Williams, 1941, pp. 63ff; 'Johnson, Francis Howe', 1942; Neel, 1942, chap. 4; and Persons, 1950a, pp. 447ff.

34. F. H. Johnson, 1891, pp. 260, 262, 264, 265, 272, 276.

35. *Ibid.*, pp. 282–3.

36. *Ibid.*, pp. 292, 312.

37. *Ibid.*, p. 493.

38. Macmillan, 1907; J. C. Tyler, 1960; Matheson, 1885, p. 184.

39. Matheson, 1885, pp. 79, 80, 86–7, 91, 157–8, 169, 171, 172, 191. See H. Spencer, 1862–96, II (1864), chap. 10 and Spencer to Youmans, March 1885, in Duncan, 1911, p. 252.

40. Savage, 1892b, p. 11. See Savage, 1887b; Lyttle, 1935; and the recollection of Henry Wilder Foote in a comment on Schlesinger, 1967 as it was first published in the Massachusetts Historical Society's *Proceedings*, 64 (1932), 547.

41. Savage, 1892*b*, p. 26; Savage, 1876, p. 59. On Spencer, see Spencer to Youmans, 8 September 1880, in Duncan, 1911, p. 212; Spencer to Savage, 9 January 1883, in Goblet d'Alviella, 1886, p. 220; Savage, 1886, p. 43; and Savage, 1887*a*, p. 13. Savage, 1880 is dedicated 'by permission to Herbert Spencer and his friend, John Fiske'.

42. Savage, 1876, p. 47; Savage, 1880, p. 166; Savage, 1886, pp. 32-4. See Persons, 1950*a*, pp. 433-6 and McCrossin, 1970, pp. 117ff.

43. From 'One law', in Savage, 1884, p. 68 (emphasis in original).

44. 'Mr Fiske's speech', in Youmans, 1887, p. 55. See Duncan, 1911, pp. 156ff; Pannill, 1957; Higgins, 1960; and Berman, 1961.

45. 'Darwinism verified' (1876), in Fiske, 1885*a*, pp. 2, 3; Fiske, 1874, II, chaps. 10-12; Fiske, 1885*b*, pp. 129-30, 150-1.

46. Cf. similar conclusions in Ellegård, 1958*a*, pp. 269-79; Neel, 1942, pp. 266ff; Overman, 1967, pp. 101-10; Boller, 1969, pp. 20-1; and Pfeifer, 1974, pp. 191-2.

47. G. D. Campbell, 1869, pp. 52ff, 71, 73-4, 128; G. D. Campbell, 1884, pp. 274, 315, 418 (cf. pp. 532ff). See Gillespie, 1977, pp. 46-7, 49-50.

48. Mivart, 1871, pp. 324, 331; Mivart, 1873, pp. 188-92; Mivart, 1889.

49. Temple, 1885, pp. 172-6.

50. *Ibid.*, pp. 177-9, 186-8.

51. Henslow, 1873, pp. 107ff, 123, 173n., 178-80; Henslow, 1871, pp. 20-1; Henslow, 1904, pp. 218-20, 360-1.

52. Beecher, 1885, pp. 26, 27, 428.

53. *Ibid.*, p. 429.

54. *Ibid.*, p. 81; cf. p. 323. Beecher's colleague, Lyman Abbott, was no more definite about human evolution. See Abbott, 1892, pp. 215, 219; Abbott, 1897, pp. 33ff, 41; and Abbott, 1915, pp. 459-60.

55. Bascom, 1871, p. 236; Bascom, 1880, p. 200.

56. Fiske, 1884*a*, p. 54. See the cool response in Darwin to J. Fiske, 9 November 1871, in F. Darwin and Seward, 1903, I, 334.

57. Fiske, 1901, p. 85; Fiske, 1884*a*, pp. 56, 117 (cf. pp. 43, 65, 110); Fiske, 1899*b*, p. 82.

58. E.g., see Le Conte, 1897, p. 77; Le Conte, 1891, pp. 302, 318-19, 330, 364; F. H. Johnson, 1891, pp. 448ff, 479; and Savage, 1876, pp. 81-92, 245, 252. Cf. the conclusions in Ellegård, 1958*a*, pp. 30-3, 311-29 and Dillenberger, 1960, p. 223.

59. 'Evolution and theology' (1898), in Pfleiderer, 1900, p. 8.

60. Beecher, 1885, pp. 3-4, 125-6; Drummond, 1894, pp. 15, 279 and *passim*; Matheson, 1885, p. 131; Savage, 1887*a*, p. 5; Fiske, 1885*b*, pp. 129-30; 'The doctrine of evolution: its scope and purport' (1891), in Fiske, 1899*a*, pp. 39-40.

61. Temple, 1885, pp. 107, 113, 122, 127, 130, 188; Henslow, 1873, pp. viii, 28-9, 124-5, 129, 156-7; Mivart, 1876, pp. 359, 361.

62. Bascom, 1880, p. 125; Bascom, 1897, p. 180; Le Conte, 1891, pp. 3, 4, 65; Abbott, 1897, p. 15; Abbott, 1892, p. 2; F. H. Johnson, 1891, pp. 366ff, 379ff, 409, 435; F. H. Johnson, 1911, p. 120.

63. Burrow, 1970, p. 99. See Bury, 1955, pp. 335, 345-6; Baillie, 1950, pp. 153-4; Buckley, 1966, pp. 38ff; and R. M. Young, 1970*a*, pp. 27ff.

64. Beecher, 1885, p. 429; Drummond, 1894, pp. 37-8, 262, 266; Matheson, 1885, pp. 240, 241, 242; Savage, 1876, pp. 64-8; 'Mr Fiske's speech', in Youmans, 1887, p. 223.

65. Temple, 1885, pp. 117-18, 146; Henslow, 1873, pp. 156-7, 213, 214 (cf.

Henslow, 1904, chap. 10); Bascom, 1897, pp. 131, 181, 187–8; Le Conte, 1891, pp. 365–75; Le Conte, 1897, p. 73; F. H. Johnson, 1891, pp. 350–1; F. H. Johnson, 1911, pp. 112, 217. Cf. the analyses in Lovejoy, 1909; Benz, 1967, chaps. 8–9; and Wagar, 1972, chap. 6. Thus if providence were rightly conceived it need not have been 'incongruous' with a belief in progress (Bury, 1955, p. 21), though the tendency in such definitions has ever been 'towards rejecting religion' in favour of a belief in those'.immanent "laws" ' by which providence is supposed to proceed (Passmore, 1972, p. 238).

66. 'Death of Dr J. S. Van Dyke', 1915.
67. Van Dyke, 1886, pp. xvii–xix.
68. *Ibid.*, pp. xvii–xviii.
69. *Ibid.*, p. xxi. Hodge's *imprimatur* was not accompanied by the *nihil obstat.* See *ibid.*, p. xxii and D. F. Johnson, 1968, chap. 4.
70. Van Dyke, 1886, p. xi.
71. *Ibid.*, pp. 24, 29, 33.
72. *Ibid.*, pp. 33–4.
73. *Ibid.*, pp. 27, 31.
74. *Ibid.*, p. 41.
75. *Ibid.*, pp. 44–5, 47.
76. *Ibid.*, pp. 51, 71.
77. *Ibid.*, pp. 77, 98, 112.
78. McCosh, 1896, p. 82. See Leslie, 1974.
79. See Veysey, 1965, chap. 1.
80. McCosh, 1896, p. 233.
81. McCosh, 1890, pp. viii–x. See Illick, 1960, pp. 234–43. In 1878 McCosh sought to persuade J. W. Dawson, the anti-Darwinian president of McGill University, to succeed Arnold Guyot in the chairs of physical geography and geology at the College. On this account C. F. O'Brien, 1971, p. 21 refers to McCosh's 'active connivance' with the board of trustees to erect an 'anti-Darwinian citadel' at Princeton. But O'Brien himself provides evidence which would modify this judgement: first, the fact that Charles Hodge, a dogmatic anti-Darwinian, was the chairman of the board and thus, in one sense, McCosh's employer; second, the striking similarity of thought and wording in the letters which Hodge and McCosh individually sent to Dawson, thus suggesting that McCosh's letter was not altogether private; and third, the telling fact that Dawson's reply went to Hodge, not to McCosh, from whom the invitation first had been received (see the letters, quoted in *ibid.*, pp. 20–1).

　　O'Brien also provides evidence to explain McCosh's inability to hire E. D. Cope in 1873 as Princeton's first professor of natural history and Joseph Le Conte in 1874 as professor of geology. He points out that there was a great desire to maintain the College as a thoroughly orthodox Presbyterian institution. 'All its Trustees are Presbyterians', Hodge boasted to Dawson (6 April 1878). 'All its Presidents have been Presbyterians' (*ibid.*, p. 21). After Cope was refused the post he was not therefore mistaken in saying that McCosh 'objected to my evolution sentiments, for those views are much condemned at Princeton' (E. D. Cope to A. Cope, 16 March 1873, in Reingold, 1964, p. 245; cf. Cope's letter of 8 August 1875, in Osborn, 1931, p. 537). It was doubtless *his* speculative sentiments regarding evolution, particularly as affected by liberal Quaker beliefs, and not evolution *per se*, which McCosh found objectionable. Likewise in the case of Le Conte, though a recommendation came from none other than

Asa Gray, he was not employed, and this due perhaps to his growing liberalism in theology (see A. Gray to J. McCosh, 16 April 1874, quoted in M. McGiffert, 1958, p. 16, n. 1 and Le Conte, 1903, pp. 17, 265).

82. Kramer, 1948, p. 235; Hicks, 1883, p. 283.

83. See McCosh and Dickie, 1857, pp. 432-6, 462, 474-8. On the 'new natural theology' which Owen inspired, see Kramer, 1948, pp. 234-88 and Bowler, 1976a, p. 99.

84. Quoted in McCosh, 1896, pp. 123, 124. Cf. Gerstner, 1945, p. 124. Abbott, 1922, p. 98 ascribes the general acceptance of evolution in America to the influence of McCosh, Beecher, and Fiske.

85. McCosh, 1871, p. 346; McCosh, 1883, pp. 12ff, 22ff. Cf. McCosh, 1885 with Spencer's response, which refers to 'Dr McBosh' (Spencer to Youmans, 4 May 1885, in Duncan, 1911, p. 246). See Cashdollar, 1978, pp. 76-7.

86. McCosh, 1871, pp. 41, 42, 64, 346, 349, 350.

87. McCosh, 1883, p. 22; McCosh, 1890, pp. 16-18; King, quoted in Wilkins, 1958, p. 211. In writing the *Religious Aspect of Evolution* McCosh relied on works by Dana, Le Conte, Geikie, Dawson, Cope, Conn, and Wallace (McCosh, 1890, p. x). Before publication he submitted it to the scrutiny of some of his former pupils who had become accomplished naturalists, among them Henry Fairfield Osborn, professor of comparative anatomy at Princeton, Cope's *protégé*, and later his biographer. (Osborn dedicated his well-known history of evolutionary thought, *From the Greeks to Darwin*, to McCosh in 1894, the year of McCosh's death. See also Osborn, 1926, p. 31.) See the evaluations in Le Conte to McCosh, 1 March 1888, in McCosh, 1896, p. 234; and in Dawson, 1890, p. 182.

88. McCosh, 1871, pp. 39, 42-3, 346, 348; McCosh, 1890, pp. 52, 55; McCosh, 1883, p. 36. Cf. Riley, 1923, pp. 204-5.

89. For McCosh's fairly traditional views on science and scripture, see McCosh, 1871, pp. 43-5; McCosh, 1890, chap. 6; Gerstner, 1945, pp. 168-76; and Osgood, 1951, p. 25.

90. McCosh, 1871, pp. 50-1 (cf. pp. 354, 361); McCosh, 1883, p. 40; McCosh, 1890, pp. 103-4.

91. McCosh, 1890, pp. 12, 18; McCosh, 1883, pp. 40-1.

92. McCosh, 1890, p. 7.

93. *Ibid.*, pp. 58ff. See Gerstner, 1945, p. 4.

94. McCosh, 1890, p. 110.

95. Gerstner, 1945, p. 188; Illick, 1960, p. 239; Persons, 1950a, p. 427; Ratner, 1936, p. 114; C. F. O'Brien, 1971, p. 19. O'Brien makes his case by linking McCosh's appeal to Dawson in 1878 (n. 81 above) with his publication in the same year of an anonymous 'anti-Darwinian article' in the *North American Review*. Actually, however, 'An advertisement for a new religion' by An Evolutionist (July 1878) was the first of a carefully planned and integrated series of four articles published by McCosh in the same journal over a period of three years. The other articles were: 'The confessions of an agnostic' by An Agnostic (September 1879); 'What morality have we left?' by A New Light Moralist (May 1881); and 'Religious conflicts of the age' by A Yankee Farmer (July 1881). Together the articles were republished, again anonymously, as *The Conflicts of the Age* (1881). McCosh intended the series, not as a covert confession of personal misgivings, but as a 'sly defense of religion' (McCosh, 1881, p. 64). The *dénouement* comes in the last article when the Evolutionist, the Agnostic, and the New Light Moralist are confronted by the Yankee Farmer, who has invited them to dinner at

his New England home. The farmer, speaking for McCosh, argues on behalf of orthodox theology and intuitionist philosophy. In conclusion he reveals the point of the preceding articles by expressing just the paternal concern for the spiritual and intellectual well-being of college students which was typical of Princeton's president: 'I mean to continue to pester these college youths who affect (there is a great deal of affection in the whole thing) to believe in nothing, while each one has a firm conviction that he is "somebody" of no mean importance.... I notice that when they marry and have several mouths to feed, they give up Nihilism' (*ibid.*, p. 90; cf. McCosh, 1896, pp. 231–5). Thus, rightly understood, none of the articles is inconsistent with McCosh's public pronouncements on evolution. To obviate misunderstandings such as O'Brien's the articles were published anonymously. As Pfeifer, 1957, pp. 141–2 suggests, McCosh wished to protect the reconciliation he had created from those who would charge him with inconsistency.

96. Gerstner, 1945, p. 154.

Chapter 11

1. Traill, 1900, pp. 342–3.
2. See Rice, 1867, p. 634.
3. Mackenzie, 1920, p. 56; Ewing, 1914, I, 56; 'Iverach, James', 1910.
4. Iverach, 1884c, p. 35.
5. *Ibid.*; Iverach, 1884a, chap. 4; Iverach, 1887; Iverach, 1894, chaps. 1–2; Iverach, 1900, chap. 9.
6. Iverach, 1894, pp. 5, 12, 32, 46–9.
7. Iverach, 1900, p. 41; Iverach, 1894, p. 68.
8. Iverach, 1894, p. 105. See T. H. Huxley, 1893–4, II, 84.
9. Iverach, 1894, pp. 86, 95, 103, 107, 109. In his later book Iverach was rather more sceptical. See Iverach, 1900, p. 76.
10. Iverach, 1894, pp. 112, 114–15, 118, 121.
11. *Ibid.*, p. 134. Cf. Iverach, 1900, p. 90.
12. Iverach, 1894, pp. 128, 129. Cf. T. H. Huxley, 1893–4, II, 223 with T. H. Huxley to G. J. Romanes, 5 January 1888, in L. Huxley, 1903, III, 57.
13. Iverach, 1900, p. 79, 80, 82. Cf. Darwin, 1874, p. 613 and Darwin, 1958, p. 92.
14. Iverach, 1900, pp. 103–4; Iverach, 1894, pp. 159, 166.
15. Iverach, 1894, pp. 161–2.
16. *Ibid.*, p. 172.
17. *Ibid.*, pp. 173, 174, 175.
18. *Ibid.*, pp. 175–6.
19. Iverach, 1884a, pp. 49, 185, 197–8.
20. Poulton, 1909, p. 17.
21. 'Memoirs', in A. L. Moore, 1890b, p. xv.
22. W. Lock, in *ibid.*, p. xxxv. See Blakiston, 1909.
23. E. S. Talbot, 'Memoirs', in A. L. Moore, 1890b, pp. xiii–xviii; Romanes, in *ibid.*, pp. xxviii–xxx. Moore's sudden and untimely death from influenza in January 1890 was a 'terrible' blow to Romanes. Not only was it 'a loss to Darwinism on its popular side' (Romanes to E. B. Poulton, 27 January 1890, in E. Romanes, 1896, p. 253); it was also, according to Romanes, the personal loss of 'one whose rich stores of knowledge and of thought had

just begun to open such large possibilities in the way of adding to my own' (Romanes to F. Paget, 18 January 1890, in Moore Autograph Collection, B/M/781, American Philosophical Society Library, Philadelphia).

24. Romanes, 'Memoirs', in A. L. Moore, 1890b, p. xxx. For the essay's influence on Romanes, see G. J. Romanes, 1895b, pp. 120–6 and E. Romanes, 1896, pp. 249–52.

25. A. L. Moore, 1889b, pp. 223, 224, 225, 226, 232.

26. Ibid., p. 99.

27. Ibid., pp. 87, 88, 105.

28. A. L. Moore, 1890b, pp. 202–17.

29. A. L. Moore, 1890a, pp. 60–1. Spencer did not relegate religion merely to the unknown but to the 'unknowable'.

30. A. L. Moore, 1889b, pp. 166, 167. The essay was commended by James Paget, W. H. Flower, Ray Lankester, and T. H. Huxley in letters to the author, now preserved in the Moore Autograph Collection (n. 23 above). Moore issued the essay under the title *Evolution and Christianity* (1889a) as one of the 'Oxford House Papers'. The tract adds to the discussion of the Fall and specifies books that Moore found 'especially helpful' (p. 2) in his research: *inter alia*, those of the Duke of Argyll, Joseph Le Conte, and Asa Gray.

31. A. L. Moore, 1889b, pp. 170, 172, 180, 184. Cf. Tennant, 1909, pp. 421–3. Among the few to note Moore's *tour de force* are Willey, 1961, p. 7 and Vidler, 1961, pp. 120–1.

32. A. L. Moore, 1889b, pp. 184–5. Cf. Tennant, 1909, pp. 435ff.

33. A. L. Moore, 1889b, pp. 193, 194, 195. Cf. Cupitt, 1975, pp. 129–30.

34. A. L. Moore, 1889b, pp. 189–93. See F. Darwin, 1887, I, 309, 312, 314; III, 189; T. H. Huxley, 1887, p. 201; T. H. Huxley, 1893–4, II, 82ff, 110ff; and A. Gray, 1963, p. 237.

35. A. L. Moore, 1889b, pp. 196, 197; cf. pp. 89–90, 234.

36. Ibid., pp. 163, 164.

37. A. L. Moore, 1890b, pp. 1, 3, 20, 21, 32, 35.

38. Ibid., pp. 18, 19.

39. A. L. Moore, 1889b, pp. 204, 206.

40. Ibid., pp. 92, 208–9.

41. Ibid., pp. 209–11. See Moore's essay 'Creation and creatianism' (A. L. Moore, 1890b, pp. 67–82), where the *infusio animae* is interpreted as the establishment of a *nova relatio* between human beings and their Creator which cannot be reduced to the terms of evolution.

42. A. L. Moore, 1889b, p. 208.

43. A. L. Moore, 1890b, pp. 38–9.

44. Ibid., pp. 45–7, 57, 59–60.

45. A. L. Moore, 1890a, pp. 58, 78, 98.

46. Ibid., pp. 99–100. A similar argument appears in Iverach, 1884a, chaps. 7–10.

47. See Darwin to A. Gray, 25 April 1855 and 5 September 1857, in F. Darwin, 1887, II, 120–5. Most of Gray's letters to Darwin before 1862 have apparently been destroyed. See J. L. Gray, 1893, II, 454, 456–8 and F. Darwin and Seward, 1903, I, 421–2, 428–33, 442–4.

48. On the American edition of the *Origin*, see Darwin to Gray, 21 December 1859 and 28 January 1860, in F. Darwin, 1887, II, 244–5, 269–70 and Gray to Darwin, 23 January 1860, in J. L. Gray, 1893, II, 456. On Gray's review, see Darwin to Gray, 18 February 1860, in F. Darwin, 1887, II, 286–7 and

Darwin to Gray, February 1860, in Loewenberg, 1973, letters 20, 22, 95, where Darwin proposes that the American edition be a 'joint publication' with Gray's review 'at the head'. On Gray's debates before the American Academy, see Dupree, 1959, chap. 15. On the pamphlet (A. Gray, 1861), see Darwin to J. D. Hooker, 11 December 1860, in F. Darwin, 1887, II, 355, 370–1 and the letters from Darwin to Gray, September 1860 to June 1861, in ibid., II, 371–4 and in Loewenberg, 1973, pp. 31–3, 41, 73, 75–6, 78, 90, 93. Darwin strongly urged that the pamphlet's title show its bearing on natural theology; he distributed no less than 100 copies to scientists, theologians, reviewers, and libraries (the list included Huxley, Hooker, Lyell, and Chambers; Henslow, Whewell, Kingsley, and Wilberforce); and he commended the pamphlet in the third edition (1861) of the *Origin* as 'an admirable, and, to a certain extent, favourable Review' (Darwin, 1959, p. 57).

49. Darwin to Gray, 22 July, 10 September, and 26 September 1860, in F. Darwin, 1887, II, 326, 338, 344. See also Darwin to Huxley, 20 July 1860, in F. Darwin and Seward, 1903, I, 157.

50. Dupree, 1959, pp. 6, 19–23, 37–8, 44–5, 120, 182.

51. J. L. Gray, 1893, I, 321. Cf. Dupree, 1959, pp. 136, 261, 359.

52. Dupree, 1959, pp. 136, 220–1. See M. McGiffert, 1958, pp. 21ff.

53. Gray to Darwin, 31 March 1862, in J. L. Gray, 1893, II, 479–80.

54. A. Gray, 1963, p. 88; cf. pp. 89–90, 98, 107, 207. See also A. Gray, 1880, pp. 61–2.

55. A. Gray, 1963, pp. 4, 11, 109, 118, 289.

56. Gray to C. L. Brace, 1861?, in J. L. Gray, 1893, II, 459. See Gray to E. Fry, 1 December 1882, in J. L. Gray, 1893, II, 740; A. Gray, 1963, p. 206; and Dupree, 1959, pp. 302–3, 364–5.

57. Darwin to Gray, 22 May 1860, in F. Darwin, 1887, II, 310.

58. A. Gray, 1963, pp. 43, 44, 119, 122, 124; cf. pp. 42, 71, 144. See also A. Gray, 1880, pp. 88–9. On Gray's analogy of the nebular hypothesis, see Numbers, 1977, pp. 113–17.

59. A. Gray, 1963, pp. 304, 308–11; A. Gray, 1880, pp. 68–9, 86. See M. McGiffert, 1958, pp. 220ff.

60. See Darwin to Gray, 22 May 1860, 26 November 1860, and 17 September [1861?], in F. Darwin, 1887, II, 312, 353, 378; Darwin to Gray, 3 July 1860 and 11 December 1861, in F. Darwin, 1887, I, 314–15 (Loewenberg, 1973, letter 27) and II, 381–2 (Loewenberg, 1973, letter 39); Darwin to Gray, 10 September 1860 and n.d., in Loewenberg, 1973, letters 30, 136. On Darwin's nose, the shape of which, he believed, could have been crucial for his entire scientific career, see Darwin, 1958, pp. 72, 76–7 and Hyman, 1967. 'Le nez de Cléopatre: s'il eust esté plus court, toute la face de la terre auroit changé' (Pascal, *Pensées*, i, 93).

61. A. Gray, 1963, pp. 49, 62.

62. *Ibid.*, pp. 121–2; cf. p. 125. See a similar analogy in Cope, 1887a, p. 16.

63. Darwin to Gray, 26 November 1860 and 5 June 1861, in F. Darwin, 1887, II, 353, 373. Cf. Wallace, 1891, pp. 149–53.

64. Darwin, 1875, II, 426–8; cf. p. 236. The metaphor was adumbrated in Darwin to Gray, 4 August 1863, in Loewenberg, 1973, letter 53 and in E. Darwin to P. Matthew, 21 November 1863, in De Beer, 1959, p. 41. It is referred to in Darwin to W. R. Greg, 31 December 1878, in Carroll, 1976, letter 557. On Darwin's difficulties with designed variation, see also Darwin to Gray, 17 September [1861?], in F. Darwin, 1887, II, 378; the letters from

Darwin to C. Lyell, who had taken up Gray's metaphor, June 1860 to August 1861, in F. Darwin and Seward, 1903, I, 154, 190–4; and Darwin to J. Herschel, 23 May 1861, in De Beer, 1959, p. 35.

65. Gray to Darwin, 25 May 1868, in J. L. Gray, 1893, II, 562.

66. A. Gray, 1963, pp. 316–17. The quotation is from John 3:8.

67. A. Gray, 1880, pp. 49, 72–7. Cf. Cupitt, 1975, p. 128.

68. See Warfield, 1932a, pp. 556–68 and Mandelbaum, 1958, pp. 368ff. Later, however, Darwin's views were still in a 'muddle'. See Darwin to Gray, 30 June 1874, in Loewenberg, 1973, letter 114; Darwin to J. Fordyce, May 1879, in De Beer, 1958, p. 88; and the recollection in G. D. Campbell, 1885b, p. 244.

69. A. Gray, 1883, p. 291.

70. G. J. Romanes, 1883b, p. 363.

71. A. Gray, 1883, pp. 527–8, 78; G. J. Romanes, 1883b, p. 529.

72. G. J. Romanes, 1883b, p. 101. Having received a personal letter from Gray (see Gray to Hooker, in J. L. Gray, 1893, II, 742), Romanes found he had mistaken the spirit in which the exchange was initiated. His response was an apology accompanied by a copy of his *Candid Examination of Theism*. He confessed that he did not now hold to all the book's arguments; he merely hoped Gray would read it to discover 'how gladly I would enter your camp if I could only see that it is on the side of Truth' (Romanes to Gray, 16 May 1883, in E. Romanes, 1896, p. 154).

73. Gray to Darwin, 7 March 1872, in J. L. Gray, 1893, II, 624; A. Gray, 1880, pp. 70–1.

74. A. Gray, 1963, pp. 25, 101, 112, 317. See Dupree, 1959, pp. 357, 381.

75. Hooker to Huxley, 27 March 1888, in L. Huxley, 1918, II, 305; *Popular Science Monthly*, 9 (September 1876), 625. Cf. Neel, 1942, p. 51 and M. McGiffert, 1958, pp. 5–6, 151, 160.

76. Dupree, 1959, p. 382. Cf. Gray to R. W. Church, 22 August 1869, in J. L. Gray, 1893, II, 592 and Gray to J. D. Dana, 22 June 1872, in J. L. Gray, 1893, II, 627.

77. A. Gray, 1963, p. 76; Darwin to Gray, [5 February 1871], in F. Darwin, 1887, III, 131; Gray to Darwin, 14 April 1871, in J. L. Gray, 1893, II, 615.

78. A. Gray, 1963, p. 296; A. Gray, 1880, pp. 44, 99–100, 101, 103.

79. A. Gray, 1880, p. 102. In a poignant letter (30 December 1883) Romanes told Gray, 'I quite agree with your view, that the doctrine of the human mind having been proximately evolved from lower minds is not incompatible with the doctrine of its having been due to a higher and supreme mind' (E. Romanes, 1896, p. 154). Cf. M. McGiffert, 1958, pp. 141–2 and Dupree, 1959, p. 376.

80. G. F. Wright, 1916, p. 128; Morison, 1971, chap. 1.

81. On Johnson and Wright, see G. F. Wright, 1916, p. 128 and Morison, 1971, p. 139.

82. G. F. Wright, 1916, pp. 132ff; Morison, 1971, chap. 5. See Vanderpool, 1971 and D. D. Williams, 1941.

83. Morison, 1971, chap. 6.

84. *Ibid.*, chaps. 6–7.

85. G. F. Wright, 1916, p. 116; M. McGiffert, 1958, chap. 3; Morison, 1971, chap. 2; Dupree, 1959, pp. 360ff.

86. Cf. Ahlstrom, 1972, p. 769.

87. Morison, 1971, pp. 56ff; Gray to G. F. Wright, 1 July 1875, in J. L. Gray, 1893, II, 655–6.

88. M. McGiffert, 1958, chaps. 4, 6.
89. G. F. Wright, 1916, p. 138.
90. G. F. Wright, 1882, pp. v–vi.
91. G. F. Wright, 1875, pp. 538–42, 543, 549. For the distance between Wright and the Baconian Charles Hodge, see *ibid.*, pp. 548–9 and G. F. Wright, 1882, pp. 213, 214.
92. G. F. Wright, 1882, pp. 7, 10, 11.
93. *Ibid.*, pp. 12, 13, 15, 16, 17–18.
94. *Ibid.*, pp. 20, 21, 23, 25 (emphasis in original).
95. *Ibid.*, pp. 33, 34–73.
96. *Ibid.*, pp. 74, 75, 77–8.
97. Quoted in Wright to Gray, 25 September 1876, in Morison, 1971, p. 105. Cf. G. F. Wright, 1916, p. 138. On the copy of the latter article, preserved in the Darwin Reprint Collection at Cambridge University Library, Darwin has marked a telling illustration of the imperfection of the geological record. Wright's other articles in the collection are unmarked.
98. G. F. Wright, 1882, pp. 89, 110, 118.
99. *Ibid.*, pp. 128, 141. See T. H. Huxley, 1893–4, II, 181ff.
100. Quoted above at the outset of chapter 8.
101. G. F. Wright, 1882, p. 153.
102. *Ibid.*, pp. 161, 164.
103. Gray to Wright, 14 August 1875, in J. L. Gray, 1893, II, 656.
104. G. F. Wright, 1882, p. 168.
105. *Ibid.*, pp. 175, 183, 185–6.
106. *Ibid.*, pp. 192, 194, 196.
107. *Ibid.*, pp. 196–7.
108. *Ibid.*, p. 202.
109. *Ibid.*, p. 201. See M. McGiffert, 1958, pp. 81–6, 223ff.
110. Quoted in M. McGiffert, 1958, p. 102.
111. G. F. Wright, 1882, pp. 219–20.
112. *Ibid.*, pp. 215–16. See McNeill, 1967.
113. *Ibid.*, p. 220.
114. *Ibid.*, pp. 225, 226.
115. *Ibid.*, pp. 230–1, 232, 241.
116. *Ibid.*, pp. 244, 246, 247, 248, 249–50.
117. *Ibid.*, pp. 251, 254, 255. Wright compares the *origin* of species with the *transmission* of revelation, both of which occur by natural means, whereas previously he compared Darwinism as an hypothesis which explains the *origin* of species with supernaturalism as an hypothesis which explains the *origin* of revelation (pp. 244–6). In the present instance the correct analogy would seem to be between *special* creation and *special* revelation. Otherwise formerly Wright should have compared the origin of *life* with the origin of revelation.
118. Morison, 1971, pp. 313ff, chaps. 9–10.
119. G. F. Wright, 1888; G. F. Wright, 1889, p. 181; G. F. Wright, 1898, pp. 42–61, 89ff; G. F. Wright, 1900, pp. 305–6, 311.
120. G. F. Wright, 1909b, pp. 332, 340, 342–3; G. F. Wright, 1909a, p. 691.
121. G. F. Wright, 1913, pp. 366–70.
122. *Ibid.*, p. 386.
123. G. F. Wright, 1888, p. 527; G. F. Wright, 1913, p. 433. See G. F. Wright, 1882, pp. 101–2, 155ff, 221–2, 225–30 and G. F. Wright, 1898, pp. 104–14.
124. G. F. Wright, 1913, p. 388.

125. *Ibid.*, p. 435. Wright was prepared to countenance the endowment of matter with the 'power of thought', which comes progressively to manifestation in life, instinct, and consciousness (pp. 414–15). Cf. M. McGiffert, 1958, pp. 141–2.
126. G. F. Wright, 1913, pp. 416–24. Cf. G. F. Wright, 1916, pp. 422–3.
127. M. McGiffert, 1958, p. 380.

Chapter 12

1. Himmelfarb, 1968*a*, pp. 445, 448–9.
2. T. H. Huxley to J. D. Hooker, 4 September 1861, in L. Huxley, 1903, I, 328.
3. Betts, 1959, p. 161. See also R. J. Elliott, 1912, pp. 49–50 and Barton, 1976, pp. 110–11.
4. May, 1949, p. 47; cf. p. 142. See Hull, 1974, p. 392 and Ruse, 1975*d*, pp. 518–22 for similar interpretations.
5. Mackenzie, 1920, pp. 58ff; Iverach, 1884*b*; A. L. Moore, 1889*b*, p. xii; A. Gray, 1963, p. 5; G. F. Wright, 1916, pp. 418ff.
6. On Mivart, see J. W. Gruber, 1960, chaps. 10–11; on Bascom, see Bates, 1929; on Le Conte, see Le Conte, 1903, p. 17; on Abbott, see I. V. Brown, 1953, p. 15; on Johnson, see D. D. Williams, 1941, pp. 44–8, 60ff and Schneider, 1945, pp. 11–14; on Matheson, see W. F. Gray, 1912, p. 588 and Macmillan, 1907, p. 124; on Beecher, see Hibben, 1942, pp. 301–2; and on Drummond, see 'Overture anent Confession of Faith', 1892, 'Professor Drummond's "Ascent of Man"', 1895, and Simmons, 1897, p. 498.
7. Powell, 1856, pp. 329ff; Powell, 1859, p. 172; Powell, 1861, p. 139. For Darwin's response, see Darwin, 1959, p. 69 (66) and his letters to Powell in De Beer, 1959, pp. 51–4. Cf. Knight, 1968, pp. 86–7; Limoges, 1970*a*, pp. 364–5; and Ruse, 1975*d*, *passim*.
8. C. Kingsley to Darwin, 18 November 1859, in F. Darwin, 1887, II, 287; Darwin, 1959, p. 748 (XIV:*183.2–3:b*); Kingsley to Darwin, 6 June 1867, in F. Kingsley, 1877, II, 247–8; Kingsley to A. R. Wallace, 22 October 1870, in F. Kingsley, 1877, II, 338 (cf. R. Smith, 1972, p. 191 and Kottler, 1974, p. 190); Kingsley to F. D. Maurice, 1863?, in F. Kingsley, 1877, II, 171 (cf. pp. 174, 424). Cf. A. Gray, 1963, pp. 231–2.
9. Kingsley to Maurice, in F. Kingsley, 1877, II, 181. Nor in fact was Maurice a radical. Welch, 1972, chap. 10 describes his theology as 'critical orthodoxy'. If Kingsley is to be held accountable for his defence of Maurice's *Theological Essays* (1853), which advanced a view of eternal damnation that cost Maurice his professorship at King's College, London, it must be in the light of his lengthy and sensitive correspondence on this and other subjects with the converted atheist Thomas Cooper (F. Kingsley, 1877, I, 377–400). On Kingsley's relationship with Maurice, see Tulloch, 1885, pp. 286–94.
10. Kingsley to A. P. Stanley, 19 February 1861, Kingsley to the Lord Bishop of Winchester (C. R. Sumner), 1861, and Kingsley to Stanley, 1 July 1863, in F. Kingsley, 1877, II, 130, 131, 182; the recollection of the Rev. William Harrison, Kingsley's curate for six years, in F. Kingsley, 1877, II, 283; C. Kingsley, 1890, pp. 315–16.

NOTES TO PAGES 308–10

NOTES TO PAGES 308–10

11. Cannon, 1961, pp. 109, 110.
12. *Ibid.*, pp. 111–12. See also Cannon, 1960, pp. 8–9; Cannon, 1968, pp. 157, 160; Cannon, 1976, pp. 117–19; Hooykaas, 1957, pp. 15–16; Hooykaas, 1963, pp. 98–100, 146–7; Hooykaas, 1966, pp. 3–19; and Hooykaas, 1970, pp. 41–2, 47. Lyell's anti-progressionism may have been 'a negligible obstruction to evolutionary thought' (Bartholomew, 1973, p. 265, n. 17; see also Bartholomew, 1976*b*, p. 170 and cf. Ospovat, 1977, pp. 321, 332) because Darwin's sense of historic development was so strong, a sense acquired from some 'framework' larger than Lyell's *Principles of Geology*, which could encompass both extended time and geologic directionality. (See De Beer *et al.*, 1967, VI, 134 [*B*, 108], 166 [*E*, 5], where Darwin's commitment to actualistic geology and 'progressive development' is already evident.) In 1856 Lyell himself thought it ironic 'how much the 6-days theories have led towards C. Darwin' (L. G. Wilson, 1970, pp. 88–9), though it has since been argued that this 'Design framework' or 'teleological time' is what Darwin destroyed (Haber, 1971, pp. 306–7). However, if the creationist framework of the natural theology tradition was progressive, not of necessity, but by virtue of the directionality of earth history (Bowler, 1976*a*, pp. 29–30, 36, 45–6), then it might well have nurtured a mechanism of creation in which progression was simply a by-product of divergence and specialisation of structure through adaptation to the changing environment.
13. On natural theology and Darwin's teachers, see Gruber and Barrett, 1974, pp. 56–60, chap. 4. For Darwin's use of the *Bridgewater Treatises*, see *ibid.*, pp. 346 (*N*, 89), 392–4 (OUN 34–6), 419 ('Essay on Theology and Natural Selection' [hereafter ETNS], 9); De Beer *et al.*, 1960–1, I, 58–60 (*B*, 141–3, 149); IV, 179–80 (*E*, 157–8); *ibid.*, 1967, VI, 145–6 (*C*, 72, 91). Among literary works *Paradise Lost* was Darwin's 'chief favourite' during his years on the *Beagle* (Darwin, 1958, p. 85). See Darwin to J. S. Henslow, 24 November 1832, in Barlow, 1967, p. 63.
14. R. M. Young, 1973, pp. 370–6 calls attention to the neglect of Paley and Malthus in the study of Darwinian backgrounds.
15. Darwin, 1958, p. 59; see also p. 87. For an earlier expression of Paley's value, see Darwin to Henslow, 2 July 1848, in Barlow, 1967, p. 161.
16. Darwin to J. Lubbock, 15 November 1859, in F. Darwin, 1887, II, 219.
17. Paley, 1803, chaps. 3 (the eye), 18 (instinct), pp. 359–61 (bees), 64, 78 (utility); Darwin, 1859, pp. 186–9 (the eye), 202–3, 224–35 (bees), chap. 7 (instinct), 199–203 (utility); Darwin, 1874, p. 61 (utility). In 1838 Darwin's belief in utility was strengthened by reading Thomas Browne's *Religio Medici*, especially section 16, headed '*Natura nihil agit frustra*' (De Beer *et al.*, 1967, VI, 159 [*D*, 54–6]).
18. Paley, 1803, pp. 283, 298; Darwin, 1959, pp. 290–7 (V:97—*137:f*). Cf. Darwin, 1875, II, 311ff, 335–7. Darwin first referred to the 'corelation of parts' in 1838 (De Beer *et al.*, 1960–1, IV, 164 [*E*, 51–4]). The concepts were perhaps more readily available in the works of Cuvier and Etienne Geoffroy Saint Hilaire. See Vorzimmer, 1972, p. 40; cf. pp. 84–9.
19. Paley, 1803, pp. 324ff, 490ff; Darwin, 1874, pt 2; Darwin, 1872. Here perhaps is another source of that 'family tradition' referred to in Gruber and Barrett, 1974, p. 298, n. 27. See also E. B. Poulton, 'The value of colour in the struggle for life', in Seward, 1909, pp. 272–5; J. A. Campbell, 1968, pp. 89–107, 122–34; J. A. Campbell, 1970, pp. 7–10; Limoges, 1970*a*, pp.

354-8; Limoges, 1970b, pp. 42-7; Yokoyama, 1971; and LeMahieu, 1976, pp. 178-80.

20. Paley, 1803, pp. 7, 76; cf. p. 447. The 'principle of order' is that of Philo in Hume, 1779, p. 125.

21. Paley, 1803, pp. 459-73. See LeMahieu, 1976, chap. 3.

22. Darwin to Huxley, 3 January 1861, and Darwin to F. Müller, 11 January 1866, in F. Darwin and Seward, 1903, I, 178, 264; Darwin to A. Hyatt, 10 October 1872, and Darwin to E. S. Morse, 23 April 1877, in F. Darwin, 1887, III, 154, 233. Cf. the misunderstanding in Cox, 1976, pp. 54-7. On Darwin's early opposition to Platonic accounts of structure, see Gruber and Barrett, 1974, pp. 417-18 (ETNS, 5). On his rejection of an 'absolute tendency to progression' as taught by his grandfather and Lamarck, see ibid., p. 339 (N, 47); De Beer et al., 1960-1, IV, 169 (E, 95); Darwin to Hooker, 11 January 1844, in F. Darwin and Seward, 1903, I, 41; and Darwin to C. Lyell, 12 March 1863, in F. Darwin, 1887, III, 14.

23. Darwin, 1958, p. 120. Herbert, 1977, pp. 191, 216-17 points out that Darwin almost certainly did not read Malthus merely 'for amusement'.

24. De Beer et al., 1967, VI, 163 (D, 135).

25. Cf. Darwin to Wallace, 6 April 1859, in F. Darwin and Seward, 1903, I, 118; Gruber and Barrett, 1974, pp. 163-74; and Ruse, 1975a with Limoges, 1970b, pp. 74-9 and Herbert, 1971. On Darwin's indebtedness to Malthus, see Vorzimmer, 1969; Ruse, 1973 and 1975c, pp. 170-2; and Bowler, 1976b.

26. See De Beer et al., 1960-1, IV, 160 (E, 3); Gruber and Barrett, 1974, pp. 332 (N, 10-11), 390 (OUN 29), 419 (ETNS, 10).

27. Darwin to A. Gray, 8 June 1860, in F. Darwin and Seward, 1903, I, 153. See also Darwin to Henslow, 23 [August/September 1855?], in Barlow, 1967, p. 185; Darwin to Hooker, 5 June 1860, in F. Darwin, 1887, II, 110; and Darwin to Wallace, 5 July 1866, in F. Darwin and Seward, 1903, I, 271. On Haughton, see Jessop, 1973. For his review of the Origin of Species, see Hull, 1973b, pp. 216-28.

28. Malthus, 1826, II, 412, 413.

29. Ibid., II, 17. See Bonar, 1924 and R. M. Young, 1969a.

30. Malthus, 1826, II, 270.

31. Ibid., II, 256. See Paley, 1803, pp. 563ff.

32. Malthus, 1826, II, 257.

33. Ibid., II, 268.

34. De Beer et al., 1967, VI, 163 (D, 135). Cf. Darwin, 1958, p. 90.

35. Malthus, 1826, II, 440-1. See Levin, 1966.

36. Darwin, 1874, pp. 146, 619. The similarities between these phrases from the concluding paragraphs of the Descent of Man and the Essay on the Principle of Population suggest that Darwin may have read the second volume of the Essay, though the second volume of his personal copy, preserved in the Cambridge University Library, is uncut.

It has been claimed that Darwin employed the struggle for existence to secure the very progress that Malthus had eschewed (Himmelfarb, 1968a, p. 163; Gale, 1972, pp. 331-2, 336, 341; Brooke and Richardson, 1974, p. 79) and, further, that in magnifying the struggle rather than absorbing it in a theodicy of moral improvement Darwin fairly removed himself from the natural theology tradition (Brooke and Richardson, 1974, pp. 44-50). On the first point, see the references above in n. 22, and in chapter 7, n. 14, where it appears that natural selection produces neither indefinite nor inexorable progression, and adaptive, rather than aesthetic or moral, perfection. On the second point, observe that the Darwinian theodicy is, by deliberate analogy with the Malthusian, one of *adaptive* improvement – natural selection works 'solely by and for the good of each being' (Darwin, 1959, p. 758 [xɪv:266]; cf. Fleming, 1959, p. 443). In the *Origin* Darwin attributed this principle to Paley (Darwin, 1959, p. 373 [vɪ:228]; cf. Paley, 1803, p. 502) and, though the 'very old argument from the existence of suffering against the existence of an intelligent first cause' seemed to him 'a strong one', still he agreed with Paley that 'happiness decidedly prevails', that 'all sentient beings have been formed so as to enjoy, as a general rule, happiness' (Darwin, 1958, pp. 88, 90; cf. Paley, 1803, pp. 488–527). Moreover, as we shall see below, Darwin's theodicy was part of a 'grander' theology of nature which lingers even in the *Origin*.

37. Mandelbaum, 1958, pp. 363–6.
38. Darwin, 1958, pp. 22, 56–7; Gruber and Barrett, 1974, p. 125.
39. Darwin, 1958, pp. 68, 85.
40. Barlow, 1967.
41. Barlow, 1933, pp. 23 (11 January 1832), 383 (18 January 1836). See Gruber and Gruber, 1962.
42. Barlow, 1933, pp. 118–19 (18 December 1832).
43. *Ibid.*, p. 171 (4–7 September 1833); cf. pp. 375–9 (12 January 1836), 388–9 (5 February 1836).
44. *Ibid.*, p. 428 (late September 1836). See 'A letter, containing remarks on the moral state of Tahiti, New Zealand, &c.' (1836), in Darwin, 1977, ɪ, 20–1.
45. Cf. Herbert, 1974, pp. 229, 233.
46. De Beer *et al.*, 1960–1, ɪɪ, 91 (*C*, 79).
47. *Ibid.*, ɪɪ, 100 (*C*, 154–5); cf. ɪ, 69 (*B*, 231), 71 (*B*, 248). On the Darwin family's opposition to slavery, see Gruber and Barrett, 1974, pp. 65–8.
48. De Beer *et al.*, 1960–1, ɪɪɪ, 134 (*D*, 49); ɪɪ, 106 (*C*, 196). See Ruse, 1977, pp. 256–7, 263–4.
49. *Ibid.*, ɪɪ, 101 (*C*, 166); cf. ɪ, 69 (*B*, 232); ɪɪ, 111 (*C*, 244). That 'man is *one* great object for which the world was brought into present state', said Darwin, '...few will dispute'. 'That it was the sole object', he added parenthetically, 'I will dispute' (*ibid.*, ɪv, 163–4 [*E*, 49]).
50. Gruber and Barrett, 1974, pp. 417, 419 (ETNS, 3, 4, 10).
51. *Ibid.*, pp. 417–18 (ETNS, 5).
52. *Ibid.*, p. 296 (*M*, 154e).
53. De Beer *et al.*, 1960–1, ɪ, 53 (*B*, 101). Cf. *ibid.*, ɪ, 47 (*B*, 45), 65 (*B*, 196) and *ibid.*, 1967, vɪ, 136 (*B*, 160). See Gruber and Barrett, 1974, p. 337 (*N*, 36); Darwin and Wallace, 1958, pp. 59 ('Sketch of 1842'), 154, 250–1 ('Essay of 1844'); and Darwin to Lyell, 17 June 1860, in F. Darwin and Seward, 1903, ɪ, 154.
54. Ruse, 1975c, p. 166.
55. De Beer *et al.*, 1960–1, ɪ, 55 (*B*, 114) with corrigendum from *ibid.*, v, 197.
56. *Ibid.*, ɪɪ, 103(*C*, 175),101 (*C*, 166).See Gruber and Barrett, 1974, pp. 213–17.
57. Gruber and Barrett, 1974, pp. 278 (*M*, 69), 279 (*M*, 74), 332 (*N*, 12)
58. *Ibid.*, p. 292 (*M*, 136). Cf. De Beer *et al.*, 1960–1, ɪɪ, 111 (*C*, 244).
59. See De Beer *et al.*, 1960–1, ɪ, 41 (*B*, 5), 47 (*B*, 49), 82 (*B*, 252); ɪɪ, 98 (*C*, 146). Cf. *ibid.*, ɪɪ, 107 (*C*, 203) and *ibid.*, 1967, vɪ, 133 (*B*, 55).
60. *Ibid.*, 1960–1, ɪɪɪ, 141 (*D*, 112); v, 194 (*D*, 114); *ibid.*, 1967, vɪ, 168 (*E*, 26); *ibid.*, 1960–1, ɪv, 172 (*E*, 112), 175 (*E*, 137). Cf. *ibid.*, ɪɪɪ, 147 (*D*, 167) and Gruber and Barrett, 1974, p. 278 (*M*, 70).
61. De Beer *et al.*, 1960–1, ɪv, 177 (*E*, 147); Paley, 1803, p. 466n.

62. Gruber and Barrett, 1974, pp. 296 (*M*, 154e), 419 (ETNS, 10). Cf. Darwin and Wallace, 1958, p. 49 ('Sketch of 1842').
63. De Beer *et al.*, 1960–1, III, 132 (*D*, 36–7). Cf. *ibid.*, 1967, VI, 160 (*D*, 74). Schweber, 1977, pp. 255–6 points out the connexion of this passage with the review of Comte which Darwin read.
64. Darwin and Wallace, 1958, pp. 86–7 ('Sketch of 1842'), 253–4 ('Essay of 1844'); Darwin, 1859, pp. 488–90.
65. Darwin, 1959, pp. 336 (VI:112), 343 (VI:131), 749 (XIV:192); Darwin, 1958, p. 93; Darwin to W. Graham, 3 July 1881, in F. Darwin, 1887, I, 316; Darwin to Lord Farrer, 28 August 1881, in F. Darwin and Seward, 1903, I, 395.
66. Darwin, 1859, p. 352; Darwin, 1958, p. 90; Darwin to Gray, 22 May 1860, in F. Darwin, 1887, II, 312. Cf. Darwin to Gray, July 1860, in F. Darwin, 1887, I, 314–15. Darwin's conception of 'chance' was identical with that of Paley and other natural theologians. The word simply expressed ignorance of causes. See Schweber, 1977, pp. 266–74.
67. Darwin to Lyell, 17 June 1860, in F. Darwin and Seward, 1903, I, 154; Darwin, 1958, p. 87 (written in June or July 1876, just before the same metaphor appeared in the concluding essay of Gray's *Darwiniana* – see Darwin to Gray, 9 August 1876, in Loewenberg, 1973, letter 74); Darwin to Gray, 11 December 1861, in F. Darwin, 1887, II, 382.
68. Darwin to Lyell, 2 August 1861, in F. Darwin and Seward, 1903, I, 192. Cf. Darwin to Lyell, 21 August 1861, in F. Darwin and Seward, 1903, I, 194 and Darwin to Lubbock, 12 November 1859, in F. Darwin, 1887, II, 218–19.
69. Darwin to Lyell, 20 October 1859, and Darwin to Gray, 5 June 1861, in F. Darwin, 1887, II, 174, 373. Cf. Darwin to Lyell, 11 October 1859, in F. Darwin, 1887, II, 210–11; Darwin, 1959, pp. 757–9 (XIV:259–70); and Mandelbaum, 1958, pp. 366–78.
70. See Gruber and Barrett, 1974, p. 127 for a diagram of 'Darwin's changing world-view'.
71. Paley, 1803, p. 563; cf. pp. 559–60.
72. *Ibid.*, pp. 443, 478–9, 480, 580; cf. pp. 54–9, 449–50, 457, 477, 559.
73. *Ibid.*, pp. 43–4.
74. Darwin and Wallace, 1958, pp. 45–6 ('Sketch of 1842').
75. Ruse, 1975a and 1975c, pp. 175–6 argue that Darwin's analogy between artificial and natural selection was no mere pedagogical device (Limoges, 1970a, pp. 370–2; Limoges, 1970b, p. 143). The analogy predated his reading of Malthus and ensured that, in Herschel's terms, natural selection was a *vera causa*.
76. Darwin and Wallace, 1958, pp. 114–16 ('Essay of 1844'). The suggested aetiology of this passage in Schweber, 1977, p. 311 is as tendentious as the article's larger interpretation of Darwin's religious views.
77. Darwin, 1975, p. 224.
78. Darwin, 1859, pp. 83–4.
79. Darwin's demiurge appears as late as 1857 however. See Darwin to Gray, 5 September 1857, in F. Darwin, 1887, II, 123.
80. Darwin, 1959, pp. 167–8 (IV:28—39:b); Gray to Darwin, 23 January 1860, in F. Darwin, 1887, II, 271. See also Darwin, 1959, pp. 753 (XIV:220:b), 759 (XIV:270:b) for added references to the 'Creator'.
81. Darwin, 1959, p. 165 (IV:14.6–9:c). Cf. *ibid.*, p. 167 (IV:29:c) and Darwin, 1875, I, 6–7.

82. Here, then, is why Ruse, 1971, pp. 329–30 finds Darwin 'so infuriatingly unclear'. Darwin wants Paley's purposeful adaptations without his Providence. He eschews 'mysterious teleological forces' but imputes their intelligence to 'an aggregate of laws'. Cf. Limoges, 1970b, p. 152n. and R. M. Young, 1971a, pp. 461ff.
83. Malthus, 1826, I, 529. Cf. Malthus, 1798, p. 362.
84. De Beer et al., 1960–1, IV, 160 (E, 3).
85. Ibid.
86. Darwin and Wallace, 1958, pp. 86–7 ('Sketch of 1842').
87. Ibid., pp. 250–1 ('Essay of 1844').
88. Darwin, 1859, p. 488.
89. By the 'orthodoxy' of Paley and Malthus we refer primarily to their doctrines of creation and providence, not to their utilitarianism. See Clarke, 1974, chap. 5; LeMahieu, 1976, chaps. 1, 5; and Bonar, 1924, bk III.
90. Gruner, 1975 justly criticises a composite version of this 'revisionist' interpretation. However, the arguments are not uniformly compelling and they seem less effective in dealing with the doctrine of creation – what we emphasise here – than with the anthropology which is supposed to have sanctioned the modern scientific enterprise. For further discussions of the subject, see Fruchtbaum, 1964 and Klaaren, 1970.
91. Hooykaas, 1972, pp. 12, 13.
92. Ibid., pp. 29–30, 51; Darwin, 1959, p. 40 ([ii]). See M. B. Foster, 1973, pp. 303–13.
93. Hooykaas, 1972, pp. 16–26, 44–52.
94. Butler, 1896, pp. 10 (intro., para. 9), 46 (1.1.31).
95. Cannon, 1960; Hooykaas, 1963, pp. 192–226; Ruse, 1975d, pp. 510–13; Ruse, 1977, pp. 250–3, 258–9.
96. Van Dyke, 1886, pp. 29, 43.
97. Van Dyke all but plagiarised sentences from Gray's Darwiniana, thereby incurring Wright's rebuke (G. F. Wright, 1888, pp. 525–6). McCosh stated in a letter printed in the New York World that Darwiniana reflected 'substantially his own position' (quoted in G. F. Wright to Gray, 31 October 1876, in Morison, 1971, pp. 74–5). Moore likewise read the book appreciatively, especially the chapter on evolutionary teleology. If Iverach was influenced by Gray it may have been indirectly through Moore. Between their books there are several similarities of emphasis and wording.
98. A. Gray, 1963, p. 47.
99. Van Dyke, 1886, pp. 461ff; McCosh, 1882, pp. 71–4.
100. Iverach, 1900, pp. 88–9; Iverach, 1894, pp. 70–1. Cf. Iverach, 1884a, pp. 205, 233.
101. A. L. Moore, 1889b, pp. 198, 199. See above, chap. 7, n. 14.
102. See n. 36 above.
103. See A. Gray, 1963, p. 312.
104. G. F. Wright, 1900, p. 314. Cf. Malthus, 1798, pp. 378–9.
105. Paley, 1803, p. 576.
106. Darwin to Lyell, April 1860, in F. Darwin, 1887, II, 303–4; Darwin to Lyell, 13 and 21 August 1861, in F. Darwin and Seward, 1903, I, 192–3, 194; Darwin to Gray, 17 September 1861? and 11 December 1861, in F. Darwin, 1887, II, 378, 382; Darwin, 1875, II, 428. See Darwin to Hooker, 12 July 1870, in F. Darwin and Seward, 1903, I, 321. See also n. 66 above.
107. Warfield, 1915, p. 208. See Hooykaas, 1963, pp. 211–12 and Hooykaas, 1972, pp. 107–9.

108. For adumbrations of this point, see D. D. Williams, 1941, pp. 54, 63, 74; Schneider, 1945, pp. 5–10; Persons, 1950a, p. 452, n. 16; Blau, 1959, p. 143; Fitch, 1959, p. 24; Passmore, 1959, pp. 47–8; Himmelfarb, 1968a, p. 395; Daniels, 1968, pp. xvii–xviii; P. A. Carter, 1971, pp. 58–61; and Bulhof, 1974, pp. 303–4. Cf. the implicit dissent in Viner, 1972, pp. 25–6.

109. McCosh, 1882, p. 164.

110. F. H. Foster, 1963, p. 547. Apart from B. B. Warfield, Wright seems to have been quite alone among American Calvinist theologians in perceiving the relevance of his theological tradition to a Christian acceptance of Darwinism. See Warfield, 1932a, pp. 556–68; F. H. Foster, 1963, pp. 550–1; and Marsden, 1970, pp. 149ff, 243.

111. Cf. the consensus view of Darwin's impact on the doctrine of providence in Gillispie, 1959a, pp. 219ff; J. Bronowski, 'Introduction', in Banton, 1961, p. xix; Peckham, 1970, pp. 195–6; and Altick, 1974, p. 228.

112. Darwin, 1874, p. 613. Darwin here refers his readers to Picton, 1870, pp. 190–204. Picton, a Congregational minister who later became a Christian pantheist, argues much as Iverach and Moore. For the same argument from a Darwinian unbeliever, see Lankester, 1880, pp. 66–70.

113. See Warfield, 1915, pp. 253–4.

114. Cf. Gray to Wright, 21 May 1876, in J. L. Gray, 1893, II, 659.

115. A. L. Moore, 1890c, p. 516; cf. pp. 389, 390–4.

116. A. L. Moore, 1890a, p. 102.

117. Ibid., pp. 96, 102.

118. Ibid., p. 95. Cf. A. L. Moore, 1889b, p. xliii with T. H. Huxley, 1887, p. 201. Although the authors of Lux Mundi were indebted as a whole to the idealism of their Oxford contemporary, T. H. Green, for a 'spiritual' as opposed to a 'materialistic' conception of the world and for an appreciation of 'personality' as a category in which mankind's relation to God and the world could be understood, there seems to be little evidence that Moore was among those most influenced. See Webb, 1923, pp. 48–54; Webb, 1933, chap. 5; A. M. Ramsey, 1960, pp. 8–10; and Langford, 1968, p. 68.

119. Kingsley to T. Cooper, 8 June 1857, in F. Kingsley, 1877, I, 398; see also p. 324.

120. Kingsley to Maurice, 1863?, in F. Kingsley, 1877, II, 171; C. Kingsley, 1890, p. 335. For the 'x' reference, see Huxley to Kingsley, 22 May 1863, in L. Huxley, 1903, I, 351–2. See also A. Johnston, 1959, pp. 218–19.

121. Kingsley to Maurice, 1863?, in F. Kingsley, 1877, II, 171. See also Kingsley's remarks in F. Kingsley, 1877, II, 175, 241, 338; and Hooykaas, 1963, pp. 215–18. For Darwin's response, see Darwin to Huxley, 28 December 1862, in F. Darwin and Seward, 1903, I, 225 and Darwin to J. Fordyce, May 1879, in De Beer, 1958, p. 88. Cf. Symonds, 1893, p. 427 and Warfield, 1932, pp. 556–8.

122. Van Dyke, 1886, pp. 444, 445; McCosh, 1882, p. 147 (cf. McCosh, 1890, chaps. 1, 3); Iverach, 1894, p. 139; Iverach, 1900, pp. 91–2.

123. A. Gray, 1963, p. 47. See A. Gray, 1880, p. 82 and M. McGiffert, 1958, pp. 214ff, 250–1. On Hodge, see A. Gray, 1963, pp. 209ff. On Dawson, see ibid., pp. 203–7 and J. L. Gray, 1893, II, 734–41. Concerning miracles, Gray's student Chauncey Wright wrote: 'To admit twenty or more (the more the better), as some geologists do, is quite enough to make them pious and safe. I would go even farther, and admit an infinite number of miracles, constituting continuous creation and the order of nature' (C. Wright to Mrs Lesley, 12 February 1860, in Thayer, 1878, p. 43).

124. A. Gray, 1887, p. 402.
125. See the sentences added to G. F. Wright, 1876a, as it appears in G. F. Wright, 1882, pp. 33–4. See also G. F. Wright, 1895.
126. Cf. G. F. Wright, 1888, p. 527 with A. Gray, 1963, pp. 47, 130. Schneider, 1945, p. 9 simply repeats Wright's error.
127. G. F. Wright, 1916, p. 419; G. F. Wright, 1913, pp. 409, 410. On the varying fortunes of the concept of divine immanence in America, see Berman, 1961, chap. 9. On British attitudes, see Webb, 1933, chaps. 5–7.
128. On evolution and liberal theology, see H. W. Clark, 1914, pp. 276–88; J. M. Wilson, 1925; Webb, 1933, chaps, 2, 5–6; H. S. Smith, 1955, chap. 8; Cauthen, 1962, pp. 1–25; and Averill, 1967, pp. 70–1.
129. M. B. Foster, 1973, pp. 303–6.
130. Cf. Hutchison, 1976, pp. 78–9.
131. M. B. Foster, 1973, p. 312.
132. Gillispie, 1959b, pp. 276, 290. Cf. M. J. S. Hodge, 1971, pp. 337, 349, 351. On Lamarck's deism, see Wilkie, 1964, pp. 266–70, 304–5 and R. Burkhardt, 1977, pp. 184–5.
133. Chambers, 1969, p. 385. Cf. ibid., 11th edn, 1860, pp. 284–5 and M. J. S. Hodge, 1972, p. 132.
134. On Spencer, see J. D. Y. Peel, 1971, pp. 8, 103ff and Burrow, 1970, pp. 211–12. On Butler, see F. M. Turner, 1974b, p. 169 and Breuer, 1975, pp. 378ff. Passmore, 1972, p. 243 and passim discusses this Lamarckian 'vacillation', which 'reflects in a new form the old Augustinian–Pelagian controversy'.

Conclusion

1. Passmore, 1972, p. 259.
2. T. H. Huxley, 1887, p. 202; Wiener, 1972, chap. 3; Hull, 1967, p. 337; Spilsbury, 1974; Wells, 1938, pp. 336–7. On Darwin's 'affective decline', cf. J. A. Campbell, 1974 with Fleming, 1961.
3. J. A. Campbell, 1968, p. 99. The evidence presented here belies the assertion in J. A. Campbell, 1974, p. 167 that 'Darwin's affective response to nature was from the first independent of belief in God'. See also J. A. Campbell, 1970, pp. 10–13; J. A. Campbell, 1975, pp. 444–8; and Cannon, 1968, p. 167.
4. There is also the curious note added to the second edition of the Descent of Man, in which Darwin seems to provide for a new doctrine of the Fall. See Darwin, 1874, p. 46, n. 62; cf. p. 145.
5. On the relationship between Darwin's metaphysical struggles and his health, see Colp, 1977, pp. 14ff, 29–30, 55–6, 105–6, 142–3. Darwin may have 'regretted' that he had 'truckled to public opinion' in using 'the Pentateuchal term of creation' – i.e. 'life, with its several powers, having been breathed into a few forms or into one' (Darwin, 1959, p. 759 [XIV:270], emphasis added) – but in this particular, unlike so many others, he neither modified the Origin nor clarified that he had meant ' "appeared" by some wholly unknown process' (Darwin to J. D. Hooker, 29 March 1863, in F. Darwin, 1887, III, 18). See 'The doctrine of heterogeny and modification of species' (1863), in Darwin, 1977, II, 78.
6. J. A. Campbell, 1968, p. 339. See also J. A. Campbell, 1970, pp. 13–14. In J. A. Campbell, 1975, pp. 446–8 the terms of 'theistic accommodation'

inferred from Darwin's view of nature are perhaps more orthodox than either Campbell or Paley might have supposed. Cf. the suggestive, if fanciful, interpretation in Hyman, 1959, pp. 545–8.

7. By now it should be clear that Darwin wrote the *Origin* without 'any relation whatever to Theology' (Darwin to C. Ridley, 28 November 1878, in F. Darwin, 1887, III, 235) only in the same sense and to the same extent that his book on orchids was 'like a Bridgewater treatise' (Darwin to J. Murray, 21 September 1861, in F. Darwin, 1887, III, 266). See Basalla, 1962, p. 973.

8. T. H. Huxley, 1887, p. 186; T. H. Huxley, 1870; T. H. Huxley to Darwin, 28 November 1880, and Huxley to J. Knowles, 29 February 1889, in L. Huxley, 1903, III, 107. See Murphy, 1973 and Hilts, 1975, p. 56.

9. T. H. Huxley, 1892, p. 567.

10. *Ibid.*, pp. 568, 569, 570.

11. *Ibid.*, p. 569.

12. Huxley to F. Dyster, 10 October 1854, in L. Huxley, 1903, I, 164. See T. H. Huxley, 1893–4, IX, 142–4 on Augustine and Jonathan Edwards; T. H. Huxley, 1887, pp. 201–3; and Helfand, 1977, p. 176.

13. D. D. Williams, 1941, p. 74; cf. p. 54. See Blau, 1952, p. 157.

14. T. H. Huxley, 1887, p. 203. The same conclusion appears in G. J. Romanes, 1892–7, I, 411–18.

15. D. D. Williams, 1941, p. 158. See Niebuhr, 1958, p. 34; Dietz, 1958, pp. 73–4; and P. A. Carter, 1971, pp. 55–61.

16. J. W. Smith, 1961, pp. 412, 422, 423.

BIBLIOGRAPHY

This bibliography contains the fullest available inventory of literature dealing with science and religion in the later nineteenth century and with the post-Darwinian controversies in particular. So far as is known, the only other published bibliography which specifically undertakes the subject appears in the pamphlet *Religion and Science* (1965) by Arthur Maltby. For critical and synoptic annotations on many of the entries, including unpublished theses and dissertations, one may consult the author's doctoral thesis, completed at the University of Manchester in 1975. This thesis also includes a catalogue of primary sources which is the most complete to date.

Entries for each author are given below in the chronological order of the editions employed in this study. Individual essays published in multi-author collections are entered by titles, date, and pagination when the collections which contain them are entered separately. Such essays are specified as appearing 'in' the collections. For essays that appear in collections by a single author the word 'in' is omitted. Dates following the titles of essays are those of first publication. Dates in square brackets are those of first editions.

Abbreviations

AS	Annals of Science
BJHS	British Journal for the History of Science
BS	Bibliotheca Sacra
CH	Church History
DAB	Dictionary of American Biography. 20 vols. London: Oxford University Press, 1928–36
DNB	Dictionary of National Biography. 22 vols. London: Smith, Elder & Co., 1908–9. 2nd supplement, 1912
JASA	Journal of the American Scientific Affiliation
JHB	Journal of the History of Biology
JHI	Journal of the History of Ideas
NR	Notes and Records of the Royal Society of London
PAPS	Proceedings of the American Philosophical Society
PSM	Popular Science Monthly
SHPS	Studies in History and Philosophy of Science
VS	Victorian Studies

402   BIBLIOGRAPHY

Abbott, Lyman. *The Evolution of Christianity.* Boston: Houghton, Mifflin & Co., 1892.
The Theology of an Evolutionist. Boston: Houghton, Mifflin & Co., 1897.
Reminiscences. Boston: Houghton, Mifflin & Co., 1915.
Silhouettes of My Contemporaries. London: George Allen & Unwin, 1922.
Abelès, F. *et al. Le XIXᵉ siècle.* Vol. 3. *Histoire générale des sciences,* ed. René Taton. Paris: Presses Universitaires de France, 1961.
Adams, Henry. 'Darwinism (1867–1868).' *The Education of Henry Adams: An Autobiography,* chap. 15. Boston: Houghton Mifflin Co., 1918.
Adeney, W. F. 'Religion and Science.' *A Century's Progress in Religious Life and Thought,* chap. 4. London: James Clarke & Co., 1901.
Agar, William M. *Catholicism and the Progress of Science.* New York: Macmillan Co., 1940.
Agassiz, Louis. *The Structure of Animal Life: Six Lectures Delivered at the Brooklyn Academy of Music in January and February, 1862.* London: Sampson Low, Son & Marston, 1866.
An Essay on Classification, ed. Edward Lurie. Cambridge, Mass.: Belknap Press of Harvard University Press, 1962 [1857].
Ahlstrom, Sydney. 'The Scottish Philosophy and American Theology.' *CH,* 24 (1955), 257–72.
'Theology in America: A Historical Survey.' In *The Shaping of American Religion,* pp. 232–321. Vol. 1. *Religion in American Life,* ed. James Ward Smith and A. Leland Jamison. Princeton, N.J.: Princeton University Press, 1961.
A Religious History of the American People. New Haven, Conn.: Yale University Press, 1972.
Aldrich, Michele L. 'United States: Bibliographical Essay.' In *Comparative Reception of Darwinism,* ed. T. F. Glick (1974), pp. 207–26.
Aliotta, Antonio. *The Idealistic Reaction against Science,* trans. Agnes McCaskill. London: Macmillan, 1914.
'Science and Religion in the Nineteenth Century.' In *Science, Religion, and Reality,* ed. Joseph Needham, pp. 149–86. London: Sheldon Press, 1925.
Allen, David Elliston. *The Naturalist in Britain: A Social History.* London: Allen Lane, 1976.
Allen, Leslie H., ed. *Bryan and Darrow at Dayton: The Record and Documents of the 'Bible-Evolution' Trial.* New York: A. Lee & Co., 1925.
Altick, Richard D. *Victorian People and Ideas.* London: J. M. Dent & Sons, 1974.
Altner, Günter. *Schöpfungsglaube und Entwicklungsgedanke in der protestantischen Theologie zwischen Ernst Haeckel und Teilhard de Chardin.* Zurich: EVZ-Verlag, 1965.
Amick, David Eldridge. 'The Rhetoric of Dogma: An Analysis of the Rhetorical Strategies of Two Representative Speakers in the Evolution Controversy of the 1920's.' Ph.D diss., University of Oregon, 1975.

Anderson, Olive. 'The Growth of Christian Militarism in Mid-Victorian Britain.' *English Historical Review*, 86 (1971), 46–72.

Andrews, Samuel J. *Christianity and Anti-Christianity in Their Final Conflict.* New York: G. P. Putnam's Sons, 1898.

Annan, Noel. 'The Strands of Unbelief.' In *Ideas and Beliefs of the Victorians* (1949), pp. 150–6.

——— *Leslie Stephen: His Thought and Character in Relation to His Time.* London: Macgibbon & Kee, 1951.

——— 'Science, Religion, and the Critical Mind: Introduction.' In *1859: Entering an Age of Crisis*, ed. P. Appleman et al. (1959), pp. 31–50.

Appleman, Philip, ed. *Darwin.* New York: W. W. Norton & Co., 1970a.

——— 'Darwin: On Changing the Mind.' In *Darwin*, ed. P. Appleman (1970b), pp. 629–51.

Appleman, Philip; Madden, William A.; and Wolff, Michael, eds. *1859: Entering an Age of Crisis.* Bloomington: Indiana University Press, 1959.

Argyll, The [Eighth] Duke of. *See* Campbell, George Douglas.

Armstrong, A. C. *Transitional Eras in Thought, with Special Reference to the Present Age.* New York: Macmillan Co., 1904.

Armstrong, Richard A. *Agnosticism & Theism in the 19th Century: An Historical Study of Religious Thought.* London: Philip Green, 1905.

Arnstein, Walter Leonard. *The Bradlaugh Case: A Study in Late Victorian Opinion and Politics.* Oxford: Clarendon Press, 1965.

Atkins, Gaius Glenn. *Religion in Our Times.* New York: Round Table Press, 1932.

Aubrey, Edwin Ewart. 'Religious Bearings of the Modern Scientific Movement.' In *Environmental Factors in Christian History*, ed. John Thomas McNeill et al., pp. 361–79. Chicago: University of Chicago Press, 1939.

Aveling, Edward B. *The Religious Views of Charles Darwin.* London: Free Thought Publishing Co., 1883.

Averill, Lloyd J. *American Theology in the Liberal Tradition.* Philadelphia: Westminster Press, 1967.

Bacon, Benjamin Wisner. 'Burr, Enoch Fitch.' In *DAB*, 3 (1929), 321–2.

——— *Theodore Thornton Munger, New England Minister.* New Haven, Conn.: Yale University Press, 1934.

Bailey, Kenneth K. 'The Antievolution Crusade of the Nineteen-Twenties.' Ph.D. diss., Vanderbilt University, 1953.

Baillie, John. *The Belief in Progress.* London: Oxford University Press, 1950.

——— *Natural Science and the Spiritual Life.* London: Oxford University Press, 1951.

Bainton, Roland H. *George Lincoln Burr.* Ithaca, N.Y.: Cornell University Press, 1942.

——— *Christian Attitudes toward War and Peace.* Nashville, Tenn.: Abingdon Press, 1960.

Baker, Ronald Duncan. 'The Influence of Charles Darwin on the Development of Theological Thought.' B.D. thesis, Trinity College (Dublin), 1962.

Baker, William J. 'Thomas Huxley in Tennessee.' *South Atlantic Quarterly*, 73 (1974), 473–86.

Baldwin, James Mark. *Darwin and the Humanities*. Baltimore, Md.: Review Publishing Co., 1909.

Bannister, R. C. ' "The Survival of the Fittest Is Our Doctrine": History or Histrionics?' *JHI*, 31 (1970), 377–98.

Banton, Michael, ed. *Darwinism and the Study of Society: A Centenary Symposium*. London: Tavistock Publications, 1961.

Barber, Bernard. 'Resistance by Scientists to Scientific Discovery.' *Science*, n.s., 134 (1961), 596–602.

Barbour, Ian G. *Issues in Science and Religion*. London: S.C.M. Press, 1966.

Barker, Ernest. *Political Thought in England from Herbert Spencer to the Present Day*. London: Williams & Norgate, 1915.

Barlow, Nora, ed. *Charles Darwin's Diary of the Voyage of H.M.S. 'Beagle'*. Cambridge University Press, 1933.

ed. *Darwin and Henslow, the Growth of an Idea: Letters, 1831–1860*. London: Bentham–Moxon Trust, John Murray, 1967.

Barnes, Harry Elmer. *A History of Historical Writing*. 2nd revised edn. New York: Dover Publications, 1962 [1937].

Barnett, S. A., ed. *A Century of Darwin*. London: Heinemann, 1958.

Barrett, Paul H. 'Darwin's "Gigantic Blunder".' *Journal of Geological Education*, 21 (1973), 19–28.

'The Sedgwick–Darwin Geologic Tour of North Wales.' *PAPS*, 118 (1974), 146–64.

Bartholomew, Michael. 'Lyell and Evolution: An Account of Lyell's Response to the Idea of an Evolutionary Ancestry for Man.' *BJHS*, 6 (1973), 261–303.

'Lyell's Conception of the History of Life.' Ph.D. thesis, University of Lancaster, 1974.

'Huxley's Defence of Darwin.' *AS*, 32 (1975), 525–35.

'The Award of the Copley Medal to Charles Darwin.' *NR*, 30 (1976a), 209–18.

'The Non-Progress of Non-progression: Two Responses to Lyell's Doctrine.' *BJHS*, 9 (1976b), 166–74.

Barton, Ruth. 'The X Club: Science, Religion, and Social Change in Victorian England.' Ph.D. diss., University of Pennsylvania, 1976.

Barzun, Jacques. *Darwin, Marx, Wagner: Critique of a Heritage*. Revised edn. Garden City, N.Y.: Doubleday & Co., Anchor Books, 1958 [1941].

Basalla, George. 'Darwin's Orchid Book.' *Actes du X⁰ Congrès International d'Histoire des Sciences Naturelles et de la Biologie* (Ithaca, 1962), II, 971–4.

Basalla, George; Coleman, William; and Kargon, Robert H., eds. *Victorian Science: A Self-Portrait from the Presidential Addresses of the British*

Association for the Advancement of Science. Garden City, N.Y.: Doubleday & Co., Anchor Books, 1970.

Bascom, John. *Science, Philosophy, and Religion.* New York: G. P. Putnam's Sons, 1871.

Natural Theology. New York: G. P. Putnam's Sons, 1880.

Evolution and Religion; or, Faith as a Part of a Complete Cosmic System. New York: G. P. Putnam's Sons, 1897.

Things Learned by Living. New York: G. P. Putnam's Sons, 1913.

Bates, Ernest Sutherland. 'Bascom, John.' In *DAB*, 2 (1929), 32–3.

Baumer, Franklin L. *Religion and the Rise of Scepticism.* New York: Harcourt, Brace & Co., 1960.

Beahm, William M. 'Factors in the Development of the Student Volunteer Movement for Foreign Missions.' Ph.D diss., University of Chicago, 1941.

Beale, Howard K. *A History of Freedom of Teaching in American Schools.* New York: Charles Scribner's Sons, 1941.

Beale, Lionel S. *Life Theories: Their Influence upon Religious Thought.* London: J. & A. Churchill, 1871.

Beecher, Henry Ward. *Evolution and Religion: Part I. Eight Sermons Discussing the Bearings of the Evolutionary Philosophy on the Fundamental Doctrines of Evangelical Christianity; Part II. Eighteen Sermons Discussing the Application of the Evolutionary Principles and Theories to the Practical Aspects of Religious Life.* New York: Fords, Howard, & Hulbert, 1885.

Begouën, Henri. *Quelques souvenirs sur le mouvement des idées transformistes dans les milieux catholiques, suivi de La mentalité spiritualiste des premiers hommes.* Paris: Bloud & Gay, 1945.

Behney, John Bruce. 'Conservatism and Liberalism in the Theology of Late Nineteenth Century American Protestantism: A Comparative Study of the Basic Doctrines of Typical Representatives.' Ph.D. diss., Yale University, 1941.

Bellah, Robert N. 'Civil Religion in America' (1967). In *Religion in America*, ed. William G. McLoughlin and Robert N. Bellah, pp. 3–23. Boston: Beacon Press, 1968.

Benn, Alfred William. *The History of English Rationalism in the Nineteenth Century.* 2 vols. London: Longmans, Green & Co., 1906.

Benz, Ernst. *Evolution and Christian Hope: Man's Concept of the Future from the Fathers to Teilhard de Chardin,* trans. Heinz G. Frank. London: Victor Gollancz, 1967 [1966].

'Theologie der Evolution im 19. Jahrhundert.' In *Biologismus im 19. Jahrhundert: Vorträge eines Symposiums vom 30. bis 31. Oktober 1970 in Frankfurt am Main,* ed. Gunter Mann, pp. 43–72. Stuttgart: Ferdinande Enke, 1973.

Berman, Milton. *John Fiske: The Evolution of a Popularizer.* Cambridge, Mass.: Harvard University Press, 1961.

Betts, John Rickards. 'Darwinism, Evolution, and American Catholic Thought, 1860–1900.' *Catholic Historical Review,* 45 (1959), 161–85.

Bezirgan, Najm A. 'The Islamic World.' In *Comparative Reception of Darwinism*, ed. T. F. Glick (1974), pp. 375–87.

Bibby, Cyril. 'Huxley and the Reception of the "Origin".' *VS*, 3 (1959a), 76–86.

T. H. Huxley: Scientist, Humanist, and Educator. London: Watts & Co., 1959b.

Scientist Extraordinary: The Life and Scientific Work of Thomas Henry Huxley, 1820–1895. Oxford: Pergamon Press, 1972.

Bill, E. G. W. 'The Declaration of Students of the Natural and Physical Sciences, 1865.' *Bodleian Library Record*, 5 (1954–6), 262–7.

Billington, Ray A. *The Protestant Crusade, 1800–1860: A Study of the Origins of American Nativism.* Reprint edn. Chicago: Quadrangle Books, 1964 [1938].

Birks, Thomas Rawson. *The Scripture Doctrine of Creation, with Reference to Religious Nihilism and Modern Theories of Development.* London: Christian Evidence Committee of the Society for Promoting Christian Knowledge, 1872.

Modern Physical Fatalism and the Doctrine of Evolution, Including an Examination of Mr H. Spencer's 'First Principles'.. .with a Preface in Reply to the Strictures of Mr H. Spencer, by C. Pritchard. 2nd edn. London: Macmillan, 1882 [1876].

Bishop, Morris. *The History of Cornell.* Ithaca, N.Y.: Cornell University Press, 1962.

Black, Max. *Models and Metaphors.* Ithaca, N.Y.: Cornell University Press, 1962.

Blakiston, H. E. D. 'Moore, Aubrey Lackington.' In *DNB*, 13 (1909), 789.

Blau, Joseph. *Men and Movements in American Philosophy.* Englewood Cliffs, N.J.: Prentice-Hall, 1952.

'The Influence of Darwin on American Philosophy.' *Bucknell Review*, 8 (1959), 141–53.

Bliakher, Leonid I. 'Die Diskussion zwischen Spencer und Weismann über die Bedeutung der natürlichen Zuchtwahl und der direkten Anpassung für die Evolution.' *Zeitschift für Geschichte der Naturwissenschaften, Technik, und Medizin*, 10 (1973), 50–8.

Blomefield, Leonard (late Jenyns). *Memoir of the Rev. John Stevens Henslow.* London: John van Voorst, 1862.

Chapters in My Life, with Appendix Containing Special Notices of Particular Incidents and Persons; also, Thoughts on Certain Subjects. Revised edn. Bath: printed for private circulation, 1889.

Bock, Kenneth E. 'Darwin and Social Theory.' *Philosophy of Science*, 22 (1955), 123–34.

Bölsche, Wilhelm. *Haeckel: His Life and Work*, trans. Joseph McCabe. London: T. Fisher Unwin, 1906.

Boller, Paul F., Jr. *American Thought in Transition: The Impact of Evolutionary Naturalism, 1865–1900.* Chicago: Rand McNally & Co., 1969.

'The New Science and American Thought.' In *The Gilded Age*, ed.

H. Wayne Morgan, pp. 239–57. Revised edn. Syracuse, N.Y.: Syracuse University Press, 1970 [1963].

Bonar, James. *Malthus and His Work.* 2nd edn. London: George Allen & Unwin, 1924 [1885].

Bordin, Ruth. 'Andrew Dickson White, Teacher of History.' *Michigan Historical Collections*, no. 8 (1958), 3–17.

Boromé, Joseph A. 'The Evolution Controversy.' In *Essays in American Historiography: Papers Presented in Honor of Allan Nevins*, ed. Donald Sheehan and Harold C. Syrett, pp. 169–92. New York: Columbia University Press, 1960.

Bourdier, Franck. 'Geoffroy Saint-Hilaire versus Cuvier: The Campaign for Paleontological Evolution (1825–1838).' In *Toward a History of Geology*, ed. Cecil J. Schneer, pp. 36–61. Cambridge, Mass.: M.I.T. Press, 1969.

Bourne, Gilbert Charles. *Herbert Spencer and Animal Evolution.* The Herbert Spencer Lecture. Oxford: Clarendon Press, 1910.

Boutroux, Emile. *Science and Religion in Contemporary Philosophy*, trans. Jonathan Nield. London: Duckworth & Co., 1909.

Bowden, Henry Warner. *Church History in the Age of Science: Historiographical Patterns in the United States, 1876–1918.* Chapel Hill: University of North Carolina Press, 1971.

Bowen, Desmond. *The Idea of the Victorian Church: A Study of the Church of England, 1833–1889.* Montreal: McGill University Press, 1968.

Bowers, David F. 'Hegel, Darwin, and the American Tradition.' In *Foreign Influences in American Life: Essays and Critical Bibliographies*, ed. David F. Bowers, pp. 146–71. Princeton, N.J.: Princeton University Press, 1944.

Bowle, John. 'Mid-Century Prospect: The Impact of Darwinism.' *Politics and Opinion in the Nineteenth Century: An Historical Introduction*, bk 2, chap. 1. London: Jonathan Cape, 1954.

Bowler, Peter J. 'Darwin's Concepts of Variation.' *Journal of the History of Medicine and Allied Sciences*, 29 (1974), 196–212.

'The Changing Meaning of "Evolution".' *JHI*, 36 (1975), 95–114.

Fossils and Progress: Paleontology and the Idea of Progressive Evolution in the Nineteenth Century. New York: Science History Publications, 1976a.

'Malthus, Darwin, and the Concept of Struggle.' *JHI*, 37 (1976b), 631–50.

'Darwinism and the Argument from Design: Suggestions for a Reevaluation.' *JHB*, 10 (1977a), 29–43.

'Edward Drinker Cope and the Changing Structure of Evolutionary Theory.' *Isis*, 68 (1977b), 249–65.

Bozeman, Theodore Dwight. *Protestants in an Age of Science: The Baconian Ideal and Antebellum American Religious Thought.* Chapel Hill: University of North Carolina Press, 1977.

Bree, Charles Robert. *Species Not Transmutable nor the Result of*

408 BIBLIOGRAPHY

Secondary Causes: Being a Critical Examination of Mr Darwin's Work Entitled 'Origin and Variation of Species'. London: Groombridge & Sons, 1860.

An Exposition of Fallacies in the Hypothesis of Mr Darwin. London: Longmans, Green & Co., 1872.

Brett, Raymond L. 'The Influence of Darwin upon His Contemporaries.' *South Atlantic Quarterly*, 59 (1960), 69–81.

Breuer, Hans-Peter. 'Samuel Butler's "The Book of the Machines" and the Argument from Design.' *Modern Philology*, 72 (1975), 365–83.

Brewster, Edwin Tenney. *Creation: A History of Non-Evolutionary Theories.* Indianapolis, Ind.: Bobbs–Merrill, 1927.

Brock, W. H. and MacLeod, R. M. 'The Scientists' Declaration: Reflexions on Science and Belief in the Wake of "Essays and Reviews", 1864–5.' *BJHS*, 9 (1976), 39–66.

Bronowski, J. 'Unbelief and Science.' In *Ideas and Beliefs of the Victorians* (1949), pp. 164–9.

Brooke, John Hedley. 'Natural Theology and the Plurality of Worlds: Observations on the Brewster–Whewell Debate.' *AS*, 34 (1977a), 221–86.

'Richard Owen, William Whewell, and the "Vestiges".' *BJHS*, 10 (1977b), 132–45.

Brooke, John Hedley and Richardson, Alan. *The Crisis of Evolution.* Science and Belief: from Copernicus to Darwin (An Inter-faculty Second Level Course in the History of Science, The Open University) block 5, units 12–14. Milton Keynes, Bucks.: Open University Press, 1974.

Brown, Alan Willard. *The Metaphysical Society: Victorian Minds in Crisis, 1869–1880.* New York: Columbia University Press, 1947.

Brown, Ira V. *Lyman Abbott, Christian Evolutionist: A Study in Religious Liberalism.* Cambridge, Mass.: Harvard University Press, 1953.

Brush, Stephen G. 'The Prayer Test.' *American Scientist*, 62 (1974), 561–3; vol. 63 (1975), 6–7.

Bryce, James. 'Personal Reminiscences of Charles Darwin and of the Reception of the "Origin of Species".' *PAPS*, 48 (1909), iii–xiv.

Bube, Richard H. 'A Case History of the Power of Public Opinion on a Subject of Particular Interest to the A.S.A.' *JASA*, 12 (1960), 24–5.

Buchsbaum, Ralph, ed. *A Book that Shook the World: Anniversary Essays on Charles Darwin's 'Origin of Species'.* Pittsburgh, Pa.: University of Pittsburgh Press, 1958.

Buckham, John Wright. *Progressive Religious Thought in America: A Survey of the Enlarging Pilgrim Faith.* Boston: Houghton Mifflin Co., 1919.

Buckley, Jerome Hamilton. *The Triumph of Time: A Study of the Victorian Concepts of Time, History, Progress, and Decadence.* Cambridge, Mass.: Belknap Press of Harvard University Press, 1966.

Budd, Susan. 'The Loss of Faith: Reasons for Unbelief among Members of the Secular Movement in England, 1850–1950.' *Past and Present*, 36 (1967), 106–25.

Bulhof, Ilse N. 'The Netherlands.' In *Comparative Reception of Darwinism*, ed. T. F. Glick (1974), pp. 269–306.

Burchfield, Joe D. 'Darwin and the Dilemma of Geological Time.' *Isis*, 65 (1974), 301–21.

Lord Kelvin and the Age of the Earth. New York: Science History Publications, 1975.

Burkhardt, Frederick. 'England and Scotland: The Learned Societies.' In *Comparative Reception of Darwinism*, ed. T. F. Glick (1974), pp. 32–74.

Burkhardt, Richard W., Jr. 'The Inspiration of Lamarck's Belief in Evolution.' *JHB*, 5 (1972), 413–38.

The Spirit of System: Lamarck and Evolutionary Biology. Cambridge, Mass.: Harvard University Press, 1977.

Burr, Enoch Fitch. *Pater Mundi; or, Modern Science Testifying to the Heavenly Father: Being in Substance Lectures Delivered to Senior Classes in Amherst College*. 1st ser. Boston: Nichols & Noyes, 1869.

Pater Mundi; or, Doctrine of Evolution: Being in Substance Lectures Delivered in Various Colleges and Theological Seminaries. 2nd ser. Boston: Noyes, Holmes & Co., 1873.

Ecce Terra; or, The Hand of God in the Earth. Philadelphia: Presbyterian Board of Publication, 1883.

Burr, George Lincoln. 'White, Andrew Dickson.' In *DAB*, 20 (1936), 88–93.

Burrow, J. W. *Evolution and Society: A Study in Victorian Social Theory*. New edn. Cambridge University Press, 1970 [1966].

Burtt, Edwin Arthur. *The Metaphysical Foundations of Modern Physical Science*. Revised edn. Garden City, N.Y.: Doubleday & Co., Anchor Books, 1954 [1924].

Bury, J. B. *The Idea of Progress: An Inquiry into Its Origin and Growth*. Reprint edn. New York: Dover Publications, 1955 [1932].

Butler, Joseph. *The Analogy of Religion Natural and Revealed to the Course and Constitution of Nature*. Vol. 1. *The Works of Joseph Butler*, ed. W. E. Gladstone. Oxford: Clarendon Press, 1896.

Bynum, William Frederick. 'Time's Noblest Offspring: The Problem of Man in the British Natural Historical Sciences, 1800–1863.' Ph.D thesis, University of Cambridge, 1974.

[Campbell, George Douglas] The [Eighth] Duke of Argyll. *The Reign of Law*. 5th edn. London: Strahan & Co., 1868 [1867].

Primeval Man: An Examination of Some Recent Speculations. London: Strahan & Co., 1869.

The Unity of Nature. London: Strahan & Co., 1884.

Geology and the Deluge. Glasgow: Wilson & McCormick, 1885a.

'What is Science?' *Good Words*, 26 (April 1885b), 236–45.

Organic Evolution Cross-Examined; or, Some Suggestions on the Great Secret of Biology. London: John Murray, 1898.

[Campbell, Ina]. The Dowager Duchess of Argyll. *George Douglas, Eighth Duke of Argyll, K.G., K.T. (1823–1900): Autobiography and Memoirs.* 2 vols. London: John Murray, 1906.

Campbell, John Angus. 'A Rhetorical Analysis of *The Origin of Species* and of American Christianity's Response to Darwinism.' Ph.D diss., University of Pittsburgh, 1968.

'Darwin and "The Origin of Species": The Rhetorical Ancestry of an Idea.' *Speech Monographs,* 37 (1970), 1–14.

'Nature, Religion and Emotional Response: A Reconsideration of Darwin's Affective Decline.' *VS,* 18 (1974), 159–74.

'Charles Darwin and the Crisis of Ecology: A Rhetorical Perspective.' *Quarterly Journal of Speech,* 60 (1975), 442–9.

Campbell, John P. *Biological Teaching in the Colleges of the United States.* Washington: Government Printing Office, 1891.

Cannon, W. F. 'The Problem of Miracles in the 1830's.' *VS,* 4 (1960), 5–32.

'The Bases of Darwin's Achievement: A Re-evaluation.' *VS,* 5 (1961), 109–34.

'The Normative Role of Science in Early Victorian Thought.' *JHI,* 25 (1964a), 487–502.

'Scientists and Broad Churchmen: An Early Victorian Intellectual Network.' *Journal of British Studies,* 4 (1964b), 65–88.

'Darwin's Vision in "On the Origin of Species".' In *The Art of Victorian Prose,* ed. George Levine and William Madden, pp. 154–76. New York: Oxford University Press, 1968.

'Charles Lyell, Radical Actualism, and Theory.' *BJHS,* 9 (1976), 104–20.

Carlyle, Gavin. *The Battle of Unbelief.* London: Hodder & Stoughton, 1878.

Carpenter, S. C. *Church and People, 1789–1889: A History of the Church of England from William Wilberforce to 'Lux Mundi'.* London: Society for Promoting Christian Knowledge, 1933.

Carroll, P. Thomas, ed. *An Annotated Calendar of the Letters of Charles Darwin in the Library of the American Philosophical Society.* Wilmington, Del.: Scholarly Resources, 1976.

Carter, G. S. 'The Reaction to "The Origin" among the General Public.' *A Hundred Years of Evolution,* chap. 5. London: Sidgwick & Jackson, 1957.

Carter, Paul A. 'The Fundamentalist Defense of the Faith.' In *Change and Continuity in Twentieth-Century America: The 1920's,* ed. John Braeman et al., pp. 179–214. Columbus: Ohio State University Press, 1968.

The Spiritual Crisis of the Gilded Age. Dekalb: Northern Illinois University Press, 1971.

Cashdollar, Charles D. 'Auguste Comte and the American Reformed Theologians.' *JHI,* 39 (1978), 61–79.

Cauthen, Kenneth. *The Impact of American Religious Liberalism.* New York: Harper & Row, 1962.

Chadwick, Owen. *The Victorian Church*. Vols. 7–8. *An Ecclesiastical History of England*, ed. J. C. Dickinson. New York: Oxford University Press, 1966–70.
The Secularization of the European Mind in the Nineteenth Century. Cambridge University Press, 1975.
[Chambers, Robert]. *Vestiges of the Natural History of Creation*. Reprint edn. Leicester University Press, 1969 [1844].
Church, Avery Milton. 'The Reaction of the American Pulpit to the Modern Scientific Movement from 1850 to 1900.' Th.D. diss., Southern Baptist Theological Seminary (Louisville, Ky.), 1943.
Church, Mary C., ed. *Life and Letters of Dean Church*. London: Macmillan, 1894.
Churchill, Frederick B. 'The Weismann–Spencer Controversy over the Inheritance of Acquired Characters.' In *Human Implications of Scientific Advance: Proceedings of the XVth International Congress of the History of Science*, ed. E. G. Forbes, pp. 451–68. Edinburgh University Press, 1978.
Clark, Edward Lassiter. 'The Southern Baptist Reaction to the Darwinian Theory of Evolution.' Th.D. diss., Southern Baptist Theological Seminary (Fort Worth, Texas), 1952.
Clark, G. Kitson. *The Making of Victorian England*. London: Methuen, 1962.
Clark, Henry W. *Liberal Orthodoxy: A Historical Survey*. London: Chapman & Hall, 1914.
Clark, Linda Loeb. 'Social Darwinism and French Intellectuals, 1860–1915.' Ph.D. diss., University of North Carolina, 1968.
Clark, Robert E. D. *Darwin, Before and After: An Evangelical Assessment*. New edn. Chicago: Moody Press, 1967 [1948].
Clarke, M. L. *Paley: Evidences for the Man*. London: S.P.C.K., 1974.
Clifford, W. K. *Lectures and Essays*, ed. Leslie Stephen and Frederick Pollock. 2 vols. London: Macmillan, 1879.
Clodd, Edward. *Thomas Henry Huxley*. Edinburgh: William Blackwood & Sons, 1902.
Pioneers of Evolution: From Thales to Huxley, with an Intermediate Chapter on the Causes of Arrest of the Movement. Revised edn. London: Cassell & Co., 1907 [1897].
Cockshut, A. O. J. *Anglican Attitudes: A Study of Victorian Religious Controversies*. London: Collins, 1959.
The Unbelievers: English Agnostic Thought, 1840–1890. London: Collins, 1964.
Religious Controversies of the Nineteenth Century: Selected Documents. London: Methuen, 1966.
Coffin, Henry Sloane. 'Evolutionary Science.' *Religion Yesterday and Today*, chap. 1. Nashville, Tenn.: Cokesbury Press, 1940.
Cole, Stewart G. *The History of Fundamentalism*. Reprint edn. Westport, Conn.: Greenwood Press, 1971 [1931].
Coleman, Brian. 'Samuel Butler, Darwin, and Darwinism.' *Journal of*

the Society for the Bibliography of Natural History, 7 (1974), 93–105.

Coleman, William. *Georges Cuvier, Zoologist: A Study in the History of Evolution Theory.* Cambridge, Mass.: Harvard University Press, 1964.

Biology in the Nineteenth Century: Problems of Form, Function, and Transformation. New York: John Wiley & Sons, 1971.

Coletta, Paolo. *William Jennings Bryan.* 3 vols. Lincoln: University of Nebraska Press, 1964–9.

Collard, Edgar Andrew. 'Lyell and Dawson: A Centenary.' *Dalhousie Review*, 22 (1942), 133–44.

Collins, James. 'Darwin's Impact on Philosophy' (1959). In *Darwin's Vision and Christian Perspectives*, ed. W. J. Ong (1960), pp. 33–103.

Colp, Ralph, Jr. *To Be an Invalid: The Illness of Charles Darwin.* Chicago: University of Chicago Press, 1977.

Commager, Henry Steele. *The American Mind: An Interpretation of American Thought and Character since the 1880's.* New Haven, Conn.: Yale University Press, 1950.

Conn, H. W. *The Method of Evolution: A Review of the Present Attitude of Science toward the Question of the Laws and Forces Which Have Brought about the Origin of Species.* New York: G. P. Putnam's Sons, 1900.

Evolution of To-day: A Summary of Evolution as Held by Scientists at the Present Time and an Account of the Progress Made by the Discussions and Investigations of a Quarter of a Century. New York: G. P. Putnam's Sons, 1907.

Conry, Yvette. *L'introduction du darwinisme en France au XIXe siècle.* Paris: Librairie Philosophique J. Vrin, 1974.

Cope, E. D. *The Origin of the Fittest: Essays on Evolution.* London: Macmillan, 1887a [1886].

Theology of Evolution: A Lecture. Philadelphia: Arnold & Co., 1887b.

The Primary Factors of Organic Evolution. Chicago: Open Court Publishing Co., 1896.

Cornish, Francis Warre. *The English Church in the Nineteenth Century.* Vol. 8, 2 pts. *A History of the English Church*, ed. W. R. W. Stephens and William Hunt. London: Macmillan, 1910.

'Correspondence between Mr Romanes and Mr Gulick.' *BS*, 53 (January 1896), 165–7.

Coulson, C. A. 'The Changing Relationship of Science and Religion.' *London Quarterly & Holborn Review*, 184 (1959), 280–3.

Cowan, Ruth Schwartz. 'Francis Galton's Contribution to Genetics.' *JHB*, 5 (1972a), 389–412.

'Francis Galton's Statistical Ideas: The Influence of Eugenics.' *Isis*, 63 (1972b), 509–28.

Cox, John D. 'Darwin's Revolution from Hume's Perspective.' *Christian Scholar's Review*, 6 (1976), 52–7.

Cramer, Frank. *The Method of Darwin: A Study in Scientific Method.* Chicago: A. C. McClurg & Co., 1896.

Cratchley, W. J. 'Influence of the Theory of Evolution on the Christian Doctrine of the Atonement.' M.A. thesis, University of Bristol, 1933.

[Creighton, Louise]. *Life and Letters of Mandell Creighton, D.D. Oxon and Cam., Sometime Bishop of London.* 2 vols in one. New York: Longmans, Green & Co., 1913.

Croce, Benedetto. *History of Europe in the Nineteenth Century*, trans. Henry Furst. New edn. New York: Harcourt, Brace & World, Harbinger Books, 1963 [1933].

Crombie, A. C. 'Darwin's Scientific Method.' *Actes du IXe Congrès International d'Histoire des Sciences* (Barcelona, 1959), II, 354–62.

Cross, F. Leslie. *Religion and the Reign of Science.* London: Longmans, Green & Co., 1930.

Crowther, M. A. *Church Embattled: Religious Controversy in Mid-Victorian England.* Newton Abbot, Devon: David & Charles, 1970.

Cupitt, Don. 'Darwinism and English Religious Thought.' *Theology*, 78 (1975), 125–31.

Curti, Merle. *The Growth of American Thought.* 3rd edn. New York: Harper & Row, 1964 [1943].

Curtis, George Ticknor. *Creation or Evolution? A Philosophical Inquiry.* New York: D. Appleton & Co., 1887.

Dampier, William Cecil. *A History of Science and Its Relations with Philosophy & Religion.* Cambridge University Press, 1929.

Daniels, George H., ed. *Darwinism Comes to America.* Waltham, Mass.: Blaisdell Publishing Co., 1968.

Science in American Society: A Social History. New York: Alfred A. Knopf, 1971.

Darlington, C. D. *Darwin's Place in History.* Oxford: Blackwell, 1960.

Darwin, Charles. *On the Origin of Species by Means of Natural Selection, or the Preservation of Favoured Races in the Struggle for Life.* London: John Murray, 1859.

The Expression of the Emotions in Man and Animals. London: John Murray, 1872.

The Descent of Man, and Selection in Relation to Sex. 2nd edn revised. London: John Murray, 1874 [1871].

The Variation of Animals and Plants under Domestication. 2 vols. 2nd edn revised. London: John Murray, 1875 [1868].

The Autobiography of Charles Darwin, 1809–1882, with Original Omissions Restored, ed. Nora Barlow. New York: Harcourt, Brace & World, 1958.

The Origin of Species by Charles Darwin: A Variorum Text, ed. Morse Peckham. Philadelphia: University of Pennsylvania Press, 1959.

Charles Darwin's 'Natural Selection': Being the Second Part of His Big Species Book Written from 1856 to 1858, ed. Richard C. Stauffer. Cambridge University Press, 1975.

The Collected Papers of Charles Darwin, ed. Paul H. Barrett. 2 vols. Chicago: University of Chicago Press, 1977.

Darwin, Charles and Wallace, Alfred Russel. *Evolution by Natural Selection*, ed. Gavin de Beer. Cambridge University Press, 1958.

Darwin, Francis, ed. *The Life and Letters of Charles Darwin, Including an Autobiographical Chapter*. 3 vols. London: John Murray, 1887.

Darwin, Francis and Seward, A. C., eds. *More Letters of Charles Darwin: A Record of His Work in a Series of Hitherto Unpublished Letters*. 2 vols. London: John Murray, 1903.

Davenport, F. Garvin. 'Scientific Interests in Kentucky and Tennessee, 1870–1890.' *Journal of Southern History*, 14 (1948), 500–21.

'Alexander Winchell: Michigan Scientist and Educator.' *Michigan History*, 35 (1951), 185–201.

Davies, Horton. *From Newman to Martineau, 1850–1900*. Vol. 4. *Worship and Theology in England*. Princeton, N.J.: Princeton University Press, 1962.

Davis, Ozora Stearns. 'The Conception of Man's Place in Nature.' In *Recent Christian Progress: Studies in Christian Thought and Work during the Last Seventy-five Years by Professors and Alumni of Hartford Theological Seminary, in Celebration of Its Seventy-fifth Anniversary, May 24–26, 1909*, ed. Lewis Bayles Paton, pp. 197–204. New York: Macmillan Co., 1909.

Dawson, John William. *Archaia; or, Studies of the Cosmogony and Natural History of the Hebrew Scriptures*. Montreal: B. Dawson & Son, 1860.

The Origin of the World according to Revelation and Science. 2nd edn. London: Hodder & Stoughton, 1880a [1877].

The Story of the Earth and Man. 6th edn. London: Hodder & Stoughton, 1880b [1873].

Modern Ideas of Evolution as Related to Revelation and Science. Revised edn. London: Religious Tract Society, 1890 [1890].

'Death of Dr J. S. Van Dyke.' *The Presbyterian*, 85 (1915), 28.

De Beer, Gavin, ed. 'Further Unpublished Letters of Charles Darwin.' *AS*, 14 (1958), 83–115.

ed. 'Some Unpublished Letters of Charles Darwin.' *NR*, 14 (1959), 12–66.

Charles Darwin: Evolution by Natural Selection. London: Thomas Nelson & Sons, 1963.

De Beer, Gavin *et al.*, eds. 'Darwin's Notebooks on Transmutation of Species': [Parts I–VI]. *Bulletin of the British Museum (Natural History)*, Historical Series, 2 (1960–1), 23–200; vol. 3 (1967), 129–76.

'Part I. First Notebook (July 1837–February 1838)' [pp. 23–73].
'Part II. Second Notebook (February–July 1838)' [pp. 75–118].
'Part III. Third Notebook (July 15th 1838–October 2nd 1838)' [pp. 119–50].
'Part IV. Fourth Notebook (October 1838–10 July 1839)' [pp. 151–83].

'[Part V]. Addenda and Corrigenda' [pp. 185–200; with M. J. Rowlands].

'Part VI. Pages Excised by Darwin' [vol. 3 (1967), 129–76; with M. J. Rowlands and B. M. Skramovsky].

De Camp, L. Sprague. *The Great Monkey Trial.* Garden City, N.Y.: Doubleday & Co., 1968.

Deely, John N. and Nogar, Raymond J. *The Problem of Evolution: A Study of the Philosophical Repercussions of Evolutionary Science.* New York: Appleton–Century–Crofts, 1973.

De Jong, John Arlo. 'American Attitudes toward Evolution before Darwin.' Ph.D diss., State University of Iowa, 1962.

Delage, Yves and Goldsmith, M. *Les théories de l'évolution.* Revised edn. Paris: Ernest Flammarion, 1920 [1909].

Delany, Selden Peabody. 'Charles Darwin and the Church.' *American Church Monthly,* 21 (1927), 134–9.

Dewey, John. 'The Influence of Darwin on Philosophy.' *The Influence of Darwin on Philosophy, and Other Essays in Contemporary Thought,* pp. 1–19. New York: Henry Holt & Co., 1910.

The Quest for Certainty: A Study of the Relation of Knowledge and Action. New York: Minton, Balch & Co., 1929.

Dickie, John. *Fifty Years of British Theology: A Personal Retrospect.* Edinburgh: T. & T. Clark, 1937.

Dietz, Reginald W. 'Eastern Lutheranism in American Society and American Christianity, 1870–1914: Darwinism – Biblical Criticism – The Social Gospel.' Ph.D. diss., University of Pennsylvania, 1958.

Dillenberger, John. *Protestant Thought and Natural Science: A Historical Interpretation.* Nashville, Tenn.: Abingdon Press, 1960.

'Discussion on Darwinism and the Doctrine of Development.' In *History, Essays, Orations and Other Documents of the Sixth General Conference of the Evangelical Alliance Held in New York, October 2–12, 1873,* ed. Philip Schaff and S. Irenaeus Prime, pp. 317–23. New York: Harper & Bros., 1874.

Döllinger, Ignaz von. *Declarations and Letters on the Vatican Decrees, 1869–1887,* ed. F. H. Reusch. Edinburgh: T. & T. Clark, 1891.

Dörpinghaus, Hermann Josef. *Darwins Theorie und der deutsche Vulgärmaterialismus im Urteil deutscher katholischer Zeitschriften zwischen 1854 und 1914.* Doctoral diss., University of Freiburg in Breisgau, 1969.

Dollar, George W. 'The Early Days of American Fundamentalism.' *BS,* 123 (1966), 115–23.

A History of Fundamentalism in America. Greenville, S.C.: Bob Jones University Press, 1973.

Dorlodot, Henri de. *Darwinism and Catholic Thought,* trans. Ernest Messenger. London: Burns Oates & Washbourne, 1922.

Dorn, Jacob Henry. *Washington Gladden: Prophet of the Social Gospel.* Columbus: Ohio State University Press, 1966.

416   BIBLIOGRAPHY

Douglas, Mary. *Natural Symbols: Explorations in Cosmology*. London: Barrie & Rockliff, The Cresset Press, 1970.
Implicit Meanings: Essays in Anthropology. London: Routledge & Kegan Paul, 1975.
Drachman, Julian M. *Studies in the Literature of Natural Science*. New York: Macmillan Co., 1930.
Draper, John William. *History of the Conflict between Religion and Science*. International Scientific Series, vol. 13. London: Henry S. King & Co., 1875a [1874].
A History of the Intellectual Development of Europe. 2 vols. Revised edn. London: George Bell & Sons, 1875b [1862].
Drummond, Henry. *Natural Law in the Spiritual World*. London: Hodder & Stoughton, 1883.
The Lowell Lectures on the Ascent of Man. London: Hodder & Stoughton, 1894.
Duncan, David. *The Life and Letters of Herbert Spencer*. New edn. London: Williams & Norgate, 1911 [1908].
Dupree, A. Hunter. 'Jeffries Wyman's Views on Evolution.' *Isis*, 44 (1953), 243–6.
Asa Gray, 1810–1888. Cambridge, Mass.: Belknap Press of Harvard University Press, 1959.
Eaton, Clement. 'Professor James Woodrow and the Freedom of Teaching in the South' (1962). In *The Pursuit of Southern History: Presidential Addresses of the Southern Historical Association, 1935–1963*, ed. George Brown Tindall, pp. 438–50. Baton Rouge: Louisiana State University Press, 1964.
Ebenstein, William. 'The Early Reception of the Doctrine of Evolution in the United States.' *AS*, 4 (1939), 306–21.
Egerton, Frank N., III. 'Darwin's Early Reading of Lamarck.' *Isis*, 67 (1976), 452–6.
[Eggleston], Sister Mary Frederick. *Religion and Evolution since 1859: Some Effects of the Theory of Evolution on the Philosophy of Religion*. Chicago: Loyola University Press, 1935.
Eiseley, Loren. *Darwin's Century: Evolution and the Men Who Discovered It*. New edn. Garden City, N.Y.: Doubleday & Co., Anchor Books, 1961 [1958].
'The Intellectual Antecedents of "The Descent of Man".' In *Sexual Selection and the Descent of Man, 1871–1971*, ed. Bernard Campbell, pp. 1–16. London: Heinemann, 1972.
Eisen, Sydney. 'Huxley and the Positivists.' *VS*, 7 (1964), 337–58.
'Frederic Harrison and Herbert Spencer: Embattled Unbelievers.' *VS*, 12 (1968), 33–56.
Elert, Werner. *Der Kampf um das Christentum: Geschichte der Beziehungen zwischen dem Evangelischen Christentum in Deutschland und dem Allgemeinen Denken seit Schleiermacher und Hegel*. Munich: C. H. Beck, 1921.

Ellegård, Alvar. 'The Darwinian Theory and Nineteenth-Century Philosophies of Science.' *JHI*, 18 (1957a), 362–93.
'The Darwinian Theory and the Argument from Design.' In *Lychnos*, 1956, pp. 173–92. Uppsala: Almquist & Wiksell, 1957b.
Darwin and the General Reader: The Reception of Darwin's Theory of Evolution in the British Periodical Press, 1859–1872. Gothenburg: Elanders Boktryckeri Aktiebolag, 1958a.
'Public Opinion and the Press: Reactions to Darwinism.' *JHI*, 19 (1958b), 379–87.
Elliott, Robert James. 'The Improved Relation between the Scientific Doctrine of Evolution and Theology in the Last Fifty Years.' Ph.D. diss., Boston University, 1912.
Elliott-Binns, L. E. *Religion in the Victorian Era.* 2nd edn. London: Lutterworth Press, 1946 [1936].
The Development of English Theology in the Later Nineteenth Century. London: Longmans, Green & Co., 1952.
English Thought, 1860–1900: The Theological Aspect. London: Longmans, Green & Co., 1956.
'End of the MacQueary Case.' *Magazine of Christian Literature*, 5 (November 1891), 176.
Eve, A. S. and Creasey, C. H. *Life and Work of John Tyndall.* London: Macmillan, 1945.
Evolution in the Light of Modern Knowledge: A Collective Work. Glasgow: Blackie & Son, 1925.
Ewing, William, ed. *Annals of the Free Church of Scotland, 1843–1900.* 2 vols. Edinburgh: T. & T. Clark, 1914.
Farber, Paul L. 'Buffon and the Concept of Species.' *JHB*, 5 (1972), 259–84.
Farley, John. 'The Initial Reactions of French Biologists to Darwin's "Origin of Species".' *JHB*, 7 (1974), 275–300.
Farmer, William Wayne. 'The Anti-Evolution Crusade in Missouri, 1922–1971.' Ph.D. diss., University of Missouri (Columbia), 1974.
Farrar, Reginald. *The Life of Frederic William Farrar, D.D., F.R.S., Sometime Dean of Canterbury.* New York: Thomas Y. Crowell & Co., 1904.
Feibleman, James K. 'Darwin and Scientific Method.' *Tulane Studies in Philosophy*, 8 (1959), 3–14.
Festinger, Leon. *A Theory of Cognitive Dissonance.* London: Tavistock Publications, 1959 [1957].
Festinger, Leon et al. *Conflict, Decision, and Dissonance.* London: Tavistock Publications, 1964a [1959].
Festinger, Leon; Riecken, Henry W.; and Schachter, Stanley. *When Prophecy Fails: A Social and Psychological Study of a Modern Group That Predicted the Destruction of the World.* New York: Harper & Row, Harper Torchbooks, 1964b [1956].
Fine, Sidney. *Laissez Faire and the General-Welfare State: A Study of Conflict in American Thought, 1865–1901.* Reprint edn. Ann

Arbor: University of Michigan Press, Ann Arbor Paperbacks, 1964 [1956].

Fisch, Max H. 'Evolution in American Philosophy.' *Philosophical Review*, 56 (1947), 357–73.

Fischer, David Hackett. *Historians' Fallacies: Toward a Logic of Historical Thought.* New York: Harper & Row, Harper Torchbooks, 1970.

Fish, Carl Russell. 'Curtis, George Ticknor.' In *DAB*, 4 (1930), 613–14.

Fisher, George P. 'The Alleged Conflict of Natural Science and Religion.' *Princeton Review*, n.s., 12 (July 1883), 29–47.

Fiske, John. *Outlines of Cosmic Philosophy Based on the Doctrine of Evolution, with Criticisms on the Positive Philosophy.* 2 vols. London: Macmillan, 1874.

The Destiny of Man Viewed in the Light of His Origin. Boston: Houghton, Mifflin & Co., 1884a.

Excursions of an Evolutionist. Boston: Houghton, Mifflin & Co., 1884b.

Darwinism, and Other Essays. Revised edn. Boston: Houghton, Mifflin & Co., 1885a [1879].

The Idea of God as Affected by Modern Knowledge. Boston: Houghton, Mifflin & Co., 1885b.

Edward Livingston Youmans: Interpreter of Science for the People. New York: D. Appleton & Co., 1894.

A Century of Science, and Other Essays. Boston: Houghton, Mifflin & Co., 1899a.

Through Nature to God. Boston: Houghton, Mifflin & Co., 1899b.

Life Everlasting. Boston: Houghton, Mifflin & Co., 1901.

Fitch, Robert E. 'Darwinism and Christianity.' *Antioch Review*, 19 (1959), 20–32.

Fleming, Donald. *John William Draper and the Religion of Science.* Philadelphia: University of Pennsylvania Press, 1950.

'The Centenary of the "Origin of Species".' *JHI*, 20 (1959), 437–46.

'Charles Darwin: The Anaesthetic Man.' *VS*, 4 (1961), 219–36.

Flint, Robert. 'Some Requirements of a Present-day Christian Apologetics.' *Sermons and Addresses*, pp. 299–333. Edinburgh: William Blackwood & Sons, 1899.

Foote, G. W. *Darwin on God.* London: Progressive Publishing Co., 1889.

Forrest, D. W. *Francis Galton: The Life and Work of a Victorian Genius.* London: Paul Elek, 1974.

Fosdick, Harry Emerson. *The Living of These Days: An Autobiography.* London: S.C.M. Press, 1957.

Foster, Frank Hugh. *The Modern Movement in American Theology: Sketches in the History of American Protestant Thought from the Civil War to the World War.* New York: Fleming H. Revell, 1939.

A Genetic History of the New England Theology. Reprint ed. New York: Russell & Russell, 1963 [1907].

Foster, M. B. 'The Christian Doctrine of Creation and the Rise of Modern

Natural Science' (1934). In *Science and Religious Belief*, ed. C. A. Russell (1973), pp. 294–315.

Fothergill, Philip G. *Historical Aspects of Organic Evolution*. London: Hollis & Carter, 1952.

'Darwinian Theory and Its Effects.' *London Quarterly & Holborn Review*, 184 (1959), 289–94.

Evolution and Christians. London: Longmans, 1961.

Frederick, Sister Mary. See Eggleston, Sister Mary Frederick.

Freeman, Derek. 'The Evolutionary Theories of Charles Darwin and Herbert Spencer.' *Current Anthropology*, 15 (1974), 211–37.

French, Richard D. 'Darwin and the Physiologists, or the Medusa and Modern Cardiology.' *JHB*, 3 (1970), 253–74.

Antivivisection and Medical Science in Victorian Society. Princeton, N.J.: Princeton University Press, 1975.

Fruchtbaum, Harold. 'The Wisdom and the Works: Natural Theology and the Rise of Science.' Ph.D. diss., Harvard University, 1964.

Fulton, James Street. 'Philosophical Adventures of the Idea of Evolution: 1859–1959.' *Rice Institute Pamphlet*, 46 (1959), 1–31.

The Fundamentals. 12 vols. Chicago: Testimony Publishing Co., [1910–15].

Furniss, Norman F. *The Fundamentalist Controversy, 1918–1931*. New Haven, Conn.: Yale University Press, 1954.

Gabriel, Ralph Henry. *The Course of American Democratic Thought*. 2nd edn. New York: Ronald Press, 1956[1940].

Gale, Barry G. 'Darwin and the Concept of a Struggle for Existence: A Study in the Extrascientific Origins of Scientific Ideas.' *Isis*, 63 (1972), 321–44.

Galton, Francis. *English Men of Science: Their Nature and Nurture*. London: Macmillan, 1874.

'A Theory of Heredity.' *Journal of the Anthropological Institute of Great Britain and Ireland*, 5 (1876), 329–48.

Memories of My Life. London: Methuen, 1908.

Gantz, Kenneth. 'The Beginnings of Darwinian Ethics, 1859–1871.' Ph.D. diss., University of Chicago, 1937.

'The Beginnings of Darwinian Ethics, 1859–1871.' In *Studies in English*, pp. 180–209. Austin: University of Texas Press, 1939.

Garrison, Winfred Ernest. *The March of Faith: The Story of Religion in America since 1865*. New York: Harper & Bros., 1933.

Gasper, Louis. *The Fundamentalist Movement*. The Hague: Mouton & Co., 1963.

Gatewood, Willard B., Jr. *Preachers, Pedagogues & Politicians: The Evolution Controversy in North Carolina, 1920–1927*. Chapel Hill: University of North Carolina Press, 1966.

ed. *Controversy in the Twenties: Fundamentalism, Modernism, and Evolution*. Nashville, Tenn.: Vanderbilt University Press, 1969.

Geikie, Archibald. 'Campbell, George Douglas.' In *DNB*, 22 (1909), 390–1.

Geison, Gerald L. 'The Protoplasmic Theory of Life and the Vitalist–Mechanist Debate.' *Isis*, 60 (1969), 273–92.

Gentle, Brian Glynn. 'The Natural Theology of Newman Smyth: A Study of a Response of Late Nineteenth-Century New England Calvinism to Darwinian Evolutionary Science.' Ph.D. diss., Duke University, 1976.

George, Wilma. *Biologist Philosopher: A Study of the Life and Writings of Alfred Russel Wallace*. London: Abelard-Schuman, 1964.

Gerstner, John H., Jr. 'Scotch Realism, Kant and Darwin in the Philosophy of James McCosh.' Ph.D. diss., Harvard University, 1945.

Ghiselin, Michael T. *The Triumph of the Darwinian Method*. Berkeley: University of California Press, 1969.

'The Individual in the Darwinian Revolution.' *New Literary History*, 3 (1972), 113–34.

Gilkey, Langdon. *Maker of Heaven and Earth: A Study of the Christian Doctrine of Creation*. Garden City, N.Y.: Doubleday & Co., Anchor Books, 1965 [1959].

Gillespie, Neal C. 'The Duke of Argyll, Evolutionary Anthropology, and the Art of Scientific Controversy.' *Isis*, 68 (1977), 40–54.

Gillispie, Charles Coulston. *Genesis and Geology: A Study in the Relations of Scientific Thought, Natural Theology, and Social Opinion in Great Britain, 1790–1850*. New edn. New York: Harper & Row, Harper Torchbooks, 1959a [1951].

'Lamarck and Darwin in the History of Science.' In *Forerunners of Darwin, 1745–1859*, ed. B. Glass et al. (1959b), pp. 265–91.

Gilman, Daniel Coit. *The Life of James Dwight Dana*. New York: Harper & Bros., 1899.

Ginger, Ray. *Six Days or Forever? Tennessee v. John Thomas Scopes*. Reprint edn. Chicago: Quadrangle Books, Quadrangle Paperbacks, 1969 [1958].

Girvetz, Harry K. 'Philosophical Implications of Darwinism.' *Antioch Review*, 19 (1959), 9–19.

Gladstone, William Ewart. ' "Robert Elsmere:" The Battle of Belief' (1888). *Later Gleanings*, pp. 77–117. Vol. 8, Theological and Ecclesiastical. *Gleanings of Past Years*, 1885–1896. London: John Murray, 1897.

Glass, Bentley. 'The Germination of the Idea of Biological Species.' In *Forerunners of Darwin*, ed. B. Glass et al. (1959a), pp. 30–48.

'Heredity and Variation in the Eighteenth Century Concept of the Species.' In *Forerunners of Darwin*, ed. B. Glass et al. (1959b), pp. 144–72.

Glass, Bentley; Temkin, Owsei; and Straus, William L., Jr., eds. *Forerunners of Darwin, 1745–1859*. Baltimore, Md.: Johns Hopkins Press, 1959.

Glick, Thomas F., ed. *The Comparative Reception of Darwinism*. Austin: University of Texas Press, 1974a.

'Spain.' In *Comparative Reception of Darwinism*, ed. T. F. Glick (1974b), pp. 307–45.

Goblet d'Alviella, Eugène. *The Contemporary Evolution of Religious Thought in England, America, and India*, trans. J. Moden. New York: G. P. Putnam's Sons, 1886.

Gode-von Aesch, Alexander. *Natural Science in German Romanticism*. New York: Columbia University Press, 1941.

Goudge, T. A. 'Philosophical Trends in Nineteenth Century America.' *University of Toronto Quarterly*, 16 (1947), 133–42.

The Ascent of Life: A Philosophical Study of the Theory of Evolution. Toronto: University of Toronto Press, 1961.

'Evolutionism.' In *Dictionary of the History of Ideas: Studies of Selected Pivotal Ideas*, ed. Philip P. Wiener *et al.*, vol. 2, pp. 174–89. New York: Charles Scribner's Sons, 1973.

Gray, Asa. *Natural Selection Not Inconsistent with Natural Theology: A Free Examination of Darwin's Treatise on the Origin of Species and of Its American Reviewers*. London: Trübner & Co., 1861.

Natural Science and Religion: Two Lectures Delivered to the Theological School of Yale College. New York: Charles Scribner's Sons, 1880.

'Natural Selection and Natural Theology.' *Nature*, 27 (25 January 1883), 281–2; (5 April 1883), 527–8; vol. 28 (24 May 1883), 78.

'Darwin's Life and Letters.' *The Nation*, 45 (17 November 1887), 399–402; (24 November 1887), 420–1.

Darwiniana: Essays and Reviews Pertaining to Darwinism, ed. A. Hunter Dupree. Cambridge, Mass.: Belknap Press of Harvard University Press, 1963 [1876].

Gray, Jane Loring, ed. *The Letters of Asa Gray*. 2 vols. Boston: Houghton, Mifflin & Co., 1893.

Gray, Virginia. 'Anti-evolution Sentiment and Behaviour: The Case of Arkansas.' *Journal of American History*, 57 (1970), 352–66.

Gray, W. Forbes. 'Matheson, George.' In *DNB*, 2nd supp. (1912), 587–9.

Greene, John C. 'Objectives and Methods in Intellectual History.' *Mississippi Valley Historical Review*, 44 (1957), 58–74.

'Biology and Social Theory in the Nineteenth Century: Auguste Comte and Herbert Spencer.' In *Critical Problems in the History of Science: Proceedings of the Institute for the History of Science at the University of Wisconsin, September 1–11, 1957*, ed. Marshall Clagett, pp. 419–46. Madison: University of Wisconsin Press, 1959a.

The Death of Adam: Evolution and Its Impact on Western Thought. Ames: Iowa State University Press, 1959b.

Darwin and the Modern World View. Baton Rouge: Louisiana State University Press, 1961.

'Darwin and Religion' (1959). In *European Intellectual History since Darwin and Marx: Selected Essays*, ed. W. Warren Wagar, pp. 12–34. New York: Harper & Row, Harper Torchbooks, 1967.

'The Kuhnian Paradigm and the Darwinian Revolution in Natural

History.' In *Perspectives in the History of Science and Technology*, ed. Duane H. D. Roller, pp. 3–25. Norman: University of Oklahoma Press, 1971.

'Science and Religion.' In *The Rise of Adventism: Religion and Society in Mid-Nineteenth-Century America*, ed. Edwin Scott Gaustad, pp. 50–69. New York: Harper & Row, 1974.

'Reflections on the Progress of Darwin Studies.' *JHB*, 8 (1975), 243–73.

'Darwin as a Social Evolutionist.' *JHB*, 10 (1977), 1–27.

Grønbech, Vilhelm. *Religious Currents in the Nineteenth Century*, trans. P. M. Mitchell and W. D. Paden. Lawrence: University of Kansas Press, 1964.

Gruber, Howard E. and Barrett, Paul H. *Darwin on Man: A Psychological Study of Scientific Creativity*. London: Wildwood House, 1974.

Gruber, Howard E. and Gruber, Valmai 'The Eye of Reason: Darwin's Development during the "Beagle" Voyage.' *Isis*, 53 (1962), 186–200.

Gruber, Jacob W. *A Conscience in Conflict: The Life of St George Jackson Mivart*. New York: Temple University Publications, Columbia University Press, 1960.

Gruner, Rolf. 'Science, Nature, and Christianity.' *Journal of Theological Studies*, n.s., 26 (1975), 55–81.

Gulick, Addison. *Evolutionist and Missionary: John Thomas Gulick, Portrayed through Documents and Discussions*. Chicago: University of Chicago Press, 1932.

Gulick, John T. 'Christianity and the Evolution of Rational Life: A Statement Made on Solicitation of the Late George H. [sic] Romanes.' *BS*, 53 (January 1896), 68–74.

Guralnick, Stanley M. 'Geology and Religion before Darwin: The Case of Edward Hitchcock, Theologian and Geologist (1793–1864).' *Isis*, 63 (1972), 529–43.

'Sources of Misconception on the Role of Science in the Nineteenth-Century American College.' *Isis*, 65 (1974), 352–66.

Science and the Ante-Bellum American College. Philadelphia: American Philosophical Society, 1976.

Gustafson, Robert. 'A Study of the Life of James Woodrow, Emphasizing His Theological and Scientific Views as They Relate to the Evolution Controversy.' Th.D. diss., Union Theological Seminary (Richmond, Va.), 1964.

Haar, Charles M. 'E. L. Youmans: A Chapter in the Diffusion of Science in America.' *JHI*, 9 (1948), 193–213.

Haber, Francis C. *The Age of the World: Moses to Darwin*. Baltimore, Md.: Johns Hopkins Press, 1959.

'The Darwinian Revolution in the Concept of Time.' *Studium Generale*, 24 (1971), 289–307.

Hales, E. E. Y. *The Catholic Church in the Modern World: A Survey from the French Revolution to the Present*. London: Eyre & Spottiswoode, 1958.

Haller, John S., Jr. *Outcasts from Evolution: Scientific Attitudes of Racial Inferiority, 1859–1900*. Reprint ed. New York: McGraw-Hill Book Co., 1975 [1971].

Halliday, R. J. 'Social Darwinism: A Definition.' *VS*, 14 (1971), 389–405.

Handy, Robert T. *A Christian America: Protestant Hopes and Historical Realities*. New York: Oxford University Press, 1971.

Hannah, J. 'The Attitude of the Clergy towards Science.' *Contemporary Review*, 6 (September 1867), 1–17.

'A Few More Words on the Relation of the Clergy to Science.' *Contemporary Review*, 9 (November 1868), 395–404.

Hardin, Garrett J. *Nature and Man's Fate*. London: Jonathan Cape, 1960 [1959].

Hardwick, John Charlton. *Religion and Science: From Galileo to Bergson*. London: Society for Promoting Christian Knowledge, 1920.

Harper, George McLean. 'Shields, Charles Woodruff.' In *DAB*, 17 (1935), 104–5.

Harrington, Carroll E. 'The Fundamentalist Movement in America, 1870–1920.' Ph.D. diss., University of California (Berkeley), 1959.

Harvie, Christopher. 'Ideology and Home Rule: James Bryce, A. V. Dicey and Ireland, 1880–1887.' *English Historical Review*, 91 (1976), 298–314.

Hauber, W. A. 'Evolution and Catholic Thought.' *American Ecclesiastical Review*, 106 (1942), 161–77.

Hayes, Carlton J. H. *A Generation of Materialism, 1871–1900*. New York: Harper & Bros., 1941.

Heimann, P. M. ' "The Unseen Universe": Physics and the Philosophy of Nature in Victorian Britain.' *BJHS*, 6 (1972), 73–9.

Helfand, Michael S. 'T. H. Huxley's "Evolution and Ethics": The Politics of Evolution and the Evolution of Politics.' *VS*, 20 (1977), 159–77.

Henry, Carl F. H. *Personal Idealism and Strong's Theology*. Wheaton, Ill.: Van Kampen Press, 1951.

Henslow, George. *Genesis and Geology: A Plea for the Doctrine of Evolution*. London: Robert Hardwicke, 1871.

The Theory of Evolution of Living Things and the Application of the Principles of Evolution to Religion, Considered as Illustrative of the 'Wisdom and Beneficence of the Almighty'. London: Macmillan, 1873.

The Origin of Floral Structures through Insect and Other Agencies. International Scientific Series, vol. 64. London: Kegan Paul, Trench & Co., 1888.

The Origin of Plant Structures by Self-Adaptation to the Environment. International Scientific Series, vol. 77. London: Kegan Paul, Trench, Trübner & Co., 1895.

Present-Day Rationalism Critically Examined. London: Hodder & Stoughton, 1904.

The Heredity of Acquired Characters in Plants. London: John Murray, 1908.

Herbert, Sandra. 'Darwin, Malthus, and Selection.' *JHB*, 4 (1971), 209–17. 'The Place of Man in the Development of Darwin's Theory of Transmutation: Part I. To July 1837.' *JHB*, 7 (1974), 217–58. 'The Place of Man in the Development of Darwin's Theory of Transmutation. Part II. *JHB*, 10 (1977), 155–227.

Herschel, John Frederick William. *Preliminary Discourse on the Study of Natural Philosophy*. London: Longman, Rees, Orme, Brown & Green, 1830.

Hibben, Paxton. *Henry Ward Beecher: An American Portrait*. New edn. New York: Press of the Readers Club, 1942 [1927].

Hicks, L. E. *A Critique of Design-Arguments: A Historical Review and Free Examination of the Methods of Reasoning in Natural Theology*. New York: Charles Scribner's Sons, 1883.

Higgins, John Edward. 'The Young John Fiske, 1842–1874.' Ph.D. diss., Harvard University, 1960.

Hilts, Victor L. 'A Guide to Francis Galton's "English Men of Science".' *Transactions of the American Philosophical Society*, n.s., 65 (1975), pt 5.

Himmelfarb, Gertrude. *Darwin and the Darwinian Revolution*. Reprint edn. New York: W. W. Norton & Co., 1968a [1959]. *Victorian Minds*. London: Weidenfeld & Nicolson, 1968b.

Himmelfarb, Samuel and Eagly, Alice Hendrickson. *Readings in Attitude Change*. New York: John Wiley & Sons, 1974.

Hinchliff, Peter Bingham. *John William Colenso*. London: Nelson, 1964.

Ho, Wing Meng. 'Methodological Issues in Evolutionary Theory, with Special Reference to Darwinism and Lamarckism.' D. Phil. thesis, University of Oxford, 1966.

Hodge, A. A. *The Life of Charles Hodge, D.D., LL.D.* New York: Charles Scribner's Sons, 1880.

Hodge, Charles. *Systematic Theology*. 3 vols. New York: Scribner, Armstrong & Co., 1872–3. *What Is Darwinism?* New York: Scribner, Armstrong & Co., 1874.

Hodge, M. J. S. 'Lamarck's Science of Living Bodies.' *BJHS*, 5 (1971), 323–52. 'The Universal Gestation of Nature: Chambers' "Vestiges" and "Explanations".' *JHB*, 5 (1972), 127–51. 'England.' In *Comparative Reception of Darwinism*, ed. T. F. Glick (1974a), pp. 3–31. 'England: Bibliographical Essay.' In *Comparative Reception of Darwinism*, ed. T. F. Glick (1974b), pp. 75–80.

Hofstadter, Richard. *Social Darwinism in American Thought*. Revised edn. Boston: Beacon Press, 1955 [1944]. *Anti-Intellectualism in American Life*. New York: Vintage Books, 1963.

Holifield, E. Brooks. 'The English Methodist Response to Darwin.' *Methodist History*, 10 (1972), 14–22.

Hooykaas, R. 'The Parallel between the History of the Earth and the History of the Animal World.' *Archives Internationales d'Histoire des Sciences*, 10 (1957), 1–18.

The Principle of Uniformity in Geology, Biology, and Theology. New edn. Leiden: E. J. Brill, 1963 [1959].

'Geological Uniformitarianism and Evolution.' *Archives Internationales d'Histoire des Sciences*, 19 (1966), 3–19.

Catastrophism in Geology, Its Scientific Character in Relation to Actualism and Uniformitarianism. Amsterdam: North-Holland Publishing Co., 1970.

Religion and the Rise of Modern Science. Grand Rapids, Mich.: William B. Eerdmans Publishing Co., 1972.

Hooykaas, R; Lawless, Clive; and Russell, Colin A. The New Outlook for Science. Science and Belief: from Copernicus to Darwin (An Inter-Faculty Second Level Course in the History of Science, The Open University), block 6, units 15–16. Milton Keynes, Bucks.: Open University Press, 1974.

Hopkins, Charles Howard. The Rise of the Social Gospel in American Protestantism, 1865–1915. New edn. New Haven, Conn.: Yale University Press, 1967 [1940].

Hort, Arthur Fenton. Life and Letters of Fenton John Anthony Hort. 2 vols. London: Macmillan, 1896.

Houghton, Walter E. The Victorian Frame of Mind, 1830–1870. New Haven, Conn.: Yale University Press, 1957.

Hovenkamp, Herbert John. 'Science and Religion in America, 1800–1860.' Ph.D. diss., University of Texas (Austin), 1976.

Hudson, William Henry. An Introduction to the Philosophy of Herbert Spencer. Revised edn. London: Watts & Co., 1904 [1894].

Hudson, Winthrop S., ed. Nationalism and Religion in America: Concepts of American Identity and Mission. New York: Harper & Row, Harper Forum Books, 1970.

Hübner, Jürgen. Theologie und biologische Entwicklungslehre: Ein Beitrag zum Gespräch zwischen Theologie und Naturwissenschaft. Munich: C. H. Beck, 1966.

Hull, David L. 'The Metaphysics of Evolution.' *BJHS*, 3 (1967), 309–37.

'Charles Darwin and Nineteenth-Century Philosophies of Science.' In Foundations of Scientific Method: The Nineteenth Century, ed. Ronald N. Giere and Richard S. Westfall, pp. 115–32. Bloomington: Indiana University Press, 1973a.

Darwin and His Critics: The Reception of Darwin's Theory of Evolution by the Scientific Community. Cambridge, Mass.: Harvard University Press, 1973b.

'Darwinism and Historiography.' In Comparative Reception of Darwinism, ed. T. F. Glick (1974), pp. 388–402.

Hume, David. Dialogues Concerning Natural Religion. 2nd edn. London: N.p., 1779 [1779].

Hunt, John. *Religious Thought in England in the Nineteenth Century.* London: Gibbings & Co., 1896.

Huntley, William B. 'David Hume and Charles Darwin.' *JHI*, 33 (1972), 457–70.

Hurst, John Fletcher. *History of Rationalism: Embracing a Survey of the Present State of Protestant Theology.* Revised edn. New York: Eaton & Mains, 1901 [1865].

Hutchison, William R. *The Modernist Impulse in American Protestantism.* Cambridge, Mass.: Harvard University Press, 1976.

Hutton, Frederick Wollaston. *Darwinism and Lamarckism, Old and New.* New York: G. P. Putnam's Sons, 1899.

Huxley, Leonard. *Life and Letters of Thomas Henry Huxley.* 3 vols. New edn. London: Macmillan, 1903 [1900].

Life and Letters of Sir Joseph Dalton Hooker. 2 vols. London: John Murray, 1918.

Thomas Henry Huxley: A Character Sketch. London: Watts & Co., 1920.

Huxley, Thomas Henry. *Evidence as to Man's Place in Nature.* London: Williams & Norgate, 1863.

Lay Sermons, Addresses, and Reviews. London: Macmillan, 1870.

Critiques and Addresses. London: Macmillan, 1873.

'On the Reception of the "Origin of Species".' In *Life and Letters of Charles Darwin*, ed. F. Darwin (1887), II, 179–204.

'An Apologetic Irenicon.' *Fortnightly Review*, n.s., 52 (1 November 1892), 557–71.

Collected Essays. 9 vols. London: Macmillan, 1893–4.
  1. *Method and Results*, 1893.
  2. *Darwiniana*, 1893.
  3. *Science and Education*, 1893.
  4. *Science and Hebrew Tradition*, 1893.
  5. *Science and Christian Tradition*, 1894.
  6. *Hume, with Helps to the Study of Berkeley*, 1894.
  7. *Man's Place in Nature, and Other Anthropological Essays*, 1894.
  8. *Discourses Biological and Geological*, 1894.
  9. *Evolution and Ethics, and Other Essays*, 1894.

'Owen's Position in the History of Anatomical Science.' In *Life of Richard Owen*, by Rev. R. Owen (1894), II, 273–332.

The Scientific Memoirs of Thomas Henry Huxley, ed. Michael Foster and E. Ray Lankester. 4 vols. London: Macmillan, 1898–1902.

Hyman, Stanley Edgar. 'The "Origin" as Scripture.' *Virginia Quarterly Review*, 35 (1959), 540–52.

'Darwin's Sidelight: The Shape of the Young Man's Nose.' *Atlantic Monthly*, 220 (1967), 96–104.

Ideas and Beliefs of the Victorians. London: Sylvan Press, 1949.

Illick, Joseph E., III. 'The Reception of Darwinism at the Theological Seminary and the College at Princeton, New Jersey.' *Journal of the Presbyterian Historical Society*, 38 (1960), 152–65, 234–43.

The Impact of Darwinian Thought on American Life and Culture:

Papers Read at the Fourth Annual Meeting of the American Studies Association of Texas at Houston, Texas. Austin: University of Texas, 1959.

Irvine, William. *Apes, Angels, and Victorians: The Story of Darwin, Huxley, and Evolution.* Reprint edn. New York: McGraw-Hill Book Co., 1972 [1955].

Iverach, James. *Is God Knowable?* London: Hodder & Stoughton, 1884a.

'Jonathan Edwards.' In *The Evangelical Succession: A Course of Lectures Delivered in St George's Free Church, Edinburgh*, pp. 109–43. 3rd ser. Edinburgh: Macniven & Wallace, 1884b.

The Philosophy of Mr Herbert Spencer Examined. London: Religious Tract Society, [1884c].

The Ethics of Evolution Examined. London: Religious Tract Society, [1887].

Christianity and Evolution. London: Hodder & Stoughton, 1894.

Theism in the Light of Present Science and Philosophy. London: Hodder & Stoughton, 1900 [1899].

'Iverach, James.' In *New Schaff-Herzog Encyclopedia of Religious Knowledge*, vol. 6, p. 71. New York: Funk & Wagnalls, 1910.

Jacks, Lawrence Pearsall. *Life and Letters of Stopford Brooke.* 2 vols. London: John Murray, 1917.

James, D. G. *Henry Sidgwick: Science and Faith in Victorian England.* London: Oxford University Press, 1970.

'James Dwight Dana.' *BS*, 52 (July 1895), 557–8.

James, Henry, ed. *The Letters of William James.* 2 vols. London: Longmans, Green & Co., 1920.

Jensen, J. Vernon. 'The Rhetoric of Thomas H. Huxley and Robert G. Ingersoll in Relation to the Conflict between Science and Theology.' Ph.D. diss., University of Minnesota, 1959.

'The Rhetorical Influence of T. H. Huxley on the United States.' *Western Speech*, 31 (1967), 29–36.

'The X Club: Fraternity of Victorian Scientists.' *BJHS*, 5 (1970), 63–72.

'Interrelationships within the Victorian "X Club".' *Dalhousie Review*, 51 (1971–2), 539–52.

Jenyns, Leonard. *See* Blomefield, Leonard.

Jessop, W. J. E. 'Samuel Haughton: A Victorian Polymath.' *Hermathena*, no. 116 (1973), 5–26.

Johnson, Deryl Freeman. 'The Attitudes of the Princeton Theologians toward Darwinism and Evolution from 1859–1929.' Ph.D. diss., University of Iowa, 1968.

Johnson, Francis Howe. *What Is Reality? An Inquiry as to the Reasonableness of Natural Religion and the Naturalness of Revealed Religion.* Boston: Houghton, Mifflin & Co., 1891.

God in Evolution: A Pragmatic Study of Theology. New York: Longmans, Green & Co., 1911.

'Johnson, Francis Howe.' In *Who Was Who in America*, vol. 1 (1897–1942), p. 638. Chicago: A. N. Marquis Co., 1942.

Johnson, LeRoy. 'The Evolution Controversy during the 1920's.' Ph.D. diss., New York University, 1954.

Johnston, Arthur, ' "The Water Babies": Kingsley's Debt to Darwin.' *English*, 12 (1959), 215–19.

Johnston, John Octavius. *Life and Letters of Henry Parry Liddon.* London: Longmans, Green & Co., 1904.

Jones, G. J. 'Darwinism and Social Thought – A Study of the Relationship between Science and the Development of Sociological Theory in Britain, 1860–1914.' Ph.D. thesis, University of London, 1974.

Judd, John W. *The Coming of Evolution: The Story of a Great Revolution in Science.* Cambridge University Press, 1910.

Källstad, Thorvald. *John Wesley and the Bible: A Psychological Study.* Stockholm: NYA Bokförlags Aktiebolaget, 1974.

Keeton, Morris T. 'Edmund Montgomery – Pioneer of Organicism.' *JHI*, 8 (1947), 309–41.

Keith, Arthur. 'Darwin's Theory of Man's Descent as It Stands Today.' *Nature*, no. 3018 (3 September 1927), supp., 14–21.

Kellogg, Vernon L. *Darwinism To-day: A Discussion of Present-day Scientific Criticism of the Darwinian Selection Theories together with a Brief Account of the Principal Other Proposed Auxiliary and Alternative Theories of Species Forming.* New York: Henry Holt & Co., 1907.

Kelvin, Lord. *See* Thomson, William.

Kennedy, Gail, ed. *Evolution and Religion: The Conflict between Science and Theology in Modern America.* Lexington, Mass.: D. C. Heath & Co., 1957.

Kent, John. *From Darwin to Blatchford: The Role of Darwinism in Christian Apologetic, 1875–1910.* Friends of Doctor Williams's Library Twentieth Lecture, 1966. London: Dr Williams's Trust, 1966.

Keynes, John Maynard. 'Alfred Marshall.' *Essays in Biography*, pp. 161–231. Vol. 10. *The Collected Writings of John Maynard Keynes.* London: Macmillan, St Martin's Press for the Royal Economic Society, 1972.

King, Agnes Gardner. *Kelvin the Man: A Biographical Sketch by His Niece.* London: Hodder & Stoughton, 1925.

King, Clarence. 'Catastrophism and Evolution.' *American Naturalist*, 11 (August 1877), 449–70.

Kingsley, Charles. 'The Natural Theology of the Future' (1871). *Scientific Lectures and Essays*, pp. 313–36. New edn. London: Macmillan, 1890 [1880].

[Kingsley, Fanny]. *Charles Kingsley: His Letters and Memories of His Life.* 2 vols. London: Henry S. King & Co., 1877.

Kinns, Samuel. *Moses and Geology; or, The Harmony of the Bible with Science.* 7th edn. London: Cassell & Co., 1884 [1881].

Kirkman, Thomas Penyngton. *Philosophy without Assumptions.* London: Longmans, Green & Co., 1876.

Klaaren, Eugene Marion. 'Belief in Creation and the Rise of Modern Natural Science.' Ph.D. diss., Harvard University, 1970.

Kline, George L. 'Darwinism and the Russian Orthodox Church.' In *Continuity and Change in Russian and Soviet Thought*, ed. E. J. Simmons, pp. 307–28. Cambridge, Mass.: Harvard University Press, 1955.

Kneller, Karl Alois. *Christianity and the Leaders of Modern Science: A Contribution to the History of Culture in the Nineteenth Century*, trans. T. M. Kettle. London: B. Herder, 1911 [1910].

Knight, D. M. 'Professor Baden Powell and the Inductive Philosophy.' *Durham University Journal*, n.s., 29 (1968), 81–7.

Knudson, Albert C. 'Townsend, Luther Tracy.' In *DAB*, 18 (1936), 618–19.

Kottler, Malcolm Jay. 'Alfred Russel Wallace, the Origin of Man, and Spiritualism.' *Isis*, 65 (1974), 145–92.

Kramer, Herbert J. 'The Intellectual Background and Immediate Reception of Darwin's "Origin of Species".' Ph.D. diss., Harvard University, 1948.

Kuhn, Thomas S. *The Structure of Scientific Revolutions*. Vol. 2, no. 2. *International Encyclopedia of Unified Science*, ed. Otto Neurath. Revised edn. Chicago: University of Chicago Press, 1970 [1962].

Kultgen, J. H. 'Biological Evolution and American Metaphysics.' In *Impact of Darwinian Thought on American Life and Culture* (1959), pp. 84–92.

Lack, David. *Evolutionary Theory and Christian Belief: The Unresolved Conflict*. New edn. London: Methuen, 1961 [1957].

Lane, Helen. 'Heredity and Environment in American Social Thought, 1900–1929: The Aftermath of Spencer.' Ph.D. diss., Columbia University, 1950.

Langford, Thomas A. *In Search of Foundations: English Theology, 1900–1920*. Nashville, Tenn.: Abingdon Press, 1968.

Lankester, E. Ray. *Degeneration: A Chapter in Darwinism*. London: Macmillan, 1880.

Latourette, Kenneth Scott. *Christianity in a Revolutionary Age: A History of Christianity in the Nineteenth and Twentieth Centuries*. 5 vols. Reprint edn. Grand Rapids, Mich.: Zondervan Publishing House, 1969 [1958–62].

Layard, George Somes. *Mrs Lynn Linton: Her Life, Letters, and Opinions*. London: Methuen, 1901.

Le Conte, Joseph. *Evolution: Its Nature, Its Evidences, and Its Relation to Religious Thought*. Revised edn. New York: D. Appleton & Co., 1891 [1888].

'God, and Connected Problems, in the Light of Evolution.' In *The Conception of God: A Philosophical Discussion Concerning the Nature of the Divine Idea as a Demonstrable Reality*, by Josiah Royce *et al.*, pp. 65–78. New York: Macmillan Co., 1897.

The Autobiography of Joseph Le Conte, ed. William Dallam Armes. New York: D. Appleton & Co., 1903.

Le Duc, Thomas. *Piety and Intellect at Amherst College, 1865–1912*. New York: Columbia University Press, 1946.

Leeds, Anthony. 'Darwinian and "Darwinian" Evolutionism in the Study of Society and Culture.' In *Comparative Reception of Darwinism*, ed. T. F. Glick (1974), pp. 437–85.

LeMahieu, D. L. *The Mind of William Paley: A Philosopher and His Age*. Lincoln: University of Nebraska Press, 1976.

Lennox, Cuthbert. *See* Napier, John H.

Lesch, John E. 'The Role of Isolation in Evolution: George J. Romanes and John T. Gulick.' *Isis*, 66 (1975), 483–503.

Leslie, W. Bruce. 'James McCosh in Scotland.' *Princeton University Library Chronicle*, 36 (1974), 47–60.

Letwin, Shirley. *The Pursuit of Certainty: David Hume, Jeremy Bentham, John Stuart Mill, Beatrice Webb*. Cambridge University Press, 1965.
'Certainty since the Seventeenth Century.' In *Dictionary of the History of Ideas: Studies of Selected Pivotal Ideas*, ed. Philip P. Wiener *et al.*, vol. 1, pp. 312–25. New York: Charles Scribner's Sons, 1973.

Leuba, James H. *The Belief in God and Immortality: A Psychological, Anthropological, and Statistical Study*. Chicago: Open Court Publishing Co., 1921 [1916].

Leuchtenberg, William E. 'The New Deal and the Analogue of War.' In *World War I*, pp. 124–52. War and Society (Arts: A Third Level Course, The Open University), block 5, units 14–17. Milton Keynes, Bucks.: Open University Press, 1973.

Levere, Trevor H. and Jarrell, Richard A., eds. *A Curious Field-Book: Science & Society in Canadian History*. Toronto: Oxford University Press, 1974.

Leverette, William E., Jr. 'Science and Values: A Study of Edward L. Youmans' *Popular Science Monthly*, 1872–1887.' Ph.D. diss., Vanderbilt University, 1963.
'E. L. Youmans's Crusade for Scientific Respectability.' *American Quarterly*, 17 (1965), 12–32.

Levin, Samuel M. 'Malthus and the Idea of Progress.' *JHI*, 27 (1966), 92–108.

Levine, Lawrence. *Defender of the Faith, William Jennings Bryan: The Last Decade, 1915–1925*. New York: Oxford University Press, 1965.

Liddon, Henry Parry. *Some Elements of Religion: Lent Lectures, 1870*. London: Rivingtons, 1872.
Sermons Preached before the University of Oxford. 2nd ser., 1868–1880. London: Rivingtons, 1880.
Life of Edward Bouverie Pusey. 4 vols. London: Longmans, Green & Co., 1893–7.

Limoges, Camille. 'Darwinisme et adaptation.' *Revue des questions scientifiques*, 141 (1970a), 353–74.

La sélection naturelle: Etude sur la première constitution d'un concept (*1837–1859*). Paris: Presses Universitaires de France, 1970b.

Lippman, Walter. *American Inquisitors: A Commentary on Dayton and Chicago.* New York: Macmillan Co., 1928.

Livingston, James C. *The Ethics of Belief: An Essay on the Victorian Religious Conscience.* Tallahassee, Fla.: American Academy of Religion, 1974.

Loetscher, Lefferts A. *The Broadening Church: A Study of Theological Issues in the Presbyterian Church since 1869.* Philadelphia: University of Pennsylvania Press, 1954.

Loewenberg, Bert James. 'The Reaction of American Scientists to Darwinism.' *American Historical Review,* 38 (1933), 687–701.

'The Impact of the Doctrine of Evolution on American Thought.' Ph.D. diss., Harvard University, 1934.

'The Controversy over Evolution in New England, 1859–1873.' *New England Quarterly,* 8 (1935), 232–57.

'Darwinism Comes to America, 1859–1900.' *Mississippi Valley Historical Review,* 28 (1941), 339–68.

Darwinism: Reaction or Reform? New York: Holt, Rinehart & Winston, 1957.

'Darwin Scholarship of the Darwin Year.' *American Quarterly,* 11 (1959a), 521–33.

'The Mosaic of Darwinian Thought.' *VS,* 3 (1959b), 3–18.

'Darwin and Darwin Studies, 1959–1963.' *History of Science,* 4 (1965), 15–54.

introd. *Calendar of the Letters of Charles Robert Darwin to Asa Gray.* Reprint edn. Wilmington, Del.: Scholarly Resources, 1973 [1939].

Loraine, Nevison. *The Battle of Belief: A Review of the Present Aspects of the Conflict.* 3rd edn. London: Society for Promoting Christian Knowledge, 1903 [1891].

Losee, John. *A Historical Introduction to the Philosophy of Science.* London: Oxford University Press, 1972.

Lovejoy, Arthur O. 'Some Aspects of Darwin's Influence upon Modern Thought.' *Bulletin of Washington University,* April 1909, pp. 85–99.

'The Argument for Organic Evolution before the "Origin of Species", 1830–1858.' In *Forerunners of Darwin,* ed. B. Glass *et al.* (1959), pp. 356–414.

The Great Chain of Being: A Study of the History of an Idea. Reprint edn. New York: Harper & Row, Harper Torchbooks, 1960 [1936].

Lunn, Arnold. *The Flight from Reason: A Study of the Victorian Heresy.* London: Eyre & Spottiswoode, 1931.

Lurie, Edward. 'Louis Agassiz and the Idea of Evolution.' *VS,* 3 (1959), 87–108.

Louis Agassiz: A Life in Science. Chicago: University of Chicago Press, 1960.

Luzzatti, Luigi. 'Sulle idee filosofiche e religiose di Darwin, sotto l'influenza delle sue dottrine naturali.' *Atti della Reale Academia dei Lincei,*

5th ser., Rendiconti classe di scienze fisiche, matematiche, e naturali, 10 (1901), 60–72.

Lyell, Charles. *The Geological Evidences of the Antiquity of Man, with Remarks on Theories of the Origin of Species by Variation.* 2nd edn. London: John Murray, 1863 [1863].

Principles of Geology; or, The Modern Changes of the Earth and Its Inhabitants as Illustrative of Geology. 2 vols. 10th edn. London: John Murray, 1867–8 [3 vols., 1830–3].

[Lyell, K. M.], ed. *Life, Letters and Journals of Sir Charles Lyell, Bart.* 2 vols. London: John Murray, 1881.

Lyon, John. 'Immediate Reactions to Darwin: The English Catholic Press' First Reviews of the "Origin of the Species" [sic].' *CH*, 41 (1972), 78–93.

Lyttle, Charles H. 'Savage, Minot Judson.' *DAB*, 16 (1935), 389–90.

McCabe, Joseph. *Life and Letters of George Jacob Holyoake.* 2 vols. London: Watts & Co., 1908.

McConnaughey, Gloria. 'Darwin and Social Darwinism.' *Osiris*, 9 (1950), 397–412.

McCosh, James. *Christianity and Positivism: A Series of Lectures to the Times on Natural Theology and Apologetics.* New York: Robert Carter & Bros., 1871.

[McCosh, James]. *The Conflicts of the Age.* New York: Charles Scribner's Sons, 1881.

McCosh, James. *The Method of the Divine Government, Physical and Moral.* 12th edn. London: Macmillan, 1882 [1850].

Development: What It Can Do and What It Cannot Do. New York: Charles Scribner's Sons, 1883.

Herbert Spencer's Philosophy as Culminated in His Ethics. New York: Charles Scribner's Sons, 1885.

The Religious Aspect of Evolution. Rev. ed. New York: Charles Scribner's Sons, 1890 [1888].

The Life of James McCosh: A Record Chiefly Autobiographical, ed. William Milligan Sloane. New York: Charles Scribner's Sons, 1896.

McCosh, James and Dickie, George. *Typical Forms and Special Ends in Creation.* 2nd edn. Edinburgh: Thomas Constable & Co., 1857 [1856].

McCrossin, G. Michael. 'World Views in Conflict: Evolution, Progress, and Christian Tradition in the Thought of John Fiske, Minot Savage, and Lyman Abbott.' Ph.D. diss., University of Chicago, 1970.

McDonagh, Eileen L. 'Attitude Changes and Paradigm Shifts: Social Psychological Foundations of the Kuhnian Thesis.' *Social Studies of Science*, 6 (1976), 51–76.

MacDonald, Frederick Charles. *Handley Carr Glyn Moule, Bishop of Durham: A Biography.* London: Hodder & Stoughton, [1922].

McElligott, John Francis. 'Before Darwin: Religion and Science as Presented in American Magazines, 1830–1860.' Ph.D. diss., New York University, 1973.

McGiffert, Arthur Cushman. *The Rise of Modern Religious Ideas.* New York: Macmillan Co., 1915.
'The Progress of Theological Thought during the Past Fifty Years.' *American Journal of Theology,* 20 (1916), 321–32.
McGiffert, Michael. 'Christian Darwinism: The Partnership of Asa Gray and George Frederick Wright, 1874–1881.' Ph.D. diss., Yale University, 1958.
McGrath, Earl. 'The Control of Higher Education in America.' *Educational Record,* 17 (1936), 259–79.
McIver, Malcolm Chester, Jr. 'The Preaching of Henry Drummond, with Reference to His Work among Students.' Ph.D. thesis, University of Edinburgh, 1958.
Mackenzie, Donald. 'James Iverach.' *Expository Times,* 32 (1920), 55–60.
McKinney, H. Lewis. *Wallace and Natural Selection.* New Haven, Conn.: Yale University Press, 1972.
Mackintosh, Robert. *From Comte to Benjamin Kidd: The Appeal to Biology or Evolution for Human Guidance.* London: Macmillan, 1899.
MacLeod, Roy M. 'Evolutionism and Richard Owen, 1830–1868: An Episode in Darwin's Century.' *Isis,* 56 (1965), 259–80.
'The X-Club: A Social Network of Science in Late-Victorian England.' *NR,* 24 (1970), 305–22.
McLoughlin, William G. *The Meaning of Henry Ward Beecher: An Essay on the Shifting Values of Mid-Victorian America.* New York: Alfred A. Knopf, 1970.
Macmillan, D. *The Life of George Matheson, D.D., LL.D., F.R.S.E.* London: Hodder & Stoughton, 1907.
McNabb, Vincent, ed. *The Decrees of the Vatican Council.* London: Burns & Oates, 1907.
McNeill, John T. *The History and Character of Calvinism.* New York: Oxford University Press, Galaxy Books, 1967 [1954].
McPheeters, Chilton Claudius. 'The Changing Apologetic Emphasis of Anglican Theology as Represented in the Bampton Lectures, 1780–1940.' Ph.D. diss., Drew University, 1948.
Macpherson, Hector. *The Church and Science: A Study of the Interrelation of Theological and Scientific Thought.* London: James Clarke & Co., [1927].
MacQueary, Thomas Howard. *The Evolution of Man and Christianity.* Revised edn. New York: D. Appleton & Co., 1891a [1890].
Topics of the Times. New York: United States Book Co., 1891b.
Ecclesiastical Liberty: Being the Defense of the Rev. Howard Mac-Queary before the Ecclesiastical Court of the Episcopal Church in Northern Ohio against the Charges of Heresy, Delivered in Cleveland, Ohio, January Seventh 1891. New York: United States Book Co., n.d.
'MacQueary, Thomas Howard.' In *Who Was Who in America,* vol. 1 (1897–1942), p. 766. Chicago: A. N. Marquis Co., 1942.

Macran, F. W. *English Apologetic Theology*. London: Hodder & Stoughton, 1905.
Maitland, Frederic William. *The Life and Letters of Leslie Stephen*. London: Duckworth & Co., 1906.
Major, H. D. A. *English Modernism: Its Origin, Methods, Aims*. Cambridge, Mass.: Harvard University Press, 1927.
Maltby, Arthur. *Religion and Science*. London: Library Association, 1965.
Malthus, T. R. *Essay on the Principle of Population as It Affects the Future Improvement of Society, with Remarks on the Speculations of W. Godwin, M. Condorcet, and Other Writers*. London: J. Johnson, 1798.
An Essay on the Principle of Population; or, A View of Its Past and Present Effects on Human Happiness, with an Inquiry into Our Prospects Respecting the Future Removal or Mitigation of the Evils Which It Occasions. 2 vols. 6th edn. London: John Murray, 1826 [1798].
Mandelbaum, Maurice. 'The Scientific Background of Evolutionary Theory in Biology.' *JHI*, 18 (1957), 342–61.
'Darwin's Religious Views.' *JHI*, 19 (1958), 363–78.
History, Man & Reason: A Study in Nineteenth Century Thought. Baltimore, Md.: Johns Hopkins Press, 1971.
Manier, Edward. *The Young Darwin and His Cultural Circle: A Study of Influences Which Helped Shape the Language and Logic of the First Drafts of the Theory of Natural Selection*. Dordrecht, The Netherlands: D. Reidel Publishing Co., 1977.
Marchant, James. *Alfred Russel Wallace: Letters and Reminiscences*. 2 vols. London: Cassell & Co., 1916.
Marsden, George M. *The Evangelical Mind and the New School Presbyterian Experience: A Case Study of Thought and Theology in Nineteenth-Century America*. New Haven, Conn.: Yale University Press, 1970.
'Defining Fundamentalism.' *Christian Scholar's Review*, 1 (1971), 141–51.
'From Fundamentalism to Evangelicalism: A Historical Analysis.' In *The Evangelicals: What They Believe, Who They Are, Where They Are Changing*, ed. David F. Wells and John D. Woodbridge, pp. 122–42. Nashville, Tenn.: Abingdon Press, 1975.
Matheson, George. *Can the Old Faith Live with the New? or, The Problem of Evolution and Revelation*. Edinburgh: William Blackwood & Sons, 1885.
Mathews, Basil. *John R. Mott, World Citizen*. New York: Harper & Row, 1934.
Maurice, Frederick, ed. *The Life of Frederick Denison Maurice*. 2 vols. London: Macmillan, 1884.
May, Henry F. *Protestant Churches and Industrial America*. New York: Harper & Bros., 1949.

Mayr, Ernst. *Evolution and the Diversity of Life: Selected Essays.* Cambridge, Mass.: Belknap Press of Harvard University Press, 1976.
Mead, G. H. *Movements of Thought in the Nineteenth Century.* Reprint edn. Chicago: University of Chicago Press, 1967 [1936].
Meadows, Milo Martin. 'Fundamentalist Thought and Its Impact in Kentucky, 1900–1928.' Ph.D. diss., Syracuse University, 1972.
Meldola, Raphael. *Evolution, Darwinian and Spencerian.* The Herbert Spencer Lecture. Oxford: Clarendon Press, 1910.
Mendelsohn, Everett. 'The Biological Sciences in the Nineteenth Century: Some Problems and Sources.' *History of Science,* 3 (1964), 39–59.
Merz, John Theodore. *A History of European Thought in the Nineteenth Century.* 4 vols. Edinburgh: William Blackwood & Sons, 1896–1914.
Messenger, Ernest C. *Evolution and Theology: The Problem of Man's Origin.* London: Burns Oates & Washbourne, 1931.
 ed. *Theology and Evolution (A Sequel to 'Evolution and Theology').* London: Sands & Co., [1949].
Metzger, Walter P. 'Darwinism and the New Regime.' *Academic Freedom in the Age of the University,* chap. 2. New York: Columbia University Press, 1961 [1955].
Mill, John Stuart. *A System of Logic, Ratiocinative and Inductive: Being a Connected View of the Principles of Evidence and the Methods of Scientific Investigation.* 2 vols. 8th edn. London: Longmans, Green, Reader & Dyer, 1872 [1843].
Millhauser, Milton. *Just before Darwin: Robert Chambers and the 'Vestiges'.* Middletown, Conn.: Wesleyan University Press, 1959.
Mivart, St George. *On the Genesis of Species.* 2nd edn. London: Macmillan, 1871 [1871].
 Man and Apes: An Exposition of Structural Resemblances and Differences Bearing upon Questions of Affinity and Origin. London: Robert Hardwicke, 1873.
 Lessons from Nature as Manifested in Mind and Matter. London: John Murray, 1876.
 The Origin of Human Reason: Being an Examination of Recent Hypotheses Concerning It. London: Kegan Paul, Trench & Co., 1889.
Molloy, John D. 'Spencer's Impact on American Conservatism.' Ph.D. diss., University of Cincinnati, 1959.
Montagu, Ashley. *Darwin: Competition & Cooperation.* New York: H. Schuman, 1952.
Montgomery, William Morey. 'Evolution and Darwinism in German Biology, 1800–1883.' Ph.D. diss., University of Texas (Austin), 1974a.
 'Germany.' In *Comparative Reception of Darwinism,* ed. T. F. Glick (1974b), pp. 81–116.
Moody, J. W. T. 'The Reading of the Darwin and Wallace Papers: An Historical "Non-Event".' *Journal of the Society for the Bibliography of Natural History,* 5 (1971), 474–6.
Moore, Aubrey Lackington. *Evolution and Christianity.* London: Rivingtons, 1889a.

Science and the Faith: Essays on Apologetic Subjects. London: Kegan Paul, Trench, Trübner & Co., 1889b.
'The Christian Doctrine of God.' In *Lux Mundi: A Series of Studies in the Religion of the Incarnation,* ed. Charles Gore. 10th edn. London: John Murray, 1890a [1889].
Essays Scientific and Philosophical. London: Kegan Paul, Trench, Trübner & Co., 1890b.
Lectures and Papers on the History of the Reformation in England and on the Continent. London: Kegan Paul, Trench, Trübner & Co., 1890c.
Moore, Edward Caldwell. *An Outline of the History of Christian Thought since Kant.* London: Duckworth & Co., 1912.
Moore, LeRoy, Jr. 'Another Look at Fundamentalism: A Response to Ernest R. Sandeen.' *CH,* 37 (1968), 195–202.
Moore, Ruth. *Man, Time, and Fossils: The Story of Evolution.* London: Jonathan Cape, 1955 [1953].
Moreno, Roberto. 'Mexico.' In *Comparative Reception of Darwinism,* ed. T. F. Glick (1974), pp. 346–74.
Morison, William James. 'George Frederick Wright: In Defense of Darwinism and Fundamentalism, 1838–1921.' Ph.D. diss., Vanderbilt University, 1971.
Morris, Francis Orpen. *Difficulties of Darwinism: Read before the British Association at Norwich and Exeter in 1868 and 1869, with a Preface and a Correspondence with Professor Huxley.* London: Longmans, Green & Co., 1869.
A Double Dilemma in Darwinism. London: William Poole, [1870].
All the Articles of the Darwin Faith. London: William Poole, 1875.
A Guard against the 'Guardian': Being Letters on Evolution Addressed to the 'Guardian' Newspaper by F.O.M., with a Few Remarks. London: William Poole, [1877].
The Darwin Craze: [A Paper] for the British Association, Swansea Meeting, 1880, on the Plumage of Birds and Butterflies. London: William Poole, 1880.
The Demands of Darwinism on Credulity. London: Partridge & Co., [1890].
Morris, M. C. F. *Francis Orpen Morris: A Memoir.* London: John C. Nimmo, 1897.
Morrison, John Lee. 'A History of American Catholic Opinion on the Theory of Evolution, 1859–1950.' Ph.D. diss., University of Missouri, 1951.
Mowat, R. B. *The Victorian Age.* London: George E. Harrap & Co., 1939.
Mozley, John Kenneth. *Some Tendencies in British Theology from the Publication of 'Lux Mundi' to the Present Day.* London: S.P.C.K., 1951.
Muckerman, H. *Attitude of Catholics to Darwinism and Evolution.* St Louis, Mo.: B. Herder, 1928.

Mudford, P. G. 'The Impact of the Theory of Evolution on the Late Nineteenth Century, with Special Reference to Thomas Henry Huxley and Samuel Butler.' B.Litt. thesis, University of Oxford, 1966.

Mullen, Pierce C. 'The Preconditions and Reception of Darwinian Biology in Germany, 1800–1870.' Ph.D. diss., University of California (Berkeley), 1964.

Mullens, W. H. and Swann, H. Kirke. *A Bibliography of British Ornithology from the Earliest Times to the End of 1912, Including Biographical Accounts of the Principal Writers and Bibliographies of Their Published Works.* London: Macmillan, 1917.

Murphy, Bruce Gordon. 'Thomas Huxley and His New Reformation.' Ph.D. diss., Northern Illinois University, 1973.

Myers, Frederic W. H. 'Charles Darwin and Agnosticism' (1888). *Science and a Future Life, with Other Essays*, pp. 51–75. London: Macmillan, 1893.

[Napier, John H.] Lennox, Cuthbert. *Henry Drummond: A Biographical Sketch (with Bibliography).* 3rd edn. London: Andrew Melrose, 1901 [1901].

Nash, J. V. 'The Religious Evolution of Darwin.' *Open Court*, 42 (1928), 449–63.

Neel, Samuel Regester, Jr. 'The Reaction of Certain Exponents of American Religious Thought to Darwin's Theory of Evolution.' Ph.D. diss., Duke University, 1942.

Neely, Alan Preston. 'James Orr: A Study in Conservative Christian Apologetics.' Th.D. diss., Southwestern Baptist Theological Seminary (Fort Worth, Texas), 1960.

Nesbitt, H. H. J., ed. *Darwin in Retrospect.* Toronto: Ryerson Press, 1960.

Nevins, Allan. *The Emergence of Modern America, 1865–1878.* Vol. 8. *A History of American Life*, ed. Arthur M. Schlesinger [Sr] and Dixon Ryan Fox. New York: Macmillan Co., 1927.

Nias, John. *Flame from an Oxford Cloister: The Life and Writings of Philip Napier Waggett, 1862–1939, Scientist, Religious Theologian, Missionary, Philosopher, Diplomat, Author, Orator, Poet.* London: Faith Press, 1961.

Niebuhr, Reinhold. 'Christianity and Darwin's Revolution.' In *A Book That Shook the World*, ed. Ralph Buchsbaum (1958), pp. 30–7.

Nordenskiöld, Erik. *The History of Biology: A Survey*, trans. Leonard Bucknall Eyre. London: Kegan Paul, Trench, Trubner & Co., 1929.

Numbers, Ronald L. 'Science Falsely So-Called: Evolution and Adventists in the Nineteenth Century.' *JASA*, 27 (1975), 18–23.

    Creation by Natural Law: Laplace's Nebular Hypothesis in American Thought. Seattle: University of Washington Press, 1977.

O'Brien, Charles F. '*Eozoön canadense*: The Dawn Animal of Canada.' *Isis*, 61 (1970), 206–23.

    Sir William Dawson: A Life in Science and Religion. Philadelphia: American Philosophical Society, 1971.

O'Brien, John A. *Evolution and Religion: A Study of the Bearing of*

438 BIBLIOGRAPHY

Evolution upon the Philosophy of Religion. New York: Century Co., 1932.

Ogilvie, Marilyn Bailey. 'Robert Chambers and the Successive Revisions of the "Vestiges of the Natural History of Creation".' Ph.D. diss., University of Oklahoma, 1973.

Olby, Robert C. *Origins of Mendelism.* New York: Schocken Books, 1966.

Omodeo, P. 'La classification et la philogenie dans l'oeuvre de Lamarck.' In *Colloque international 'Lamarck', tenu au Muséum National d'Histoire Naturelle, Paris, les 1–2 et 3 juillet 1971...*, ed. Joseph Schiller, pp. 11–27. Paris: Librairie Scientifique et Technique A. Blanchard, 1971.

Ong, Walter J., ed. *Darwin's Vision and Christian Perspectives.* New York: Macmillan Co., 1960.

Orr, James. *The Christian View of God and the World as Centring in the Incarnation.* 3rd edn. New York: Charles Scribner's Sons, 1897 [1893].
God's Image in Man and Its Defacement in the Light of Modern Denials. London: Hodder & Stoughton, 1905.
Sin as a Problem of Today. London: Hodder & Stoughton, 1910.

Osborn, Henry Fairfield. *From the Greeks to Darwin: An Outline of the Development of the Evolution Idea.* 2nd edn. New York: Macmillan Co., 1905 [1894].
Evolution and Religion in Education: Polemics of the Fundamentalist Controversy of 1922 to 1926. New York: Charles Scribner's Sons, 1926.
Cope, Master Naturalist: The Life and Writings of Edward Drinker Cope, with a Bibliography of His Writings Classified by Subject. Princeton, N.J.: Princeton University Press, 1931.

Osgood, Charles G. *Lights in Nassau Hall: A Book of the Bicentennial, Princeton, 1746–1946.* Princeton, N.J.: Princeton University Press, 1951.

Ospovat, Dov. 'Lyell's Theory of Climate.' *JHB*, 10 (1977), 317–39.

Ostoya, Paul. *Les théories de l'évolution: Origines et histoire du transformisme et des idées qui s'y rattachent.* Paris: Payot, 1951.

Overman, Richard H. *Evolution and the Christian Doctrine of Creation: A Whiteheadian Interpretation.* Philadelphia: Westminster Press, 1967.

Overton, Grant. *Portrait of a Publisher and the First Hundred Years of the House of Appleton.* New York: D. Appleton & Co., 1925.

'Overture anent Confession of Faith.' In *Proceedings and Debates of the General Assembly of the Free Church of Scotland, Held at Edinburgh, May 1892*, ed. Thomas Crerar, pp. 172–9, 189–203. Edinburgh: Ballantyne, Hanson & Co., 1892.

Owen, Richard. *On the Archetype and Homologies of the Vertebrate Skeleton.* London: John van Voorst, 1848.
On the Nature of Limbs: A Discourse Delivered on Friday, February 9,

at an Evening Meeting of the Royal Institution. London: John van
Voorst, 1849.
On the Anatomy of Vertebrates. 3 vols. London: Longmans, Green &
Co., 1866–8.
Owen, Rev. Richard. The Life of Richard Owen. 2 vols. London: John
Murray, 1894.
Packard, Alpheus, S., Jr. 'Rapid as Well as Slow Evolution.' The Indepen-
dent, 29 (23 August 1877), 6–7.
'The Law of Evolution.' The Independent, 33 (5 February 1880), 10.
Lamarck, The Founder of Evolution: His Life and Work, with Trans-
lations of His Writings on Organic Evolution. New York: Longmans,
Green & Co., 1901.
Paley, William. Natural Theology; or, Evidences of the Existence and
Attributes of the Deity Collected from the Appearances of Nature.
5th edn. London: R. Faulder, 1803 [1802].
Pannill, H. Burnell. The Religious Faith of John Fiske. Durham, N.C.:
Duke University Press, 1957.
Pantin, C. F. A. 'Darwin's Theory and the Causes of Its Acceptance.'
School Science Review, 32 (1950–1), 75–83.
Parkinson, George H. 'Charles Darwin's Influence on Religion and
Politics of the Present Day.' Ph.D. diss., University of Chicago, 1942.
Parrington, Vernon L. The Beginnings of Critical Realism in America.
Vol. 3. Main Currents in American Thought: An Interpretation of
American Literature from the Beginnings to 1920. New York: Har-
court, Brace & Co., 1958.
Passmore, John. A Hundred Years of Philosophy. London: Gerald Duck-
worth & Co., 1957.
'Darwin's Impact on British Metaphysics.' VS, 3 (1959), 41–54.
'The Objectivity of History' (1958). In Philosophical Analysis and
History, ed. William H. Dray, pp. 75–94. New York: Harper & Row,
1966.
The Perfectibility of Man. 2nd edn. London: Gerald Duckworth & Co.,
1972 [1970].
Patton, Carl S. 'The American Theological Scene: Fifty Years in Retro-
spect.' Journal of Religion, 16 (1936), 445–62.
Paul, Harry W. 'Religion and Darwinism: Varieties of Catholic Reaction.'
In Comparative Reception of Darwinism, ed. T. F. Glick (1974), pp.
403–36.
The Edge of Contingency: French Catholic Reaction to Scientific Change
from Darwin to Duhem. Gainesville: University Presses of Florida,
1979.
Pearson, W. W. 'The Theory of Evolution in Science, Ethics, and
Religion, with Special Reference to Its Teleological Nature, Tracing
the Development of the Theory in Question from Lamarck to Our
Own Day.' B.Sc. thesis, University of Oxford, 1907.
Peckham, Morse. 'Darwinism and Darwinisticism' (1959). The Triumph of
Romanticism: Collected Essays, pp. 176–201. Columbia: University of
South Carolina Press, 1970.

Peel, George. 'Campbell, George Douglas.' In *DNB*, 22 (1909), 385–90.

Peel, J. D. Y. *Herbert Spencer: The Evolution of a Sociologist*. London: Heinemann, 1971.

Pelikan, Jaroslav. 'Creation and Causality in the History of Christian Thought.' In *Issues in Evolution*, pp. 24–40. Vol. 3. *Evolution after Darwin: The University of Chicago Centennial*, ed. Sol Tax and Charles Callender. Chicago: University of Chicago Press, 1960.

Perry, Ralph Barton. *The Thought and Character of William James as Revealed in Unpublished Correspondence and Notes, together with His Published Writings*. Vol. 1, *Inheritance and Vocation*. Vol. 2, *Philosophy and Psychology*. Boston: Little, Brown & Co., 1935.

Persons, Stow. 'Evolution and Theology in America.' In *Evolutionary Thought in America*, ed. S. Persons (1950a), pp. 422–53.

ed. *Evolutionary Thought in America*. New Haven, Conn.: Yale University Press, 1950b.

American Minds: A History of Ideas. New York: Henry Holt & Co., 1958.

'Darwinism and American Culture.' In *Impact of Darwinian Thought on American Life and Culture* (1959), pp. 1–10.

'Religion and Modernity, 1865–1914.' In *The Shaping of American Religion*, pp. 369–401. Vol. 1. *Religion in American Life*, ed. James Ward Smith and A. Leland Jamison. Princeton, N.J.: Princeton University Press, 1961.

Peterson, Clifford Harold. 'The Incorporation of the Basic Evolutionary Concepts of Charles Darwin in Selected American College Biology Programs in the Nineteenth Century.' Ed.D. diss., Columbia University, 1970.

Peterson, George E. *The New England College in the Age of the University*. Amherst, Mass.: Amherst College Press, 1964.

Peterson, Houston. *Huxley, Prophet of Science*. London: Longmans, Green & Co., 1932.

Pfeifer, Edward J. 'The Reception of Darwinism in the United States, 1859–1880.' Ph.D. diss., Brown University, 1957.

'The Genesis of American Neo-Lamarckism.' *Isis*, 56 (1965), 156–67.

'United States.' In *Comparative Reception of Darwinism*, ed. T. F. Glick (1974), pp. 168–206.

Pfleiderer, Otto. *The Development of Theology in Germany since Kant and Its Progress in Great Britain since 1825*, trans. J. Frederick Smith. London: Swan Sonnenschein & Co., 1890.

Evolution and Theology, and Other Essays, ed. Orello Cone. London: Adam & Charles Black, 1900.

Picton, J. Allanson. *New Theories and the Old Faith: A Course of Lectures on Religious Topics of the Day, Delivered in St Thomas's Square Chapel, Hackney*. London: Williams & Norgate, 1870.

Plochmann, George Kimball. 'Darwin or Spencer?' *Science*, n.s., 130 (1959), 1452–6.

Poulton, Edward B. *Charles Darwin and the Theory of Natural Selection.* London: Cassell & Co., 1896.

'Thomas Henry Huxley and Natural Selection' (1905). *Essays on Evolution, 1889–1907,* pp. 193–219. Oxford: Clarendon Press, 1908.

'Fifty Years of Darwinism.' In *Fifty Years of Darwinism: Modern Aspects of Evolution,* pp. 8–56. New York: Henry Holt & Co., 1909.

Powell, Baden. *The Unity of Worlds and of Nature: Three Essays on the Spirit of the Inductive Philosophy, the Plurality of Worlds, and the Philosophy of Creation.* 2nd edn revised. London: Longman, Brown, Green, Longmans & Roberts, 1856 [1855].

The Order of Nature Considered in Reference to the Claims of Revelation: A Third Series of Essays. London: Longman, Brown, Green, Longmans & Roberts, 1859.

'On the Study of the Evidences of Christianity.' In *Essays and Reviews.* 4th edn. London: Longman, Green, Longmans & Roberts, 1861 [1860].

Price, H. H. *Some Aspects of the Conflict between Science & Religion.* Cambridge University Press, 1953.

'Professor Drummond's "Ascent of Man".' In *Proceedings and Debates of the General Assembly of the Free Church of Scotland, Held at Edinburgh, May 1895,* ed. Thomas Crerar, pp. 116–32. Edinburgh: Ballantyne, Hanson & Co., 1895.

Provine, William B. *The Origins of Theoretical Population Genetics.* Chicago: University of Chicago Press, 1971.

'Psychosis'. *Our Modern Philosophers, Darwin, Bain, and Spencer; or, The Descent of Man, Mind and Body: A Rhyme with Reasons, Essays, Notes, and Quotations.* London: T. Fisher Unwin, 1884.

Pusey, E. B. *Unscience, Not Science, Adverse to Faith: A Sermon Preached before the University of Oxford on the Twentieth Sunday after Trinity, 1878.* Oxford: James Parker & Co., 1878.

Quillian, William F., Jr. 'Evolution and Moral Theory in America.' In *Evolutionary Thought in America,* ed. S. Persons (1950), pp. 398–419.

Rádl, Emanuel. *The History of Biological Theories,* trans. and ed. E. J. Hatfield. London: Oxford University Press, 1930.

Ramm, Bernard. *The Christian View of Science and Scripture.* London: Paternoster Press, 1955.

'Theological Reactions to the Theory of Evolution.' *JASA,* 15 (1963), 71–7.

Ramsey, A. M. *From Gore to Temple: The Development of Anglican Theology between 'Lux Mundi' and the Second World War.* London: Longmans, 1960.

Ramsey, Ian T. *Religion and Science, Conflict and Synthesis: Some Philosophical Reflections.* London: S.P.C.K., 1964.

442   BIBLIOGRAPHY

Randall, John Herman, Jr. 'The Changing Impact of Darwin on Philosophy.' *JHI*, 22 (1961), 435–62.
Philosophy after Darwin: Chapters for 'The Career of Philosophy', Vol. 3, and Other Essays, ed. Beth J. Singer. New York: Columbia University Press, 1977.
Randel, William Pierce. 'Huxley in America.' *PAPS*, 114 (1970), 73–99.
Ratner, Sidney. 'Evolution and the Rise of the Scientific Spirit in America.' *Philosophy of Science*, 3 (1936), 104–22.
Raven, Charles E. *Science, Religion, and the Future*. Cambridge University Press, 1943.
'Man and Nature.' In *Ideas and Beliefs of the Victorians* (1949), pp. 173–9.
'Darwin and His Universe.' In *The History of Science: Origins and Results of the Scientific Revolution*, pp. 139–47. London: Cohen & West, 1951.
Natural Religion and Christian Theology. The Gifford Lectures, 1951–2. Vol. 1, *Science and Religion*, 1st ser., 1951. Vol. 2, *Experience and Interpretation*, 2nd ser., 1952. Cambridge University Press, 1953.
Reardon, Bernard M. G. *Religious Thought in the Nineteenth Century, Illustrated from Writers of the Period*. Cambridge University Press, 1966.
From Coleridge to Gore: A Century of Religious Thought in Britain. London: Longmans, 1971.
Rehbock, Philip F. 'Huxley, Haeckel, and the Oceanographers: The Case of *Bathybius haeckelii*.' *Isis*, 66 (1975), 504–33.
Reingold, Nathan, ed. *Science in Nineteenth-Century America: A Documentary History*. New York: Hill & Wang, 1964.
Reist, Irwin. 'Augustus Hopkins Strong and William Newton Clarke: A Study in Nineteenth-Century Evolutionary and Eschatological Thought.' *Foundations*, 13 (1970), 26–43.
'William Newton Clarke: Nineteenth-Century Evolutionary and Eschatological Immanentism.' *Foundations*, 18 (1975), 5–25.
'Rev. George Henslow.' *Nature*, 117 (1926), 130.
Rice, William North. 'The Darwinian Theory of the Origin of Species.' *The New Englander*, 26 (October 1867), 603–35.
Christian Faith in an Age of Science. London: Hodder & Stoughton, 1904 [1903].
Riddle, Oscar. *The Unleashing of Evolutionary Thought*. New York: Vantage Press, 1954.
Riley, Isaac Woodbridge. *American Thought from Puritanism to Pragmatism and Beyond*. 2nd edn. New York: Henry Holt & Co., 1923 [1915].
Ritterbush, Philip C. 'Organic Form: Aesthetics and Objectivity in the Study of Form in the Life Sciences.' In *Organic Form: The Life of an Idea*, ed. G. S. Rousseau, pp. 25–59. London: Routledge & Kegan Paul, 1972.
Roberts, Windsor Hall. 'The Reaction of the American Protestant

Churches to the Darwinian Philosophy, 1860–1900.' Ph.D. diss., University of Chicago, 1936.
Robertson, J. M. *A History of Freethought in the Nineteenth Century.* 2 vols. London: Watts & Co., 1929.
'Biographical Sketch.' In *Champion of Liberty: Charles Bradlaugh*, ed. J. P. Gilmour, pp. 1–25. London: Watts & Co., 1933.
Robinson, Sanford. *John Bascom, Prophet.* New York: G. P. Putnam's Sons, 1922.
Roger, Jacques. 'Darwin en France.' *AS*, 33 (1976), 481–4.
Rogers, James Allen. 'Darwinism and Social Darwinism.' *JHI*, 33 (1972), 265–80.
'The Reception of Darwin's "Origin of Species" by Russian Scientists.' *Isis*, 64 (1973), 484–503.
'Russia: Social Sciences.' In *Comparative Reception of Darwinism*, ed. T. F. Glick (1974a), pp. 256–68.
'Russian Opposition to Darwinism in the Nineteenth Century.' *Isis*, 65 (1974b), 487–505.
Rogers, Walter P. *Andrew D. White and the Modern University.* Ithaca, N.Y.: Cornell University Press, 1942.
[Romanes, Ethel]. *The Life and Letters of George John Romanes.* London: Longmans, Green & Co., 1896.
Romanes, George John. *Christian Prayer and General Laws: Being the Burney Prize Essay for the Year 1873, with an Appendix, 'The Physical Efficacy of Prayer'.* London: Macmillan, 1874.
[Romanes, George John]. Physicus. *A Candid Examination of Theism.* London: Trübner & Co., 1878.
Romanes, George John. *Animal Intelligence.* International Scientific Series, vol. 41. London: Kegan Paul, Trench & Co., 1882a.
'Natural Selection and Natural Theology.' *Contemporary Review*, 42 (October 1882b), 536–43.
Mental Evolution in Animals, . . .with a Posthumous Essay on Instinct by Charles Darwin. London: Kegan Paul, Trench & Co., 1883a.
'Natural Selection and Natural Theology.' *Nature*, 27 (15 February 1883b), 362–4; (5 April 1883b), 528–9; vol. 28 (31 May 1883b), 100–1.
'Physiological Selection: An Additional Suggestion on the Origin of Species.' *Journal of the Linnean Society*, Zoology, 19 (1886), 337–411.
Mental Evolution in Man: Origin of Human Faculty. London: Kegan Paul, Trench & Co., 1888.
'Mr Wallace on Darwinism.' *Contemporary Review*, 56 (August 1889), 244–58.
'Darwin's Latest Critics.' *Nineteenth Century*, 27 (May 1890), 823–32.
Darwin, and after Darwin: An Exposition of the Darwinian Theory and a Discussion of Post-Darwinian Questions. 3 vols. London: Longmans, Green & Co., 1892–7.
1. *The Darwinian Theory*, 1892.
2. *Post-Darwinian Questions: Heredity and Utility*, 1894.

3. *Post-Darwinian Questions: Isolation and Physiological Selection*, 1897.

An Examination of Weismannism. London: Longmans, Green & Co., 1893.

Mind and Motion, and Monism. London: Longmans, Green & Co., 1895a.

Thoughts on Religion, ed. Charles Gore. London: Longmans, Green & Co., 1895b.

Ronayne, Maurice. *Religion and Science: Their Union Historically Considered*. New York: Peter F. Collier, 1879.

Roome, P. 'The Darwin Debate in Canada: 1860–1880.' In *Science, Technology, and Culture in Historical Perspective*, ed. L. A. Knafla et al., pp. 183–205. Calgary, Alberta: University of Calgary, 1976.

Root, John David. 'Catholics and Science in Mid-Victorian England.' Ph.D. diss., Indiana University, 1974.

Rosen, Edward. 'Kepler and the Lutheran Attitude towards Copernicanism in the Context of the Struggle between Science and Religion.' In *Kepler, Four Hundred Years: Proceedings of Conferences Held in Honour of Johannes Kepler*, ed. Arthur Beer and Peter Beer, pp. 317–37. Oxford: Pergamon Press, 1975.

Ross, Frederic R. 'Philip Gosse's "Omphalos", Edmund Gosse's "Father and Son", and Darwin's Theory of Natural Selection.' *Isis*, 68 (1977), 85–96.

Routh, H. V. *Towards the Twentieth Century: Essays in the Spiritual History of the Nineteenth*. Cambridge University Press, 1937.

Rowell, Geoffrey. *Hell and the Victorians: A Study of Nineteenth-Century Theological Controversies Concerning Eternal Punishment and the Future Life*. Oxford: Clarendon Press, 1974.

Rudwick, Martin J. S. *The Meaning of Fossils: Episodes in the History of Paleontology*. London: Macdonald, 1972.

'Darwin and Glen Roy: A "Great Failure" in Scientific Method?' *SHPS*, 5 (1974), 97–185.

Rudy, S. Willis. 'The "Revolution" in American Higher Education – 1865–1900.' *Harvard Educational Review*, 21 (1951), 155–74.

Rumney, J. *Herbert Spencer's Sociology: A Study in the History of Social Theory, to Which is Appended a Bibliography of Spencer and His Work*. London: Williams & Norgate, 1934.

Rupke, Nicholas A. '*Bathybius Haeckelii* and the Psychology of Scientific Discovery: Theory Instead of Observed Data Controlled the Late 19th Century "Discovery" of a Primitive Form of Life.' *SHPS*, 7 (1976), 53–62.

Ruse, Michael. 'Natural Selection in "The Origin of Species".' *SHPS*, 1 (1971), 311–51.

'The Nature of Scientific Models: Formal *v* Material Analogy.' *Philosophy of the Social Sciences*, 3 (1973), 63–80.

'Charles Darwin and Artificial Selection.' *JHI*, 36 (1975a), 339–50.

'Charles Darwin's Theory of Evolution: An Analysis.' *JHB*, 8 (1975b), 219–41.
'Darwin's Debt to Philosophy: An Examination of the Influence of the Philosophical Ideas of John F. W. Herschel and William Whewell on the Development of Charles Darwin's Theory of Evolution.' *SHPS*, 6 (1975c), 159–81.
'The Relationship between Science and Religion in Britain, 1830–1870.' *CH*, 44 (1975d), 505–22.
'William Whewell and the Argument from Design.' *The Monist*, 60 (1977), 244–68.
Russell, A. J. 'Darwin.' *Their Religion*, pp. 271–97. London: Hodder & Stoughton, 1934.
Russell, Bertrand. *Religion and Science*. London: Thornton Butterworth, 1935.
Russell, Colin A., ed. *Science and Religious Belief: A Selection of Recent Historical Studies*. London: University of London Press and Open University Press, 1973.
Russell, Colin A.; Hooykaas, R.; and Goodman, David C. *The 'Conflict Thesis' and Cosmology*. Science and Belief: from Copernicus to Darwin (An Inter-faculty Second Level Course in the History of Science, The Open University), block 1, units 1–3. Milton Keynes, Bucks.: Open University Press, 1974.
Russell, E. S. *Form and Function: A Contribution to the History of Animal Morphology*. London: John Murray, 1916.
Russett, Cynthia Eagle. *Darwin in America: The Intellectual Response, 1865–1912*. San Francisco: W. H. Freeman & Co., 1976.
Saffin, N. W. *Science, Religion & Education in Britain, 1804–1904*. Kilmore, Victoria, Australia: Lowden Publishing Co., 1973.
Sandeen, Ernest R. 'The Princeton Theology: One Source of Biblical Literalism in American Protestantism.' *CH*, 31 (1962), 307–21.
The Origins of Fundamentalism: Toward a Historical Interpretation. Philadelphia: Fortress Press, 1968.
The Roots of Fundamentalism: Britain and American Millenarianism, 1800–1930. Chicago: University of Chicago Press, 1970.
Sandford, E. G., ed. *Memoirs of Archbishop Temple by Seven Friends*. 2 vols. London: Macmillan, 1906.
Sanford, William F., Jr. 'Dana and Darwinism.' *JHI*, 26 (1965), 531–46.
Sarno, Ronald A. 'A Sixteenth-Century War of Ideas: Science against the Church.' *AS*, 25 (1969), 209–27.
Savage, Minot Judson. *The Religion of Evolution*. Boston: Lockwood, Brooks & Co., 1876.
The Morals of Evolution. London: Trübner & Co., 1880.
Poems of Modern Thought. London: Williams & Norgate, 1884.
Evolution and Religion from the Standpoint of One Who Believes in Both: A Lecture Delivered in the Philadelphia Academy of Music, Seventh December 1885. Philadelphia: George H. Buchanan & Co., 1886.

Herbert Spencer: His Influence on Religion and Morality. Liverpool: W. & J. Arnold, 1887a.

My Creed. Boston: George H. Ellis, 1887b.

The Evolution of Christianity. Boston: George H. Ellis, 1892a.

The Irrepressible Conflict between Two World-Theories: Five Lectures Dealing with Christianity and Evolutionary Thought, to Which Is Added 'The Inevitable Surrender of Orthodoxy'. Boston: Arena Publishing Co., 1892b.

Schlesinger, Arthur M., [Sr]. *A Critical Period in American Religion, 1875-1900.* Philadelphia: Fortress Press, 1967 [1932].

Schmidt, George Paul. *The Old Time College President.* New York: Columbia University Press, 1930.

'Colleges in Ferment.' *American Historical Review,* 59 (1953), 19–42.

The Liberal Arts College: A Chapter in American Cultural History. New Brunswick, N.J.: Rutgers University Press, 1957.

Schneewind, J. B. *Sidgwick's Ethics and Victorian Moral Philosophy.* Oxford: Clarendon Press, 1977.

Schneider, Herbert W. 'The Influence of Darwin and Spencer on American Philosophical Theology.' *JHI,* 6 (1945), 3–18.

A History of American Philosophy. 2nd edn. New York: Columbia University Press, 1963 [1946].

Schott, Kenneth Ronald. 'An Analysis of Henry Drummond and His Rhetoric of Reconciliation.' Ph.D. diss., Ohio State University, 1972.

Schweber, Silvan S. 'The Origin of the "Origin" Revisited.' *JHB,* 10 (1977), 229–316.

Scopes, John T. and Presley, James. *Center of the Storm: Memoirs of John T. Scopes.* New York: Holt, Rinehart & Winston, 1967.

Scott, Wilson L. *The Conflict between Atomism and Conservation Theory, 1644–1860.* London: Macdonald, 1970.

Sears, Paul B. *Charles Darwin: The Naturalist as a Cultural Force.* New York: Charles Scribner's Sons, 1950.

'The Sentence of Howard MacQueary.' *Magazine of Christian Literature,* 4 (May 1891), 111.

Seward, A. C., ed. *Darwin and Modern Science: Essays in Commemoration of the Centenary of the Birth of Charles Darwin and of the Fiftieth Anniversary of the Publication of 'The Origin of Species'.* Cambridge University Press, 1909.

Sharlin, Harold Issadore. 'Herbert Spencer and Scientism.' *AS,* 33 (1976), 457–65.

Shaw, Bernard. *Back to Methuselah: A Metabiological Pentateuch.* London: Constable & Co., 1921.

Sheldon, Henry C. *Unbelief in the Nineteenth Century: A Critical History.* New York: Eaton & Mains, 1907.

Sherwood, Morgan B. 'Genesis, Evolution, and Geology in America before Darwin: The Dana-Lewis Controversy, 1856–1857.' In *Toward a History of Geology,* ed. Cecil J. Schneer, pp. 305–16. Cambridge, Mass.: M.I.T. Press, 1969.

Shideler, Emerson W. 'Darwin and the Doctrine of Man.' *Journal of Religion*, 40 (1960), 198–211.

Shields, Charles Woodruff. *Philosophia Ultima*. Philadelphia: J. B. Lippincott & Co., 1861.

The Final Philosophy; or, System of Perfectible Knowledge Issuing from the Harmony of Science and Religion. New York: Scribner, Armstrong & Co., 1877.

Philosophia Ultima; or, Science of the Sciences. 3 vols. New York: Charles Scribner's Sons, 1888–1905.

Shipley, Maynard. *The War on Modern Science: A Short History of the Fundamentalist Attacks on Evolution and Modernism*. New York: A. A. Knopf, 1927.

Shriver, George H., ed. *American Religious Heretics: Formal and Informal Trials*. Nashville, Tenn.: Abingdon Press, 1966.

Sidgwick, A. and Sidgwick, E. M. *Henry Sidgwick: A Memoir*. London: Macmillan, 1906.

Simmons, Henry M. 'Henry Drummond and His Books.' *New World*, 6 (September 1897), 485–98.

Simonsson, Tord. *Face to Face with Darwinism: A Critical Analysis of the Christian Front in Swedish Discussion of the Later Nineteenth Century*. Lund: C. W. K. Gleerup, 1958.

Logical and Semantic Structures in Christian Discourses, trans. Agnes George. Oslo: Universitetsforlaget, 1971.

Simpson, George Gaylord. *The Meaning of Evolution: A Study of the History of Life and of Its Significance for Man*. New Haven, Conn.: Yale University Press, 1949.

'The Concept of Progress in Organic Evolution.' *Social Research*, 41 (1974), 28–51.

Simpson, James Young. *Henry Drummond*. Edinburgh: Oliphant Anderson & Ferrier [1901].

Landmarks in the Struggle between Science and Religion. London: Hodder & Stoughton, 1925.

Simpson, Patrick Carnegie. *The Life of Principal Rainy*. 2 vols. London: Hodder & Stoughton, 1909.

Singer, Charles. *Religion & Science Considered in Their Historical Relations*. New York: Jonathan Cape & Harrison Smith, 1929 [1928].

Skoog, Gerald Duane. 'The Topic of Evolution in Secondary School Biology Textbooks, 1900–1968.' Ph.D. diss., University of Nebraska, 1969.

Smallwood, W. M. 'How Darwinism Came to the United States.' *Scientific Monthly*, 52 (1941), 342–9.

Smith, B. A. *Dean Church: The Anglican Response to Newman*. London: Oxford University Press, 1958.

Smith, Crosbie. 'Natural Philosophy and Thermodynamics: William Thomson and "The Dynamical Theory of Heat".' *BJHS*, 9 (1976), 293–319.

Smith, G. Barnett. 'Birks, Thomas Rawson.' In *DNB*, 2 (1908), 546–7.

Smith, George Adam. *The Life of Henry Drummond*. London: Hodder & Stoughton, 1899.

Smith, H. Shelton. *Changing Conceptions of Original Sin: A Study in American Theology since 1750*. New York: Charles Scribner's Sons, 1955.

Smith, Hay Watson. *Evolution and Presbyterianism*. Little Rock, Ark.: Allsopp & Chapple, 1922.

Smith, Homer W. *Man and His Gods*. London: Jonathan Cape, 1953.

Smith, James Ward. 'Religion and Science in American Philosophy.' In *The Shaping of American Religion*, pp. 402–42. Vol. 1. *Religion in American Life*, ed. James Ward Smith and A. Leland Jamison. Princeton, N.J.: Princeton University Press, 1961.

Smith, Preserved. 'The Controversy over Evolution in New England: A Footnote to Bert J. Loewenberg's Article.' *New England Quarterly*, 8 (1935), 398–9.

Smith, Roger. 'Alfred Russel Wallace: Philosophy of Nature and Man.' *BJHS*, 6 (1972), 177–99.

Smith, Sydney. 'The Origin of the "Origin" as Discerned from Charles Darwin's Notebooks and His Annotations in the Books He Read between 1837 and 1842.' *Advancement of Science*, 16 (1960), 391–401.

Smith, Timothy L. *Revivalism and Social Reform: American Protestantism on the Eve of the Civil War*. Reprint edn. New York: Harper & Row, Harper Torchbooks, 1965 [1957].

Smith, Warren Sylvester. *The London Heretics, 1870–1914*. London: Constable & Co., 1967.

Smith, William Holt. 'The Influence of Bishop Butler's "Analogy" in American Apologetic Thinking.' Ph.D. diss., University of Chicago, 1925.

Somervell, D. C. *English Thought in the Nineteenth Century*. London: Methuen, 1929.

Sopka, Katherine. 'An Apostle of Science Visits America: John Tyndall's Journey of 1872–73.' *Physics Teacher*, 10 (1972), 369–75.

Spencer, Herbert. *A System of Synthetic Philosophy*. 10 vols. London: Williams & Norgate, 1862–96.

　　1. *First Principles* (1862; 2nd edn, 1867; 3rd edn, 1875; 4th edn, 1880; 5th edn, 1884; 6th edn, 1900).

　　2–3. *The Principles of Biology* (2 vols., 1864–7; revised edn, 1898–9).

　　4–5. *The Principles of Psychology* (1 vol., 1855; 2nd edn, 2 vols., 1870–2; 3rd edn, 1880; 4th edn, 1899).

　　6–8. *The Principles of Sociology* (Vol. 1, 1876; 2nd edn, 1877; 3rd edn, 1885. Vol. 2, pt 4, 1879; pt 5, 1882. Vol. 3, pt 6, 1885; pts 7–8, 1896).

　　9–10. *The Principles of Ethics* (Vol. 1, pt 1, 1879; pts 2–3, 1892. Vol. 2, pt 4, 1891; pts 5–6, 1893).

Illustrations of Universal Progress: A Series of Discussions,...with a

449

of Spencer's 'New System of Philosophy'. New York: D.
Appleton & Co., 1864.
Social Statics; or, The Conditions Essential to Human Happiness Speci-
fied and the First of Them Developed. New edn. London: Williams
& Norgate, 1868 [1851].
The Study of Sociology. International Scientific Series, vol. 5. London:
Henry S. King & Co., 1873.
The Factors of Organic Evolution. London: Williams & Norgate, 1887.
Essays: Scientific, Political, and Speculative. 3 vols. Revised edn.
London: Williams & Norgate, 1890 [3 series: 1857, 1863, 1874].
An Autobiography. 2 vols. London: Williams & Norgate, 1904.
Spencer, Philip. ' "Barbarian Assault": The Fortunes of a Phrase.' *JHI*,
16 (1955), 232–9.
Spilsbury, Richard. *Providence Lost: A Critique of Darwinism.* London:
Oxford University Press, 1974.
Stackhouse, Reginald. 'Darwin and a Century of Conflict.' *Christian
Century*, 76 (1959), 944–6.
Stark, W. 'Natural and Social Selection.' In *Darwinism and the Study
of Society*, ed. Michael Banton (1961), pp. 49–61.
Stebbins, Robert E. 'French Reactions to Darwin, 1859–1882.' Ph.D. diss.,
University of Minnesota, 1965.
'France.' In *Comparative Reception of Darwinism*, ed. T. F. Glick
(1974a), pp. 117–63.
'France: Bibliographical Essay.' In *Comparative Reception of Darwin-
ism*, ed. T. F. Glick (1974b), pp. 164–7.
Stecher, Robert M. 'The Darwin–Innes Letters: The Correspondence of
an Evolutionist with His Vicar, 1848–1884.' *AS*, 17 (1961), 201–58.
Stephen, Leslie. 'Darwinism and Divinity.' *Essays on Freethinking and
Plainspeaking*, pp. 72–109. London: Longmans, Green & Co., 1873.
ed. *Letters of John Richard Green.* London: Macmillan, 1901.
Stephens, Lester D. 'Evolution and Woman's Rights in the 1890s: The
Views of Joseph Le Conte.' *Historian*, 38 (1976), 239–52.
Stern, Bernhard J. 'Darwin on Spencer.' *Scientific Monthly*, 26 (1928),
180–1.
Stevenson, Robert Louis. *Memoir of Fleeming Jenkin.* New York: Charles
Scribner's Sons, 1904.
Stirling, James Hutchison. *Philosophy and Theology.* Edinburgh: T. & T.
Clark, 1890.
Stocking, George W., Jr. 'Lamarckianism in American Social Science,
1890–1915' (1962). *Race, Culture, and Evolution: Essays in the
History of Anthropology*, pp. 234–69. New York: Free Press, 1968.
Stoughton, John. *Religion in England, 1800–1880.* 2 vols. London:
Hodder & Stoughton, 1884.
Street, T. Watson. 'The Evolution Controversy in the Southern Presby-
terian Church with Attention to the Theological and Ecclesiastical
Issues Raised.' *Journal of the Presbyterian Historical Society*, 37
(1959), 232–50.

Struik, Dirk J. 'Science and Religion.' *Yankee Science in the Making*, chap. 11. Revised edn. New York: Collier Books, 1962 [1948].

Stubbe, Hans. *History of Genetics from Prehistoric Times to the Rediscovery of Mendel's Laws*, trans. T. R. W. Walters. Cambridge, Mass.: M.I.T. Press, 1972.

Summerton, N. W. 'Dissenting Attitudes to Foreign Relations, Peace and War, 1840–1890.' *Journal of Ecclesiastical History*, 28 (1977), 151–78.

Symonds, John Addington. 'Darwin's Thoughts about God.' *Essays Speculative and Suggestive*, pp. 425–8. New edn. London: Chapman & Hall, 1893 [1890].

Taylor, F. Sherwood. *A Short History of Science*. London: Heinemann, [1939].

'Geology Changes the Outlook.' In *Ideas and Beliefs of the Victorians* (1949), pp. 189–96.

Teidman, S. J. 'Darwin's Reverend Friend.' *Modern Churchman*, n.s., 6 (1963), 286–90.

Temple, Frederick. *The Present Relations of Science to Religion: A Sermon Preached on Act Sunday, July 1, 1860, before the University of Oxford, during the Meeting of the British Association*. Oxford: J. H. & Jas. Parker, 1860.

The Relations between Religion and Science: Eight Lectures Preached before the University of Oxford in the Year 1884 on the Foundation of the Late Rev. John Bampton.... London: Macmillan, 1885.

Tennant, F. R. 'The Influence of Darwinism upon Theology.' *Quarterly Review*, 211 (1909), 418–40.

Thayer, James Bradley. *Letters of Chauncey Wright, with Some Account of His Life*. Cambridge, Mass.: Press of John Wilson & Son, 1878.

Thoday, J. M. 'Natural Selection and Biological Progress.' In *A Century of Darwin*, ed. S. A. Barnett (1958), pp. 313–33.

Thomson, J. Arthur. *The Science of Life: An Outline of the History of Biology and Its Recent Advances*. London: Blackie & Son, 1899.

'The Influence of Darwinism on Thought and Life.' In *Science and Civilization*, ed. F. S. Marvin, pp. 203–20. London: Oxford University Press, 1923.

The Great Biologists. London: Methuen, 1932.

Thomson, William. *Popular Lectures and Addresses*. 3 vols. London: Macmillan, 1889–94.

1. *Constitution of Matter*, 1889.
2. *Geology and General Physics*, 1894.
3. *Navigational Affairs*, 1891.

Tillich, Paul. *Systematic Theology*. 3 vols. Chicago: University of Chicago Press, 1951–63.

Theology of Culture, ed. Robert C. Kimball. New York: Oxford University Press, 1959.

Toulmin, Stephen and Goodfield, June. *The Discovery of Time*. London: Hutchinson & Co., 1965.

Towers, Bernard. 'The Impact of Darwin's "Origin of Species" on

Medicine and Biology.' In *Medicine and Science in the 1860s: Proceedings of the Sixth British Congress on the History of Medicine, University of Sussex, 6–9 September 1967*, pp. 45–55. London: Wellcome Institute of the History of Medicine, 1968.

[Townsend, Luther Tracy]. *Credo*. Boston: Lee & Shepard, 1869.

Townsend, Luther Tracy. *The Mosaic Record and Modern Science*. Boston: Howard Gannett, 1881.

Bible Theology and Modern Thought. Boston: Lee & Shepard, 1883.

Evolution or Creation: A Critical Review of the Scientific and Scriptural Theories of Creation and Certain Related Subjects. New York: Fleming H. Revell, 1896.

Traill, H. D. 'Lucretius, Paley, and Darwin.' *The New Lucian: Being a Series of Dialogues of the Dead*, pp. 322–47. Revised edn. London: Chapman & Hall, 1900 [1884].

Trevelyan, George Macaulay. *British History in the Nineteenth Century (1782–1901)*. London: Longmans, Green & Co., 1922.

Tristram, Henry Baker. 'Address to the Members of the Tyneside Naturalists' Field Club.' *Transactions of the Tyneside Naturalists' Field Club*, 4 (1858–60), 191–228.

'On the Ornithology of Northern Africa. ...Part III. The Sahara, continued.' *The Ibis*, 1 (October 1859), 415–35.

Tufts, J. H. 'Darwin and Evolutionary Ethics.' *Psychological Review*, 16 (1909), 195–206.

Tulloch, John. *Movements of Religious Thought in Britain during the Nineteenth Century*. London: Longmans, Green & Co., 1885.

Turbayne, Colin Murray. *The Myth of Metaphor*. Revised edn. Columbia: University of South Carolina Press, 1970 [1962].

Turner, Frank Miller. *Between Science and Religion: The Reaction to Scientific Naturalism in Late Victorian England*. New Haven, Conn.: Yale University Press, 1974a.

'Rainfall, Plagues, and the Prince of Wales: A Chapter in the Conflict of Religion and Science.' *Journal of British Studies*, 13 (1974b), 46–65.

Turner, John Mills, Jr. 'The Response of Major American Writers to Darwinism, 1859–1910.' Ph.D. diss., Harvard University, 1944.

Tyler, John Crew. *The Blind Seer: George Matheson*. London: Vision Press, 1960 [1959].

Tyler, William S. *A History of Amherst College during the Administration of Its First Five Presidents, from 1821 to 1891*. New York: Frederick H. Hitchcock, 1895.

Tyndall, John. *Address Delivered before the British Association Assembled at Belfast, with Additions*. London: Longmans, Green & Co., 1874.

Faraday as a Discoverer. 5th edn. London: Longmans, Green & Co., 1894.

Fragments of Science: A Series of Detached Essays, Addresses, and Reviews. 2 vols. 6th edn. New York: D. Appleton & Co., 1899 [1879].

452   BIBLIOGRAPHY

Vanderlaan, Eldred D., ed. *Fundamentalism and Modernism.* New York: H. W. Wilson, 1925.

Vanderpool, Harold Y. 'The Andover Conservatives: Apologetics, Biblical Criticism, and Theological Change at the Andover Theological Seminary, 1808–1880'. Ph.D. diss., Harvard University, 1971.

'Charles Darwin and Darwinism: A Naturalized World and a Brutalized Man?' In *Critical Issues in Modern Religion*, by Roger A. Johnson *et al.* Englewood Cliffs, N.J.: Prentice-Hall, 1973*a*.

ed. *Darwin and Darwinism: Revolutionary Insights Concerning Man, Nature, Religion, and Society.* Lexington, Mass.: D. C. Heath & Co., 1973*b*.

Van Dyke, Joseph S. *Theism and Evolution: An Examination of Modern Speculative Theories as Related to Theistic Conceptions of the Universe.* London: Hodder & Stoughton, 1886.

Vernon, Ambrose White. 'Later Theology.' In *A History of American Literature*, ed. William Peterfield Trent, vol. 3, pp. 201–25. Cambridge University Press, 1921.

Veysey, Laurence R. *The Emergence of the American University.* Chicago: University of Chicago Press, 1965.

Vidler, Alec R. *The Church in an Age of Revolution, 1789 to the Present Day.* Harmondsworth, Middx: Penguin Books, 1961.

Viner, Jacob. *The Role of Providence in the Social Order: An Essay in Intellectual History.* The Jayne Lectures for 1966. Philadelphia: American Philosophical Society, 1972.

Von Hofsten, Nils. 'Ideas of Creation and Spontaneous Generation prior to Darwin.' *Isis*, 25 (1936), 80–94.

Vorzimmer, Peter J. 'Darwin, Malthus, and the Theory of Natural Selection.' *JHI*, 30 (1969), 527–42.

Charles Darwin, The Years of Controversy: The 'Origin of Species' and Its Critics, 1859–82. London: University of London Press, 1972 [1970].

Vucinich, Alexander. 'Russia: Biological Sciences.' In *Comparative Reception of Darwinism*, ed. T. F. Glick (1974), pp. 227–55.

Wagar, W. Warren. *Good Tidings: The Belief in Progress from Darwin to Marcuse.* Bloomington: Indiana University Press, 1972.

Waggett, P. N. 'The More General Effect of Evolutionary Doctrine.' *The Scientific Temper in Religion, and Other Addresses*, pp. 82–104. London: Longmans, Green & Co., 1905.

'The Influence of Darwin upon Religious Thought.' In *Darwin and Modern Science*, ed. A. C. Seward (1909), pp. 477–93.

Wallace, Alfred Russel. *Darwinism: An Exposition of the Theory of Natural Selection, with Some of Its Applications.* London: Macmillan, 1889.

Natural Selection and Tropical Nature: Essays on Descriptive and Theoretical Biology. London: Macmillan, 1891 [1870, 1878].

Studies Scientific and Social. 2 vols. London: Macmillan, 1900.

My Life: A Record of Events and Opinions. 2 vols. London: Chapman & Hall, 1905.

Wallace, William. *Lectures and Essays on Natural Theology and Ethics.* Oxford: Clarendon Press, 1898.

Walsh, James J. *The Popes and Science: The History of the Papal Relations to Science during the Middle Ages and down to Our Own Time.* New York: Fordham University Press, 1908.

Ward, Henshaw. *Charles Darwin: The Man and His Warfare.* Indianapolis, Ind.: Bobbs–Merrill, 1927.

Ward, Wilfrid. *William George Ward and the Catholic Revival.* London: Macmillan, 1893.

Warfield, Benjamin Breckinridge. 'Darwin's Arguments against Christianity and against Religion.' *Homiletic Review,* 17 (January 1889), 9–16.

'Calvin's Doctrine of the Creation.' *Princeton Theological Review,* 13 (1915), 190–255.

'Charles Darwin's Religious Life: A Sketch in Spiritual Biography' (1888). *Studies in Theology,* pp. 541–82. New York: Oxford University Press, 1932a.

'On the Antiquity and Unity of the Human Race' (1911). *Biblical and Theological Essays,* pp. 235–58. New York: Oxford University Press, 1932b.

Warren, Sidney. *American Freethought, 1860–1914.* New York: Columbia University Press, 1943.

Webb, Clement C. J. *A Century of Anglican Theology, and Other Lectures.* Oxford: Basil Blackwell, 1923.

A Study of Religious Thought in England from 1850. Oxford: Clarendon Press, 1933.

Weisenburger, Francis P. *Ordeal of Faith: The Crisis of Church-going America, 1865–1900.* New York: Philosophical Library, 1959.

Weismann, August. *Studies in the Theory of Descent,* trans. Raphael Meldola. 2 vols. London: Sampson Low, Marston, Searle & Rivington, 1882.

Essays upon Heredity and Kindred Biological Problems, trans. Edward B. Poulton *et al.* 2 vols. Oxford: Clarendon Press, 1891–2 (vol. 1, 2nd edn, 1891 [1889]; vol. 2, 1892).

Welch, Claude. *Protestant Thought in the Nineteenth Century.* Vol. 1, 1799–1870. New Haven, Conn.: Yale University Press, 1972.

[Wells, Geoffrey Harry] West, Geoffrey. *Charles Darwin: A Portrait.* New Haven, Conn.: Yale University Press, 1938.

Wernham, J. C. S. 'The Religious Controversy.' In *Darwin in Retrospect,* ed. H. H. J. Nesbitt (1960), pp. 17–34.

West, Geoffrey. *See* Wells, Geoffrey Harry.

Whewell, William. *The Philosophy of the Inductive Sciences.* Vols. 5–6. *The Historical and Philosophical Works of William Whewell,* ed. G. Buchdahl and L. L. Laudan. London: Frank Cass & Co., 1967 [1840].

White, Andrew Dickson. 'First of the Course of Scientific Lectures –

Prof. White on "The Battle-fields of Science".' *New York Daily Tribune*, 18 December 1869, p. 4.

'The Warfare of Science.' *PSM*, 8 (February 1876a), 385–409; (March 1876a), 553–70.

The Warfare of Science. London: Henry S. King & Co., 1876b.

A History of the Warfare of Science with Theology in Christendom. 2 vols. London: Macmillan, 1896.

Autobiography of Andrew Dickson White. 2 vols. New York: Century Co., 1905.

Seven Great Statesmen in the Warfare of Humanity with Unreason. London: T. Fisher Unwin, 1910.

White, Edward A. *Science and Religion in American Thought: The Impact of Naturalism.* Stanford, Calif.: Stanford University Press, 1952.

White, Morton. *Pragmatism and the American Mind: Essays and Reviews in Philosophy and Intellectual History.* London: Oxford University Press, Galaxy Books, 1975 [1973].

White, S. S. 'The Reception in Russia of Darwinian Doctrines Concerning Evolution.' Ph.D. thesis, Imperial College of Science and Technology (London), 1968.

Whitehead, Alfred North. *Science and the Modern World.* Cambridge University Press, 1926.

Wichler, Gerhard. *Charles Darwin: The Founder of the Theory of Evolution and Natural Selection.* New York: Pergamon Press, 1961.

Wiener, Philip P. 'Chauncey Wright's Defense of Darwin and the Neutrality of Science.' *JHI*, 6 (1945), 19–45.

Evolution and the Founders of Pragmatism. Reprint edn. Philadelphia: University of Pennsylvania Press, 1972 [1949].

Wilkie, J. S. 'Buffon, Lamarck and Darwin: The Originality of Darwin's Theory of Evolution.' In *Darwin's Biological Work: Some Aspects Reconsidered*, ed. P. R. Bell, pp. 262–307. Reprint edn. New York: John Wiley & Sons, Science Editions, 1964 [1959].

Wilkins, Thurman. *Clarence King: A Biography.* New York: Macmillan Co., 1958.

Willey, Basil. *More Nineteenth Century Studies: A Group of Honest Doubters.* London: Chatto & Windus, 1956.

'Darwin and Clerical Orthodoxy.' In *1859: Entering An Age of Crisis*, ed. P. Appleman et al. (1959), pp. 51–62.

Darwin and Butler: Two Versions of Evolution. London: Chatto & Windus, 1960.

'Darwin's Place in the History of Thought.' In *Darwinism and the Study of Society*, ed. M. Banton (1961), pp. 1–16.

Williams, C. M. *A Review of the Systems of Ethics Founded on the Theory of Evolution.* London: Macmillan, 1893.

Williams, Daniel Day. *The Andover Liberals: A Study in American Theology.* Morningside Heights, N.Y.: King's Crown Press, 1941.

Wilson, David B. 'Kelvin's Scientific Realism: The Theological Context.'

Philosophical Journal: Transactions of the Royal Philosophical Society of Glasgow, 11 (1974), 41–60.
'Victorian Science and Religion.' *History of Science*, 15 (1977), 52–67.
Wilson, James Maurice. 'The Religious Effect of the Idea of Evolution.' In *Evolution in the Light of Modern Knowledge* (1925), pp. 477–516.
Wilson, John B. 'Darwin and the Transcendentalists.' *JHI*, 26 (1965), 286–90.
Wilson, Leonard G., ed. *Sir Charles Lyell's Scientific Journals on the Species Question*. New Haven, Conn.: Yale University Press, 1970.
Wilson, R. J., ed. *Darwinism and the American Intellectual: A Book of Readings*. Homewood, Ill.: Dorsey Press, 1967.
Wiltshire, David. *The Social and Political Thought of Herbert Spencer*. Oxford University Press, 1978.
Windle, Bertram C. A. *The Church and Science*. 3rd edn revised. London: Catholic Truth Society, 1924 [1917].
The Catholic Church and Its Reactions with Science. London: Burns & Oates, 1927.
Wood, Herbert G. *Belief and Unbelief since 1850*. Cambridge University Press, 1955.
'Contemporary Religious Trends: Science and Religion.' *Expository Times*, 67 (1956), 283–6.
Wright, Conrad. 'The Religion of Geology.' *New England Quarterly*, 14 (1941), 335–58.
Wright, George Frederick. 'The Ground of Confidence in Inductive Reasoning.' *The New Englander*, 30 (October 1871), 601–15.
'Recent Books Bearing upon the Relation of Science to Religion: No. I – The Nature and Degree of Scientific Proof.' *BS*, 32 (July 1875), 537–55.
'Recent Works Bearing on the Relation of Science to Religion: No. II – The Divine Method of Producing Living Species.' *BS*, 33 (July 1876a), 448–93.
'Recent Works Bearing on the Relation of Science to Religion: [No.] III – Objections to Darwinism, and the Rejoinders of Its Advocates.' *BS*, 33 (October 1876b), 656–94.
Review of *History of the Conflict between Religion and Science*, by John William Draper. *BS*, 33 (July 1876c), 584–5.
'Recent Works Bearing on the Relation of Science to Religion: No. IV – Concerning the True Doctrine of Final Cause or Design in Nature.' *BS*, 34 (April 1877), 355–85.
'Recent Works Bearing on the Relation of Science to Religion: No. V – Some Analogies between Calvinism and Darwinism.' *BS*, 37 (January 1880), 48–76.
The Logic of Christian Evidences. London: Richard D. Dickinson, 1881 [1880].
Studies in Science and Religion. Andover, Mass.: W. F. Draper, 1882.
'The Debt of the Church to Asa Gray.' *BS*, 45 (July 1888), 523–30.
'Darwin on Herbert Spencer.' *BS*, 46 (January 1889), 181–4.

'Bad Philosophy Going to Seed.' *BS*, 52 (July 1895), 559–61.
Scientific Aspects of Christian Evidences. New York: D. Appleton & Co., 1898.
'The Evolutionary Fad.' *BS*, 57 (April 1900), 303–16.
'Calvinism and Darwinism.' *BS*, 66 (October 1909a), 685–91.
'The Mistakes of Darwin and His Would-Be Followers.' *BS*, 66 (April 1909b), 332–43.
The Origin and Antiquity of Man. London: John Murray, 1913 [1912].
Story of My Life and Work. Oberlin, O.: Bibliotheca Sacra Co., 1916.
The Year of Preparation for the Vatican Council: Including the Original and English of the Encyclical and Syllabus, and of the Papal Documents Connected with Its Convocation. London: Burns, Oates & Co., 1869.
Yokoyama, Toshiaki. '[The Influence of Theological Thought on Charles Darwin – Consideration of the Relation between William Paley and Charles Darwin.]' *Kagakusi Kenkyu*, 10 (1971), 49–59.
Youmans, Edward Livingston. 'The Conflict of Religion and Science.' *PSM*, 6 (January 1875a), 361–4.
'Draper and His Critics.' *PSM*, 7 (June 1875b), 230–3.
'The "Conflict" and the "Warfare".' *PSM*, 9 (October 1876), 757–8.
[Youmans, Edward Livingston], ed. *Herbert Spencer on the Americans and the Americans on Herbert Spencer: Being a Full Report of His Interview and of the Proceedings of the Farewell Banquet of Nov. 9, 1882*. Reprint edn. New York: D. Appleton & Co., 1887 [1882].
Young, G. M. *Victorian England: Portrait of an Age*. 2nd edn. London: Oxford University Press, Oxford Paperbacks, 1960 [1936].
Young, Robert M. 'The Development of Herbert Spencer's Concept of Evolution.' *Actes du XIᵉ Congrès International d'Histoire des Sciences* (Warsaw, 1967), II, 273–8.
'Malthus and the Evolutionists: The Common Context of Biological and Social Theory.' *Past and Present*, no. 43 (1969a), 109–45.
'Natural Theology, Victorian Periodicals, and the Fragmentation of a Common Context.' Unpublished paper presented at King's College Research Centre Seminar in Science and History, July 1969b.
'The Impact of Darwin on Conventional Thought.' In *The Victorian Crisis of Faith*, ed. John Symondson, pp. 13–35. London: S.P.C.K., 1970a.
Mind, Brain, and Adaptation in the Nineteenth Century: Cerebral Localization and Its Biological Context from Gall to Ferrier. Oxford: Clarendon Press, 1970b.
'Darwin's Metaphor: Does Nature Select?' *The Monist*, 55 (1971a), 442–503.
'"Non-Scientific" Factors in the Darwinian Debate.' *Actes du XIIᵉ Congrès International d'Histoire des Sciences Naturelles et de la Biologie* (Paris, 1971b), VIII, 221–6.
'The Historiographic and Ideological Contexts of the Nineteenth-Century Debate on Man's Place in Nature.' In *Changing Perspectives*

in the History of Science: Essays in Honour of Joseph Needham, ed. Mikuláš Teich and Robert Young, pp. 344–438. London: Heinemann, 1973.

Yule, John David. 'The Impact of Science on British Religious Thought in the Second Quarter of the Nineteenth Century.' Ph.D. thesis, University of Cambridge, 1976.

Zahm, J. A. *Evolution and Dogma*. Chicago: D. H. McBride & Co., 1896.

Zimmerman, Paul A., ed. *Darwin, Evolution, and Creation*. St Louis, Mo.: Concordia Publishing House, 1959.

Zirkle, Conway. 'The Inheritance of Acquired Characters and the Provisional Hypothesis of Pangenesis.' *American Naturalist*, 69 (1935), 417–45.

'Further Notes on Pangenesis and the Inheritance of Acquired Characters.' *American Naturalist*, 70 (1936), 529–46.

'The Early History of the Ideas of the Inheritance of Acquired Characters and of Pangenesis.' *Transactions of the American Philosophical Society*, n.s., 35 (1946), 91–151.

Zöckler, O. *Geschichte der Beziehungen zwischen Theologie und Naturwissenschaft, mit besondrer Rücksicht auf Schöpfungsgeschichte*. 2 vols. Gütersloh: C. Bertelsmann, 1877–9.

ADDENDUM

This addendum contains titles overlooked in the main bibliography or published since the book first went to press. It incorporates the addendum from the original edition.

Abbott, Lawrence Fraser. 'Charles R. Darwin, the Saint.' In *Twelve Great Modernists*, pp. 225–51. New York: Doubleday, Page, 1927.

Allan, Mea. *Darwin and His Flowers: The Key to Natural Selection*. London: Faber & Faber, 1977.

Alszeghi, Zoltan. 'Development in the Doctrinal Formulations of the Church Concerning the Theory of Evolution.' *Concilium*, 6 (1967), 14–17.

Altner, Günter. *Charles Darwin und Ernst Haeckel: Ein Vergleich nach theologischen Aspekten*. Zurich: EVZ-Verlag, 1966.

Altschuler, Glenn C. 'From Religion to Ethics: Andrew D. White and the Dilemma of a Christian Rationalist.' *CH*, 47 (1978), 308–24.

Archer, R. L. *Secondary Education in the Nineteenth Century*. Cambridge University Press, 1921.

Armstrong, A. MacC. 'Samuel Wilberforce v. T.H. Huxley: A Retrospect.' *Quarterly Review*, 256 (1958), 426–37.

Bailey, Kenneth K. 'The Enactment of Tennessee's Antievolution Law.' *Journal of Southern History*, 16 (1950), 472–510.

Bannister, Robert C. *Social Darwinism: Science and Myth in Anglo-American Social Thought.* Philadelphia: Temple University Press, 1979.

Barker, Eileen. 'In the Beginning: The Battle of Creationist Science against Evolutionism.' In *On the Margins of Science: The Social Construction of Rejected Knowledge,* ed. Roy Wallis, pp. 179–200. Sociological Review Monograph 27. Keele, Staffs.: University of Keele, 1979

Bartholomew, Michael. 'The Singularity of Lyell.' *History of Science,* 17 (1979), 276–93.

Bartholomew, Michael; Norton, Bernard; and Young, Robert M. *Problems in the Biological and Human Sciences.* Science and Belief: from Darwin to Einstein (Arts: A Third Level Course, The Open University), block 6, units 12–14. Milton Keynes, Bucks.: Open University Press, 1981.

Bartov, H. 'A fortiori Arguments in the Bible, in Paley's Writings, and in the "Origin of Species".' *Janus,* 64 (1977), 131–45.

Beimfohr, Herman Nelson. 'The Doctrine of Creation in Recent Theology as Influenced by the Theory of Evolution.' M.A. thesis, Northwestern University, 1929.

Bevis, Richard. 'Spiritual Geology: C.M. Doughty and the Land of the Arabs.' *VS,* 16 (1972–3), 163–81.

Biddiss, Michael D. *The Age of the Masses: Ideas and Society in Europe since 1870.* Harmondsworth, Middx: Penguin Books, 1977.

Boesiger, Ernest. 'Evolutionary Theories after Lamarck and Darwin.' In *Studies in the Philosophy of Biology: Reductionism and Related Problems,* ed. Francisco Jose Azala and Theodosius Dobzhansky, pp. 21–44. London: Macmillan, 1974.

Boller, Paul F., Jr. *Freedom and Fate in American Thought from Edwards to Dewey.* Dallas, Tex.: SMU Press, 1978.

Bozeman, Theodore Dwight. 'Inductive and Deductive Politics: Science and Society in Antebellum Presbyterian Thought.' *Journal of American History,* 64 (1977), 704–22.

Brod, Donald F. 'The Scopes Trial: A Look at Press Coverage after Forty Years.' *Journalism Quarterly,* 42 (1965), 219–26.

Brooke, John Hedley. 'The Natural Theology of the Geologists: Some Theological Strata.' In *Images of the Earth: Essays in the History of the Environmental Sciences,* ed. L.J. Jordanova and Roy S. Porter, pp. 39–64. Chalfont St Giles, Bucks.: British Society for the History of Science, 1979.

'Nebular Contraction and the Expansion of Naturalism' [essay review of *Creation by Natural Law: Laplace's Nebular Hypothesis in American Thought* by Ronald L. Numbers]. *BJHS,* 12 (1979), 200–211.

Browne, Elizabeth Janet. 'C.R. Darwin and J.D. Hooker: Episodes in the History of Plant Geography, 1840–1860.' Ph.D. thesis, University of London (Imperial College), 1979.

Budd, Susan. *Varieties of Unbelief: Atheists and Agnostics in English Society, 1850–1960*. London: Heinemann Educational Books, 1977.

Burggraaff, Winfield. *The Rise and Development of Liberal Theology in America*. New York: Board of Publication and Bible-School Work of the Reformed Church in America, 1928.

Bystrom, Robert E. 'The Earliest Methodist Response to Evolution, 1870–1880.' M.A. thesis, Northwestern University, 1966.

Cannon, W. Faye. 'The Whewell-Darwin Controversy.' *Journal of the Geological Society of London*, 132 (1976), 377–84.

Cannon, Susan Faye. *Science in Culture: The Early Victorian Period*. New York: Science History Publications, 1978.

Cashdollar, Charles D. 'The Social Implications of the Doctrine of Divine Providence: A Debate in Nineteenth-Century American Theology.' *Harvard Theological Review*, 71 (1978), 265-84.

Centore, F.F. 'Darwin on Evolution: A Re-estimation.' *The Thomist*, 33 (1969), 456–96.

Chant, Colin and Fauvel, John, eds. *Darwin to Einstein: Historical Studies on Science and Belief*. London: Longman and Open University Press, 1980.

Clark, Clifford E., Jr. *Henry Ward Beecher: Spokesman for a Middle-Class America*. Urbana: University of Illinois Press, 1978.

Coley, Noel G. and Hall, Vance M.D., eds. *Darwin to Einstein: Primary Sources on Science and Belief*. London: Longman and Open University Press, 1980.

Colp, Ralph, Jr. 'Charles Darwin: Slavery and the American Civil War.' *Harvard Library Bulletin*, 26 (1978), 471–89.

Corsi, P. 'Natural Theology, Methodology of Science and the Question of Species in the Works of the Rev. Baden Powell.' D. Phil. thesis, University of Oxford, 1980.

Cotkin, George Bernard. 'Working-class Intellectuals and Evolutionary Thought in America, 1870–1915.' Ph.D. diss., Ohio State University, 1978.

Cowles, Thomas. 'Malthus, Darwin, and Bagehot: A Study in the Transference of a Concept.' *Isis*, 26 (1936–7), 341–8.

Cravens, Hamilton, *The Triumph of Evolution: American Scientists and the Heredity-Environment Controversy, 1900–1941*. Philadelphia: University of Pennsylvania Press, 1978.

Crocker, Arna Ruth. 'The Religious Response to Scientific Certainty: A Study of the Metaphysical Society and Some Representative Members.' M. Litt. thesis, University of Cambridge, 1977.

Curti, Merle. *Human Nature in American Thought: A History*. Madison: University of Wisconsin Press, 1980.

Daub, Edward E. 'Demythologizing White's "Warfare of Science and [sic] Theology".' *American Biology Teacher,* 40 (1978), 553–6.

Davis, Dennis R. 'The Impact of Evolutionary Thought on Walter Rauschenbusch.' *Foundations,* 21 (1978), 254–71.

Dawson, Marshall, *Nineteenth Century Evolution and After: A Study of Personal Forces Affecting the Social Process in the Light of the Life-Sciences and Religion.* New York: Macmillan Co., 1923.

Dean, Dennis R. 'The Influence of Geology on American Literature and Thought.' In *Two Hundred Years of Geology: Proceedings of the New Hampshire Bicentennial Conference on the History of Geology,* ed. Cecil J. Schneer, pp. 289–303. Hanover, N.H.: University Press of New England, 1979.

Di Gregorio, Mario Aurelio. 'On the Side of the Apes: T.H. Huxley and the Method and Results of Science.' Ph.D. thesis, University of London (University College), 1980.

Durant, John R. 'The Meaning of Evolution: Post-Darwinian Debates on the Significance for Man of the Theory of Evolution, 1858–1908.' Ph.D. thesis, University of Cambridge, 1977.

'Scientific Naturalism and Social Reform in the Thought of Alfred Russel Wallace.' *BJHS,* 12 (1979), 31–58.

Dysart, Marjorie May. 'Darwinism versus Southern Orthodoxy: A Survey of Southern Thought about Evolution as Reflected in Periodicals of Four Protestant Churches, 1865–1900.' M.A. thesis, University of Kentucky, 1954.

Ellis, William Elliott. 'The Kentucky Evolution Controversy.' M.A. thesis, Eastern Kentucky University, 1967.

Eng, Erling. 'Thomas Henry Huxley's Understanding of "Evolution".' *History of Science,* 16 (1978), 291–303.

Farley, John. *The Spontaneous Generation Controversy from Descartes to Oparin.* Baltimore, Md.: Johns Hopkins University Press, 1977.

Freeman, R.B. *The Works of Charles Darwin: An Annotated Bibliographical Handlist.* 2nd edn revised. Folkestone, Kent: Wm Dawson & Sons, 1977 [1965].

Charles Darwin: A Companion. Folkestone, Kent: Wm Dawson & Sons, 1978.

Freeman, R.B. and Wertheimer, Douglas. *Philip Henry Gosse: A Bibliography.* Folkestone, Kent: Wm Dawson & Sons, 1980.

Garland, Martha McMackin. *Cambridge before Darwin: The Ideal of a Liberal Education, 1800–1860.* Cambridge University Press, 1980.

Geison, Gerald L. 'Darwin and Heredity: The Evolution of His Hypothesis of Pangenesis.' *Journal of the History of Medicine and Allied Sciences,* 24 (1969), 375–411.

Gilbert, Scott F. 'Altruism and Other Unnatural Acts: T.H. Huxley on

Nature, Man, and Society.' *Perspectives in Biology and Medicine*, 2 (1979), 346–58.

Gillespie, Neal C. *Charles Darwin and the Problem of Creation.* Chicago: University of Chicago Press, 1979.

Gillispie, Charles Coulston. *The Edge of Objectivity: An Essay in the History of Scientific Ideas.* Princeton, N.J.: Princeton University Press, 1960.

Goodman, David C. and Olby, Robert C. *The Mystery of Life.* Science and Belief: from Darwin to Einstein (Arts: A Third Level Course, The Open University), block 5, units 10-11. Milton Keynes, Bucks.: Open University Press, 1981.

Goodwin, Craufurd D. 'Evolution Theory in Australian Social Thought.' *JHI*, 25 (1964), 393–416.

Gould, Stephen Jay. *Ever Since Darwin: Reflections in Natural History.* New York: W.W. Norton & Co., 1977.

Ontogeny and Phylogeny. Cambridge, Mass.: Harvard University Press, 1977.

'Agassiz' Later, Private Thoughts about Evolution: His Marginalia in Haeckel's "Natürliche Schöpfungsgeschichte" (1868).' In *Two Hundred Years of Geology in America*, ed. Cecil J. Schneer, pp. 277–82. Hanover, N.H.: University Press of New England, 1979.

Grebstein, Sheldon Norman, ed. *Monkey Trial: The State of Tennessee vs. John Thomas Scopes.* Boston: Houghton Mifflin Co., 1960.

Greene, John C. 'Protestantism, Science, and American Enterprise: Benjamin Silliman's Moral Universe.' In *Benjamin Silliman and His Circle: Studies on the Influence of Benjamin Silliman on Science in America*, ed. Leonard G. Wilson, pp. 11–28. New York: Science History Publications, 1979.

Science, Ideology, and World View: Essays in the History of Evolutionary Ideas. Berkeley: University of California Press, 1981.

Gregory, Frederick. *Scientific Materialism in Nineteenth Century Germany.* Dordrecht, The Netherlands: D. Reidel Publishing Co., 1977.

Gruber, Howard E. 'Darwin's "Tree of Nature" and Other Images of Wide Scope.' In *On Aesthetics in Science*, ed. Judith Wechsler, pp. 121–40. Cambridge, Mass.: MIT Press, 1978.

Halliburton, R., Jr. 'The Adoption of Arkansas' Anti-Evolution Law.' *Arkansas Historical Quarterly*, 23 (1964), 271–83.

'Kentucky's Anti-evolution Controversy.' *Register of the Kentucky Historical Society*, 66 (1968), 97–107.

Hart, Nelson Hodges. 'The True and the False: The Worlds of an Emerging Evangelical Protestant Fundamentalism in America, 1890–1920.' Ph.D. diss., Michigan State University, 1976.

Hattiangadi, J.N. 'Alternatives and Incommensurables: The Case of Darwin and Kelvin.' *Philosophy of Science*, 38 (1971), 502–7.

Hegenbarth, Hans. *Darwin, die Bibel und die Tatsachen.* Graz, Austria: Steiermärkische Landesregierung, Steiermärkische Landesbibliothek, 1972.

Herbert, Sandra, ed. *The Red Notebook of Charles Darwin.* Ithaca, N.Y.: Cornell University Press, 1980.

Himrod, David Kirk. 'Cosmic Order and Divine Activity: A Study in the Relation of Science and Religion, 1850–1950.' Ph.D. diss., University of California (Los Angeles), 1977.

Hinds, J.I.D. *Charles Darwin: A Sketch of His Life, Writings, Theory, Character, Mental Characteristics, and Religious Views.* Revised edn. Nashville, Tenn.: Cumberland Presbyterian Publishing House, 1900.

Hoeveler, J. David, Jr. *James McCosh and the Scottish Intellectual Tradition: from Glasgow to Princeton.* Princeton, N.J.: Princeton University Press, 1981.

Holt, Niles. 'The Challenge of Darwinism.' In *Problems in European History*, ed. Harold T. Parker, pp. 124–37. Durham, N.C.: Moore Publishing Co., 1979.

Hovenkamp, Herbert. *Science and Religion in America, 1800–1860.* Philadelphia: University of Pennsylvania Press, 1978.

Hutchinson, Gov. 'Robert Chambers's Vision of Science: The Diffusion of Scientific Ideas to the General Reader in Early-Victorian Britain.' Ph.D. diss., Temple University, 1980.

Jacyna, Leon Stephen. 'Scientific Naturalism in Victorian Britain: An Essay in the Social History of Ideas.' Ph.D. thesis, University of Edinburgh, 1980.

Jaki, Stanley L. *Science and Creation: From Eternal Cycles to an Oscillating Universe.* Edinburgh: Scottish Academic Press, 1974.
The Road of Science and the Ways to God. Edinburgh: Scottish Academic Press, 1978.

James, Patricia. *Population Malthus: His Life and Times.* London: Routledge & Kegan Paul, 1979.

Janet, Paul. *Final Causes,* trans. William Affleck. 2nd edn. Edinburgh: T. & T. Clark, 1883 [1878].

Jones, Greta. 'The Social History of Darwin's "Descent of Man".' *Economy and Society*, 7 (1978), 1–23.
Social Darwinism and English Thought: The Interaction between Biological and Social Theory. Brighton, Sussex: Harvester Press, 1980.

Kass, Leon R. 'Teleology and Darwin's "The Origin of Species": Beyond Chance and Necessity?' In *Organism, Medicine, and Metaphysics: Essays in Honor of Hans Jonas on his 75th Birthday, May 10,*

1978, ed. Stuart F. Spicker, pp. 97–120. Dordrecht, The Netherlands: D. Reidel Publishing Co., 1978.

Kelly, Alfred. *The Descent of Darwin: The Popularization of Darwinism in Germany, 1860–1914.* Chapel Hill: University of North Carolina Press, 1981.

Klaaren, Eugene M. *Religious Origins of Modern Science: Belief in Creation in Seventeenth-Century Thought.* Grand Rapids, Mich.: William B. Eerdmans Publishing Co., 1977.

Klubertanz, George. 'The Influence of Evolutionary Theory upon American Thought (Conspectus Bibliographicus).' *Gregorianum*, 32 (1951), 582–90.

Kohn, David. 'Theories to Work By: Rejected Theories, Reproduction, and Darwin's Path to Natural Selection.' *Studies in History of Biology*, 4 (1980), 67–170.

Kottler, Malcolm J. 'Charles Darwin's Biological Species Concept and Theory of Geographic Speciation: the Transmutation Notebooks.' *AS*, 35 (1978), 275–97.

Krause, David J. 'Apparent Age and Its Reception in the 19th Century.' *JASA*, 32 (1980), 146–50.

Kuklick, Bruce. *The Rise of American Philosophy: Cambridge, Massachusetts, 1860–1930.* New Haven, Conn.: Yale University Press, 1977.

Landucci, Giovanni. *Darwinismo a Firenze: tra scienza e ideologia (1860–1900).* Florence: Leo S. Olschki Editore, 1977.

LeMahieu, D.L. 'Malthus and the Theology of Scarcity.' *JHI*, 40 (1979), 467–74.

Lester, Jacob Franklin. 'John Fiske's Philosophy of Science: The Union of Science and Religion through the Principle of Evolution.' Ph.D. diss., Oregon State University, 1979.

Levinson, Henry S. *Science, Metaphysics, and the Chance of Salvation: An Interpretation of the Thought of William James.* Missoula, Mont.: Scholars Press for the American Academy of Religion, 1978.

Lightman, Bernard Vise. 'Henry Longueville Mansel and the Genesis of Victorian Agnosticism.' Ph.D. diss., Brandeis University, 1978.

Livingstone, David N. 'Evolution versus Religion?' *Third Way*, March 1980, pp. 16–17.

Loades, Ann L. 'Analogy, and the Indictment of the Deity: Some Inter-related Themes.' *Studia Theologica*, 33 (1979), 25–43.

Lucas, J.R. 'Wilberforce and Huxley: A Legendary Encounter.' *Historical Journal*, 22 (1979), 313–30.

Manier, Edward, 'Darwin's Language and Logic.' *SHPS*, 11 (1980), 305–23.

'History, Philosophy and Sociology of Biology: A Family Romance.' *SHPS*, 11 (1980), 1–24.

Marchant, P.D. 'Darwin and Social Theory.' *Australian Journal of Politics and History*, 5 (1959), 213–17.

Marsden, George M. *Fundamentalism in American Culture: The Shaping of Twentieth-Century Evangelicalism, 1870–1925*. New York: Oxford Univesity Press, 1981.

Meyer, D.H. 'American Intellectuals and the Victorian Crisis of Faith.' In *Victorian America*, ed. Daniel Walker Howe, pp. 59–77. Philadelphia: University of Pennsylvania Press, 1976.

Miller, William L. 'Herbert Spencer's Factors in Social Evolution.' *Sociological Analysis and Theory*, 7 (1977), 99–115.

Moore, James R. 'Charles Darwin and the Doctrine of Man.' *Evangelical Quarterly*, 44 (1972), 196–217.

'Charles Lyell and the Noachian Deluge.' *Evangelical Quarterly*, 45 (1973), 141–60.

'Evolutionary Theory and Christian Faith: A Bibliographical Guide to the Post-Darwinian Controversies.' *Christian Scholar's Review*, 4 (1975), 211–30.

'Could Darwinism Be Introduced in France?' [essay review of *L'introduction du darwinisme en France au XIXe siècle* by Yvette Conry]. *BJHS*, 10 (1977), 246–51.

'On the Education of Darwin's Sons: The Correspondence between Charles Darwin and the Reverend G.V. Reed, 1857–1864.' *NR*, 32 (1977), 51–70.

'Varieties of Social Darwinism.' In *Conflict and Stability in the Development of Modern Europe, 1789–1970*. Arts: A Third Level Course, The Open University, block 2 (1870–1918 or 1939), part 1 (Ideas), pp. 27–41. Milton Keynes, Bucks.: Open University Press, 1980.

Beliefs in Science: An Introduction. Science and Belief: from Darwin to Einstein (Arts: A Third Level Course, The Open University), block 1, unit 1. Milton Keynes, Bucks.: Open University Press, 1981.

'Creation and the Problem of Charles Darwin' [essay review of *Charles Darwin and the Problem of Creation* by Neal C. Gillespie]. *BJHS*, 14 (1981), in press.

The Future of Science and Belief: Theological Views in the Twentieth Century. Science and Belief: from Darwin to Einstein (Arts: A Third Level Course, The Open University), block 7, unit 15. Milton Keynes, Bucks.: Open University Press, 1981.

Moore, James R.; Chant, Colin; Coley, Noel G; and Roberts, Gerrylynn K. *Science and Metaphysics in Victorian Britain*. Science and Belief: from Darwin to Einstein (Arts: A Third Level Course,

The Open University), block 2, units 2-3. Milton Keynes, Bucks.: Open Univesity Press, 1981.

Morrison, John L. 'American Catholics and the Crusade against Evolution.' *Records of the American Catholic Historical Society of Philadelphia*, 64 (1953), 59-71.

Mozley, Ann. 'Evolution and the Climate of Opinion in Australia, 1840-76.' *VS*, 10 (1966-7), 411-30.

Nelkin, Dorothy. 'Creation vs. Evolution: The Politics of Science Education.' In *The Social Production of Scientific Knowledge*, ed. Everett Mendelsohn *et al.*, pp. 265-87. Dordrecht, The Netherlands: D. Reidel Publishing Co., 1977.

Noble, David W. *The Paradox of Progressive Thought*. Minneapolis: University of Minnesota Press, 1958.

Numbers, Ronald L. 'Arnold Guyot and the Harmony of Science and the Bible.' *Proceedings of the XIVth International Congress of the History of Science* (Tokyo, 1975), III, 239-42.

'Sciences of Satanic Origin: Adventist Attitudes toward Evolutionary Biology and Geology.' *Spectrum*, 9 (1979), 17-30.

Núñez, Diego, ed. *El Darwinismo en España*. Madrid: Castalia Editorial, 1977.

O'Connor, Daniel and Oakley, Francis, eds. *Creation: The Impact of an Idea*. New York: Charles Scribner's Sons, 1969.

Oldroyd, D.R. *Darwinian Impacts: An Introduction to the Darwinian Revolution*. Milton Keynes, Bucks.: Open University Press, 1980.

Ospovat, Dov. 'Perfect Adaptation and Teleological Explanation: Approaches to the Problem of the History of Life in the Mid-nineteenth Century.' *Studies in History of Biology*, 2 (1978), 33-56.

'Darwin after Malthus.' *JHB*, 12 (1979), 211-30.

'God and Natural Selection: The Darwinian Idea of Design.' *JHB*, 13 (1980), 169-94.

The Development of Darwin's Theory: Natural History, Natural Theology, and Natural Selection, 1838-59. Cambridge University Press, 1981.

Pancaldi, Giuliano, *Charles Darwin's 'storia' ed 'economia' della natura*. Florence: La Nuova Italia, 1977.

Paradis, James G. *T.H. Huxley: Man's Place in Nature*. Lincoln: University of Nebraska Press, 1978.

Passmore, J.A. 'Darwin and the Climate of Opinion.' *Australian Journal of Science*, 22 (1959), 8-15.

Petersen, William. *Malthus*. Cambridge, Mass.: Harvard University Press, 1979.

Phillips, Walter. 'The Defense of Christian Belief in Australia, 1875-1914: The Responses to Evolution and Higher Criticism.' *Journal of Religious History*, 9 (1976-7), 402-23.

Popkin, Richard H. 'Pre-Adamism in 19th Century American Thought: "Speculative Biology" and Racism.' *Philosophia*, 8 (1978–9), 205–39.

Prendergast, Michael Laurent. 'James Dwight Dana: The Life and Thought of an American Scientist.' Ph.D. diss., University of California (Los Angeles), 1978.

Reardon, Michael F. 'Science and Religious Modernism: The New Apologetic in France, 1890–1913.' *Journal of Religion*, 57 (1977), 48–63.

Richardson, R. Alan. 'Biogeography and the Genesis of Darwin's Ideas on Transmutation.' *JHB*, 14 (1981), 1–41.

Royle, Edward, *Radicals, Secularists and Republicans: Popular Freethought in Britain, 1866–1915*. Manchester University Press, 1980.

Ruse, Michael. 'Darwin and Herschel.' *SHPS*, 9 (1978), 323–31.

The Darwinian Revolution: Science Red in Tooth and Claw. Chicago: University of Chicago Press, 1979.

'Charles Darwin and Group Selection.' *AS*, 37 (1980), 615–30.

Russell, Colin A. *Time, Chance and Thermodynamics*. Science and Belief: from Darwin to Einstein (Arts: A Third Level Course, The Open University), block 3, units 4–5. Milton Keynes, Bucks.: Open University Press, 1981.

Russell-Gebbett, Jean P. *Henslow of Hitcham: Botanist, Educationalist, and Clergyman*. Lavenham, Suffolk: T. Dalton, 1977.

Sandow, Alexander. 'Social Factors in the Origin of Darwinism.' *Quarterly Review of Biology*, 13 (1938), 315–26.

Sankey, Derek E. 'A Comparison of the Role of Analogy in Early Nineteenth Century Science and Natural Theology in Britain.' M.A. diss., University of Kent (Canterbury), 1979.

Schweber, Silvan S. 'Darwin and the Political Economists: Divergence of Character.' *JHB*, 13 (1980), 195–289.

Shapin, Steven and Barnes, Barry. 'Darwin and Social Darwinism: Purity and History.' In *Natural Order: Historical Studies of Scientific Culture*, ed. Barry Barnes and Steven Shapin, pp. 125–42. Beverly Hills, Calif.: Sage Publications, 1979.

Sharlin, Harold Issadore and Sharlin, Tiby. *Lord Kelvin: The Dynamic Victorian*. University Park, Pa.: Pennsylvania State University Press, 1979.

Shimao, Eikoh. 'Darwinism in Japan, 1877–1927.' *AS*, 38 (1981), 93–102.

Simms, Beatrice L. 'The Anti-evolution Conflict in the 1920's.' M.A. thesis, University of Kentucky, 1953.

Smith, C.U.M. 'Charles Darwin, the Origin of Consciousness, and Panpsychism.' *JHB*, 11 (1978), 245–67.

Smith, Roger. 'The Human Significance of Biology: Carpenter, Darwin and the *vera causa.*' In *Nature and the Victorian Imagination,* ed. U.C. Knoepflmacher and G.B. Tennyson, pp. 216–30. Berkeley: University of California Press, 1977.

Smith, Willard H. *The Social and Religious Thought of William Jennings Bryan.* Lawrence, Kan.: Coronado Press, 1975.

Stanley, Oma. 'T.H. Huxley's Treatment of "Nature".' *JHI,* 18 (1957), 120–7.

Steen, Franklin David. 'Taylor Lewis on Scripture: A Defense of Revelation and Creation in Nineteenth Century America.' Th.D. diss., Westminster Theological Seminary, 1971.

Stephens, Lester D. 'Joseph Le Conte on Evolution, Education, and the Structure of Knowledge.' *Journal of the History of the Behavioural Sciences,* 12 (1976), 103–19.

'Joseph Le Conte's Evolutional Idealism: A Lamarckian View of Cultural History.' *JHI,* 39 (1978), 465–80.

'Joseph Le Conte and the Development of the Physiology and Psychology of Vision in the United State.' *AS,* 37 (1980), 303–21.

Sulloway, Frank J. 'Geographical Isolation in Darwin's Thinking: The Vicissitudes of a Crucial Idea.' *Studies in History of Biology,* 3 (1979), 23–65.

Super, R.H. 'The Humanist at Bay: The Arnold–Huxley Debate.' In *Nature and the Victorian Imagination,* ed. U.C. Knoepflmacher and G.B. Tennyson, pp. 231–45. Berkeley: University of California Press, 1977.

Swanston, Hamish F.G. *Ideas of Order: Anglicans and the Renewal of Theological Method in the Middle Years of the Nineteenth Century.* Assen, The Netherlands: Van Gorcum & Co., 1974.

Swoboda, Merrily Kodis. 'The American Rhetorical Career of Louis Agassiz: A Case Study of Transformations in American Science, 1846–1860.' Ph.D. diss., University of Pittsburgh, 1977.

Szasz, Ferenc M. 'William B. Riley and the Fight against Teaching of Evolution in Minnesota.' *Minnesota History,* 41 (1969), 201–16.

'The Scopes Trial in Perspective.' *Tennessee Historical Quarterly,* 30 (1971), 288–98.

'William Jennings Bryan, Evolution, and the Fundamentalist–Modernist Controversy.' *Nebraska History,* 56 (1975), 259–78.

Tallmadge, John. 'From Chronicle to Quest: The Shaping of Darwin's "Voyage of the Beagle".' *VS,* 23 (1979–80), 325–45.

Taylor, Robert J. 'The Darwinian Revolution: The Responses of Four Canadian Scholars.' Ph.D. diss., McMaster University, 1976.

Thagard, Paul. 'Darwin and Whewell.' *SHPS,* 8 (1977), 353–6.

Thompson, James J., Jr. 'Southern Baptists and the Antievolution

Controversy of the 1920's.' *Mississippi Quarterly*, 29 (1975–6), 65–81.

Tierney, Brian; Kagan, Donald; and Williams, L. Pearce, eds. *Social Darwinism—Law of Nature or Justification of Repression?* 3rd edn. New York: Random House, 1977.

Tompkins, Jerry R., ed. *D-Days at Dayton: Reflections on the Scopes Trial.* Baton Rouge: Louisiana State University Press, 1965.

Turner, Frank M. 'Victorian Scientific Naturalism and Thomas Carlyle.' *VS*, (1974–5), 325–43.

'The Victorian Conflict between Science and Religion: A Professional Dimension.' *Isis*, 69 (1978), 356–76.

Vander Stelt, John C. *Philosophy and Scripture: A Study in Old Princeton and Westminster Theology.* Marlton, N.J.: Mack Publishing Co., 1978.

Vile, John Ralph. 'Science, Faith, and Philosophy in William James: A Theoretical Analysis with Application to James's Political Thought.' Ph.D. diss., University of Virginia, 1977.

Vorzimmer, Peter J. 'An Early Darwin Manuscript: The "Outline and Draft of 1839".' *JHB*, 8 (1975), 191–217.

'Darwin's Reading Notebooks (1838–1860).' *JHB*, 10 (1977), 107–53.

Wace, Henry. 'Some of the Relations between Science and Religion as Affected by the Work of the Last Fifty Years.' *Journal of the Transactions of the Victoria Institute, or Philosophical Society of Great Britain*, 49 (1917), 269–80.

Watanabe, Masao. 'John Thomas Gulick: American Evolutionist and Missionary in Japan.' *Japanese Studies in the History of Science*, no. 5 (1966), 140–9.

'Darwinism in Japan in the Late 19th Century.' *Actes du XIIe Congrès International d'Histoire des Sciences Naturelles et de la Biologie*, (Paris, 1971), XI, 149–54.

Watanabe, Masao and Ose, Yoko. 'General Academic Trend and the Evolution Theory in Late Nineteenth Century Japan: A Statistical Analysis of the Contemporary Periodicals.' *Japanese Studies in the History of Science*, no. 7 (1968), 129–42.

Watson, Elbert L. 'Oklahoma and the Anti-evolution Movement of the 1920's.' *Chronicles of Oklahoma*, 42 (1964–5), 396–407.

Wertheimer, Douglas. 'Philip Henry Gosse: Science and Revelation in the Crucible.' Ph.D. diss., University of Toronto, 1977.

Wetzels, Walter D. 'Aspects of Natural Science in German Romanticism.' *Studies in Romanticism*, 10 (1971), 44–59.

Whalen, Matthew Daniel. 'American Science, Society, and Civilization in the Age of Energy: An Investigation of the Relationships among Neo-Lamarckism, Social Evolutionism, and the Myth of

Atlantis between 1860 and 1920.' Ph.D. diss., University of Maryland, 1978.

Winsor, Mary P. *Starfish, Jellyfish, and the Order of Life: Issues of Nineteenth-Century Science.* New Haven, Conn.: Yale University Press, 1976.

'Louis Agassiz and the Species Question.' *Studies in History of Biology*, 3 (1979), 89–117.

The World's Most Famous Court Trial, Tennessee Evolution Case: A Word-for-Word Report of the Famous Court Test of the Tennessee Anti-evolution Act, at Dayton, July 10 to 21, 1925, Including Speeches and Arguments of Attorneys, Testimony of Noted Scientists, and Bryan's Last Speech. Enlarged reprint edn. Dayton, Tenn.: Rhea County Historical Society, 1978.

Worster, Donald E., ed. *American Environmentalism: The Formative Period, 1860–1915.* New York: John Wiley & Sons, 1973.

Wrangham, Richard W. 'The Bishop of Oxford: Not So Soapy.' *New Scientist*, 83 (9 August 1979), 450–1.

'Bishop Wilberforce: Natural Selection and the Descent of Man.' *Nature*, 287 (18 Sept. 1980), 192.

Wyllie, Irvin G. 'Bryan, Birge, and the Wisconsin Evolution Controversy, 1921–1922.' *Wisconsin Magazine of History*, 35 (1952), 294–301.

Yeo, Richard. 'Natural Theology and the Philosophy of Knowledge in Britain, 1819–1869.' Ph.D. thesis, University of Sydney, 1977.

'William Whewell, Natural Theology, and the Philosophy of Science in Mid Nineteenth Century Britain.' *AS*, 36 (1979), 493–516.

Young, David. 'The Impact of Darwinism on the Concept of God in the Nineteenth Century.' *Faith and Thought*, 100 (1972–3), 17–44.

Young, Robert M. 'Natural Theology, Victorian Periodicals and the Fragmentation of a Common Context' (1969). In *Darwin to Einstein*, ed. C. Chant and J. Fauvel (1980), pp. 69–107.

INDEX

The numbers given in brackets are note numbers.

472   INDEX

American Journal of Science and Arts
270
American Naturalist 146, 150, 372 (33)
American Philosophical Society, Library
of the 367 (26), 388 (23)
Amherst College 197
Amick, David Eldridge 362 (73, 82)
Ampère, André Marie 28
Anderson, Olive 51–2, 359 (4, 8, 11), 360
(14)
Anderson, Robert; *Doubter's Doubts
about Science and Religion* 380 (32)
Andover, Massachusetts
Association of Congregational Ministers 281
Free Christian Church 281
Theological Seminary 198, 227, 281,
304
Andrews, Samuel J. 359 (10)
Angell, James Burrill 30–1
Anglicanism 52, 63
of Brooke, S. 103
of Butler, S. 105, 344
of Darwin, C. 315
of Kingsley, C. 306
of Lankester, E. R. 110
of Malthus, T. R. 309
of Paley, W. 309
responses to evolution of 11
of Stephen, L. 104
see also Anglo-Catholicism, Broad
Church, evangelicalism, Low Church
and evolution
Anglo-Catholicism 52, 60, 305
of Clifford, W. K. 110
and divine immanence 337, 398 (118)
of Liddon, H. P. 90
of Moore, A. L. 259–60, 303
of Pusey, E. B. 90
restated in developmental terms 92
revival of 117
Annan, Noel 4, 366 (8)
anthropology *See* mankind
anthropometry 176
Antichrist 53
anti-communism 74
anti-Darwinism
of Agassiz, L. 87, 141, 207–9
Christian 122, 193–216, 300; and the
Bible *See main entry*; causes of 14–
15, 113–15, 219; historical distortion of 193; philosophy of 213–16
and legal profession 380 (32)
at Princeton University 385 (81)
among scientists 86–8
see also anti-evolutionism
Anti-Evolution League of America 74, 75
anti-evolutionism

and Fundamentalism 68–76
among theologians 88–9, 93
see also anti-Darwinism
Apocalypse of St John 45, 53
Applebaum, Wilbur x
Appleman, Philip
1859: Entering an Age of Crisis 102
Madden, W. A. and Wolff, M. 359 (5)
Appleton, Daniel (publisher) 21, 271,
282, 357 (52), 358 (58)
and International Scientific Series 20
and *Origin of Species* 270
and Spencer, H. 167
Aquinas, Thomas 45, 62
archaesthetism, hypothesis of 149
Argentina 316
Argyll, The Dowager Duchess of *See*
Campbell, Ina
Argyll, The Duke of *See* Campbell,
George Douglas
Aristotelian philosophy of nature 206,
327, 342, 344
Arminianism 293
Armstrong, A. C. 372 (1)
Armstrong, Richard A. 166, 375 (42)
Arneson, Kenneth x
Arnstein, Walter Leonard 359 (5)
artificial selection 396 (75)
Darwin, C. on *See main entry*
Dawson, J. W. on 205
Johnson, F. H. on 227
Wallace, A. R. on 186
Wright, G. F. on 292, 335
see also natural selection
astrology 189
atheism 72, 150, 242, 311, 319
of Cooper, T. 392 (9)
and Darwinism *See main entry*
of Romanes, G. J. 107–8
Athenaeum 363 (21)
Atkins, Gaius Glenn 355 (24), 382 (3)
Atkinson, John Christopher 83
Atlantic cable 57
Atlantic Monthly 38, 169, 270, 375 (49)
Augustine of Hippo, St 62, 64, 280, 293,
294, 334, 337, 400 (12)
Augustinian–Pelagian controversy 399
(134)
Aulie, Richard x
Aveling, Edward 95, 365 (57), 367 (25)
Averill, Lloyd J. 399 (128)
Aymaras 137–8

Bacon, Benjamin Wisner 365 (51), 379
(14)
Bacon, Francis 45
Advancement of Learning 328
Novum Organum 162, 194

484   INDEX

divine intervention – *cont.*
Wright, G. F. on 297–8, 340
see also miracles, providence, secondary causes, supernaturalism
Dixon, Amzi C. 361 (68), 362 (70)
Döllinger, Johann Joseph Ignaz von 27, 356 (19)
Dörpinghaus, Hermann Josef 354 (18), 355 (27)
Dollar, George W. 362 (72)
Dorlodot, Henri de 355 (27)
Dorn, Jacob Henry 383 (19)
Douglas, Mary 101, 366 (2)
Dowey, Edward, Jr x
Drachman, Julian M. 42, 358 (7)
Draper, Elizabeth 23
Draper, John William 44, 57, 356 (8–30 *passim*), 358 (62), 362 (75)
  anti-Catholicism of 24–8, 48, 102
  biographical details of 22–4
  History of the Conflict between Religion and Science 58, 63, 68; criticised 26–8; influence of 28–9, 40; origin of 24; Shields, C. W. on 45; and White, A. D. 37; Wright, G. F. on 356 (29)
  History of the Intellectual Development of Europe 21, 23
  Human Physiology 23
  and Huxley–Wilberforce confrontation 61
  and military metaphor 13, 43, 48, 50, 102
  religious views of 23, 102
  and White, A. D. 35, 37, 38
  Zöckler, O. on 46
Draper Seminary for Girls, Misses 22
Drummond, Henry 11, 235, 354 (14), 383 (23), 384 (60, 64), 392 (6)
  and Abbott, L. 227
  in America 7
  Ascent of Man 224, 227, 305, 392 (6)
  and Calvinism 343
  as Christian Darwinist 224, 341
  as evolutionist 92
  liberal theology of 305
  and Moody, D. L. 224
  Natural Law in the Spiritual World 7, 224
  on progress 239
  and Spencer, H. 224, 237
  on universal evolution 237
Dublin, University of 87
Duffield, George 55
Dumas, Jean Baptiste André 28
Duncan, David 359 (79), 366 (20), 374 (25, 30, 32–3, 35), 375 (38, 40, 49), 376 (9, 53), 378 (33), 379 (28), 382

(10), 383 (22–3, 39), 384 (41, 44), 386 (85)
Dupree, A. Hunter 354 (14), 361 (63), 367 (26), 389 (48, 52, 56), 390 (74, 76, 79, 85)
Durant, John x
Duryea, Joseph T. 381 (64)
Dynamite 74
dysteleology 113
  Darwin, C. on 273–4, 275, 309, 316, 318, 321, 333–4
  Gray, A. on 273, 331–2
  Iverach, J. on 331
  McCosh, J. on 249
  Moore, A. L. on 331
  Paley, W. on 309
  Wright, G. F. on 290, 330, 331–2
  see also design in nature, final causes, natural theology, teleology, theodicy
Dyster, Frederick 400 (12)

Eagly, Alice Hendrickson, Samuel Himmelfarb and 367 (29)
Eaton, Clement 354 (14), 355 (25), 383 (27)
Ebenstein, William 2–3, 355 (27)
Edinburgh
  New College 92, 253
  St Bernard's Church 92
  University of 87, 204, 245, 253, 315
education
  evolution in 29–30, 34, 73, 75, 78–9, 93, 176, 362 (75)
  higher, reform of American 30–1, 78–9
Edwards, Jonathan (the elder) 290, 294, 303, 335, 337, 400 (12)
Egerton, Frank N., III 372 (38)
Eggleston, Mary Frederick 2–3, 5, 7, 11, 355 (18, 27), 361 (64), 366 (6)
Eiseley, Loren 4, 370 (46), 373 (6), 377 (11), 380 (45), 382 (68)
Eisen, Sydney 360 (43), 375 (43)
Elert, Werner 359 (84)
Eliot, Charles William 30–1, 245
Ellegård, Alvar 3–11 *passim*, 84, 93–4, 115, 196, 213, 218, 354 (15–17), 355 (26–8), 360 (34), 361 (64), 363 (22), 364 (35), 365 (53), 367 (34, 36), 369 (6), 371 (7), 378 (5), 379 (9), 382 (5, 7–8, 68–9), 384 (46, 58)
Ellicott, Charles John 98
Elliott, Robert James 353 (5), 359 (89), 392 (3)
Elliott-Binns, L. E. 353 (8), 359 (89), 365 (54), 366 (5)
  Religion in the Victorian Era 47
embryology and evolution

490 INDEX